IUTAM SYMPOSIUM ON DYNAMICS OF SLENDER VORTICES

FLUID MECHANICS AND ITS APPLICATIONS
Volume 44

Series Editor: R. MOREAU
MADYLAM
Ecole Nationale Supérieure d'Hydraulique de Grenoble
Boîte Postale 95
38402 Saint Martin d'Hères Cedex, France

Aims and Scope of the Series

The purpose of this series is to focus on subjects in which fluid mechanics plays a fundamental role.

As well as the more traditional applications of aeronautics, hydraulics, heat and mass transfer etc., books will be published dealing with topics which are currently in a state of rapid development, such as turbulence, suspensions and multiphase fluids, super and hypersonic flows and numerical modelling techniques.

It is a widely held view that it is the interdisciplinary subjects that will receive intense scientific attention, bringing them to the forefront of technological advancement. Fluids have the ability to transport matter and its properties as well as transmit force; therefore fluid mechanics is a subject that is particulary open to cross fertilisation with other sciences and disciplines of engineering. The subject of fluid mechanics will be highly relevant in domains such as chemical, metallurgical, biological and ecological engineering. This series is particularly open to such new multidisciplinary domains.

The median level of presentation is the first year graduate student. Some texts are monographs defining the current state of a field; others are accessible to final year undergraduates; but essentially the emphasis is on readability and clarity.

For a list of related mechanics titles, see final pages.

IUTAM Symposium on

Dynamics of Slender Vortices

Proceedings of the IUTAM Symposium
held in Aachen, Germany,
31 August – 3 September 1997

Edited by

E. KRAUSE
Aerodynamisches Institut,
Rheinisch-Westfälische Technische Hochschule Aachen,
Germany

and

K. GERSTEN
Institut für Thermo- und Fluiddynamik,
Ruhr-Universität Bochum,
Germany

SPRINGER SCIENCE+BUSINESS MEDIA, B.V.

A C.I.P. Catalogue record for this book is available from the Library of Congress.

ISBN 978-94-010-6117-9 ISBN 978-94-011-5042-2 (eBook)
DOI 10.1007/978-94-011-5042-2

Printed on acid-free paper

Coverpicture:
Visualization of the numerical solution of vortex breakdown for Re = 3220: Vortex lines and color-coded pressure distribution by M. Weimer in Diss. Aerodynamisches Institut, RWTH Aachen, 1997

Originally published by Kluwer Academic Publishers in 1998
Softcover reprint of the hardcover 1st edition 1998

福 湯 章 夫

Dr. Akio Fukuyu, Professor of Mathematics, †1997
Dean, School of Science and Engineering, Tokyo Denki University

Scientific Committee

E. Krause, Chairman
K. Gersten, Co-Chairman
Hussain, USA
K. Kuwahara, Japan
M. Lesieur, France
C.-H. Liu, USA
H. K. Moffatt, UK
L. Ting, USA

Local Organizing Committee

E. Krause, Germany
K. Gersten, Germany
W. Limberg, Germany
W. Althaus, Germany
L. Ting, USA

Sponsors of Symposium

Deutsche Forschungsgemeinschaft
Ministerium für Wissenschaft und Forschung des Landes Nordrhein-Westfalen
Rheinisch - Westfälische Technische Hochschule Aachen
Stadt Aachen
International Union of Theoretical and Applied Mechanics
United States Air Force European Office of Aerospace Research and Development
Kluwer academic publishers

TABLE OF CONTENTS

Session 1: Asymptotic Theories

Session 2: Numerical Methods

Session 3: Vortices in Shear Layers

Session 4: Interaction of Vortices

Session 5: Vortex Breakdown

Session 6: Vortex Sound

Session 7: Aircraft and Helicopter Vortices

PREFACE

The decision of the General Assembly of the International Union of Theoretical and Applied Mechanics to organize a Symposium on Dynamics of Slender Vortices was greeted with great enthusiasm. The acceptance of the proposal, forwarded by the Deutsches Komitee für Mechanik (DEKOMECH) signalized, that there was a need for discussing the topic chosen in the frame the IUTAM Symposia offer. Also the location of the symposium was suitably chosen: It was decided to hold the symposium at the RWTH Aachen, where, years ago, Theodore von Kármán had worked on problems related to those to be discussed now anew. It was clear from the beginning of the planning, that the symposium could only be held in the von Kármán-Auditorium of the Rheinisch-Westfälische Technische Hochschule Aachen, a building named after him. The symposium was jointly organized by the editors of this volume, strongly supported by the local organizing committee.

The invitations of the scientific committee brought together scientists actively engaged in research on the dynamics of slender vortices. It was the aim of the committee to have the state of the art summarized and also to have the latest results of specific problems investigated communicated to the participants of the symposium. The topics chosen were asymptotic theories, numerical methods, vortices in shear layers, interaction of vortices, vortex breakdown, vortex sound, and aircraft and helicopter vortices.

The presentations on the first two topics focussed on explaining the present capability of asymptotic theories and numerical methods for analyzing flows containing slender vortices. These presentations were complemented by reports on new developments and specific applications of various methods. For the third topic recent results of investigations of the dynamics of slender vortices in shear layers were presented. Among other interesting findings the possibility of substantial drag reduction in turbulent boundary layers by manipulating the vortex structures in them was discussed. Of equal interest were the presentations on vortex interaction and breakdown. Also the problem of sound generation by slender vortices was vividly discussed. Finally the presentations of the last topic gave insight into problems related to trailing vortices of aircraft and to helicopter vortices. The symposium clearly showed the many aspects of the dynamics of slender vortices and the progress in the various areas. It also pointed out future research and results to be expected. There were strong interactions between the participants working on fundamental problems and those focussing on applied research.

About 60 scientists including a group of doctoral students from the RWTH Aachen followed the invitation of the scientific committee. The participants came from 9 countries, and 39 papers were presented in invited lectures and invited contributions. The presentations comprised a wide variety of fundamental investigations and of advanced engineering applications in the problem areas mentioned. Lively discussion aroused on several problems, as for example on the problem of vortex breakdown. Numerous new contacts were established. The presentations and discussions also stimulated further theoretical and experimental research with respect to all problems mentioned.

The assistance by the members of the scientific committee before and during the symposium is gratefully acknowledged. Sincere thanks are also extended to the chairpersons, the invited lecturers and contributors and all other participants for their active cooperation. The support offered by the local organizing committee and the staff of the Aerodynamisches Institut of the RWTH Aachen was one of the main sources of the success of the symposium.

Financial support for the symposium was generously provided. The following sponsors are gratefully acknowledged: The Deutsche Forschungsgemeinschaft, the Ministerium für Wissenschaft und Forschung des Landes Nordrhein-Westfalen, the Stadt Aachen, the International Union for Theoretical and Applied Mechanics, the RWTH Aachen, the United States Air Force European Office of Aerospace Research and Development, and Kluwer academic publishers.

Finally the efficient cooperation of the Kluwer academic publishers in the publication of the proceedings of the symposium in this volume is very much appreciated.

Aachen, December 1997 E. Krause K. Gersten

PARTICIPANTS

W. Althaus
Aerodynamisches Institut
RWTH Aachen
Wüllnerstr. zw. 5 u. 7
52062 Aachen
Germany
Tel.: +49 (241) 805417
Fax: +49 (241) 8888.257
wolfgang@aia.rwth-aachen.de

Chr. Blohm
ZARM
University of Bremen
Am Fallturm
28359 Bremen
Germany
Tel.: +49 (421) 218.4844
Fax: +49 (421) 218.2521
blohm@zarm.uni-bremen.de

P. Comte
L.E.G.I. - I.M.G.
Domaine Universitaire
B.P. 53
38041 Grenoble Cedex 9
France
Tel.: +33 (4) 768.25121
Fax: +33 (4) 768.25271
pierre.comte@hmg.inpg.fr

K. Bajer
Institute of Geophysics
University of Warsaw
al. Pasteura 7
02-093 Warszawa
Poland
Tel.: +48 (22) 235281
Fax: +48 (22) 6219712
kbajer@euw.edu.pl

Ch. Brücker
Aerodynamisches Institut
RWTH Aachen
Wüllnerstr. zw. 5 u. 7
52062 Aachen
Germany
Tel.: +49 (241) 805430
Fax: +49 (241) 8888.257
bruecker@aia.rwth-aachen.de

U. Ch. Dallmann
DLR-Institut für Strömungsmechanik
Bunsenstraße 10
37073 Göttingen
Germany
Tel.: +49 (551) 709.2442
Fax: +49 (551) 709.2404
uwe.dallmann@dlr.de

Daniel Margerit
Laboratoire d'Énergétique et de
Mécanique Appliquée (LEMTA)
URA CNRS 875
2, Av. de la Foret de Haye, BP 160
54054 Vandoeuvre Les Nancy
France
Tel.: +33 (383) 59.5730
Fax: +33 (383) 59.5551
daniel.margerit@ensem.u-nancy.fr

K. Ehrenfried
DLR-Institut für Strömungsmechanik
Bunsenstraße 10
37073 Göttingen
Germany
Tel.: +49 (551) 709.2813
klaus.ehrenfried@dlr.de

Marie Farge
LMD - CNRS
École Normale Supérieure
24, Rue Lhmonond
75231 Paris Cedex 5
France
Tel.: +33 (144) 32.2235
Fax: +33 (143) 36.8392
farge@lmd.ens.fr

Y. Fukumoto
Graduate School of Mathematics
Kyushu University 33
Fukuoka 812-81
Japan
Tel.: +81 (92) 642.2762
Fax: +81 (92) 642.2776
yashuhide@math.kyushu-u.ac.jp

A. Fukuyu
Tokyo Denki University
Ishizaka, Hatoyama, Hiki
Saitama-ken 350-03
Japan
Tel.: +81 (492) 96.2911
Fax: +81 (492) 96.7072
fukuyu@r.dendai.ac.jp

K. Gersten
Institut für Thermo- und Fluiddynamik
Ruhr-Universität Bochum
Hofleite 15
44795 Bochum
Germany
Tel. + Fax: +49 (234) 433388

P. Huerre
Laboratoire d'Hydrodynamique
École Polytechnique-CNRS
91128 Palaiseau Cedex
France
Tel.: +33 (1) 6933.4990
Fax: +33 (1) 6933.3030
huerre@ladhyx.polytechnique.fr

F. Hussain
Dept. of Mechanical Engineering
University of Houston
Houston, TX 77204-4792
USA
Tel.: +1 (713) 743.4545
Fax: +1 (713) 743.4503
fhussain@uh.edu

O. Inoue
Institute of Fluid Science
Tohoku University
2-1-1 Katahira, Aoba-ku
Sendai 980-77
Japan
Tel.+ Fax: +81 (22) 217.5256
inoue@ifs.tohoku-ac.jp

K. Ishii
Dept. of Computational Science
and Engineering
Nagoya University
Nagoya, Aichi 464-01
Japan
Tel.+ Fax: +81 (52) 789.4660
ishii@nuap.nagoya-u.ac.jp

O. A. Kandil
Aerospace Engineering Dept.
Old Dominion University
Norfolk, VA 23529-0247
USA
Tel.: +1 (757) 683.4913
Fax: +1 (757) 683.3200
kandil@aero.odu.edu

S. Kida
National Institute for Fusion Science
Oroshi-cho 322-6
Toki City
Toki-shi 509-52
Japan
Tel.: +81 (572) 58.2248
Fax: +81 (572) 58.262
kida@toki.theory.nifs.ac.jp

R. Klein
FB 14 - Sicherheitstechnik
Bergische Universität
Gauß-Str. 20
42097 Wuppertal
Germany
Tel.: +49 (202) 439.2070
Fax: +49 (202) 439.2047
rupert@uni-wuppertal.de

O. M. Knio
Dept. of Mechanical Engineering
The John Hopkins University
Baltimore, MD 21218
USA
Tel.: +1 (410) 516.7736
Fax: +1 (410) 516.7254
knio@flame.me.jhu.edu

N. Kornev
Dept. of Hydromechanics
Marine Technical University
Lotsmanskaya Str. 3
St. Petersburg 190008
Russia
Tel. + Fax: +7 (812) 1572500
kornev@mtu-mic.spb.su

E. Krause,
Aerodynamisches Institut
RWTH Aachen
Wüllnerstr. zw. 5 u. 7
52062 Aachen
Germany
Tel.: +49 (241) 805410
Fax: +49 (241) 8888257
ek@aia.rwth-aachen.de

A. N. Kudryavtsev
Inst. of Theoretical & Applied Mechanics
Russian Academy of Sciences
Institutskaya st., 4/1
Novosibirsk 630090
Russia
Tel.: +7 (383) 235.3169
Fax: +7 (383) 235.2268
aleks@itam.nsc.ru

St. Le Dizès
Institut de Recherche sur les
Phénomènes Hors Équilibre
12, Ave. Général Leclerc
13003 Marseille
France
Tel.: +33 (491) 505439
Fax: +33 (491) 081637
ledizes@marius.univ-mrs.fr/-ledizes

Thomas Leweke
IRPHÉ, CNRS
Universités Aix-Marseille I et II
12, Avenue Général Leclerc
13003 Marseille
France
Tel.: +33 (491) 505439
Fax: +33 (491) 081637
leweke@marius.univ-mrs.fr

F. Lund
Universidad de Chile
Fac. de Ciencas Fisicas Y Mathematicas
Dep. de Fisica
Avda. Blanco Encalada 2008
Casilla 487-3, Santiago
Chile
Tel.: +56 (2) 6784339
Fax: +56 (2) 6967359
flund@camilo.dfi.uchile.cl

K. Kuwahara
Institute of Space and
Astronautical Science
1-22-3, Haramachi
Meguro-ku, Tokyo 152
Japan
Tel.: +81 (3) 3714-6322
Fax: +81 (3) 3714-7430
kuwahara@icfd.co.jp

J. G. Leishman
Dept. of Aerospace Engineering
University of Maryland at
College Park
Maryland 20742
USA
Tel.: +1 (301) 4051126
Fax: +1 (301) 3149001
leishman@eng.umd.edu

C. H. Liu
Theoretical Flow Physics Branch
NASA Langley Research Center
Mail Stop 128
Hampton, VA 23665-5225
USA
Tel.: +1 (757) 220-8808
Fax: +1 (757) 220-3611
cfdinc@erols.com

J. E. Martin
Christopher Newport University
50 Shoe Lane
Newport News, VA 23606-299
USA
Tel.: +1 (757) 594.7945
Fax: +1 (757) 594.7919
jamie@pcs.cnu.edu

E. Meiburg
Dept. of Aerospace & Engineering
University of Southern California
Los Angeles, CA 90089-1191
USA
Tel.: +1 (213) 740.4303
Fax: +1 (213) 740.7774
eckart@spock.usc.edu

Dr. G. E. A. Meier
DLR-Institut für Strömungsmechanik
Bunsenstraße 10
37073 Göttingen
Germany
Tel.: +49 (551) 709.2177
Fax: +49 (551) 709.2889
g.e.a.meier@dlr.de

M. Meinke
Aerodynamisches Institut
RWTH Aachen
Wüllnerstr. zw. 5 u. 7
52062 Aachen
Germany
Tel.: +49 (241) 804821
Fax: +49 (241) 8888257
meinke@aia.rwth-aachen.de

H. K. Moffatt
Isaac Newton Institute for
Mathematical Sciences
Clarkson Road
Cambridge CB3 9EW
United Kingdom
Tel.: +44 (1223)-335980
hkm2@newton.cam.ac.uk

M. Nitsche
100 Mathematics Building
Ohio State University
231 West 18th Avenue
Columbus, OH 43210-1174
USA
Tel.: +1 (614) 292-5710
Fax: +1 (614) 292-7173
nitsche@math.ohio-state.edu

V. L. Okulov
Institute of Thermophysics
Siberian Branch of RAS
Lavrentyev Ave., 1
630090 Novosibirsk
Russia
Tel.: +7 (3832) 357128
Fax: +7 (3832) 357880
aleks@otani.thermo.nsk.su

Lakshmi N. Sankar
School of Aerospace Engineering
Georgia Techn. University
Mail Stop 0150
Atlanta, GA 30332-0150
USA
Tel.: +1 (404) 894.3014
Fax: +1 (404) 894.2760
lakshmi.sankar@aerospace.gatech.edu

T. Sarpkaya
Mechanical Engineering
Naval Postgraduate School
700 Dyer Road, Rm: 339
Monterey, CA 93943-5100
USA
Tel.: +1 (408) 656.3425
Fax: +1 (408) 656.2238
sarp@nps.navy.mil

Wade Schoppa
Dept. of Mechanical Engineering
University of Houston
Houston, TX 77204-4792
USA
Tel.: +1 (713) 743.4551
Fax: +1 (713) 743.4503
mece18l@jetson.uh.edu

B. Yu. Scobelev
Institute of Theoretical and
Applied Mechanics
Russian Academy of Sciences
Novosibirsk 630090
Russia
Tel.: +7 (3832) 352346
Fax: +7 (3832) 352268
adm@itam.nsk.su

P. M. Sforza
Mech. & Aerospace Engineering Dept.
Polytechnic University of Brooklyn
Route 110
Farmingdale, New York 11735
USA
Tel.: +1 (516) 755.4205
Fax: +1 (516) 755.4526
sforza@rama.poly.edu

V. N. Shtern
Dept. of Mechanical Engineering
University of Houston
Houston, TX 77204-4792
USA
Tel.: +1 (713) 743.4547
Fax: +1 (713) 743.4503
mece21w@jetson.uh.edu

S. Shtork
Institute of Thermophysics
Siberian Branch of RAS
Lavrentyev Ave., 1
Novosibirsk 630090
Russia
Tel.: +7 (3832) 357128
Fax: +7 (3832) 357880
alex@otani.thermo.nsk.su

R. Spall
Dept. of Mechanical and
Aerospace Engineering
Utah State University
Logan, UT 84322-4130
USA
Tel.: +1 (801) 797.2878
Fax: +1 (801) 797.2417
spall@fluids.me.usu.edu

R. Stuff
DLR-Institut für Strömungsmechanik
Bunsenstraße 10
37073 Göttingen
Germany
Tel.: +49 (551) 709.2865
Fax: +49 (551) 709.2830

Lu Ting
Courant Institute of Mathematical Scien
New York University
251, Mercer Street
New York, N.Y. 10012
USA
Tel.: +1 (212) 998.3139
Fax: +1 (212) 995.4121
ting@ting.cims-nyu.edu

CONTRIBUTORS

A. Abdelfattah
Aerodynamisches Institut
RWTH Aachen
Wüllnerstr. zw. 5 u. 7
52062 Aachen
Germany

S. Adachi
Institute of Computational
Fluid Dynamics
1-22-3 Haramachi
Meguro-ku
Japan

S. V. Alekseenko
Institute of Thermophysics
Lavrentyev Ave. 1
630090 Novosibirsk
Russia

W. Althaus
Aerodynamisches Institut
RWTH Aachen
Wüllnerstr. zw. 5 u. 7
52062 Aachen
Germany

M. Baffico
Departamento de Física
Fac. de Ciencas Físicas y Matematicas
Universidad de Chile
Casilla 487-3
Santiago
Chile

K. Bajer
Institute of Geophysics
University of Warsaw
al. Pasteura 7
02-093 Warszawa
Poland

P. Billant
Ladhyx
École Polytechnique
91128 - Palaiseau
France

Chr. Blohm
ZARM
University of Bremen
Am Fallturm
28359 Bremen
Germany

D. Boyer
Departamento de Física
Fac. de Ciencas Físicas y Matematicas
Universidad de Chile
Casilla 487-3
Santiago
Chile

Ch. Brücker
Aerodynamisches Institut
RWTH Aachen
Wüllnerstr. zw. 5 u. 7
52062 Aachen
Germany

J. M. Chomaz
Ladhyx Lab. d'Hydrodynamique
Ecole Polytechnique
91128 Palaiseau Cedex
France

P. Comte
L.E.G.I. - I.M.G.
Domaine Universitaire
B.P. 53
38041 Grenoble Cedex 9
France

I. Delbende
Ladhyx Lab. d'Hydrodynamique
Ecole Polytechnique
91128 Palaiseau Cedex
France

K. Ehrenfried
DLR-Institut für Strömungsmechanik
Bunsenstraße 10
37073 Göttingen
Germany

C. Eloy
Institut de recherches sur les
Phénomènes Hors Équilibre
12, Av. Général Leclerc
13003 Marseille
France

Marie Farge
LMD - CNRS
École Normale Supérieure
24, Rue Lhmonond
75231 Paris Cedex 5
France

Y. Fukumoto
Graduate School of Mathematics
Kyushu University 33
Fukuoka 812-81
Japan

A. Fukuyu
Tokyo Denki University
Ishizaka, Hatoyama, Hiki
Saitama-ken 350-03
Japan

T. B. Gatski
Aerodynamic and Acoustic Methods
Branch
NASA Langley Research Center
Mail Stop 128
Hampton, VA 23681-0001
USA

Y. Hattori
Institute of Fluid Science
Tohoku University
2-1-1 Katahira, Aoba-ku
Sendai 980-77
Japan

J. Hofhaus
BMW AG
Knorrstrasse 147
80788 München
Germany

P. Huerre
Laboratoire d'Hydrodynamique
École Polytechnique-CNRS
91128 Palaiseau Cedex
France

F. Hussain
Dept. of Mechanical Engineering
University of Houston
Houston, TX 77204-4792
USA

O. Inoue
Institute of Fluid Science
Tohoku University
2-1-1 Katahira, Aoba-ku
Sendai 980-77
Japan

K. Ishii
Dept. of Computational Science
and Engineering
Nagoya University
Nagoya, Aichi 464-01
Japan

O. A. Kandil
Aerospace Engineering Dept.
Old Dominion University
Norfolk, VA 23529-0247
USA

G. Kawahara
Ehime University
Matsuyama 790-77
Japan

N. K.-R. Kevlahan
LMD-CNRS
École Normale Supérieure
24, Rue Lhmonond
75231 Paris cedex 05
France

D. V. Khotyanosky
Inst.of Theoretical & Applied Mechanics
Novosibirsk 630090
Russia

S. Kida
National Institute for Fusion Science
Oroshi-cho 322-6
Toki City
Toki-shi 509-52
Japan

R. Klein
FB 14 - Sicherheitstechnik
Bergische Universität
Gauß-Str. 20
42097 Wuppertal
Germany

O. M. Knio
Dept. of Mechanical Engineering
The John Hopkins University
Baltimore, MD 21218
USA

N. Kornev
Dept. of Hydromechanics
Marine Technical University
Lotsmanskaya Str. 3
St. Petersburg 190008
Russia

A. N. Kudryavtsev
Inst. of Theoretical & Applied Mechanics
Russian Academy of Sciences
Institutskaya st., 4/1
Novosibirsk 630090
Russia

H. C. Kuhlmann
ZARM
Universität Bremen
Am Fallturm
28359 Bremen
Germany

P. A. Kuibin
Inst. of Thermophysics
Russian Academy of Sciences
Institutskaya st.,4/1
Novosibirsk 630090
Russia

K. Kuwahara
Institute of Space and Astronautical Science
1-22-3, Haramachi
Meguro-ku, Tokyo 152
Japan

St. Le Dizès
Institut de Recherche sur les Phénomènes Hors Équilibre
12, Ave. Général Leclerc
13003 Marseille
France

J. G. Leishman
Dept. of Aerospace Engineering
University of Maryland at College Park
Maryland 20742
USA

M. Lesieur
Lab. des Écoulements Geophysiques et Industriels
Inst. de Mécanique de Grenoble
38041 Grenoble-cedex 9
France

Thomas Leweke
IRPHÉ, CNRS
Universités Aix-Marseille I et II
12, Avenue Général Leclerc
13003 Marseille
France

C. H. Liu
Theoretical Flow Physics Branch
NASA Langley Research Center
Mail Stop 128
Hampton, VA 23665-5225
USA

T. Loiseleux
LadHyX
Lab. d'Hydrodynamique
École Polytechnique
91128 Palaiseau Cedex
France

F. Lund
Universidad de Chile
Fac. de Ciencas Fisicas Y Matematicas
Dep. de Fisica
Avda. Blanco Encalada 2008
Casilla 487-3, Santiago
Chile

Daniel Margerit
Laboratoire d'Énergétique et de Mécanique Appliquée (LEMTA)
URA CNRS 875
2, Av. de la Foret de Haye, BP 160
54054 Vandoeuvre Les Nancy
France

J. E. Martin
Christopher Newport University
50 Shoe Lane
Newport News, VA 23606-299
USA

H. Maru
Dept. of Computational Science
and Engineering
Nagoya University
Nagoya 464-01
Japan

E. Meiburg
Dept. of Aerospace & Engineering
University of Southern California
Los Angeles, CA 90089-1191
USA

G. E. A. Meier
DLR-Institut für Strömungsmechanik
Bunsenstr. 10
37073 Göttingen
Germany

M. Meinke
Aerodynamisches Institut
RWTH Aachen
Wüllnerstr. zw. 5 u. 7
52062 Aachen
Germany

H. K. Moffatt
Isaac Newton Institute for
Mathematical Sciences
Clarkson Road
Cambridge CB3 9EW
United Kingdom

M. Nitsche
100 Mathematics Building
Ohio State University
231 West 18th Avenue
Columbus, OH 43210-1174
USA

F. Novak
Naval Postgraduate School
Monterey, CA 93943
USA

V. L. Okulov
Institute of Thermophysics
Siberian Branch of RAS
Lavrentyev Ave., 1
630090 Novosibirsk
Russia

C. Olendraru
LadHyX
Lab. D'Hydrodynamique
École Polytechnique
91128 Palaiseau Cedex
France

M. Schmitz
Inst. für Technische Mechanik
RWTH Aachen
Templergraben 64
52056 Aachen
Germany

H. J. Rath
ZARM
University of Bremen
Am Fallturm
28359 Bremen
Germany

G. Reichert
Technical University Braunschweig
Schleinitzstr. 20
38023 Braunschweig
Germany

J. W. Russell
School of Aerospace Engineering
Georgia Institute of Technology
Atlanta, GA 30332-0150
USA

Lakshmi Sankar
School of Aerospace Engineering
Georgia Techn. University
Mail Stop 0150
Atlanta, GA 30332-0150
USA

T. Sarpkaya
Dept. of Mechanical Engineering
Naval Posatgraduate School
700 Dyer Rd., Rm. 339
Monterey, CA 93943-5100
USA

K. Schneider
Centre de Physique Théorique
CNRS-Luminy, Case 907
13288 Marseille Cedex 09
France

Wade Schoppa
Dept. of Mechanical Engineering
University of Houston
Houston, TX 77204-4792
USA

B. Yu. Scobelev
Institute of Theoretical and
Applied Mechanics
Russian Academy of Sciences
Novosibirsk 630090
Russia

A. Sellier
LadHyX
Lab. d'Hydrodynamique
École Polytechnique
91128 Palaiseau cedex
France

P. M. Sforza
Mech. & Aerospace Engineering Dept.
Polytechnic University of Brooklyn
Route 110
Farmingdale, New York 11735
USA

E. F. Sheta
Aerospace Engineering Dept.
Old Dominion University
Norfolk, VA 23529-0247
USA

O. A. Shmagunov
Inst. of Theoretical & Applied Mechanics
of the Russian Academy of Sciences
4/1 Institutskaya st.,
630090 Novosibirsk
Russia

V. N. Shtern
Dept. of Mechanical Engineering
University of Houston
Houston, TX 77204-4792
USA

S. Shtork
Institute of Thermophysics
Siberian Branch of RAS
Lavrentyev Ave., 1
Novosibirsk 630090
Russia

R. Spall
Dept. of Mechanical and
Aerospace Engineering
Utah State University
Logan, UT 84322-4130
USA

M. Tanaka
Kyoto Institute of Technology
Kyoto 606
Japan

Lu Ting
Courant Institute of Mathematical Sciences
New York University
251, Mercer Street
New York, N.Y. 10012
USA

V. Treshkov
Marine Technical University
St. Petersburg
Lotsmanskaya Str. 3
190008 St. Petersburg
Russia

Chee Tung
Aeroflight Dynamics Directorate
US Army Aviation and Troop Command
Ames Research Center
Moffet Field, CA 94035
USA

M. Wanschura
Center of Applied Space Technology
and Microgravity
University of Bremen
Am Fallturm
28359 Bremen
Germany

M. Weimer
Aerodynamisches Institut
RWTH Aachen
Wüllnerstr. zw. 5 u. 7
52062 Aachen
Germany

C. H. K. Williamson
Sibley School of Mechanical and
Aerospace and Engineering
Cornell University
Ithaca, NY 14853
USA

S. Yanase
Okayama University
Okayama 700
Japan

Session 1

Asymptotic Theories

ASYMPTOTIC THEORY OF SLENDER VORTEX FILAMENTS – OLD AND NEW

L. TING
Courant Institute of Mathematical Sciences
New York University, New York, NY 10012, USA

R. KLEIN
Fachbereich Sicherheitstechnik
Bergische Universität, D-42097 Wuppertal, Germany

AND

O. M. KNIO
Department of Mechanical Engineering
The Johns Hopkins University, Baltimore, MD 21218, USA

Abstract. We give a brief review of the asymptotic theory of slender vortex filaments to emphasize i) the choices of scalings, small parameters and the distinguished limit, ii) the consistency conditions, iii) the optimum similar and non-similar viscous vortical core structures and iv) their applications to complement experimental investigations. We present highlights of several extensions of the asymptotic theory: the analyses for core structures with axial variation, for the interaction of filaments with a solid body and sound generation and for a filament in a background rotational flow. We then outline the vortical flow problems currently under investigation.

1. Introduction

We consider an incompressible flow induced by an initial vorticity distribution concentrated in a slender tube-like region forming a torus, known as a slender vortex filament. Let $\mathbf{x} = \mathbf{X}(t, s)$ be the equation of the filament centerline $\mathcal{C}$ for time $t \geq 0$, with the parameter s being its initial arc length. The unit vectors $\hat{\tau}, \hat{n}$ and $\hat{b}$ denote the tangent, normal and binormal vectors of $\mathcal{C}$. Let σ, κ and T denote its linear strain, curvature and torsion. With its initial length S_0, we have

E. Krause and K. Gersten (eds.), IUTAM Symposium on Dynamics of Slender Vortices, 3-20.

the periodicity condition

$$\mathbf{X}(t, s + S_0) = \mathbf{X}(t, s) \ . \tag{1}$$

We choose the direction of $ds > 0$ or that of $\hat{\tau}$ so that the strength or the circulation of the filament $\Gamma > 0$. It is well known that the velocity of a filament depends on the vorticity distribution in the core and the velocity is undefined when the filament is modeled by a vortex line of zero core radius (Lamb, (1932)). The velocity $\mathbf{Q}$ induced by a vortex line is given by the Biot-Savart formula

$$\mathbf{Q}(t, \mathbf{x}) = [\Gamma/(4\pi)] \int_C [\mathbf{X}' - \mathbf{x}] \times d\mathbf{X}'/|\mathbf{X}' - \mathbf{x}|^3. \tag{2}$$

For a point $\mathbf{x}$ in the neighborhood $\mathcal{C}$, i. e., in the normal plane of point $\mathbf{X}(t, s)$ on $\mathcal{C}$, we write $\mathbf{x} = \mathbf{X} + r\hat{r}$ with $\hat{r} = \hat{n}\cos\phi + \hat{b}\sin\phi$ where r and ϕ are the polar coordinates. In approaching the vortex line $\mathcal{C}$, i. e., $r = |\mathbf{x} - \mathbf{X}| \to 0$, the formula gives three singular terms (Callegari & Ting, (1978)),

$$\mathbf{Q} = [\Gamma/(2\pi r)]\hat{\theta} + [\Gamma\kappa/(4\pi)]\ln[S/r]\hat{b} + [\Gamma\kappa/(4\pi)]\hat{\theta}\cos\phi + \mathbf{Q}^f \ , \tag{3}$$

where $\mathbf{Q}^f$ denotes the remainder of the Biot-Savart integral, and $\hat{\theta} = \hat{\tau} \times \hat{r}$ denotes the unit circumferential vector. The first term with $1/r$ singularity corresponds to the circumferential velocity of a 2-D vortex point in the normal plane. The second term, the binormal velocity with $\ln r$ singularity, represents the curvature effect. The third term, a circumferential velocity depending on ϕ, does not have a limit as $r \to 0$. After subtracting these three singular terms from $\mathbf{Q}$, the remainder $\mathbf{Q}^f$ has a limit, known as the finite part of the integral. These singular behaviors are not valid for a real fluid because with viscosity the flow field has to have continuous velocity gradient. Thus the Biot-Savart formula is valid away from the vortex line or slender filament. Near the filament these singular terms shall be matched with or removed by the *inner solution* for the core structure using the method of matched asymptotics. For a slender filament, there are two typical length scales, the core size δ and the length ℓ of the flow field. With the slenderness ratio $\epsilon = \delta/\ell$, as the small parameter, matched asymptotic analyses were carried out in a sequence of papers, (Ting & Tung, (1965); Tung & Ting, (1967); Ting, (1971); Callegari & Ting, (1978)), from the two-dimensional cases to the three dimensional cases with large circumferential and axial velocity components in the core. A comprehensive account of those asymptotic analyses was given by Ting and Klein (1991).

In Section **2**, we presented highlights of those published or *old* investigations, with emphases on the motivation and/or the physical meaning of the distinguished limit, the expansion scheme, the consistency conditions, and the optimum similar and non-similar core structures, and their applications. Results obtained recently are presented in three papers in this volume, One implements the theory of C-T, (Callegari & Ting, (1978)), into the vortex elements method to render the numerical scheme more accurate and faster by an order of magnitude (Klein & Knio,

(1997)). The second one applies the equations of Klein and Ting (1995) for filament with axial core structure variation to analyze vortex breakdown (Schmitz & Klein, (1997)). The third one applies the formulas obtained recently by Knio and Ting (1997) to the numerical simulations of the interaction of a filament with a rigid sphere and sound generation (Knio & Ting, (1997a)) and (Knio, Ting & Klein, (1998)). Section **3** gives a brief account of our research in progress, namely the classification of the interactions of three coaxial rings and the dynamics of long filaments.

2. Highlights of the asymptotic analyses

In Subsection **2.1**, we explain the choice of time and length scales characterizing a slender filament and the *distinguished limit* so that the solution accounts for both the effect of slenderness and viscous diffusion. The asymptotic analysis reduces the Navier-Stokes (N-S) equations to a simpler system of equations. The reduced system in turn imposes some restrictions or consistency conditions on its solution and hence on the initial conditions, i. e., some initial conditions for the N-S equations will be *lost*. In Subsection **2.2**, we state the expansion scheme and show where these consistency conditions come from, explain their physical meaning and indicate how to introduce multi-time scales to take care of those *lost* initial conditions. In Subsection **2.3**, we describe the solution of C-T for a filament core structure without axial variation, and explain the meaning of the *optimum* similarity solutions. In Subsection **2.4**, we explain how theoretical and experimental investigations can complement each other to achieve a better understanding of the dynamics of vortex filament (Klein & Ting, (1995)). In Subsection **2.5**, we outline the analyses extending the asymptotic theory for a background potential flow to a background flow with $O(1)$ vorticity (Liu & Ting, (1987); Ishii & Liu, (1987)).

2.1. CHOICE OF SCALINGS AND THE DISTINGUISHED LIMIT

For a slender filament, there are two distinct length scales, the typical core size δ^*, i. e., the length scale for the inner region, and the typical length scale for the background flow ℓ, the length scale for the outer region. We consider the flow to be incompressible with density ρ and kinematic viscosity ν. Let U be the reference background velocity, we consider the strength of the filament $\Gamma = O(U\ell)$ and the radius of curvature of $\mathcal{C}$, $1/\kappa = O(\ell)$. In case there is no background flow, we define $U = \Gamma/\ell$ and ℓ as the typical radius of curvature. We use ρ, U and ℓ as the unit density, velocity and length scales, and $t^* = \ell/U$ as the unit time scale. Thus, we can use the same symbols $\mathbf{v}$, $\mathbf{x}$ and t as the physical and dimensionless velocity, position vector and time. We can say $\delta^* = O(\epsilon\ell)$ or $O(\epsilon)$ etc. Let the pressure deviation from the ambient pressure be scaled by ρU^2 and denote it by p. To simplify the formulation of the asymptotic theory, we assume that (1) the background flow is irrotational and (2) the filament length S is finite $O(\ell)$ forming a torus. Assumptions (1 & 2) will be removed in Subsections **2.5** & **3.2**.

Since a real fluid is viscous, our problem is characterized by two small parameters, the slenderness ratio $\epsilon = \delta^*/\ell$ and the physical parameter, the inverse of the background Reynolds number, $1/R_e = \nu/(U\ell)$. We consider $R_e \gg 1$ so that the viscous effect in the background flow is of higher order while in the core structure the viscous effect is important when the core size is very small initially and the effect is accumulated over a long time. We choose the *distinguished limit*,

$$1/\sqrt{R_e} = O(\epsilon) \quad \text{or} \quad 1/\sqrt{R_e} \leq K\epsilon \ , \tag{4}$$

so that the equations for the evolution of the core structure retain the nonlinear convection (inviscid stretching) and viscous diffusion terms. The constant K represents a typical ratio of viscous to inviscid effect. The inviscid limit is $K \to 0$.

From the balance of the unsteady and the diffusion terms, we get the diffusion time scale, $t_d = (\delta^*)^2/\nu$, over which the effective core size changes by a finite factor. Then K^2 is the ratio of the inviscid and viscous time scales,

$$K^2 = \ell^2\nu/(\delta^2 U\ell) = t^*/t_d. \tag{5}$$

Consider the duration of observation of a filament in nature or in an experiment to be $O(t^*)$. If the duration of observation is much shorter than the diffusion time, $t^*/t_d \ll 1$, the diffusion effect is ignorable, i. e., the solution in the duration is related to the initial data by the inviscid theory.

2.2. EXPANSION SCHEME AND THE CONSISTENCY CONDITIONS

To concentrate on the development of the asymptotic theory, we consider the filament to be submerged in a background potential flow and write the velocity as

$$\mathbf{v} = \nabla\Phi + \mathbf{Q} \ . \tag{6}$$

Here Φ is the velocity potential without the filament, but contains the outer solutions of all the other filaments, if any, and $\mathbf{Q}$ is the velocity induced by the filament. The latter is given by (2) in the outer region, i. e., away from the filament. We consider the background velocity, $\nabla\Phi$, to be $O(1)$. Since $\delta^2\Omega = O(\Gamma)$ and $\delta\mathbf{v} = O(\Gamma)$, we have the order of magnitude of the vorticity and velocity:

$$|\Omega| = O(\epsilon^{-2}) \qquad \text{and} \quad |\mathbf{v}| = O(\epsilon^{-1}) \ . \tag{7}$$

To construct the inner solution, i. e., the core structure near a point $\mathbf{X}$ on the centerline, we introduce the intrinsic coordinates, r, ϕ, s used in (3) and then replace the polar angle ϕ in the normal plane by the angle

$$\theta = \phi - \theta_0(t,s) \ , \qquad \text{with} \quad \partial\theta_0/\partial s = -\sigma T \ , \tag{8}$$

so that the new coordinates, r, θ, s, are orthogonal with

$$d\mathbf{x} = \hat{r}\ dr + \hat{\theta}\ r d\theta + \hat{\tau}\ \sigma[1 - r\kappa\cos(\theta + \theta_0)]ds\ . \tag{9}$$

The rotation θ_0 accounts for the torsion $T(t, s)$ of $\mathcal{C}$. It is interesting to note that θ_0 is the argument of the complex "filament function" introduced by Hasimoto (1972) in his local induction approximation. The argument plays the same role of rotation of the axes $\hat{n}$ and $\hat{b}$ to $\hat{n}'$ and $\hat{b}'$ in the normal plane.

For the inner region, i. e., $r = O(\epsilon)$, we introduce the stretched radial variable $\bar{r} = r/\epsilon$. With $|\mathbf{v}| = O(\epsilon^{-1})$, (7), and the assumption that the velocity of the filament $\dot{\mathbf{X}}/U = O(1)$, the relative velocity $\mathbf{V} = u\hat{r} + v\hat{\theta} + w\hat{\tau} = \mathbf{v} - \dot{\mathbf{X}}$ is $O(\epsilon^{-1})$. To characterize a slender vortex filament, we need a large circumferential velocity, $v = O(\epsilon^{-1})$, and allow for a large axial flow w, but the radial component u remains $O(1)$. Note that the inner solutions are periodic in θ with period 2π and in s with period S_0 on account of (1). Using this physical model, we arrive at the expansion scheme,

$$f(t, \bar{r}, \theta, s, \epsilon) = \epsilon^{-m}\{f^{(0)}(t, \bar{r}, \theta, s) + \epsilon f^{(1)}(t, \bar{r}, \theta, s) + O(\epsilon^2)\}\ . \tag{10}$$

We consider $\ln(1/\epsilon)$ to be $O(1)$ with respect to ϵ, therefore, $f^{(j)}, j = 0, \cdots$, could be functions of $\ln(1/\epsilon)$. If f stands for $\mathbf{v}$, we have $m = 1$, with $u^{(0)} \equiv 0$. If f stands for p, we obtain $m = 2$ from the radial momentum equation. We substitute the power series (10) into the N-S equations, equate the coefficients of like powers of ϵ, and obtain systematically the sets of leading and higher order equations.

With $u^{(0)} \equiv 0$, we obtain from the leading order continuity equation, $v^{(0)}_{\theta} = 0$, the axial momentum equation, $w^{(0)}_{\theta} = 0$, and the radial momentum equation, $p^{(0)}_{\bar{r}} = \rho[v^{(0)}]^2/\bar{r}$. Then the circumferential momentum equation becomes an identity. The solution, independent of θ with t, s as parameters, represents a quasi steady two-dimensional flow in the normal plane of $\mathcal{C}$. Thus we say:

C.1) The leading core structure is axi-symmetric with respect to the centerline $\mathcal{C}$. Note that the dependence of the core structure, $v^{(0)}$ and $w^{(0)}$ on $t, \bar{r}$, and s are not yet defined. They shall be defined later by the compatibility conditions of the second order equations in Subsection **2.3**.

For the next order (the first order) system of equations, the pressure, $p^{(1)}$, appears only in the radial momentum equation, which is considered as the equation for $p^{(1)}$, similar to that in the leading order system. The remaining three first order equations are linear equations for the velocity components, $u^{(1)}, v^{(1)}$ and $w^{(1)}$, with inhomogeneous terms nonlinear in the leading order solutions. Because of the distinguished limit (4) and our assumption of only one time scale t^*, the viscous terms and the time derivatives do not appear in the first order equations (but will appear in the second order equations). Because of the linear equations, the first order solutions can be written as the sum of terms symmetric and asymmetric in θ,

and the θ-averages of those equations yield a system of quasi-steady inviscid equations for the axi-symmetric parts of the solutions. Consequently, the core structure for $t \geq 0$ has to be consistent with the following two classical relationships:
C.2) The circulation $\mathcal{G} = 2\pi\bar{r}v^{(0)}$ around a axi-symmetric stream tube is independent of s, i. e., remains constant along the tube, (The Helmholtz Theorem).
C.3) The total head, $\mathcal{H} = [(v^{(0)})^2 + (w^{(0)})^2]/2 + p^{(0)}/\rho$, on an axi-symmetric stream tube remains constant, (The Bernoulli's Equation).
If we denote the mass flux through an axi-symmetric stream tube by $\psi(t, r, s) = 2\pi \int_0^{\bar{r}} w^{(0)}(t, \tilde{r}, s)\tilde{r}d\tilde{r}$, then *C.2)* and *C.3)* say that the circulation $\mathcal{G}(t, \bar{r}, s)$ and the total head $\mathcal{H}(t, \bar{r}, s)$ can be expressed as functions two variables t and ψ., (Klein & Ting, (1992)).

By matching the asymmetric inner solution of the core structure with the outer solution for $r \to 0$, *the second and the third singular terms* in the Biot-Savart integral (3) are removed and the velocity of the centerline, $\dot{\mathbf{X}}$, is defined by

$$\dot{\mathbf{X}}(t, s) = \hat{n}[\mathbf{Q}_0 \cdot \hat{n}] + \hat{b}[\mathbf{Q}_0 \cdot \hat{b}] + \hat{b}[\Gamma\kappa/(4\pi)][\ln(1/\epsilon) + C_v + C_w] \,, \tag{11}$$

where $\mathbf{Q}_0(t, \mathbf{X}) = \nabla\Phi + \mathbf{Q}^f$ denotes the background velocity without the filament plus the finite part of $\mathbf{Q}$ in (3). $C_v(t, s)$ and $C_w(t, s)$ denote respectively the contributions of the circumferential and axial velocity in the core to $\dot{\mathbf{X}}$. They are,

$$C_v = \lim_{\bar{r} \to \infty} \left(\frac{4\pi^2}{\Gamma^2} \int_0^{\bar{r}} \bar{r}' v^2 \, d\bar{r}' - \ln \bar{r}\right) + \frac{1}{2} \,, \quad C_w = \frac{-8\pi^2}{\Gamma^2} \int_0^{\infty} \bar{r}' w^2 \, d\bar{r}' \,. \tag{12}$$

With $\ln(1/\epsilon)$ considered $O(1)$, we treat the terms in (11) as $O(1)$ while terms $O(\epsilon)$ have been omitted. Their omission implies that the superscripts (0) for $\mathbf{X}$ and its geometrical entities, $\sigma, \kappa, T, \hat{r}$ etc have been suppressed, i. e., we are considering $\mathbf{X}(t, s, \epsilon = 0)$. The terms omitted can come from the higher order core structure and from the effect of nonzero core size not accounted for in the Biot Savart formula (2). We recall that the formula defines the outer velocity induced by a vortex line, i. e., a filament of zero core size. Here we have a small but nonzero core size. We should derive the outer solution for a slender filament with large axial vorticity distribution of total strength $\Gamma > 0$ and show that the leading term $O(1)$ agrees with the Biot-Savart formula while the next order contribution is $O(\epsilon^2)$ because of axi-symmetry. Instead of proving the preceding statement, we use the same procedure to derive a formula for the contribution to the outer solution induced by the large axial flow, or rather the circumferential vorticity component $\epsilon^{-2}\omega(t, \bar{r}) = -\epsilon^{-2}[w^{(0)}(t, \bar{r})]_{\bar{r}}$. We need to show that this additional contribution to (11) is small, $O(\epsilon)$. Since the vector potential is related linearly to the vorticity field, we evaluate the vector potential $\mathbf{A}$ induced by the large circumferential vorticity vector $\epsilon^{-2}\omega\hat{\theta}$ separately from that by the axial vorticity. For a point $\mathbf{x}'$ in the vortical core, we write $\mathbf{x}'(s) = \mathbf{X}(s) + r\hat{r}(\phi, s)$, with $\phi = \theta + \theta_0(s)$. The dependence of the variables on t, being a parameter here, has been suppressed. Then the vector potential at point $\mathbf{x}$, using (9) for the orthogonal variables, becomes:

$$\mathbf{A} = \frac{1}{4\pi}\int_0^{S_0}\int_0^{\infty} \mathbf{F}(\bar{r},s)\, d\bar{r}ds \ , \quad \text{with } \mathbf{F} = \sigma\omega\bar{r}\int_0^{2\pi}\frac{\hat{\theta}[1-\epsilon\kappa\cos\phi]\, d\theta}{|\mathbf{R}-\epsilon\bar{r}\hat{r}(\phi,s)|} , \qquad (13)$$

where $\mathbf{R}(s) = \mathbf{x}-\mathbf{X}$. For point $\mathbf{x}$ in the outer region, where $|\mathbf{R}| \gg \epsilon$, we express the integrand of $\mathbf{F}$ in a power series of ϵ as, $[1-\epsilon(\kappa\cos\phi - \bar{r}\hat{r}\cdot\mathbf{R}/|\mathbf{R}|^{-2}) + O(\epsilon^2)]\ \hat{\theta}/|\mathbf{R}|$, We use $< f >$ to denote $\int_0^{2\pi} f d\theta$ and note that: $< \hat{\theta} >= 0$, $< \cos\phi\ \hat{\theta} > = \pi\hat{n}$, and $< (\mathbf{R}\cdot\hat{r})\hat{\theta} > = - < (\mathbf{R}\cdot\hat{\theta})\hat{r} > = \frac{1}{2} < (\mathbf{R}\cdot\hat{\theta})\hat{\theta} - (\mathbf{R}\cdot\hat{\theta})\hat{r} > = -\pi\mathbf{R}\times\hat{\tau}$. Eq. (13) then becomes

$$\mathbf{A}(\mathbf{x}) = \frac{\epsilon}{4}\int_0^{S_0}\frac{\kappa\hat{n} + \mathbf{R}\times\hat{\tau}|\mathbf{R}|^{-2}}{|\mathbf{R}|}\{\int_0^{\infty}\omega\bar{r}\ d\bar{r}\}\sigma\ ds\ + O(\epsilon^2)\ . \qquad (14)$$

The integral of $\omega\bar{r}$ in the curly brackets is equal to that of the axial velocity, $\int_0^{\infty} w^{(0)}\ d\bar{r}$. Eq. (14) says that,
C.4) The outer velocity induced by the circumferential vorticity $\epsilon^{-2}\omega$ in the core is one order smaller than that induced by the axial vorticity of strength Γ.

Since the unsteady and viscous terms do not appear in the leading and first order equations, the consistency conditions, *C.1), C.2)* and *C.3)* and Eq. (11) are valid for a fluid viscous or inviscid, i. e., for all $K \geq 0$. Early on, Callegari and Ting (1978) found a special condition on the core structure, fulfilling both *C.2)* and *C.3)*. The condition says that,
C.5) If the circumferential velocity is independent of s, then the axial component should also be independent of s. The converse is also true. This condition holds for a core structure viscous or inviscid.
They pointed out that a patched inviscid core structure (Moore & Saffman, (1972)) with a circumferential velocity independent of s but an axial velocity depending on s is not admissible for a slender filament.

The asymptotic theory requires that the initial core structure should be consistent with conditions, *C.1), C.2)* and *C.3)* and initial velocity of the filament centerline, if assigned, has to agree with (11). These restrictions on the initial data are the results of having only one time scale. In case the initial data violate *C.2)* and/or *C.3)* we should introduce a short time scale ϵt^* so that unsteady terms will appear in the first order equations. If the assigned initial velocity of the filament differs from (11) we need an even shorter time scale $\epsilon^2 t^*$, (Ting & Tung, (1965); Ting, (1971); Ting & Klein, (1991)).

2.3. THE OPTIMUM SIMILAR AND NON-SIMILAR CORE STRUCTURES

The second order equations involve the first and second order unknowns, $u^{(1)}, u^{(2)}$, etc. To remove those unknowns, we make use of the periodicity of the solution in θ and s, integrate those equations with respect to θ over the period 2π and then with respect to s over the period S_0, and arrive at two *compatibility* conditions on the inhomogeneous terms. They are the evolution equations of the core structure, $v^{(0)}(t, \bar{r}, s)$ and $w^{(0)}(t, \bar{r}, s)$. These two equations coupled with Eq. (11) for $\dot{\mathbf{X}}$ form a closed system of intro-differential equations in $t, s, \bar{r}$ for the filament dynamics. This system is extremely complex (Klein & Ting, (1992)). Theoretical analyses of the solutions are feasible only for some special cases (Schmitz & Klein, (1997)). Here, we shall summarize the analysis of Callegari and Ting (1978) for the special case that the core structure does not vary along the filament.

With the core structure depending only on t and $\bar{r}$, the the evolution equations of the axial velocity $w^{(0)}$ and vorticity $\zeta^{(0)}$ are:

$$w_t^{(0)} = (K^2/\bar{r})[\bar{r}w_{\bar{r}}^{(0)}]_{\bar{r}} + (\bar{r}^3/2)(\dot{S}/S)[w^{(0)}/\bar{r}^2]_{\bar{r}} \,, \tag{15}$$

$$\zeta_t^{(0)} = (K^2/\bar{r})[\bar{r}\zeta_{\bar{r}}^{(0)}]_{\bar{r}} + (\dot{S}/S)(\bar{r}^2\zeta^{(0)})_{\bar{r}}/(2\bar{r}) \,. \tag{16}$$

From these two equations we obtain two invariants respectively; one relating the filament length $S(t)$ and the axial mass flux $\mathcal{M}(t) = 2\pi \int_0^\infty w^{(0)}(t, \bar{r})\bar{r}d\bar{r}$, and one for the conservation of total axial vorticity. They are:

$$S^2(t)\mathcal{M}(t) = S_0^2\mathcal{M}_0 \,, \quad \text{and} \quad 2\pi \int_0^\infty \zeta^{(0)}(t, \bar{r})\bar{r}d\bar{r} = \Gamma \,. \tag{17}$$

Here the subscript 0 denotes the initial value. We use the axial vorticity as a primary variable instead of the circumferential velocity, because the formal decays exponentially in $\bar{r}$, while the latter, related to the axial vorticity by, $\bar{r}v^{(0)} = \int_0^{\bar{r}} \zeta^{(0)}(t, \tilde{r})\tilde{r}d\tilde{r}$, becomes $\Gamma/(2\pi\bar{r})$ as $\bar{r} \to \infty$. Thus the inner solution $v^{(0)}$ matches or removes the leading singular term of the Biot-Savart integral (3). Recall that the second and third singular terms were removed by the asymmetric part of the first order solutions in Subsection **2.2**, therefore, we have shown that *the leading and first order inner solutions have matched or removed all three singular terms of the outer solution* (3).

The terms in (15 & 16) with the factor K^2 account for the effect of viscous diffusion. As $K \to 0$, they reduce to the inviscid equations and their solutions are:

$$w^{(0)}(t, \bar{r}) = w_0(\beta)S_0/S(t), \quad \zeta^{(0)}(t, \bar{r}) = \zeta_0(\beta)S(t)/S_0 \quad \text{with} \quad \beta = \bar{r}\sqrt{S(t)/S_0}. \tag{18}$$

Equations (15 & 16), are linear parabolic equations in t and $\bar{r}$ for $w^{(0)}$ and $\zeta^{(0)}$ respectively with coefficients depending on the filament length $S(t)$, which in turn requires the solution of (11), a nonlinear equation for $\mathbf{X}$ in t and s. Therefore, we need to decouple these three equations. Before listing the transformations of the independent and dependent variables to new ones introduced in (Callegari & Ting, (1978)), we explain how did we conceive the feasibility of the decoupling and the transformations by tracing the earlier analyses for simpler cases.

The asymptotic analysis for a two-dimensional vortex (Ting & Tung, (1965)) shows that the evolution equation of the core structure, the axial vorticity, is governed by the two-dimensional axi-symmetric heat conduction equation,

$$[\partial_\tau - (K^2/\alpha)\partial_\alpha(\alpha\partial_\alpha)]\chi(\tau,\alpha) = 0 \tag{19}$$

where τ and α denote the time and radial variable. We now give a brief summary of the solution of (19) relevant to our analysis.

If the initial data, $\chi_0(\alpha) = \chi(0,\alpha)$ is a delta function, representing a "heat source" or point vortex of strength $\mathcal{S}$ at the origin, we have the similarity solution,

$$\chi^*(\tau,\alpha) = [\mathcal{S}/(\pi\xi^2)]\exp(-\eta^2), \quad \text{with } \xi = \sqrt{4K\tau},\ \eta = \alpha/\xi\ . \tag{20}$$

Here η is called the similarity variable and ξ the effective core radius. If the initial data is the similarity solution with radius $\xi_0 = \sqrt{4K\tau_0}$, i. e., a point vortex created at $\tau = -\tau_0$, the solution is given by (20) with a time shift of τ_0, i. e., $\chi^*(\tau+\tau_0,\alpha)$.

If the initial data $\chi_0(\alpha)$ does not correspond to a similarity solution but decays exponentially in α, the solution of (19) can be expressed as a descending power series in $(\tau+\tau_0)$, with a positive "time shift", $\tau_0 > 0$, to be chosen later. The solution is

$$\chi(\tau,\alpha) = \frac{e^{-\eta^2}}{\tau_0+\tau}\{\sum_{n=0}^{N} c_n L_n(\eta^2)[\frac{\tau_0}{\tau+\tau_0}]^n + R_N\}\ . \tag{21}$$

Here, L_n, $n = 0, 1, \cdots$ denote the Laguerre polynomials and the coefficients, $c_n = \int_0^\infty \chi_0(\alpha)L_n(\lambda)\,d\lambda$, are defined by the first n-moments of the initial profile $\chi_0(\alpha)$ with $\alpha = 2K\sqrt{\lambda\tau_0}$.

From (21), we see that as $\tau \to \infty$, χ approaches to the first term, which is the similarity solution (20) with initial age τ_0. An optimum choice of the initial age τ_0 was proposed by Kleinstein and Ting (1971) by the condition that the second term in the series (21) vanishes, i. e., $c_1 = 0$. Then the first term becomes the optimum similarity solution, the best approximation to χ, because the error is reduced from $O(\tau^{-1})$ to $O(\tau^{-2})$. The condition of $c_1 = 0$ also renders the optimum similarity solution having the strength and polar moment of χ not only for $\tau = 0$ but also for $\tau > 0$, because they are time invariant. For engineers, the approximation becomes good after a finite duration, say $\tau = 2\tau_0$ (Kleinstein & Ting, (1971)). This says that: *the solution with a non-similar initial profile will soon approach the optimum similarity solution in the diffusion time scale,* τ_0.

For a point vortex, we identify χ, τ, α and $\mathcal{S}$ as ζ, t, $\bar{r}$ and Γ and obtain the classical Lamb vortex created at $t = 0$ (Ting & Tung, (1965)). For a nonsimilar initial profile, the optimum time shift τ_0, and the series solution for the core structure were presented in (Ting, (1971)).

For a slender circular vortex ring submerged in an axi-symmetric flow (Tung & Ting, (1967)), the axial vorticity ζ obeys (16) with $S(t) = 2\pi R(t)$ where R denotes the ring radius. Note that the ratio $\dot{R}/R$ or $\dot{S}(t)/S(t)$ is independent of s

and represents not only the global but also the local rate of strain of the centerline $\mathcal{C}$. For a ring of zero initial core radius, there is no length scale in the initial core structure. Hence we follow the standard procedure to seek a similarity variable η and a solution separable in t and η, i. e., $\zeta = g(t)f(\eta)$, (Tung & Ting, (1967)). The similarity variable and solution are then identified as that for a two dimensional vortex with $g(t) = \Gamma/[\pi\xi^2]$, $f(\eta) = \exp(-\eta^2)$, $\eta = \bar{r}/\xi$ and the core size $\xi = \sqrt{4K\tau}$ using the new time variable, $\tau = \int_0^t R(t')dt'/R(t)$.

For a filament with only large circumferential flow, the core structure has to be independent of s, i. e., the same along the centerline (Ting, (1971)). The equation of motion of $\mathcal{C}$ is (11) with the contribution of the large axial flow $C_w = 0$. The evolution equation for the axial vorticity is (16) and the similarity solution for a vortex line created at $t = 0$ is identified as that of a two dimensional solution (20) in the same manner as the axi-symmetric case with $\dot{R}/R$ replaced by $\dot{S}/S$, i. e., replacing t by $\tau = \int_0^t S(t')dt'/S(t)$. Now $\dot{S}/S$ is the rate of strain of $\mathcal{C}$, with the local rate $\dot{\sigma}/\sigma$ s-dependent. Again the solution in (τ, η) is independent of $S(t)$.

If the initial axial vorticity ζ_0 is nonsimilar, then it is simpler to seek transformations of independent and dependent variables (Callegari & Ting, (1978)), so that (16) for $\zeta^{(0)}$ is transformed to the simple heat conduction equation (19). The transformations of the variables are guided by the invariant condition on $\zeta^{(0)}$ in (17) and that the leading term in the series solution (21) is the similarity solution for a filament with initial age τ_0. Likewise, we get the series solution for the axial velocity $w^{(0)}$ governed by (15).

For both equations (15) and (16), we change the independent variables t and $\bar{r}$ to the variables $\tau = \int_0^t S(t')dt'/S_0$ and α in (19). For the axial velocity, we identify $\chi(\tau, \alpha)$ as $w^{(0)}(t, \bar{r})[S(t)/S_0]$, with the initial condition, $\chi_0(\alpha) = w_0^{(0)}(\alpha)$ and for the axial vorticity, we identify $\chi(\tau, \alpha)$ as $\zeta^{(0)}(t, \bar{r})[S_0/S(t)]$, with $\chi_0(\alpha) = \zeta_0^{(0)}(\alpha)$.

For both series solutions, we choose the optimum time shift τ_0 making the similarity solution for the axial vorticity the optimum, because the leading outer solution induced by the filament, given by the Biot-Savart formula, comes from the axial vorticity of the filament with total strength Γ. See (11) and (14).

Since the initial axial velocity and vorticity decay exponentially in $\bar{r}$, the corresponding two $\chi_0(\alpha)$'s decay exponentially in α. The series solutions (21) for the two $\chi(\tau, \alpha)$'s are independent of $S(t)$, with the coefficients defined by the moments of the corresponding initial profiles, $w_0^{(0)}$ and $\zeta_0^{(0)}$. Consequently the global contributions C_v and C_w in (11) are given functions of t, $S(t)/S_0$ and the coefficients of those two series solutions respectively (Liu, Tavantzts & Ting, (1986)). Eq. (11) then becomes a partial differential equation for the motion of the centerline $\mathbf{X}(t, s)$.

For a large t relative to the optimum initial age τ_0, the series solution (21) with $K > 0$ approaches the leading term, the optimum similarity solution and the global contributions of the core structure to the filament velocity (11) are,

$$C_w(t) = -2\mathcal{M}_0^2 S_0^4/[\Gamma^2\xi^2(t)S^4(t)] \quad \text{and} \quad C_v(t) = -\ln\xi(t) + 0.442\,, \tag{22}$$

where $\xi^2(t) = 4K[\tau + \tau_0]$ and $\tau(t) = S_0^{-1} \int_0^t S(t')dt' + \tau_0$ (Ting & Klein, (1991)).

If the duration t^* of an observation or experiment is much shorter than the diffusion time scale t_d, i. e., in the inviscid limit, $K \to 0$, the shape of the core structure is frozen with the amplitude and size varying according to the inviscid stretching rules (18). On the other hand, there is no need to specify an inviscid core structure, because its global contributions (12) to the filament velocity (11),

$$C_w(t) = [S_0/S(t)]^3 C_w(0), \quad \text{and} \quad C_v(t) = C_v(0) + \ln\sqrt{S(t)/S_0}\ , \tag{23}$$

are defined by their initial values (Klein & Ting, (1995)). Thus the motion of filaments with the same initial centerline and the same two constants $C_v(0)$ and $C_w(0)$ will be the same regardless of their initial profiles. Furthermore, from the differences between (22) and (23), we see that the global contributions C_w and C_v in the inviscid limit differ from those for a similar diffusive core structure.

2.4. COMPLEMENTING THEORY AND EXPERIMENTS

We note that the asymptotic solution, (Callegari & Ting, (1978)), requires the initial data; the centerline, $\mathbf{X}_0(s)$, and the axial vorticity and velocity profiles. The profiles can be approximated the first M and N terms of their series solutions, (Liu, Tavantzts & Ting, (1986)) and those coefficients are defined by the first N and M moments of the initial profiles. In a real problem or in an experiment, a slender vortex filament begins with a small but finite core size. From the finite core observed at an instant t_i, it centerline, i. e., the position vectors $\mathbf{X}_{ij}$ of its j-th point, $j = 1, \cdots J$ can be determined but not the core profiles. Here we shall elaborate on the scheme of Klein and Ting (1995), using both the theory (Callegari & Ting, (1978)) and experimental data on the centerline, $\mathbf{X}_{i,j}$, for several i's, to study the filament dynamics, in particular, to identify the initial profiles of a filament generated by an experimental device.

Klein and Ting (1995) proposed to express the contributions $C_v(t)$ and $C_w(t)$ to $\dot{\mathbf{X}}$,(12), as functions of t and the first N and M coefficients and determine these coefficients for the best fit to the experiment data. These coefficients in turn define the initial profiles via the first N and M moments of $\zeta_0^{(0)}$ and $w_0^{(0)}$. Here we shall do it in two steps. In the first step, we find the two functions $C_v(t)$ and $C_w(t)$ for the best fit to the experimental data. We could define the best fit at an instant t_i by minimizing the mean $\bar{d}_i$ of the distances $d_{i,j}$ from j-th point $\mathbf{X}_{i,j}$ to the theoretical centerline $\mathbf{X}(t_i, s)$ for $j = 1, \cdots J$. The accuracy of the fit is measured by the maximum of the mean $\bar{d}_i$ for $i = 1, \cdots, I$. In the second step we determine from the functions $C_v(t)$ and $C_w(t)$ the two sets of coefficients characterizing the initial profiles. To extend the theory to a turbulent core, we could consider K as an unknown parameter $K_e = \sqrt{\nu_e/U\ell}/\epsilon$ with the kinematic viscosity ν in K replaced by an unknown ν_e. From the fit to C_v and C_w, we get also K_e and then ν_e. If $\nu_e \approx \nu$, the core is laminar. If $\nu_e \gg \nu$, the core is turbulent with eddy viscosity ν_e.

After the first step, We can draw the following conclusions on the core structure: *(1)* If the theory and experiments fit well with $C_v(t)$ and $C_w(t)$ restricted by the inviscid theory (23) involving only two free constants $C_v(0)$ and $C_w(0)$, we know that the diffusion effect is negligible. *(2)* If instead they can be fit well with $C_v(t)$ and $C_w(t)$ restricted by (22) containing only two adjustable constants $\mathcal{M}_0$ and $\xi(0)$ for the initial similar profiles while the parameter K being adjustable for a turbulent core. *(3)* If they can be fit well by finding two functions $C_v(t)$ and $C_w(t)$ without the restriction of (23) or (22), then the core structure is nonsimilar and diffusive. We need to carry out the second step to determine the N and M coefficients in the two series solutions and the parameter K_e. From the value of K_e, we then know whether the core is laminar or turbulent. *(4)* If we can not find two functions of t for C_v and C_w such that the centerlines predicted by the theory fit well with those observed in the experiment, we conclude that the C-T solution is not applicable, because the core structure has an axial variation and/or the turbulence model by an eddy viscosity coefficient is no good.

2.5. FILAMENT IN A BACKGROUND ROTATIONAL FLOW

In formulating the asymptotic theory, we assume the background flow to be a potential flow in order to simplify the analysis of the outer solution. Since a potential flow in absence of a filament and boundary is an exact solution of the N-S equations, the outer solution is expressed in (6) as the superposition of the potential flow plus the velocity $\mathbf{Q}$ induced by the filament.

For filaments in a background rotational flow, we consider the background vorticity $\mathbf{W}(t,\mathbf{x})$ to be $O(U/\ell)$, or $O(1)$, with length scale ℓ. Near a filament, i. e., in the scale δ^*, the background vorticity in the normal plane of $\mathcal{C}$ is nearly uniform, $\mathbf{W}(t.\mathbf{x}) = \mathbf{W}(t,\mathbf{X})[1+O(\epsilon)]$. The redistribution of the background vorticity by convection will change it by at most $O(1)$, which is two orders smaller than the vorticity $\Omega = O(\epsilon^{-2})$ in the core (7). With this simple order of magnitude analysis, we expect that the evolution of the core structure and the equations of motion of the filament obtained before for a background potential flow remain valid for a rotational flow. On the other hand, the variation of the background vorticity $\mathbf{W}(t,\mathbf{x})$ is coupled with the motion of the filament, because the velocity $\mathbf{v}$ in the vorticity evolution equation, $\mathbf{W}_t + \nabla\cdot(\mathbf{v}\mathbf{W}) - \nabla\cdot(\mathbf{W}\mathbf{v}) = \nu\Delta\mathbf{W}$, contains the velocity $\mathbf{Q}$ induced by the moving filament. The viscous term is included in the equation so that the background flow can include a boundary layer along a surface. Away from the filament, $\mathbf{v}$ is $O(1)$, when we use the composite solution which removes the singular parts of $\mathbf{Q}$. For a filament with only large circumferential velocity $\epsilon^{-1}v^{(0)}$, the velocity changes $\mathbf{W}$ in the core by $O(\epsilon)$. Therefore, we can replace $\mathbf{v}$ by its average $<\mathbf{v}>$ in the sense of two-length analysis, so that the evolution equation of $\mathbf{W}$ can be solved numerically with grid size and time step independent of δ^*.

The above expectation is confirmed by carrying out the matched asymptotic analysis by Liu and Ting (1987) for a two-dimensional unsteady flow and numerical examples were presented to simulate the interaction of a pair of trailing vortices in a spanwise shear flow near the ground. Ishii and Liu (1987) carried out the analysis for axi-symmetric flows and presented numerical examples to simulate the interaction of vortex rings in a coaxial pipe and that of a ring approaching a wall normal to the axis of symmetry.

3. Research in progress

Currently we are studying the interaction of filaments with multiple length scales in the axial variation of a filament and/or in the background flow. These length scales are much larger than the typical core size δ^*. In Subsection **3.1**, we study the interaction of three coaxial rings to show how to classify the initial data for a systematic numerical investigation and physical understanding of the problem. In Subsection **3.2**, we describe multi-length scale problems for which the analyses can be carried out without the prior knowledge of whether the filament forms a torus, extends to infinity, or ends on a boundary.

3.1. INTERACTION OF THREE COAXIAL VORTEX RINGS

Due to symmetry, it suffices to study the motion of the three vortex centers in a meridian plane, the RZ plane, or the motion of the triangle formed by the vortex centers, (R_j, Z_j), $j = 1, 2, 3$.

Figure 1 shows the interaction of three rings with the same strength, $\Gamma_j = \Gamma$. Initially they are coplanar with $Z_j = 0$ and $R_1 - R_2 = R_2 - R_3 = d = 0.5$ and have the same core radius $\delta_j = \delta_0 = 0.025$. The Reynolds number is large so that in the duration of the numerical simulation of the interaction, the viscous diffusion effect is ignorable. The left column shows the trajectories in the meridian plane while the right column shows the trajectories in the coordinates $\bar{R}, \bar{Z}$ relative to the weighted center R^*, Z^* of the triangle, i. e., $\bar{R} = R - R^*$, $\bar{Z} = Z - Z^*$, with

$$R^* = \sum_{j=1}^{3} \Gamma_j R_j(0)/[\Gamma_1 + \Gamma_2 + \Gamma_3] \text{ and } Z^*(t) = \sum_{j=1}^{3} \Gamma_j Z_j(t)/[\Gamma_1 + \Gamma_2 + \Gamma_3]\,. \quad (24)$$

In the first row, we have $R^* = R_2 = 4.0$, and hence the distance to ring radius ratio is small, $d/R_2 = 0.125$. The trajectories of the ring centers are aperiodic and intermittent with strong or weak pairing of two rings. In the second row, we have $R_2 = 0.75$ with a moderate ratio $d/R_2 = 2/3$. The trajectories of the ring centers are qualitatively similar to those shown in the first row, but the intermittency is more pronounced. Qualitatively similar trajectories were reported by (Klein, Knio & Ting, (1996)) for the rings with same initial configuration as that for the second row except that the middle ring center is slightly ahead of the other two, with $Z_2(0) = 0.2$. These examples are interesting but they show only a few

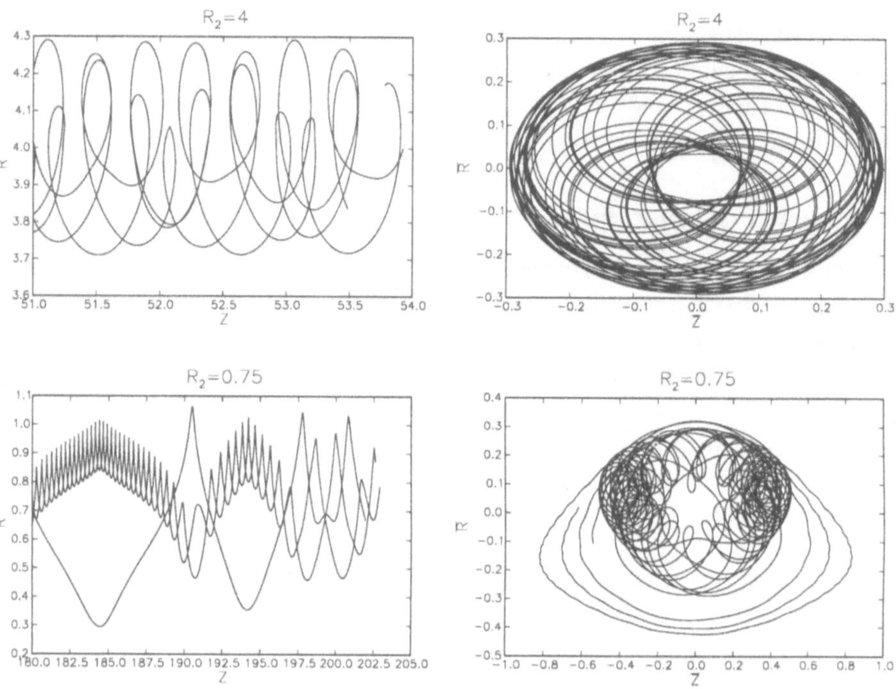

Figure 1. Interaction of three coaxial vortex rings. Initial configuration is coplanar.

special cases of the interaction problems. In general, the problem depends on six parameters, or initial data, for example, R_j, Z_j and Γ_j, $j = 2, 3$, with $Z_1 = 0$ and $R_1 = \ell = 1$ and $\Gamma_1 = 1$. In addition there are three parameters, the initial core sizes, assuming that the cores have similarity profiles. We need some guidance in the selection or classification of the initial data in order to begin a systematic numerical investigation of the interaction problem. We propose to begin with the restricted three ring problem, for which the leading order velocity near a ring center is that of a two-dimensional vortex, i. e, the effects of the curvature and core size are of higher order. This implies that in the local expansions of the Biot-Savart integrals (3) of two adjacent rings, the leading term of an adjacent ring is much larger than the curvature term, i. e., $\Gamma/[2\pi d_{ij}] \gg \Gamma/[2\pi\ell \ln \epsilon]$ with $R_j = O(\ell)$ and $\epsilon = \delta_0/\ell$. To be in this regime, the two outer length scales, d for the sides of the triangle, $d_{ij} = O(d)$, and ℓ for the typical ring radius, $R_j = O(\ell)$, have to observe the inequality, $\delta_0 \ll d \ll \ell$. Thus the condition for the regime of the restricted three ring problem is

$$\epsilon \ll \lambda = d/\ell \ll 1 . \tag{25}$$

With the configuration of, or the triangle formed by, the three ring centers in the RZ plane identified as that of three vortices in the xy plane, the equations of motions of the ring centers can be identified as perturbations of the equations for three planar vortices, with $R_j > 0$ insured by (25). The dynamics of three vortices was studied by Synge (1949) using the trilinear coordinates $\xi_{ij} = d_{ij}/$[the sum of three sides], thus $\xi_{12}+\xi_{23}+\xi_{31} = 1$. There are only two independent coordinates say $\xi_{12} \in [0,1]$ and $\xi_{23} \in [0, 1-\xi_{12}]$. Triangles with identical trilinear coordinates are similar. The equilibrium points or critical points were then identified. In particular the stability of the critical point where the configuration is an equilateral triangle, i. e., the trilinear coordinates equal to 1/3, was analyzed. Synge's study was continued by Tavantzis and Ting (1988). They obtained an integral invariant and then the global behavior of the trajectories in the trilinear coordinates. For all combinations of Γ_j, the critical points are shown to be either *centers* or *saddle points*. We shall infer the dynamics of three restricted rings from the corresponding two dimensional case with the same initial configuration.

If the initial configuration of the ring centers in trilinear coordinates corresponds to that of three 2-D vortices at a *saddle point* or on a *separatrix*, we expect that the trajectories of the ring centers will be aperiodic and become chaotic in long time. The initial configurations of the two examples shown in Fig. 1 have the same trilinear coordinates $(0.25, 0.25, 0.5)$, which is a saddle point of three 2-D vortices with the same strength. The corresponding trilinear coordinates for the example shown in (Klein, Knio & Ting, (1996)) are $(0.2808, 0.2808, 0.4384)$, which is is collinear with the saddle point and the *center*, $(1/3, 1/3, 1/3)$ but is closer to the saddle point with distance $0.5 - 0.4384 = 0.0616$ than the *center* with distance 0.1051. Thus the trajectories of the ring centers become aperiodic with intermittency.

If the initial configuration is at a *center* or on a periodic orbit around the *center*, we expect that quasiperiodic trajectories of the restricted problem exist for a sufficiently small $\lambda = d/\ell$. Figure 2 shows the interaction of three rings with the same strength Γ. Their centers form an equilateral triangle at $t = 0$, with side $d = 1$, $Z_1 = Z_3 = 0$ and $Z_2 = d\sqrt{3}/2$, $R_{2\pm1} = R_2 \pm d/2$. The initial configuration corresponds to a *center* in trilinear coordinates. For the first row, we have $R_2(0) = 5 = \ell$ and hence a small $\lambda = 0.2$. The trajectories remain quasiperiodic. For the second row, we have $R_2(0) = 1$ and a moderate $\lambda = 1$. The trajectories are aperiodic showing intermittency. Here we see that mathematical study of this restricted three ring problem is needed to establish a bound for λ such that quasiperiodic trajectories exist.

3.2. MULTI-LENGTH SCALE PROBLEMS

In the analysis of C-T 1978 and the extension (Klein & Ting, (1992)), we assume that the solution has one length scale δ^* for the inner region and one scale ℓ for the

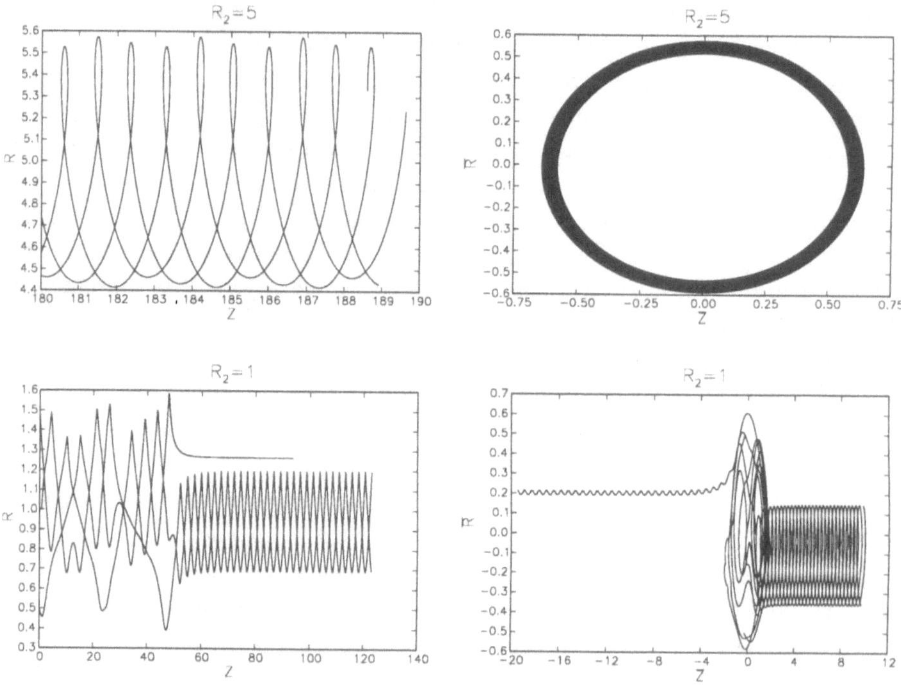

Figure 2. Interaction of three coaxial vortex rings, initially at equal distance.

outer region with $\delta^*/\ell = \epsilon$. In both analyses, we had to make use of the fact that the centerline C forms a loop of length $S = O(\ell)$, so that all the physical entities in the inner region around C are periodic function of s with period S_0. Here we shall present two classes of problems requiring length scales in addition to ℓ for C and/or the background flow so that the analysis can be carried out without using the periodicity condition.

3.2.1. *An intermediate length scale λ between δ^* and ℓ*

With ℓ as the unit length scale, we can treat λ as a small parameter with $\epsilon \ll \lambda \ll 1$. In this case, we have a double series expansion in λ and ϵ. For each order of ϵ we are looking for the series in λ. For example, we consider $\dot{\mathbf{X}}$ given by (11) to be $O(1)$ with respect to ϵ but it can be of different orders of λ. The leading order term in (11) with respect to λ depends only on the *local* solution, i. e., in the length scale $\lambda \ll \ell$. This fact renders the analysis *local*, avoiding using the periodicity condition of s with period $S_0 = O(\ell)$.

In our asymptotic analysis, we treated $\ln(1/\epsilon)$ to be $O(1)$. If $1/\ln(1/\epsilon)$ is treated as the small parameter λ, then the leading term in (11) is $\hat{b}[\Gamma\kappa/4\pi]\ln(1/\epsilon)$, depending only on the local geometry of $\mathcal{C}$. Thus a simple theory for small amplitude short wavelength distortion of a nearly straight filament was obtained (Klein & Majda, (1991)), generalizing the analysis of Hasimoto (1972).

If there are two filaments with distance $d \ll \ell$, e. g., two nearly parallel filaments, we have $d/\ell = \lambda$, the leading term in $\dot{\mathbf{X}}$ of one filament is $O(1/\lambda)$ due to the velocity induced by the other filament included in $\mathbf{Q}^f$. Analysis for this case was carried out (Klein, Majda & Damodaran, (1995)).

We are studying the dynamics of a nearly helical filament with small pitch $O(\lambda\ell)$ and the interaction of several co-axial helical filaments with the distance between two adjacent filaments $O(\lambda\ell)$. For the leading outer solution in the scale ℓ, the filament(s) becomes a cylindrical vortex sheet of radius ℓ with constant circumferential and axial vorticity strengths. The sheet with constant circumferential vorticity strength induces an uniform jet inside the cylinder with no flow outside. The sheet with constant axial vorticity strength induces a potential flow outside as that of a rotating cylinder with no flow inside. The outer solutions for the above two problems in the intermediate length scale $\lambda\ell$ will be different but the analysis does not need the end conditions of the filament(s) in the length scale ℓ or larger.

3.2.2. *A long filament with another length scale $L \gg \ell$*

For a long filament, the scale L could be the scale for the gradual axial variation of the core structure, the background flow, and/or the slow variation of the geometry of the centerline. We now have another small parameter $\varsigma = \ell/L \ll 1$. For the *distinguished limit*, we set $\varsigma = O(\epsilon)$. We express the dependence of a variable f on s as a function of two independent variables, the microscopic variable, s, and the macroscopic variable $\tilde{s}$ with $s = \varsigma\tilde{s}$. We apply the standard two-length analysis in s and $\tilde{s}$ to the vortex filament problem, including the dependence of f on t, θ and $\bar{r}$ or r and carry out the matching of inner and outer solutions in $\bar{r}$ and r. We shall then replace the periodic condition in the analysis for a filament with length $O(\ell)$, (Callegari & Ting, (1978); Klein & Ting, (1992)) by applying the s-averaging in the two-length analysis to the governing equations. We shall arrive at a set of equations for canonical microscopic or cell problems with $\tilde{s}$ as a parameter and at a system of equations in the macroscopic scale involving the bulk contributions of the canonical cell problems. The end condition(s) of the filament(s) are then needed for the solution of the macroscopic problem.

Acknowledgements

The research of Ting was partially supported by the Air Force Office of Scientific Research, USA, and by the Alexander von Humboldt Foundation, Germany. The research of Klein was partially supported by the Deutsche Forschungsgemainschaft.

References

Callegari, A. and Ting, L. (1978) Motion of a Curved Vortex Filament with Decaying Vortical Core and Axial Velocity, *SIAM J. Appl. Math.*, **35**, pp. 148–175.

Hasimoto, H., (1972) A Soliton on a Vortex Filament, *J. Fluid Mech.*, **51**, pp. 477–485.

Ishii, K. and Liu, C. H., (1987) Interaction of Decaying Vortex Ring with a Rotational Background Flow Bounded by a Solid Wall, *AIAA paper no. 87-1342*, 19th Fluid and Plasma Dynamics and Lasers Conference, Hawaii.

Klein, R. and Knio, O. (1995) Asymptotic Vorticity Structure and Numerical Simulation of Slender Vortex Filaments, *J. Fluid Mech.*, **284**, pp. 275–321, 1995.

Klein, R. and Knio, O. (1997) Optimized Vortex Element Schemes for Slender Vortex Simulation, *Proceedings of IUTAM Symposium on the Dynamics of Slender Vortices, Aachen*, Kluwer Academic Publishers, Dordrecht.

Klein, L., Knio, O. and Ting, L. (1996) Representation of Core Dynamics in Slender Vortex Filament Simulations, *Phys. Fluids*, **8**, pp. 2415–2425.

Klein R. and Majda, A. (1991) Self-stretching of a Perturbed Vortex Filament, I & II, *Physica D*, **49**, pp. 323–352, & **53**, pp. 267–294.

Klein, R., Majda, A. and Damodaran, K. (1995) Simplified Equations for the Interaction of Nearly Parallel Vortex Filaments, *J. Fluid Mech.*, **288**, pp. 201–248.

Klein, R. and Ting, L, (1992) Vortex Filament with Axial Core Structure Variation, *Appl. Math. Letters*, **5**, pp. 99–103.

Klein, R. and Ting, L. (1995) Theoretical and Experimental Studies of Slender Vortex Filaments, *Appl. Math. Letters*, **8**, pp. 45–50.

Kleinstein, G. and Ting, L. (1971) Optimum Solutions for Heat Conduction Problems, *ZaMM*, **51**, pp. 1–16.

Knio, O. and Ting, L. (1997) Vortical Flow outside a Sphere and Sound Generation, *SIAM J. Appl. Math.*, **57**, pp. 972–981.

Knio, O. and Ting, L. (1997a) Noise Emission due to Slender Vortex - Solid Body Interactions, *Proceedings of IUTAM Symposium on the Dynamics of Slender Vortices, Aachen*, Kluwer Academic Publishers, Dordrecht.

Knio, O., Ting, L. and Klein, R. (1998) Interaction of a Slender Vortex with a Rigid Sphere: Dynamics and Far-field Sound, to appear in *J. Acoust. Soc. Amer.*, **103**.

Lamb, H. (1932) *Hydrodynamics*, Dover Publ., New York.

Liu, C. H., Tavantzis, J and Ting, L. (1986) Numerical Studies of Motion and Decay of Vortex Filaments, *AIAA J.*, **24**, pp. 1290–1297.

Liu, C. H. and Ting, L. (1987) Interaction of Decaying Trailing Vortices in Spanwise Shear Flow, *J. Computer and Fluids*, **15**, pp. 77–92.

Moore, D. M. and Saffman, P. G. (1972) The motion of a vortex filament with an axial flow, *Philos. Trans. Roy. Soc. London Ser. A* **272**, pp. 403–429.

Schmitz, M. and Klein, R. (1997) Recent Developments in the Asymptotic Theory of Vortex Breakdown, *Proceedings of IUTAM Symposium on the Dynamics of Slender Vortices, Aachen*, Kluwer Academic Publishers, Dordrecht.

Synge, J. L. (1949) On the Motion of Three Vortices *Can. J. Math.*, **1**, pp. 257–270.

Tavantzis, J. and Ting, L. (1988) The Dynamics of Three Vortices Revisited, *Phys. Fluids*, **31**, pp. 1392–1409.

Ting, L. (1971) Studies in the Motion and Decay of Vortices, *Aircraft Wake Turbulence and its Detection*, Eds.: Olsen, J. H., Goldburg, A. and Rogers, M., Plenum Publ., New York, pp. 11-39.

Ting, L. and Klein, R. (1991) *Viscous Vortical Flows*, Lecture Notes in Physics **374**, Springer-Verlag, New York.

Ting, L. and Tung, C. (1965) Motion and Decay of a Vortex in a Nonuniform Stream, *Phys. Fluids*, **8**, pp. 1039–1051.

Tung, C. and Ting, L. (1967) Motion and Decay of a Vortex Ring, *Physics of Fluids*, **10**, pp. 901–910.

MOTION OF A THIN VORTEX RING IN A VISCOUS FLUID: HIGHER-ORDER ASYMPTOTICS

YASUHIDE FUKUMOTO
Graduate School of Mathematics,
Kyushu University 33, Fukuoka 812–81, Japan

AND

H. K. MOFFATT
Isaac Newton Institute for Mathematical Sciences,
20 Clarkson Road, Cambridge CB3 0EH, UK

Abstract. The motion of an axisymmetric vortex ring of small cross-section in a viscous incompressible fluid is investigated using the method of matched asymptotic expansions. A general formula for the ring speed is obtained up to third order in $\epsilon = \delta/R_0$ $(= (\nu/\Gamma)^{1/2})$, the ratio of core to curvature radii, which takes account of the influence of the self-induced strain. Here Γ is the circulation and ν is the kinematic viscosity of fluid. It is pointed out that the dipole distributed along the centerline of the ring plays a vital role in its movement. Its strength needs be specified at the initial instant in order to remove the indeterminacy of the theory. A new asymptotic development of the Biot-Savart law enables us to calculate the non-local induction velocity at $O(\epsilon^3)$ from the dipole. In a special case, we recover Dyson's inviscid formula (1893). It is demonstrated that the viscosity acts, at $O(\epsilon^3)$, to expand the radius of the loop consisting of the stagnation points in the core, when viewed from a certain comoving frame.

1. Introduction

The motion of a vortex ring is one of the most classical and fundamental problems of vortex dynamics. Extending Kelvin's result, Dyson (1893) (see also Fraenkel 1972) obtained the speed U of an axisymmetric vortex ring, embedded in an inviscid incompressible fluid, up to third (virtually fourth) order in a small parameter:

$$U = \frac{\Gamma}{4\pi R_0}\left\{\log\left(\frac{8}{\epsilon}\right) - \frac{1}{4} - \frac{3\epsilon^2}{8}\left[\log\left(\frac{8}{\epsilon}\right) - \frac{5}{4}\right] + O(\epsilon^4 \log\epsilon)\right\}, \quad (1)$$

where Γ is the circulation, R_0 is the ring radius and $\epsilon = \delta/R_0$ is the ratio of core radius δ to R_0. The vorticity distribution in the core is proportional to the

E. Krause and K. Gersten (eds.), IUTAM Symposium on Dynamics of Slender Vortices, 21-34.

distance from the symmetry axis. We consider Kelvin's formula as the first order and ϵ^2-term as the third. The vortex ring induces a local straining field on itself which deforms the core into an ellipse at second order:

$$r = \delta\left\{1 + \epsilon^2\left[\frac{3}{8}\log\left(\frac{8}{\epsilon}\right) - \frac{17}{32}\right]\cos 2\theta + \cdots\right\}, \tag{2}$$

where (r, θ) are local cylindrical coordinates about the core center which will be introduced in §2. It is remarkable that the inclusion of the third-order term in the propagation velocity achieves a great improvement in approximation: equation (1) exhibits fair agreement even with the exact value for the "fat" limit of Hill's spherical vortex ($\epsilon = \sqrt{2}$).

The viscosity acts to diffuse the vorticity. Its influence on the propagation speed, at large Reynolds number, was calculated by Tung and Ting (1967) (Callegari and Ting 1978) and Saffman (1970), up to $O((\nu/\Gamma)^{1/2})$, as

$$U = \frac{\Gamma}{4\pi R_0}\left\{\log\left(\frac{8R_0}{2\sqrt{\nu t}}\right) - \frac{1}{2}(1 - \gamma + \log 2) + \cdots\right\}, \tag{3}$$

where t is the time measured from a virtual instant at which the vorticity is 'δ-function' concentrated, and $\gamma = 0.57721566\cdots$ is Euler's constant. The vorticity distribution has a diffusing Gaussian profile with circular symmetry.

Recent direct numerical simulations of fully developed turbulence have revealed that the small-scale structure is dominated by high-vorticity regions concentrated in tubes (see, for example, Siggia 1981; Kerr 1985; Hosokawa and Yamamoto 1989). These occupy a small fraction of the total volume, but are responsible for a much larger fraction of viscous dissipation. This observation led Moffatt *et al.* (1994, 1996) to develop a large-Reynolds-number asymptotic theory to solve the Navier-Stokes equations for a vortex subjected to uniform non-axisymmetric irrotational strain. The solution satisfactorily accounts for the structure of the dissipation field previously obtained by numerical computation (Kida and Ohkitani 1992). The viscosity is an agent to pick out vorticity distribution. At leading order, the Burgers vortex is obtained, and at the next order ($O(\nu/\Gamma)$), a quadrupole component emerges, reflecting an elliptical vorticity distribution. The salient feature is that the major axis of the ellipse is aligned at 45° to the principal axis of the external strain. This fact leads us to the belief that the strained crosssection of a propagating vortex ring, commonly observed in nature, is established as an equilibrium between self-induced strain and viscous diffusion.

The aim of our study is to elucidate the structure of this strained core and its influence on the translation speed of an axisymmetric vortex ring. As a first step, we present, in this paper, a general framework to address this problem. A partial answer is given as to how viscosity affects the radial drift of vorticity.

The method of matched asymptotic expansions has been previously developed to derive the velocity of a slender curved vortex tube in a fluid both with and without viscosity (Tung and Ting 1967; Widnall *et al.* 1971; Callegari and Ting 1978; Klein and Majda 1991). However these studies have been limited to the second-order curvature effect (Moore and Saffman 1972; Fukumoto and Miyazaki 1991). The self-induced straining field of a vortex ring makes its appearance at second order in $\epsilon = (\nu/\Gamma)^{1/2}$, and the translation speed is affected at the next order. We make an attempt to extend asymptotic expansions to a higher order and to calculate the speed of a vortex ring up to $O(\epsilon^3)$.

The existing asymptotic formula for the potential flow caused by a circular vortex loop is not sufficient to carry through this program. After a brief statement about the general setting of asymptotic expansions in §2, we obtain an asymptotic expression of the Biot-Savart integral accommodating an arbitrary vorticity distribution in § 3. In §4, the inner expansions are recalled and extended to second order. Based on theses, we establish, in § 5, a general formula for the translation velocity of a vortex ring, valid up to third order. Dyson's formula (1) is recovered in a special case. Moreover, it is revealed that the radius of the loop consisting of the stagnation points in the core, when viewed from the frame moving with the core, expands linearly in time owing to the action of viscosity. Our procedure pursuing higher-order asymptotics highlights the significance of the dipole distributed along the ring. Its strength must be prescribed at the initial instant and thereby the problem of undetermined constants at $O(\epsilon)$ is remedied.

2. Formulation of the matched asymptotic expansions

Two length scales are available, namely, the (typical) core radius δ and the ring radius R_0. We assume that their ratio is very small. We retain only the slow mode of core dynamics, suppressing fast waves on the core. Then, in view of (1), the time-scale is of order $R_0/(\Gamma/R_0) = R_0^2/\Gamma$. In the presence of viscosity, the core radius grows as $\delta \sim (\nu t)^{1/2} \sim (\nu/\Gamma)^{1/2} R_0$ during this time. Thus our assumption requires that

$$\epsilon = \delta/R_0 = \sqrt{\nu/\Gamma} \ll 1 . \tag{4}$$

Let us introduce cylindrical coordinates (ρ, ϕ, z) with z-axis along the axis of symmetry and ϕ along the vortex lines. The vorticity distribution $\boldsymbol{\omega}$ is axisymmetric but otherwise arbitrary:

$$\boldsymbol{\omega} = \zeta(\rho, z)\boldsymbol{e}_\phi , \tag{5}$$

where $\boldsymbol{e}_\phi$ is the unit vector in the azimuthal direction. The Stokes streamfunction ψ for the flow produced by (5) is given, via the Biot-Savart law, in the form

$$\psi(\rho, z) = -\frac{\rho}{4\pi}\int_{-\infty}^{\infty}\int_0^{2\pi}\int_0^{\infty} \frac{\zeta(\rho', z')\cos\phi' d\rho' d\phi' dz'}{\sqrt{\rho^2 - 2\rho\rho'\cos\phi' + \rho'^2 + (z - z')^2}} . \tag{6}$$

As is well known, the expression (6) for an infinitely thin core is not convergent near the core center. A way out is to connect the outer flow to an inner viscous flow which decays rapidly with distance from the core center. Thus we are led to inner and outer expansions (Tung and Ting 1967). The inner region has length scale of order the core radius δ and there we seek the solution of the Navier-Stokes equations matched to the outer solution given by (6).

It is expedient to choose a coordinate frame moving with the core center $(R(t), Z(t))$ in which we introduce local cylindrical coordinates (r, θ) such that

$$\rho = R(t) + r\cos\theta\,, \qquad z = Z(t) + r\sin\theta\,. \tag{7}$$

Introduce dimensionless variables:

$$\left.\begin{array}{l} r^* = r/\epsilon R_0\,, \quad t^* = t/\frac{R_0}{\Gamma}\,, \quad \psi^* = \frac{\psi}{\Gamma R_0}\,, \quad \zeta^* = \zeta/\frac{\Gamma}{R_0^2\epsilon^2}\,, \\ \boldsymbol{v}^* = \boldsymbol{v}/\frac{\Gamma}{R_0\epsilon}\,, \quad (\dot{R}^*, \dot{Z}^*) = (\dot{R}, \dot{Z})/\frac{\Gamma}{R_0}\,. \end{array}\right\} \tag{8}$$

Here $\boldsymbol{v}$ is the velocity relative to the moving coordinates and the difference in normalization between the last two of (8) should be noted. The equation handled in the inner region are the vorticity equation and the relation between ζ and ψ. Dropping the stars, these take the following form:

$$\begin{aligned} &\frac{\partial\zeta}{\partial t} + \frac{1}{\epsilon^2}\left(u\frac{\partial\zeta}{\partial r} + \frac{v}{r}\frac{\partial\zeta}{\partial\theta}\right) - \frac{1}{\epsilon\rho^2}\left(\frac{\partial\psi}{\partial r}\sin\theta + \frac{1}{r}\frac{\partial\zeta}{\partial\theta}\cos\theta\right) \\ &= \Delta\zeta + \frac{\epsilon}{\rho}\left(\cos\theta\frac{\partial}{\partial r} - \frac{\sin\theta}{r}\frac{\partial}{\partial\theta}\right)\zeta - \frac{\epsilon^2}{\rho^2}\zeta\,, \end{aligned} \tag{9}$$

$$\zeta = \frac{1}{\rho}\Delta\psi - \frac{\epsilon}{\rho^2}\left(\cos\theta\frac{\partial}{\partial r} - \frac{\sin\theta}{r}\frac{\partial}{\partial\theta}\right)\psi\,. \tag{10}$$

where

$$\rho = R + \epsilon r\cos\theta\,, \tag{11}$$

Δ is the two-dimensional Laplacian, and u and v are the r- and θ-components of the relative velocity $\boldsymbol{v}$:

$$u = \frac{1}{r\rho}\frac{\partial\psi}{\partial\theta} - \epsilon(\dot{Z}\sin\theta + \dot{R}\cos\theta)\,, \tag{12a}$$

$$v = -\frac{1}{\rho}\frac{\partial\psi}{\partial r} - \epsilon(\dot{Z}\cos\theta - \dot{R}\sin\theta)\,. \tag{12b}$$

We look for the solution of (9) and (10) in the form

$$\zeta = \zeta^{(0)} + \epsilon\zeta^{(1)} + \epsilon^2\zeta^{(2)} + \epsilon^3\zeta^{(3)} + \cdots\,, \tag{13a}$$

$$\psi = \psi^{(0)} + \epsilon\psi^{(1)} + \epsilon^2\psi^{(2)} + \epsilon^3\psi^{(3)} + \cdots\,, \tag{13b}$$

$$R = R^{(0)} + \epsilon R^{(1)} + \epsilon^2 R^{(2)} + \cdots\,, \tag{13c}$$

$$Z = Z^{(0)} + \epsilon Z^{(1)} + \epsilon^2 Z^{(2)} + \cdots\,, \tag{13d}$$

where $\zeta^{(i)}$ and $\psi^{(i)}$ $(i = 0, 1, 2, \cdots)$ are taken to be functions of r, θ and t.

The permissible solution must satisfy the condition[1],

$$u \text{ and } v \text{ are finite at } r = 0\,; \tag{14}$$

the requirement that it smoothly matches the outer solution will determine the values of $\dot{R}^{(i)}$ and $\dot{Z}^{(i)}$ $(i = 0, 1, 2, \cdots)$.

3. Outer solution

The streamfunction ψ_m for the flow induced by a circular vortex loop $\zeta = \delta(\rho - R)\delta(z)$ of unit strength is obtainable from (6):

$$\psi_m(\rho, z, R) = -\frac{\rho}{4\pi}\int_0^{2\pi} \frac{R\cos\phi' d\phi'}{\sqrt{\rho^2 - 2\rho R\cos\phi' + R^2 + z^2}}\,. \tag{15}$$

We call (15) the monopole field. So far, this has been exclusively employed as the outer solution.

It turns out however that, when going into higher orders, (15) is not enough to be qualified as the outer solution. The elaboration of the detailed structure of (6) is unavoidable. To this aim, it is advantageous to adapt Dyson's technique to an arbitrary distribution of vorticity:

$$\begin{aligned}
\psi &= -\frac{\rho}{4\pi}\iint_{-\infty}^{\infty} dx' dz' \zeta(x', z') e^{x'\frac{\partial}{\partial R} - z'\frac{\partial}{\partial z}} \int_0^{2\pi} \frac{R\cos\phi' d\phi'}{\sqrt{\rho^2 - 2\rho R\cos\phi' + R^2 + z^2}} \\
&= \iint_{-\infty}^{\infty} dx' dz' \zeta(x', z') \left\{ 1 + x'\frac{\partial}{\partial R} - z'\frac{\partial}{\partial z} + \frac{1}{2!}\left(x'\frac{\partial}{\partial R} - z'\frac{\partial}{\partial z}\right)^2 \right. \\
&\quad + \frac{1}{3!}\left(x'\frac{\partial}{\partial R} - z'\frac{\partial}{\partial z}\right)^3 + \frac{1}{4!}\left(x'\frac{\partial}{\partial R} - z'\frac{\partial}{\partial z}\right)^4 + \frac{1}{5!}\left(x'\frac{\partial}{\partial R} - z'\frac{\partial}{\partial z}\right)^5 \\
&\quad \left. + \frac{1}{6!}\left(x'\frac{\partial}{\partial R} - z'\frac{\partial}{\partial z}\right)^6 + \cdots \right\} \psi_m\,.
\end{aligned} \tag{16}$$

The expected spatial dependence of vorticity distribution is

$$\zeta^{(0)} = \zeta^{(0)}\,, \tag{17a}$$

$$\zeta^{(1)} = \zeta_{11}^{(1)} \cos\theta\,, \tag{17b}$$

$$\zeta^{(2)} = \zeta_0^{(2)} + \zeta_{21}^{(2)} \cos 2\theta\,, \tag{17c}$$

$$\zeta^{(3)} = \zeta_{11}^{(3)} \cos\theta + \zeta_{12}^{(3)} \sin\theta + \zeta_{31}^{(3)} \cos 3\theta\,. \tag{17d}$$

For $\zeta_{jk}^{(i)}$, i denotes the order of perturbation, j the Fourier mode, and $k = 1$ and 2 correspond to $\cos j\theta$ and $\sin j\theta$ respectively. Substituting (17a)–(17d) into (16),

[1]This condition is better than the restrictive one that $u = v = 0$ at $r = 0$.

integrating with respect to x' and z', and taking the derivatives of ψ_m, substituted from (15), with respect to R and z, we obtain the asymptotic form of the outer solution valid at $r \ll R$, which is expressed in dimensionless form as

$$\begin{aligned}
\psi &= -\frac{R}{2\pi}\Gamma \log\left(\frac{8R}{\epsilon r}\right) + \epsilon\left(-\frac{\Gamma}{4\pi}\left[\log\left(\frac{8R}{\epsilon r}\right) - 1\right] r\cos\theta + d\frac{\cos\theta}{r}\right) \\
&+ \epsilon^2\left(-\frac{\Gamma}{2^5\pi R}\left\{\left[2\log\left(\frac{8R}{\epsilon r}\right) + 1\right] r^2 - \left[\log\left(\frac{8R}{\epsilon r}\right) - 2\right] r^2\cos\theta\right\}\right. \\
&\quad \left. + \frac{d}{2R}\left[\log\left(\frac{8R}{\epsilon r}\right) + \frac{\cos 2\theta}{2}\right] + q\frac{\cos 2\theta}{r^2}\right) \\
&+ \epsilon^3\left(\frac{3\Gamma}{2^7\pi R^2}\left\{\left[\log\left(\frac{8R}{\epsilon r}\right) - \frac{1}{3}\right] r^3\cos\theta - \left[\log\left(\frac{8R}{\epsilon r}\right) - \frac{7}{3}\right] r\cos 3\theta\right\}\right. \\
&\quad - \frac{d}{8R^2}\left\{\left[\log\left(\frac{8R}{\epsilon r}\right) - \frac{7}{4}\right] r\cos\theta + \frac{r\cos 3\theta}{4}\right\} \\
&- \frac{1}{8\pi}\left\{\left[2\pi\int_0^\infty r^3\zeta_0^{(2)}dr\right] + 4R\left[\pi\int_0^\infty r^2\zeta_{11}^{(3)}dr\right] + \left[\pi\int_0^\infty r^3\zeta_{21}^{(2)}dr\right]\right\}\frac{\cos\theta}{r} \\
&\quad + \frac{q}{4R}\left(\frac{\cos\theta}{r} + \frac{\cos 3\theta}{r}\right) - \frac{1}{\pi R}\left\{\frac{1}{3\cdot 2^8}\left[2\pi\int_0^\infty r^5\zeta^{(0)}dr\right]\right. \\
&\quad - \frac{R}{8\cdot 4!}\left[\pi\int_0^\infty r^6\zeta_{11}^{(1)}dr\right] + \frac{R^2}{4!}\left[\pi\int_0^\infty r^5\zeta_{21}^{(2)}dr\right] \\
&\quad \left.\left. + \frac{R^3}{6}\left[\pi\int_0^\infty r^4\zeta_{31}^{(3)}dr\right]\right\}\frac{\cos 3\theta}{r^3} - \frac{R}{2\pi}\left[\pi\int_0^\infty r^2\zeta_{12}^{(3)}dr\right]\frac{\sin\theta}{r}\right) + \cdots, \qquad (18)
\end{aligned}$$

where

$$\Gamma = 2\pi\int_0^\infty r\zeta^{(0)}dr, \qquad (19)$$

($\Gamma = 1$ when nondimensionalised), and d and q are related to the strength of low-order dipole and quadrupole:

$$d = -\frac{1}{2\pi}\left\{\frac{1}{4}\left[2\pi\int_0^\infty r^3\zeta^{(0)}dr\right] + R\left[\pi\int_0^\infty r^2\zeta_{11}^{(1)}dr\right]\right\}, \qquad (20)$$

$$q = \frac{1}{2\pi R}\left\{\frac{1}{2^6}\left[2\pi\int_0^\infty r^5\zeta^{(0)}dr\right] - \frac{R}{8}\left[\pi\int_0^\infty r^4\zeta_{11}^{(1)}dr\right] - \frac{R^2}{2}\left[\pi\int_0^\infty r^3\zeta_{21}^{(2)}dr\right]\right\}. \qquad (21)$$

The terms multiplied by Γ stem from $\Gamma\psi_m$, which are augmented by the induction velocities due to the dipole, quadrupole, hexapole distributed along the center $r = 0$ of the core. Parts of (18) supply the matching conditions on the inner solution. The distributions of $\zeta_{11}^{(1)}$, $\zeta_0^{(2)}$, $\zeta_{21}^{(2)}$, $\zeta_{11}^{(3)}$, $\zeta_{12}^{(3)}$ and $\zeta_{31}^{(3)}$ are as yet unknown, but will be fixed by the inner expansions and the matching procedure. It will be clarified that the dipole components $\zeta_{11}^{(1)}$, $\zeta_{11}^{(3)}$, $\zeta_{12}^{(3)}$ are distinctive. In the subsequent sections we investigate the flow field inside the core.

4. Inner expansions up to second order

Before going to third order, we give a brief outline of the inner perturbations up to second order.

Collecting like powers of ϵ in (9) and (10), along with (11)–(12b), substituted from (13a)–(13d), the Navier-Stokes equations at each order are deduced successively.

At $O(\epsilon^0)$, we obtain the Jacobian form of the Euler equation:

$$[\zeta^{(0)}, \psi^{(0)}] = 0\,, \tag{22}$$

where we define $[\zeta^{(0)}, \psi^{(0)}] = \partial(\zeta^{(0)}, \psi^{(0)})/\partial(r,\theta)/r$. Hence $\zeta^{(0)} = \mathcal{F}(\psi^{(0)})$ for some function $\mathcal{F}$. Suppose that the flow $\psi^{(0)}$ has a single stagnation point at $r = 0$, the streamlines being all closed around that point. Then it is probable that the solution of (22), coupled with $\zeta^{(0)} = \Delta\psi^{(0)}/R^{(0)}$ (see (10)), is radial $\psi^{(0)} = \psi^{(0)}(r)$, that is, the streamlines are circles (Moffatt *et al.* 1994)[2]. The functional form of $\psi^{(0)}(r)$ and $\zeta^{(0)}(r)$ remain undetermined at this level of approximation, but is determined through the axisymmetric part of the vorticity equation at $O(\epsilon^2)$:

$$\frac{\partial \zeta^{(0)}}{\partial t} = \left(\zeta^{(0)} + \frac{r}{2}\frac{\partial \zeta^{(0)}}{\partial r}\right)\frac{\dot{R}^{(0)}}{R^{(0)}} + \left(\frac{\partial^2 \zeta^{(0)}}{\partial r^2} + \frac{1}{r}\frac{\partial \zeta^{(0)}}{\partial r}\right). \tag{23}$$

where a dot stands for the differentiation with respect to time. We focus our attention on the case for which, at the initial instant, the vorticity is concentrated in the circle of radius R_0:

$$\zeta^{(0)} = \delta(\rho - R_0)\delta(z) \quad \text{at} \;\; t = 0\,. \tag{24}$$

Anticipating that $R^{(0)}$ is constant at $O(\epsilon)$ (see (31)), we obtain the decaying circular vortex

$$\zeta^{(0)} = \frac{1}{4\pi t} e^{-\frac{r^2}{4t}}\,. \tag{25}$$

(Tung and Ting 1967; Jiménez *et al.* 1996). Interestingly, viscosity selects the distribution of vorticity, even in the limit of $\nu \to 0$.

The first-order perturbation $\psi^{(1)}$ satisfies

$$\Delta\psi^{(1)} - a\psi^{(1)} = -\cos\theta v^{(0)} + aR_0 r(\dot{Z}^{(0)}\cos\theta - \dot{R}^{(0)}\sin\theta) + 2r\zeta^{(0)}\cos\theta\,, \tag{26}$$

[2]This result may be derived from the theorem proved by Gidas *et al.* (1979).

where $R_0 = R^{(0)}$ (with some abuse of notation) and

$$v^{(0)} = -\frac{1}{R_0}\frac{\partial \psi^{(0)}}{\partial r}, \qquad a = -\frac{1}{v^{(0)}}\frac{\partial \zeta^{(0)}}{\partial r}. \tag{27}$$

Here we have used the fact that the axisymmetric part of $\zeta^{(1)}$ is suppressed from the result of (31) and the analysis of the vorticity equation at $O(\epsilon^3)$. The solution meeting the condition that the relative velocity $(u^{(1)}, v^{(1)})$ is finite at $r = 0$ is

$$\psi^{(1)} = \psi_{11}^{(1)} \cos\theta + \psi_{12}^{(1)} \sin\theta; \tag{28a}$$

where

$$\psi_{11}^{(1)} = \tilde{\psi}_{11}^{(1)} - R_0 r \dot{Z}^{(0)}, \qquad \text{with } \tilde{\psi}_{11}^{(1)} = \Psi_{11}^{(1)} + c_{11}^{(1)} v^{(0)}, \tag{28b}$$

$$\psi_{12}^{(1)} = c_{12}^{(1)} v^{(0)}, \tag{28c}$$

$c_{11}^{(1)}$ and $c_{12}^{(1)}$ are constants, and $\Psi_{11}^{(1)}$ is a particular solution:

$$\Psi_{11}^{(1)} = v^{(0)} \int_0^r \frac{dr'}{r'[v^{(0)}(r')]^2} \left\{ \int_0^{r'} \eta v^{(0)}(\eta)[-v^{(0)}(\eta) + 2\eta\zeta^{(0)}(\eta)] d\eta \right\}, \tag{29}$$

(Widnall *et al.* 1971; Callegari and Ting 1978).

Irrespective of any choice of the parameter values $c_{11}^{(1)}$ and $c_{12}^{(1)}$, the matching condition

$$\psi^{(1)} \sim -\frac{1}{4\pi}\left[\log\left(\frac{8R_0}{\epsilon r}\right) - 1\right] r\cos\theta \quad \text{as } r \to \infty, \tag{30}$$

results in (3) and

$$\dot{R}^{(0)} = 0. \tag{31}$$

To have an idea on the constants, we revisit the discrete model in an inviscid flow studied by Dyson. At leading order, it is the Rankine vortex, that is, the vorticity is constant in the circular core of unit radius surrounded by an irrotational flow:

$$\zeta^{(0)} = \begin{cases} 1/\pi, \\ 0, \end{cases} \qquad v^{(0)} = \begin{cases} -r/2\pi, & (r \leq 0) \\ -1/2\pi r, & (r0) \end{cases} \tag{32}$$

Continuity of velocity across the core boundary $r = 1$ gives

$$c_{11}^{(1)} = 5/8, \quad c_{12}^{(1)} = 0. \tag{33}$$

(Widnall *et al.* 1971). However a difficulty arises when the discrete distribution is replaced by a continuous one, because the continuity condition is no longer of help. To make matters worse, both $c_{11}^{(1)}$ and $c_{12}^{(1)}$ admit arbitrary time dependence as long as we stick to the matching condition (30). This is true also for the discrete model, and therefore (33) is merely one possibility.

We can show that $c_{11}^{(1)}$ and $c_{12}^{(1)}$ serve as the parameters placing the circular core in the moving frame, to an accuracy of $O(\epsilon)$ in terms of the inner spatial scale. Increase of $c_{11}^{(1)}$ and $c_{12}^{(1)}$ by c amounts to the shift of the core-center by $\epsilon c/R_0$ in the ρ- and z-directions respectively. Without loss of generality, we may assume that $c_{12}^{(1)} = 0$. Still, a freedom of the choice of the location of the center in the radial direction is at our disposal. We realise that fixing the initial location of the core is equivalent to giving the strength of dipole at $t = 0$, and (30) is superseded by

$$\psi^{(1)} \sim \left\{ -\frac{1}{4\pi} \left[\log\left(\frac{8R_0}{\epsilon r}\right) - 1 \right] r + \frac{d_0(t)}{r} \right\} \cos\theta \quad \text{as } r \to \infty . \tag{34}$$

Comparison of (34) with (18) gives rise to the following identity:

$$d_0 = -\frac{1}{2\pi} \left\{ \left[2\pi \int_0^\infty r^3 \zeta^{(0)} dr \right] + R_0 \left[\pi \int_0^\infty r^2 \zeta_{11}^{(1)} dr \right] \right\} . \tag{35}$$

With the specification of $d_0(0)$, a proper formulation of the initial-value problem is completed. Yet, we suffer from arbitrariness of the temporal evolution of $d_0(t)$. We can verify that this is consistently absorbed into the third-order radial velocity $\dot{R}^{(2)}$ as exemplified at the end of §5. It implies that the perturbation solution is unique, while it has an infinite variety of representations.

Next, we proceed to the second-order perturbation $\psi^{(2)}$. It is shown to have the following θ-dependence:

$$\psi^{(2)} = \psi_0^{(2)} + \psi_{21}^{(2)} \cos 2\theta , \tag{36}$$

meaning that a quadrupole is produced in conjunction with the elliptical core deformation. The governing equations and matching conditions are

$$\begin{aligned} \left(\frac{\partial^2}{\partial r^2} + \frac{1}{r}\frac{\partial}{\partial r} \right) \psi_0^{(2)} &= R_0 \zeta_0^{(2)} + \dot{R}^{(2)} \zeta^{(0)} \\ &+ \frac{ra}{2R_0} \tilde{\psi}_{11}^{(1)} + \frac{1}{2R_0} \left[r v^{(0)} + r^2 \zeta^{(0)} + \frac{\partial \psi_{11}^{(1)}}{\partial r} + \frac{\psi_{11}^{(1)}}{r} \right] , \end{aligned} \tag{37}$$

with

$$\psi_0^{(2)} \sim -\frac{1}{2^4 \pi R_0} \left[\log\left(\frac{8R_0}{\epsilon r}\right) + \frac{1}{2} \right] r^2 + \left(\frac{d_0}{2R_0} - \frac{\dot{R}^{(2)}}{2\pi} \right) \log\left(\frac{8R_0}{\epsilon r}\right) \quad \text{as } r \to \infty , \tag{38}$$

and

$$\begin{aligned} \left(\frac{\partial^2}{\partial r^2} + \frac{1}{r}\frac{\partial}{\partial r} - \frac{4}{r^2} - a \right) \psi_{21}^{(2)} &= \frac{r^2 a}{4} \dot{Z}^{(0)} + \frac{b}{4R_0} \left[\tilde{\psi}_{11}^{(1)} \right]^2 \\ &+ \frac{ra}{R_0} \tilde{\psi}_{11}^{(1)} + \frac{1}{2R_0} \left[r v^{(0)} + r^2 \zeta^{(0)} + \frac{\partial \psi_{11}^{(1)}}{\partial r} - \frac{\psi_{11}^{(1)}}{r} \right] , \end{aligned} \tag{39}$$

with

$$\psi_{21}^{(2)} \sim \frac{1}{2^5 \pi R_0}\left[\log\left(\frac{8R_0}{\epsilon r}\right) - 2\right] r^2 + \frac{d_0}{4R_0} \quad \text{as } r \to \infty , \tag{40}$$

where $\zeta_0^{(2)}$ is the axisymmetric part of the second-order vorticity perturbation and

$$b = -\frac{1}{v^{(0)}}\frac{\partial a}{\partial r} . \tag{41}$$

Finding $\zeta_0^{(2)}$ requires us to analyse the vorticity equation at $O(\epsilon^4)$.

5. Third-order velocity of a vortex ring

We are now ready to tackle the third-order problem. The dipole field again shows up as the result of nonlinear interactions among the mono-, di- and quadru-poles up to $O(\epsilon^2)$. It is this field that takes part in the correction to the ring speed at $O(\epsilon^3)$. The streamfunction $\psi^{(3)}$ at $O(\epsilon^3)$ consists of three terms:

$$\psi^{(3)} = \psi_{11}^{(3)} \cos\theta + \psi_{12}^{(3)} \sin\theta + \psi_{31}^{(3)} \cos 3\theta , \tag{42}$$

only $\cos\theta$ and $\sin\theta$ components being relevant to the speed.

After lengthy but tedious algebra, the Navier-Stokes equations collapse to the following equation for $\psi_{11}^{(3)}$:

$$\frac{1}{r}\left(\frac{\partial \zeta^{(0)}}{\partial r}\psi_{11}^{(3)} + R_0 v^{(0)} \zeta_{11}^{(3)}\right) + R_0 \dot{Z}^{(2)} \frac{\partial \zeta^{(0)}}{\partial r} + R^{(2)}\left(\dot{Z}^{(0)}\frac{\partial \zeta^{(0)}}{\partial r} + \frac{v^{(0)}\zeta^{(0)}}{R_0}\right)$$
$$= f(r) , \tag{43}$$

where

$$\begin{aligned}\zeta_{11}^{(3)} = {} & \frac{1}{R_0}\Delta\psi_{11}^{(3)} - \frac{r}{R_0}\left\{\zeta_0^{(2)} + \frac{a}{2R_0}\tilde{\psi}_{21}^{(2)} + \frac{b}{8R_0^2}\left[a\tilde{\psi}_{11}^{(1)}\right]^2 + \frac{ra}{4R_0}\tilde{\psi}_{11}^{(1)}\right\} \\ & - \frac{1}{R_0^2}\left(\frac{\partial \psi_0^{(2)}}{\partial r} + \frac{1}{2}\frac{\partial \psi_{21}^{(2)}}{\partial r} + \frac{\psi_{21}^{(2)}}{r}\right) + \frac{r}{4R_0^3}\left(3\frac{\partial \psi_{11}^{(1)}}{\partial r} + \frac{\psi_{11}^{(1)}}{r}\right) \\ & + \frac{3r^2}{4R_0^3}v^{(0)} - \frac{R^{(2)}}{R_0^2}\left(a\tilde{\psi}_{11}^{(1)} + r\zeta^{(0)}\right) ,\end{aligned} \tag{44}$$

and

$$\begin{aligned} f(r) = {} & \frac{1}{2R_0}\left(\frac{b}{r}\tilde{\psi}_{11}^{(1)} + a\right) v^{(0)} \tilde{\psi}_{21}^{(2)} + \frac{1}{4R_0^2}\left\{2a\tilde{\psi}_{21}^{(2)}\frac{\partial \tilde{\psi}_{11}^{(1)}}{\partial r}\right. \\ & \left. + \frac{2b}{r}\left[\tilde{\psi}_{11}^{(1)}\right]^2 \frac{\partial \tilde{\psi}_{11}^{(1)}}{\partial r} + \frac{1}{2r}\frac{\partial b}{\partial r}\left[\tilde{\psi}_{11}^{(1)}\right]^3 + \left(\frac{2a}{r} - \frac{3bv^{(0)}}{2}\right)\left[\tilde{\psi}_{11}^{(1)}\right]^2\right\} \\ & + \left(\frac{\dot{Z}^{(0)}}{2R_0} + \frac{1}{R_0 r}\frac{\partial \psi_0^{(2)}}{\partial r} - \frac{r v^{(0)}}{2R_0^2}\right) a\tilde{\psi}_{11}^{(1)} + \zeta_0^{(2)} v^{(0)} - \frac{1}{r}\frac{\partial \zeta_0^{(2)}}{\partial r}\psi_{11}^{(1)} .\end{aligned} \tag{45}$$

The boundary conditions are

$$\psi_{11}^{(3)} \propto r \quad \text{as } r \to 0\,, \tag{46}$$

and, from (18),

$$\psi_{11}^{(3)} \sim \frac{3}{2^7 \pi R_0^2}\left[\log\left(\frac{8R_0}{\epsilon r}\right) - \frac{1}{3}\right] r^3 - \frac{d_0}{8R_0^2}\left[\log\left(\frac{8R_0}{\epsilon r}\right) - \frac{7}{4}\right] r - \frac{R^{(2)}}{4\pi R_0} r + \frac{d^{(3)}}{r} \quad \text{as } r \to \infty\,, \tag{47}$$

with $d^{(3)}$ being the strength of the third-order dipole. The last term of (47) pertains to fixing the location of the core center with an accuracy of $O(\epsilon^3)$, but may be ignored for determining the speed at the present order. To deduce $\dot{Z}^{(2)}$, we can skip the full solution of (43)–(47). It suffices to multiply (43) by r^2 and to integrate from 0 to some large value with respect to r. To simplify the expression, (37)–(41) is invoked. Taking the limit $r \to \infty$, we eventually arrive at the desired formula:

$$\begin{aligned} \dot{Z}^{(2)} &= \frac{\pi}{4R_0^3}\int_0^\infty \left[\frac{17}{8} r v^{(0)} - \frac{3}{R_0}\psi^{(0)}\right] r^3 \zeta^{(0)} dr - \frac{5\pi}{4R_0^3} B \\ &- \frac{\pi}{R_0^2}\int_0^\infty \left[ra + \frac{b}{2}\tilde{\psi}_{11}^{(1)}\right] r v^{(0)} \tilde{\psi}_{21}^{(2)} dr - \frac{R^{(2)}}{4\pi R_0}\left[\log\left(\frac{8R_0}{\epsilon}\right) - \frac{3}{2} + A\right] \\ &+ \frac{\pi}{8R_0^3}\int_0^\infty \left\{ ra\left[\tilde{\psi}_{11}^{(1)} - 3r\frac{\partial \tilde{\psi}_{11}^{(1)}}{\partial r}\right]\tilde{\psi}_{11}^{(1)} + b\left[\tilde{\psi}_{11}^{(1)} - r\frac{\partial \tilde{\psi}_{11}^{(1)}}{\partial r}\right]\left[\tilde{\psi}_{11}^{(1)}\right]^2 \right\} dr \\ &- \frac{\pi}{2R_0^4}\int_0^\infty r^2 \psi^{(0)} a \tilde{\psi}_{11}^{(1)} dr - \frac{\pi}{R_0}\int_0^\infty \left[2 r v^{(0)} + \frac{\psi^{(0)}}{R_0}\right] r \zeta_0^{(2)} dr \\ &+ \frac{\pi}{R_0}\int_0^\infty \left[\frac{\partial \zeta_0^{(2)}}{\partial r} - \frac{a}{r}\int_0^r r' \zeta_0^{(2)} dr'\right] r \tilde{\psi}_{11}^{(1)} dr\,, \end{aligned} \tag{48a}$$

where definitions (27) and (41) of a and b should be remembered, and

$$A = \lim_{r\to\infty}\left\{4\pi^2 \int_0^r r'[v^{(0)}(r')]^2 dr' - \log r\right\}, \tag{48b}$$

$$B = \lim_{r\to\infty}\left\{\int_0^r r' v^{(0)} \tilde{\psi}_{11}^{(1)} dr' + \frac{r^2}{4}\left(\int_0^r r'[v^{(0)}]^2 dr'\right) - \frac{d_0}{2\pi}\left[\log\left(\frac{8R_0}{\epsilon r}\right) - \frac{7}{10}\right]\right\}. \tag{48c}$$

The fact that (48c) includes the parameter d_0 brings out the contribution of the dipole distributed along the core-centerline to the induction velocity at $O(\epsilon^3)$, which has so far gone unnoticed. This is traced back to the matching condition (47), essentially non-local in its nature. It must be emphasised that the asymptotic formula (18) of the Biot-Savart integral is essential to make the systematic evaluation of multi-pole induction feasible.

In order to find $\dot{Z}^{(2)}$, it remains to numerically calculate $\psi_0^{(2)}$ and $\psi_{21}^{(2)}$. Fortunately, the explicit solution is at hand for the Rankine vortex (32). In this case,

$$B = \frac{3}{2^5\pi^2}\log\left(\frac{8R_0}{\epsilon}\right) - \frac{71}{15\cdot 2^5\pi^2}\,. \tag{49}$$

Noting that $a = -2\delta(r-1)$ and (41), the last four integrals of (48a) vanish and we are left with

$$\dot{Z}^{(2)} = -\frac{3}{2^5\pi^2 R_0^3}\left[\log\left(\frac{8R_0}{\epsilon}\right) - \frac{5}{4}\right], \tag{50}$$

in accordance with (1). Otherwise stated, (48a) is a generalisation of Dyson's result to an arbitrary distribution of leading-order vorticity in the presence or absence of viscosity.

The rest of this section concerns the third-order radial velocity $\dot{R}^{(2)}$. The equation for $\psi_{12}^{(3)}$ is reducible to

$$\begin{aligned} &\frac{1}{r}\left(\frac{\partial\zeta^{(0)}}{\partial r}\psi_{12}^{(3)} + v^{(0)}\frac{\partial}{\partial r}\left[\frac{1}{r}\frac{\partial}{\partial r}\left(r\psi_{12}^{(3)}\right)\right]\right) - R_0\dot{R}^{(2)}\frac{\partial\zeta^{(0)}}{\partial r} \\ &= R_0\left(-\frac{\partial\zeta_{11}^{(1)}}{\partial t} + \Delta\zeta_{11}^{(1)} + \frac{1}{R_0}\frac{\partial\zeta^{(0)}}{\partial r}\right), \end{aligned} \tag{51}$$

subject to the matching condition that $\psi_{12}^{(3)} \propto 1/r$ as $r \to \infty$. As before, we implement the integration of (51) with respect to r after multiplication by r^2. The diffusion equation (23) of $\zeta^{(0)}$ helps to simplify the result, and thus we obtain the speed of the origin $r = 0$ of the local moving coordinates in the ρ-direction:

$$\dot{R}^{(2)} = \frac{2\pi}{R_0}\dot{d}_0\,. \tag{52}$$

It is noteworthy that (52) is consistent with the conservation law of the fluid impulse. Recall that the impulse is constant even in the presence of viscosity. Only the z-component P_z is nontrivial for the axisymmetric flow, giving, in dimensionless form,

$$P_z \cong \pi R_0^2 + \epsilon^2\pi\left\{2R_0R^{(2)} + \pi\int_0^\infty r^3\zeta^{(0)}dr + 2R_0\pi\int_0^\infty r^2\zeta_{11}^{(1)}dr\right\}. \tag{53}$$

In the light of (35), the constancy of $O(\epsilon^2)$-term gives rise to (52). This observation implies that the first-order solution, combined with impulse conservation, is sufficient to get $\dot{R}^{(2)}$ and therefore that we may skip the third-order solution $\psi_{12}^{(3)}$. Notice that the initial value of d_0 defined by (35) sets that of P_z up to second order. This manifests a remarkable aspect that our formulation of the initial-value problem rests upon the fundamental laws of conservation of both circulation and impulse.

Finally we illustrate how the vorticity distribution radially evolves starting from a delta-function core (24). In this case, $P_z = \pi R_0^2$ identically with the $O(\epsilon^2)$ correction term being absent. The particular solution $\Psi_{11}^{(1)}$ given by (29) corresponds to the dipole field whose stagnation point is permanently sitting at $r = 0$ (Klein and Knio 1995). The evaluation of the behaviour of $\Psi_{11}^{(1)}$, at large values of r, is carried out with ease to yield

$$\Psi_{11}^{(1)} = \frac{r}{4\pi}\left\{\log r + \lim_{r\to\infty}\left(4\pi^2\int_0^r r'[v^{(0)}(r')]^2 dr' - \log r\right) + \frac{1}{2}\right\} + \frac{D_0}{r} + \cdots,$$

$$D_0 \cong 0.41225489 \times \frac{t}{2\pi}. \tag{54}$$

Comparing with (52), we reach the conclusion that, given initially a circular line vortex of radius R_0, the stagnation point $\rho_s(t)$ in the core drifts outward linearly in time owing to the action of viscosity:

$$\rho_s \cong R_0 + 0.41225489\nu t/R_0. \tag{55}$$

References

Callegari, A. J. and Ting, L. (1978) Motion of a curved vortex filament with decaying vortical core and axial velocity, *SIAM J. Appl. Maths* **35**, pp. 148–175.

Dyson, F. W. (1893) The potential of an anchor ring – part II, *Phil. Trans. Roy. Soc. Lond. A* **184**, pp. 1041–1106.

Fraenkel, L. E. (1972) Examples of steady vortex rings of small cross-section in an ideal fluid, *J. Fluid Mech.* **51**, pp. 119–135.

Fukumoto, Y. and Miyazaki, T. (1991) Three-dimensional distortions of a vortex filament with axial velocity, *J. Fluid Mech.* **222**, pp. 369–416.

Gidas, B., Ni, W.-M., and Nirenberg, L. (1979) Symmetry and related properties via the maximum principle, *Commun. Math. Phys.* **68**, pp. 209–243.

Hosokawa, I. and Yamamoto, K. (1989) Fine structure of a directly simulated isotropic turbulence, *J. Phys. Soc. Japan* **59**, pp. 401–404.

Jiménez, J., Moffatt, H. K., and Vasco, C. (1996) The structure of vortices in freely decaying two-dimensional turbulence, *J. Fluid Mech.* **313**, pp. 209–222.

Kerr, R. M. (1985) Higher-order derivative correlation and the alignment of small-scale structure in isotropic turbulence, *J. Fluid Mech.* **153**, pp. 31–58.

Kida, S. and Ohkitani, K. (1992) Spatiotemporal intermittency and instability of a forced turbulence, *Phys. Fluids A* **4**, pp. 1018–1027.

Klein, R. and Knio, O. M. (1995) Asymptotic vorticity structure and numerical simulation of slender vortex filaments, *J. Fluid Mech.* **284**, pp. 275–321.

Klein, R. and Majda, A. J. (1991) Self-stretching of a perturbed vortex filament. I. The asymptotic equation for deviations from a straight line, *Physica D* **49**, pp. 323–352.

Moffatt, H. K., Kida, S., and Ohkitani, K. (1994) Stretched vortices – the sinews of turbulence; large-Reynolds-number asymptotics, *J. Fluid Mech.* **259**, pp. 241–264.

Moore, D. W. and Saffman, P. G. (1972) The motion of a vortex filament with axial flow, *Phil. Trans. R. Soc. Lond. A* **272**, pp. 403–429.

Saffman, P. G. (1970) The velocity of viscous vortex rings, *Stud. Appl. Math.* **49**, pp. 371–380.

Siggia, E. D. (1981) Numerical study of small scale intermittency in three-dimensional turbulence, *J. Fluid Mech.* **107**, pp. 375–406.

Tung, C. and Ting, L. (1967) Motion and decay of a vortex ring, *Phys. Fluids* **10**, pp. 901–910.

Widnall, S. E., Bliss, D. B., and Zalay, A. (1971) Theoretical and experimental study of the stability of a vortex pair, In *Aircraft Wake Turbulence and its Detection* (eds, Olsen, Goldberg, Rogers), Plenum, pp. 305–338.

RECENT DEVELOPMENT IN THE ASYMPTOTIC THEORY OF VORTEX BREAKDOWN

M. SCHMITZ
Institut für Technische Mechanik, RWTH Aachen
Templergraben 64, D-52056 Aachen, Germany

AND

R. KLEIN
FB Sicherheitstechnik, BUGH Wuppertal
Gauß Str. 20, D-42097 Wuppertal, Germany

1. Introduction

Slender vortex dynamics is dominated by two very different sources of nonlinearity. There is first the nonlinearity inherent in the motion of a vortex filament due to nonlocal self-induction and local curvature effects. Secondly, there is a nonlinear core flow, which produces local axisymmetric breakdown under suitable boundary and initial conditions. Both these aspects of slender vortex dynamics have been studied in the past using tools of matched asymptotic analysis. The non-axisymmetric spiral mode of vortex breakdown appears to be due to an interaction of the core flow and the vortex centerline motion. Thus, it is a challenge to derive a unified formulation in the slender vortex limit which includes both the filament dynamics and nontrivial core structures.

One interesting issue in this context concerns axial core structure variations. Obviously, the axisymmetric mode of vortex breakdown would likely be included in such an extended theory. But, in addition there is a strong sensitivity of three-dimensional vortex filament motion with respect to changes of the core structure through the well-known curvature-binormal term in the equation of motion. Thus, in the presence of axial variations of the vortex core, any large scale bend of the vortex centerline will lead to an additional deformation in the binormal direction. A concrete example would be a circular vortex ring, originally embedded in a plane: Axial core structure variations would lead to a bending of the filament out of the plane.

E. Krause and K. Gersten (eds.), IUTAM Symposium on Dynamics of Slender Vortices, 35-44.

Ting & Klein (1991) and Klein & Ting (1992) have outlined the asymptotic theory for slender, three-dimensional vortex filaments with axial core structure variation, including vortex stretching and viscosity effects. While the equation of motion for the filament centerline is very similar to that for a vortex with axially homogeneous core, the core structure evolution equations reveal new insight. At leading order, one finds the core flow to be axisymmetric. At the first order one recovers essentially the well-known slender vortex approximation for an inviscid, steady, axisymmetric and (locally) columnar vortex. These equations determine the axial structure of the leading order core at fixed times. An immediate consequence is that the total head and circulation of the core flow are conserved along concentric surfaces of constant axial mass flow, Ψ. Unsteady terms, the diffusion of vorticity and vortex stretching appear at the second order as observed ealier in (Callegari & Ting, 1978) for vortices with axially homogeneous cores. For axially varying cores these terms attain a non-standard form: The net effect of molecular transport is obtained as the axial average of the viscous terms along the stream surfaces. Thus, in the present one-time scale asymptotics, viscosity is nonlocal with respect to the axial coordinate.

The first sections of this paper summarize this extended asymptotic theory for three-dimensional slender vortices with axial variation of the core structure.

In preparation of a comprehensive model that will include all the effects listed above, Schmitz (1995) re-considered the first order equations for the special case of a straight columnar vortex. One important issue that needs to be resolved in this context is the occurance of "Benjamin's critical point" (Benjamin, 1967): For certain distributions of total head and circulation as functions of the axial mass flux variable there is a boundary condition singularity. Generally the vortex core structure is determined in any vortex crossection by the distributions of total head and circulation vs. the mass flux variable, plus two radial boundary conditions. Within the critical crossection a compatibility condition connecting the radial boundary data exists. For incompatible boundary data a solution to the slender vortex equations cannot be determined. Studies of the phenomenon using different numerical and analytical approaches can be found, e.g., in (Reyna & Menne, 1988), (Krause, 1989) and the references therein.

The introduction of the standard slender vortex approximation amounts to neglecting the second axial derivative term $\frac{1}{2y}\Psi_{ss}$ in the Bragg-Hawthorne or Squire-Long equation

$$\Psi_{yy} + \frac{1}{2y}\Psi_{ss} = \frac{dH}{d\Psi} - \frac{1}{2\,y}\frac{dI}{d\Psi} \tag{1}$$

for the mass flux $\Psi(s, y)$ of an axisymmetric columnar steady vortex. Here $y = r^2/2$ is proportional to the square of the radial coordinate, the axial coordinate s is scaled - like r - by a typical core diameter and $H(\Psi)$ and $I(\Psi)$ are essentially the total head and the square of the circulation on stream surfaces $\Psi(s, y) = const.$.

When approching the critical point, the solution of the slender vortex equation

$$\Psi_{yy} = \frac{dH}{d\Psi} - \frac{1}{2y}\frac{dI}{d\Psi} \tag{2}$$

behaves as $\sqrt{s - s_{\text{crit}}}$, so that the neglected second derivative is s grows without bounds. Hence, in an extended theory this second axial derivative term needs to be included at least in the immediate vicinity of the critical point. The second part of this paper is concerned with such a critical layer analysis. Perturbations about the critical state are considered and a new scaling of coordinates and perturbation amplitudes is found that allows one to include the effects of the second axial derivative without having to numerically solve the full Bragg-Hawthorne / Squire-Long equation.

The major result of the critical layer analysis is the perturbation amplitude equation

$$\alpha_{\zeta\zeta} + a\,\alpha^2 = b\zeta\ , \tag{3}$$

with coefficients a, b that depend on the distributions $H(\Psi), I(\Psi)$ and ζ a rescaled axial coordinate. Generally, a strong singularity develops inside the critical layer at a finite location in terms of the rescaled inner coordinate ζ. In this case, the critical layer terminates with a flow region whose structure precludes the application of any slenderness approximation. Only under very specific circumstances does the solution pass through criticality without locally leaving the slender vortex regime.

Note that the above conclusions stictly hold only for flows in tubes with radii comparable to the vortex core size. For flows with bounding walls that are far away from the core or for free vortices, a modified analysis would have to be developed. See also the remarks near the end of Section 4.

2. Slender Vortex Filaments with Axial Core Structure Variation

2.1. THE ASYMPTOTIC REGIME AND THE EXPANSION SCHEME

Here we consider slender vortices with asymptotic scalings analogous to those outlined in (Callegari & Ting, 1978). That is, we assume an incompressible flow in three space dimensions with the vorticity concentrated in a thin tube-like region about a dynamically evolving centerline $\mathcal{L}(t)$. The vorticity field is, to leading order, axisymmetric with respect to the $\mathcal{L}(t)$ and, also to leading order, has no radial component in orthogonal crossections of the vortex tube. The vortex core size d is large compared to a typical centerline radius of curvature R, so that

$$\delta = \frac{d}{R} \ll 1\,. \tag{4}$$

The viscosity is assumed small, such that the Reynolds $Re = \Gamma/\nu$, defined as the ratio of the total vortex circulation Γ and the kinematic viscosity ν, satisfies the estimate

$$\frac{1}{Re} \leq O(\delta^2). \tag{5}$$

We emphasize that the theory presented here, just as that of Callegari & Ting (1978) and follow-ups of this work, is uniformly valid for all the following cases:

- the distinguished limit $1/Re = \mu\delta^2$ with $\mu = O(1)$ as $\delta \to 0$,
- the case of much smaller viscosity $1/Re = o(\delta^2)$ as $\delta \to 0$ and
- the case of vanishing viscosity $\nu \equiv 0$.

The asymptotic regime is thus exactly that also considered by Callegari & Ting (1978), yet, as is Klein & Ting (1992) we allow a non-zero variation of the leading order vorticity distribution in the axial direction. Thus, we consider an asymptotic expansion of the vorticity and velocity fields according to

$$\omega = \frac{1}{\delta^2}\left(\eta^{(0)}\boldsymbol{\theta} + \zeta^{(0)}\boldsymbol{t}\right) + \frac{1}{\delta}\left(\xi^{(1)}\boldsymbol{r} + \eta^{(1)}\boldsymbol{\theta} + \zeta^{(1)}\boldsymbol{t}\right) + \dots \tag{6}$$

$$\boldsymbol{v} = \frac{1}{\delta}\left(v^{(0)}\boldsymbol{\theta} + w^{(0)}\boldsymbol{t}\right) + \left(u^{(1)}\boldsymbol{r} + v^{(1)}\boldsymbol{\theta} + w^{(1)}\boldsymbol{t}\right) + \dots \tag{7}$$

Here $\boldsymbol{r}, \boldsymbol{\theta}, \boldsymbol{t}$ are the radial, circumferential and tangential unit basis vectors in a centerline-attached coordinate system (Callegari & Ting, 1978). The expansion functions have the general dependencies

$$\phi^{(0)} = \phi^{(0)}(r, \xi, t), \quad \phi^{(i)} = \phi^{(i)}(r, \theta, \xi, t), \quad (i = 1, 2, ...), \quad (\xi = \delta s), \tag{8}$$

where (r, θ, ξ) are the centerline-attached coordinates and t denotes time.

It is shown in (Klein & Ting, 1992) that the centerline equation of motion is exactly the same as for the case of an axially homogeneous core, i.e, one has

$$\dot{\boldsymbol{X}} = \frac{\Gamma}{4\pi}\kappa\big(\ln(\frac{1}{\delta}) + (C_v + C_w)(\xi, t)\big)\,\boldsymbol{b} + \boldsymbol{Q}^f(\xi, t)\,, \tag{9}$$

for any point $\boldsymbol{X}(\xi, t)$ on the filament centerline. Here κ is the centerline curvature, $\boldsymbol{b}$ the local binormal and $\boldsymbol{Q}^f$ the so-called finite part of the line Biot-Savart integral. We have explicitly indicated the ξ-dependence of the core structure coefficients

$$C_v = \lim_{r\to\infty}\big(\frac{4\pi^2}{\Gamma^2}\int_0^r r' {v^{(0)}}^2\,dr' - \ln r\big) + \frac{1}{2}, \qquad C_w = \frac{-8\pi^2}{\Gamma^2}\int_0^\infty r' {w^{(0)}}^2\,dr'\,. \tag{10}$$

which will introduce the geometrical distortions in the binormal direction mentioned in the introduction.

2.2. KNOWN RESULTS ON THE LEADING ORDER CORE STRUCTURE

Ting & Klein (1991) show that the leading order velocity field in the core satisfies, at each fixed time t, the conservation constraints on total head and circulation, well known from axisymmetric columnar vortices in inviscid flows. Thus, let

$$\Psi(r, s, t) = \int_0^r w^{(0)} r'\,dr' \tag{11}$$

denote a stream function for the axial mass flux through a circle of radius r centered on $\mathcal{L}(t)$, then the total head and square of the circulation

$$H = H(\Psi, t) = p^{(0)} + \frac{1}{2}\left(v^{(0)^2} + w^{(0)^2}\right) \quad \text{and} \quad I = I(\Psi, t) = \frac{1}{2}\left(rv^{(0)}\right)^2 \tag{12}$$

are functions of (Ψ, t) only. Notice that $H = H(\Psi, t)$ and $I = I(\Psi, t)$ are evolving in time under the influence of vortex stretching and, if applicable, viscous diffusion. See (Klein & Ting, 1992) for the relevant evolution equations. The mass flux Ψ satisfies the Bragg-Hawthorne / Squire-Long equation (1).

3. The Axial Flow Structure at Fixed Time

The slenderness condition has been accounted for by a stretched axial coordinate $\xi = \delta s$ with $\delta = d/R \ll 1$. Using the expansion

$$\Psi = \Psi^{(0)} + \delta\Psi^{(1)} + \delta^2\Psi^{(2)} + \ldots; \tag{13}$$

we derive the following hierarchy of perturbation equations from (1):

$$\Psi^{(0)}_{yy} - \left[H' - \frac{1}{2y}I'\right] = 0 \tag{14}$$

$$\Psi^{(1)}_{yy} - \Psi^{(1)}\left[H'' - \frac{1}{2y}I''\right] = 0 \tag{15}$$

$$\Psi^{(2)}_{yy} - \Psi^{(2)}\left[H'' - \frac{1}{2y}I''\right] - \frac{\Psi^{(1)^2}}{2}\left[H''' - \frac{1}{2y}I'''\right] = -\frac{1}{2y}\Psi^{(0)}_{\xi\xi}. \tag{16}$$

Here $H' = H'(\Psi^{(0)})$, $I' = I'(\Psi^{(0)})$ etc. and primes denote differentiation.

On the filament axis the mass flux vanishes ($\Psi \equiv 0$). At a given outer wall $y - a(\xi)$ the total mass flux Ψ_{max} must be achieved:

$$\Psi^{(0)} = \Psi^{(1)} = \Psi^{(2)} = \ldots \equiv 0. \quad \text{at} \quad (y = 0). \tag{17}$$

$$\Psi^{(0)} = \Psi_{max}, \quad \Psi^{(1)} = \Psi^{(2)} = \ldots = 0 \quad \text{at} \quad (y = a(\xi)) \tag{18}$$

Schmitz (1995) derives a boundary condition for given farfield pressure $p_\infty(\xi)$:

$$\Psi^{(0)}_y \to \sqrt{2\left(\lim_{y\to\infty} H(\Psi^{(0)}) - p_\infty(\xi)\right)}, \quad \Psi^{(i)}_y \to 0 \quad \text{as} \quad (y \to \infty) \tag{19}$$

Either set of conditions leads to $\Psi^{(1)} \equiv 0$, so that the correct expansion, away from the critical point, is

$$\Psi = \Psi^{(0)} + \delta^2\Psi^{(2)} + \ldots \tag{20}$$

3.1. BEHAVIOUR NEAR THE CRITICAL POINT

The axial derivative $\Psi_\xi^{(0)}$ of the leading order stream function obeys the first order perturbation equation from (15). The boundary conditions at $y = 0, a(\xi)$ yield

$$\Psi_\xi^{(0)} \equiv 0 \;\text{ at }\; (y = 0)\,; \qquad \Psi_\xi^{(0)} + \frac{da}{d\xi}\Psi_y^{(0)} = 0 \;\text{ at }\; (y = a(\xi))\,. \tag{21}$$

Generally, $da/d\xi \neq 0$ and also $\Psi_y^{(0)}(a(\xi), \xi) \neq 0$. As a consequence, the problem is to find solutions to (15) that are zero on the axis and unity at the outer boundary. A suitable re-scaling then would yield the desired function $\Psi_\xi^{(0)}$. Benjamin (1967) shows that for certain values of a and distributions $H(\Psi), I(\Psi)$ no such solution to (15) exists. That is, any non-trivial solution will be exactly zero at the outer boundary. In approaching this crossection $\Psi_\xi^{(0)}$ develops a singularity (the exceptions for $\Psi_y^{(0)}(y = a) = 0$ and $da/d\xi = 0$ will not be considered here):

Only $\Psi_\xi^{(0)}$ diverges, while $\Psi^{(0)}$ stays bounded. Thus, we seek a coordinate transformation $\eta = f(\xi)$ such that $\Psi^{(0)}(y, \xi) = \Phi^{(0)}(y, \eta)$ with $\Phi_\eta^{(0)}$ non-singular and $f(\xi_{\rm cr}) = 0$. Recalling that $\Psi^{(0)}(a(\xi), \xi) \equiv \Psi_{\rm max} = \text{const.}$ we find

$$0 = \Psi^{(0)}(a(\xi), \xi) - \Psi^{(0)}(a_{\rm cr}, \xi_{\rm cr}) \tag{22}$$

$$= \Psi_{y,\rm cr}^{(0)}(\xi - \xi_{\rm cr})a'(\xi) + \Phi_{\eta,\rm cr}^{(0)} f(\xi) + \Phi_{\eta\eta,\rm cr}^{(0)} \frac{1}{2} f(\xi)^2 + \dots \tag{23}$$

where the subscript $_{,\rm cr}$ denotes evaluation at $y = a(\xi), \xi = \xi_{,\rm cr}, \eta = 0$. The derivative $\Phi_\eta^{(0)}$ is subject to the same constraints as $\Psi_\xi^{(0)}$, so that $\Phi_{\eta,\rm cr}^{(0)} = 0$. As a consequence,

$$f(\xi) \sim \sqrt{\xi - \xi_{\rm cr}}\,, \tag{24}$$

unless also $\Phi_{\eta\eta,\rm cr}^{(0)} = 0$. This is a special case that we do not consider here. Under most circumstances, the stream function should thus vary like $\sqrt{\xi - \xi_{\rm cr}}$ near the critical point, and $\Psi_\xi^{(0)}$ and $\Psi_{\xi\xi}^{(0)}$ should diverge as $(\xi - \xi_{\rm cr})^{-\frac{1}{2}}$ and $(\xi - \xi_{\rm cr})^{-\frac{3}{2}}$, respectively. Since the second derivative would grow without bounds, at some point its neglection in the full stream function equation (1) is no longer acceptable. A narrow layer around the critical crossection exists in which this term representing axial interactions regains its importance.

4. Critical Layer Analysis

In the vicinity of the critical point $(\xi - \xi_{\rm cr})$ we introduce the new expansion

$$\tilde{\Psi} = \tilde{\Psi}^{(0)}(\hat{y}) + \delta^{\frac{2}{5}}\tilde{\Psi}^{(1)}(\hat{y}, \zeta) + \delta^{\frac{4}{5}}\tilde{\Psi}^{(2)}(\hat{y}, \zeta) + \dots\,, \qquad \zeta = \delta^{-\frac{4}{5}}(\xi - \xi_{\rm cr}). \tag{25}$$

In describing the boundary conditions, the following y-coordinate transformation is found to be useful in addition:

$$\hat{y} = y\frac{a_c}{a(\xi)} \quad \text{with} \quad a(\xi) = a_c + \delta^{4/5}\zeta a'_{\rm cr}\,, \tag{26}$$

where a_{cr} denotes the boundary at the critical point, a'_{cr} its axial derivative. The outer boundary then coincides with $\hat{y} = 1$ for all ζ.

The equations for the leading and first order mass fluxes are identical to (14), (15). Obviously, the leading order solution matches the critical flow state from the outer analysis. The first order solution is

$$\tilde{\Psi}^{(1)}(\hat{y}, \zeta) = \alpha(\zeta)\tilde{\Phi}(\hat{y}), \tag{27}$$

where $\Phi(y)$ is a normalized solution to (15), which is zero at both boundaries and has unit slope at $\hat{y} = 0$. The yet unknown amplitude $\alpha(\zeta)$ is obtained from a solvability condition for the second order equation:

$$\begin{aligned} \tilde{\Psi}^{(2)}_{\hat{y}\hat{y}} - \left[H'' - \tfrac{1}{2\hat{y}}I''\right]\Psi^{(2)} &= \left[H''' - \tfrac{1}{2\hat{y}}I'''\right]\frac{(\alpha\tilde{\Phi})^2}{2} \\ &\quad + \left[H' - \tfrac{1}{4\hat{y}}I'\right]2\zeta\left(\tfrac{a_\xi}{a}\right)_{\text{cr}} - \frac{1}{2\hat{y}}\alpha_{\zeta\zeta}\tilde{\Phi}\,. \end{aligned} \tag{28}$$

Multiplying by $\tilde{\Phi}$, integrating in y and using $\int_0^{a_{\text{cr}}} \tilde{\Phi}\tilde{\Psi}^{(2)}_{\hat{y}\hat{y}}dy = \int_0^{a_{\text{cr}}} \tilde{\Psi}^{(2)}\tilde{\Phi}_{\hat{y}\hat{y}}dy$, one obtains a nonlinear second order differential equation for α:

$$\alpha_{\zeta\zeta} - \frac{\hat{b}}{N}\alpha^2 - \frac{\hat{c}}{N}\zeta = 0. \tag{29}$$

with coefficients

$$\hat{b} = \int_0^{a_{\text{cr}}} \frac{\tilde{\Phi}^3}{2}\left[H''' - \frac{1}{2\hat{y}}I'''\right]dy, \quad \hat{c} = \int_0^{a_{\text{cr}}} \frac{2a_\xi}{a}\Big|_{\text{cr}}\tilde{\Phi}\left[H' - \frac{1}{4\hat{y}}I'\right]dy, \quad N = \int_0^{a_{\text{cr}}} \frac{\tilde{\Phi}^2}{2\hat{y}}dy\,.$$

Eqn. (29) defines Painlevé's transcendent functions (Ince (1956)). For $\hat{b}\hat{c} > 0$ and $\zeta \to -\infty$ the relevant solution branch of (29) behaves like $\sqrt{-\zeta}$ (unstable approach (Bender & Orzag (1978)). This matches the behavior of the outer solution when $\xi \to \xi_{\text{cr}}^-$. With the opposite sign of $\hat{b}\hat{c}$, the square root behavior of the outer solution can be matched for $\zeta \to +\infty$.

Without restriction of generality we assume the first case, $\hat{b}\hat{c} > 0$. Then for $\zeta > 0$, $\alpha_{\zeta\zeta}$ is always positive and α is rapidly driven to ∞. In fact, the Painlevé functions asymptotically diverge like the Weierstrass elliptic function at some finite distance in ζ. As a consequence, it is not possible to match this inner critical layer solution to the outer flow for $\zeta \to +\infty$, when $\hat{b}\hat{c} > 0$ or for $\zeta \to -\infty$, when $\hat{b}\hat{c} < 0$. Instead, there must be a narrow region near the critical point, where the axial scales of the flow field become comparable with the radial extension of the core structure. In this layer, the full axisymmetric flow equation from (1) needs to be solved.

Fig. 1 shows a typical solution and a sketch of the associated flow field. The amplitude α typically has two zeroes, so that the solution passes through criticality twice. This is reminiscent of the two-stage transition proposed earlier by Escudier & Keller (1983).

Schmitz (1995) shows that the above conclusions strictly hold only when the flow tube radius is comparable to the vortex core radius. When the ratio of flow tube radius vs. vortex core radius is large, then in the appropriate scalings $\hat{b}/N$ in (29) becomes negligible and the solution behaves as ζ^3 for large ζ. In principle, matching to some adjacent layer with a different axial scaling and a transition through criticality without leaving the slender vortex regime would then be conceivable. This is, however, "asymptotics on top of asymptotics". A more systematic approach would from the start employ a new distinguished for axial scales and the tube to vortex core radius and then re-expand the solution near the critical point. We leave this analysis for future work.

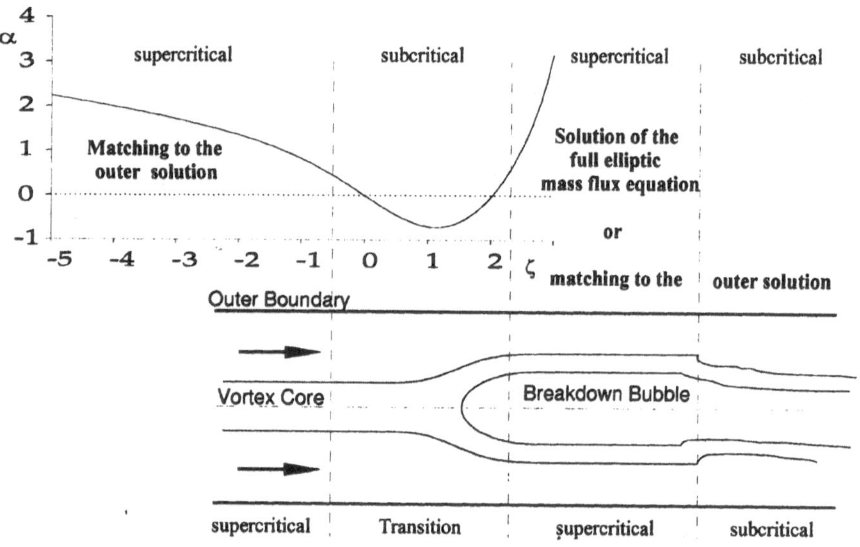

Figure 1. Two-stage transition through the critical flow state

5. Comparison with Direct Numerical Simulation

In this section we compare solutions of the asymptotic model with a solution of a direct numerical simulation for a bubble type breakdown of a free vortex (Breuer 1991), Weimer (1994), see also Althaus' contribution in the present volume!). This 3-D simulation was performed on a relatively coarse carthesian grid with $35 \times 35 \times 65$ gridpoints in $x-, y-, z-$direction. On the order of 20 gridpoints between the inflow crossection and the onset of reverse flow could be used for comparison with the asymptotics in the radial and axial directions, respectively.

The direct simulation essentially represents a free vortex. Thus, one way of treating this problem in the framework of the slender vortex asymptotics is to prescribe the axial velocity $w^{(0)}(0,\xi)$ on the vortex axis in agreement with the direct simulation and to obtain the outer flow field by solving the asymptotic mass flux equation with two initial data imposed at $y = 0$: $\Psi\big|_{y=0} = 0$ and $\Psi_y\big|_{y=0} = w^{(0)}(0,\xi)$. The radial distributions of axial and circumferential velocities in a single reference crossection (here at $\xi \approx 0.35$) can be used to derive the structure functions $H(\Psi)$ and $I(\Psi)$. Away from the critical state the leading order equation (14) will provide an accurate approximation, while near the critical state (27) and (3) must be used in addition.

In fig. 2 the axial velocities of the leading order outer solution $w^{(0)}$ and the axial velocity of the direct numerical simulation w_{AIA} are shown at different axial cross sections of the vortex tube. Close to the critical point near $\xi = 0.7$ the deviation of $w^{(0)}$ from w_{AIA} increases. In this region the critical layer solution yields an improvement, w_{inner}, over the leading order solution, albeit not a dramatic one.

The deviations may be due to the coarse mesh in the direct simulation, the fact that the Reynolds number was $Re = 500$ in the simulation, whereas the asymptotic model is inviscid to the order considered, and that the slenderness parameter δ was be estimated to be about $\delta = 0.2$, which is too large to truly guarantee the success of an asymptotic analysis. Nevertheless, we observe that the asymptotic theory yields reasonable solutions even near the limits of its validity.

Acknowledgements

This work was supported by the Deutsche Forschungsgemeinschaft in the framework of SFB 25 at RWTH Aachen.

References

Benjamin T.B. (1967) Some developments in the theory of vortex breakdown, *J. Fluid Mech.***28**, 65–84.

Callegari, A. and Ting, L. (1978) Motion of a Curved Vortex Filament with Decaying Vortical Core and Axial Velocity, *SIAM J. Appl. Math.***35**, pp. 148-175.

Escudier M.P., Keller J.J. (1983) Vortex Breakdown: A two Stage Transition, *AGARD CP*, **342**

Klein R., Ting L. (1992) Vortex filaments with axial core structure variation, *Appl. Math. Lett.***5**, pp. 99–103.

Krause E. (1989) Pressure Variation in Axially Symmetric Breakdown, Conf. Proc. "Colloquium on Vortex Breakdown", RWTH Aachen, Febr. 11./12., (1989).

Reyna L.G., Menne S. (1988) Numerical Prediction of Flow in Slender Vortices, *Computers and Fluids***16**, pp. 239-256.

Schmitz M. (1995) Axiale Entwicklung der Kernstruktur schlanker Wirbelfäden, Dissertation, RWTH-Aachen.

Ting L., Klein R. (1991) *Viscous Vortical Flows*, Lecture Notes in Physics, **374**, Springer-Verlag.

Weimer M. (1995), Personal communication, Aerodyn. Institut, RWTH Aachen, Germany (See also Althaus' contribution to this volume).

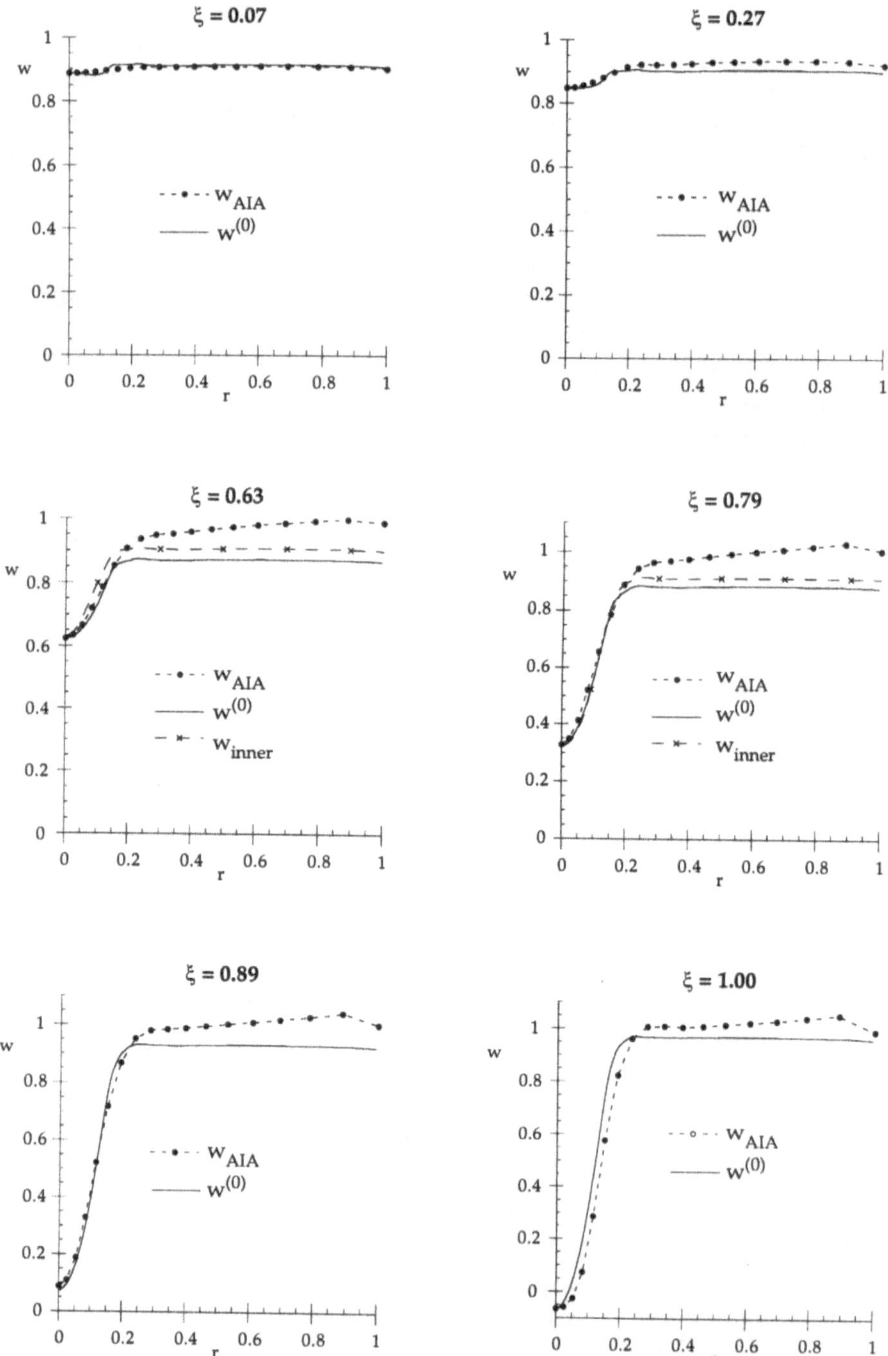

Figure 2. Axial velocity along the vortex axis

THE COMPLETE FIRST ORDER EXPANSION OF A SLENDER VORTEX RING

D. MARGERIT
LEMTA, CNRS URA 875, Nancy, France

Abstract. Equations for the axisymmetric part of the velocity field and for the equation of motion of a *non circular* slender vortex ring are given at first order. This is the correction to the known leading order given by Callegari and Ting [2].

1. Definitions and Notations

The length scales of the vortex ring that are different from its thickness δ , for example : the radius of curvature, the ring length, are of the same order L with $\delta/L = O(\varepsilon) << 1$. The central curve is described parametricaly with the use of a function $\vec{\mathbf{X}} = \vec{\mathbf{X}}(s,t)$. A local curvilinear co-ordinate system (r,φ,s), with a frame $(\vec{\mathbf{r}},\vec{\theta},\vec{\tau})$, is introduced near this central curve [2]. There is an *outer problem* defined by the *outer limit* : $\varepsilon \to 0$ with r fixed, which describes the situation far from the central line and an *inner problem* defined by the *inner limit* : $\varepsilon \to 0$ with $\bar{r} = r/\varepsilon$ fixed, which describes the situation near the central line.

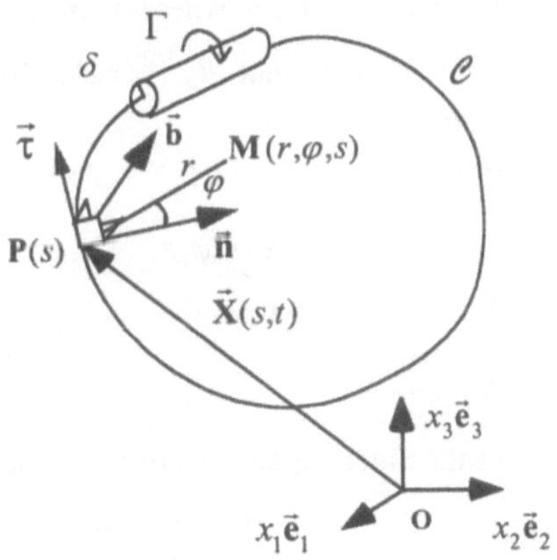

Figure 1 : The central curve and the local co-ordinates of the vortex ring.

E. Krause and K. Gersten (eds.), IUTAM Symposium on Dynamics of Slender Vortices, 45-54.

The change between Cartesian co-ordinates $\mathbf{M}(x_1,x_2,x_3)$ and local co-ordinates $\mathbf{M}(r,\varphi,s)$ satisfies :

$$\vec{\mathbf{x}} = \overrightarrow{\mathbf{OM}} = \vec{\mathbf{X}}(s,t) + r\vec{\mathbf{r}}(\varphi,s,t)$$

We have :

$$\sigma(s,t) = \left|\vec{\mathbf{X}}_s\right| \quad \vec{\mathbf{X}}_s = \sigma\vec{\tau} \quad \vec{\tau}_s = \sigma K\vec{\mathbf{n}}$$
$$\vec{\mathbf{n}}_s = \sigma(T\vec{\mathbf{b}} - K\vec{\tau}) \qquad \vec{\mathbf{b}}_s = -\sigma T\vec{\mathbf{n}}$$
$$\vec{\mathbf{r}} = \vec{\mathbf{r}}(\varphi,s) = \quad \vec{\mathbf{n}}(s)\cos\varphi + \vec{\mathbf{b}}(s)\sin\varphi$$
$$\vec{\theta} = \vec{\theta}(\varphi,s) = -\vec{\mathbf{n}}(s)\sin\varphi + \vec{\mathbf{b}}(s)\cos\varphi$$

where $|\;|$ is the usual norm of $\mathbf{R}^3$, T the local *torsion* of $\mathcal{C}$ and K the local *curvature* of $\mathcal{C}$. Notice that here and throughout this paper, the differentiation $\partial f/\partial x$ of a function f with respect to its variable x is denoted f_x ; $\times$ is the cross-product and $\bullet$ is the dot-product.

The small parameter ε is defined by $\varepsilon = \delta_0/L = \delta(t=0)/L$. Dimensionless variables :

$$r^* = r/L \quad \vec{\mathbf{X}}^* = \vec{\mathbf{X}}/L \quad \sigma^* = \sigma/L \quad K^* = LK$$
$$T^* = LT \quad S^* = S/L \quad \delta^* = \delta/L$$
$$t^* = t/(L^2/\Gamma) \quad \vec{\mathbf{v}}^* = \vec{\mathbf{v}}/(\Gamma/L) \quad \vec{\omega}^* = \vec{\omega}/(\Gamma/L^2)$$

are introduced, where S is the length of the ring and Γ is his circulation. Here, $\vec{\mathbf{v}}$ and $\vec{\omega}$ are respectively the velocity and the vorticity fields. From here on, all quantities are dimensionless and the asterisks are omitted.

The Reynolds number R_e is defined by $R_e = \Gamma/\nu$ where ν is the kinematic viscosity of the fluid. Let us define the number α such that $R_e^{-1/2} = \alpha\varepsilon$. Both inviscid : $\alpha = 0$ and viscous : $\alpha = O(1)$ vortex rings are studied.

The velocity is decomposed as follows :

$$\vec{\mathbf{v}}(r,\varphi,s,t,\varepsilon) = \dot{\vec{\mathbf{X}}}(s,t,\varepsilon) + \vec{\mathbf{V}}(r,\varphi,s,t,\varepsilon) \tag{1}$$

where
$$\vec{\mathbf{V}} = u\vec{\mathbf{r}} + v\vec{\theta} + w\vec{\tau} \tag{2}$$

and
$$\dot{\vec{\mathbf{X}}} = \frac{\partial\vec{\mathbf{X}}}{\partial t} \tag{3}$$

The following forms are chosen for the inner expansions of the velocity field :

$$\begin{aligned} u^{\text{inn}} &= & & u^{(1)}(\bar{r},\varphi,s,t)+\ldots \\ v^{\text{inn}} &= \varepsilon^{-1}v^{(0)}(\bar{r},s,t) &+& v^{(1)}(\bar{r},\varphi,s,t)+\ldots \\ w^{\text{inn}} &= \varepsilon^{-1}w^{(0)}(\bar{r},s,t) &+& w^{(1)}(\bar{r},\varphi,s,t)+\ldots \end{aligned} \tag{4}$$

with an expression of the central curve of the form :

$$\vec{\mathbf{X}} = \vec{\mathbf{X}}^{(0)}(s,t) + \varepsilon\,\vec{\mathbf{X}}^{(1)}(s,t)+\ldots \tag{5}$$

2. Limit of $\vec{\mathbf{v}}^{inn}$ at $\bar{r} \to \infty$ up to Order ε through Biot and Savart Law :

Let us have a vorticity field of the form :

$$\vec{\omega} = \frac{1}{\varepsilon^2} \vec{\omega}^{(0)}(\bar{r}, \varphi, s) \tag{6}$$

The Biot and Savart law is given on local co-ordinates by the formula :

$$\vec{\mathbf{v}}\ (r,\varphi,s,t,\varepsilon)$$
$$= \frac{1}{4\pi} \iiint \frac{\varepsilon^2 \vec{\omega}(\bar{r}', \varphi', s', t, \varepsilon) \times \left[(\vec{\mathbf{X}}(s,t,\varepsilon) + r\,\vec{\mathbf{r}}(\varphi,s,t,\varepsilon)) - (\vec{\mathbf{X}}(s',t,\varepsilon) + \varepsilon \bar{r}'\,\vec{\mathbf{r}}') \right]}{\left| (\vec{\mathbf{X}}(s,t,\varepsilon) + r\,\vec{\mathbf{r}}(\varphi,s,t,\varepsilon)) - (\vec{\mathbf{X}}(s',t,\varepsilon) + \varepsilon \bar{r}'\,\vec{\mathbf{r}}') \right|^3} h_3' \bar{r}'\, d\bar{r}'\, d\varphi'\, ds' \tag{7}$$

where $h_3' = \sigma(s',t)\ (1 - K(s',t)\varepsilon \bar{r}' \cos(\varphi'))$.

Next, in this section, s will be an arc length parameter.

The outer expansion of velocity is :

$$\vec{\mathbf{v}}^{out}(r,\varphi,s,\varepsilon) = \vec{\mathbf{v}}^{out^{(0)}}(r,\varphi,s) + \varepsilon \vec{\mathbf{v}}^{out^{(1)}}(r,\varphi,s) + O(\varepsilon^2)\,.$$

If

$$\iint (\vec{\omega} - [\vec{\omega} \bullet \vec{\tau}]\vec{\tau}\ )\bar{r} d\bar{r} d\varphi = 0 \tag{8}$$

one obtains :

$$\vec{\mathbf{v}}^{out(0)}(r,\varphi,s,\varepsilon) = \frac{1}{4\pi} \int_{\mathcal{C}} \frac{\vec{\tau}(s') \times (\vec{\mathbf{x}} - \vec{\mathbf{X}}(s'))}{\left| (\vec{\mathbf{x}} - \vec{\mathbf{X}}(s') \right|^3} ds' \tag{9a}$$

$$\vec{\mathbf{v}}^{out(1)}(r,\varphi,s) = \frac{1}{4\pi} \iiint \frac{\vec{\omega}^{(0)'} \times (\vec{\mathbf{x}} - \vec{\mathbf{X}}(s'))}{\left| \vec{\mathbf{x}} - \vec{\mathbf{X}}(s') \right|^3} \bar{r}'^2 K(s') \cos\varphi' d\bar{r}'\, d\varphi' ds' - \frac{1}{4\pi} \iiint \frac{\left(\vec{\mathbf{r}}' \times \vec{\omega}^{(0)'} \right)}{\left| \vec{\mathbf{x}} - \vec{\mathbf{X}}(s') \right|^3} \bar{r}'^2\, d\bar{r}'\, d\varphi' ds'$$
$$- \frac{1}{4\pi} \iiint 3 \frac{\left[\vec{\omega}^{(0)'} \times (\vec{\mathbf{x}} - \vec{\mathbf{X}}(s')) \right] \left(\vec{\mathbf{r}}' \bullet (\vec{\mathbf{x}} - \vec{\mathbf{X}}(s')) \right)}{\left| \vec{\mathbf{x}} - \vec{\mathbf{X}}(s') \right|^5} \bar{r}'^2\, d\bar{r}'\, d\varphi' ds' \tag{9b}$$

with : $\vec{\mathbf{x}} = \vec{\mathbf{X}}(s,t) + r\vec{\mathbf{r}}(\varphi,s,t)$.

Thus at leading order in outer co-ordinates, the velocity field exactly correspond to the Dirac delta distribution $\delta_{\mathcal{C}} \vec{\tau}$ on the central line.

In case $\vec{\omega}^{(0)} = \omega_2(\bar{r})\vec{\theta} + \omega_3(\bar{r})\vec{\tau}$, when $r = \varepsilon\ \bar{r}$ is put in $\vec{\mathbf{v}}^{out}(r \to 0, \varphi, s)$, one obtains :

$$\begin{aligned} \vec{\mathbf{v}}^{inn}(\bar{r} \to \infty,\varphi,s) &= \frac{1}{\varepsilon} \vec{\mathbf{v}}^{inn(0)}(\bar{r} \to \infty,\varphi,s) + \ln \varepsilon \vec{\mathbf{v}}^{inn(01)}(\bar{r} \to \infty,\varphi,s) + \vec{\mathbf{v}}^{inn(1)}(\bar{r} \to \infty,\varphi,s) \\ &+ \varepsilon \ln \varepsilon \vec{\mathbf{v}}^{inn(12)}(\bar{r} \to \infty,\varphi,s) + \varepsilon\ \vec{\mathbf{v}}^{inn(2)}(\bar{r} \to \infty,\varphi,s) + O(\varepsilon^2 \ln \varepsilon) \end{aligned} \tag{10}$$

with :

$$\vec{\mathbf{v}}^{\,\mathrm{inn}(0)}(\bar{r}\to\infty,\varphi,s)=\frac{1}{2\pi}\frac{\vec{\theta}}{\bar{r}}+\frac{\vec{\mathbf{I}}_1}{\bar{r}^2}+O(\frac{1}{\bar{r}^3}) \tag{11a}$$

$$\vec{\mathbf{v}}^{\,\mathrm{inn}(01)}(\bar{r}\to\infty,\varphi,s)=-\frac{K}{4\pi}\vec{\mathbf{b}} \tag{11b}$$

$$\vec{\mathbf{v}}^{\,\mathrm{inn}(1)}(\bar{r}\to\infty,\varphi,s)=\frac{K}{4\pi}\left[\ln\frac{S}{\bar{r}}-1\right]\vec{\mathbf{b}}+\frac{K}{4\pi}\cos\varphi\vec{\theta}+\vec{\mathbf{A}}+\frac{\vec{\mathbf{I}}_2}{\bar{r}}+O(\frac{1}{\bar{r}^2}) \tag{11c}$$

$$\vec{\mathbf{v}}^{\,\mathrm{inn}(12)}(\bar{r}\to\infty,\varphi,s)=\vec{\mathbf{I}}_3+\vec{\mathbf{I}}_5\bar{r} \tag{11d}$$

$$\vec{\mathbf{v}}^{\,\mathrm{inn}(2)}(\bar{r}\to\infty,\varphi,s)=(\vec{\mathbf{I}}_3+\vec{\mathbf{I}}_5\bar{r})\ln\bar{r}+\left(\vec{\mathbf{I}}_6+\vec{\mathbf{E}}_2(\varphi,s)\right)\bar{r}+\vec{\mathbf{I}}_4+\vec{\mathbf{E}}_1(s) \tag{11e}$$

$$\vec{\mathbf{E}}_2(\varphi,s)=\frac{1}{4\pi}\left(\vec{\mathbf{B}}(\varphi,s)-3\vec{\mathbf{C}}(\varphi,s)\right) \tag{11f}$$

where expressions of $\vec{\mathbf{A}}$, $\vec{\mathbf{B}}$, $\vec{\mathbf{C}}$, $\vec{\mathbf{E}}_i$, $\vec{\mathbf{I}}_i$ (i=1...6) are given in Appendix.

This expression (10) can be compared with that of Fukumoto and Miyazaki [4] (page 373) and Callegari et Ting [2] (page 173). It is the same all but here $\vec{\mathbf{A}}$, $\vec{\mathbf{B}}$, $\vec{\mathbf{C}}$ and order ε are given. Besides here the derivation was performed in an algorithmic way with formal calculus (Maple) and with the matched asymptotic expansion of singular integral method following François [3] or Bender and Orszag [1].

Let us notice that the same result would have be obtained if $\bar{r}\to\infty$ were put in the inner expansion of Biot and Savart law [6].

This result will be used in the following when the asymptotic matching will be performed.

3. Results at Order 0

Callegari and Ting [2] considered the case where $v^{(0)}, w^{(0)}$ are independent of s so that some compatibility conditions are satisfied. They deduced the following equations for $v^{(0)}, w^{(0)}$ from Navier Stokes second order equations :

$$\bar{r}\frac{\partial v^{(0)}(\bar{r},t)}{\partial t}-\alpha^2\frac{\partial v^{(0)}(\bar{r},t)}{\partial\bar{r}}+\frac{\alpha^2}{\bar{r}}v^{(0)}(\bar{r},t)-\alpha^2\bar{r}\frac{\partial^2 v^{(0)}(\bar{r},t)}{\partial\bar{r}^2}-\frac{1}{2}\bar{r}\frac{\partial\,\bar{r}v^{(0)}(\bar{r},t)}{\partial\bar{r}}\frac{\overset{\bullet}{S}{}^{(0)}}{S^{(0)}}=0 \tag{12a}$$

$$\bar{r}\frac{\partial w^{(0)}(\bar{r},t)}{\partial t}-\alpha^2\frac{\partial w^{(0)}(\bar{r},t)}{\partial\bar{r}}-\alpha^2\bar{r}\frac{\partial^2 w^{(0)}(\bar{r},t)}{\partial\bar{r}^2}-\frac{1}{2}\bar{r}^4\left(\frac{w^{(0)}(\bar{r},t)}{\bar{r}^2}\right)_{\bar{r}}\frac{\overset{\bullet}{S}{}^{(0)}}{S^{(0)}}=0 \tag{12b}$$

where $S^{(0)}$ is the length of the ring.

Through matching, they found the following equation for $\vec{\mathbf{X}}^{(0)}(s,t)$:

$$\dot{\vec{\mathbf{X}}}^{(0)} - (\dot{\vec{\mathbf{X}}}^{(0)} \bullet \vec{\tau})\vec{\tau} = \left(\frac{K^{(0)}}{4\pi}\left[\ln\frac{S^{(0)}}{\varepsilon} - 1\right] + K^{(0)}C^{*}\right)\vec{\mathbf{b}} + \vec{\mathbf{A}} - (\vec{\mathbf{A}} \bullet \vec{\tau})\vec{\tau} \qquad (13a)$$

where

$$C^{*}(t) = \frac{1}{4\pi}\left\{+\frac{1}{2} + \lim_{\bar{r}\to\infty}\left(4\pi^2\int_0^{\bar{r}}\xi\left(v^{(0)}\right)^2 d\xi - \ln(\bar{r})\right) - 8\pi^2\int_0^{\infty}\xi\left(w^{(0)}\right)^2 d\xi\right\} \qquad (13b)$$

$$\lambda(s,\bar{s},t) = \int_s^{s+\bar{s}} \sigma^{(0)}(s^{*},t)ds^{*} \qquad (13c)$$

$$\vec{\mathbf{A}}(s,t) = \frac{1}{4\pi}\int_{-\pi}^{+\pi}\left[-\sigma^{(0)}(s+\bar{s},t)\frac{\vec{\tau}^{(0)}(s+\bar{s},t)\times(\vec{\mathbf{X}}^{(0)}(s+\bar{s},t)-\vec{\mathbf{X}}^{(0)}(s,t))}{\left|\vec{\mathbf{X}}^{(0)}(s+\bar{s},t)-\vec{\mathbf{X}}^{(0)}(s,t)\right|^3} - \frac{K^{(0)}(s,t)}{2}\frac{\vec{\mathbf{b}}^{(0)}(s,t)\sigma^{(0)}(s+\bar{s},t)}{\left|\lambda^{(0)}(s,\bar{s},t)\right|}\right]d\bar{s} \qquad (13d)$$

4. Results at Order 1

In the same way that first order Navier Stokes equations give compatibility equations for $v^{(0)}, w^{(0)}$, second order Navier Stokes equations give compatibility equations for the axisymmetric part $v_c^{(1)}, w_c^{(1)}$ of $v^{(1)}, w^{(1)}$. These equations are automatically satisfied if $v_c^{(1)}$ is assumed to be independent of s and if $w_c^{(1)}$ is such that :

$$\frac{\partial w_c^{(1)}(\bar{r},s,t)}{\partial s} = -\dot{\sigma}^{(0)} + a(s,t)\sigma^{(0)} \qquad (14a)$$

$$a(s,t) = \dot{S}^{(0)} / S^{(0)} \qquad (14b)$$

We write : $$w_c^{(1)}(\bar{r},s,t) = w_{cc}^{(1)}(s,t) + w_c^{(1)}(\bar{r},t) \qquad (14c)$$

Equations for $v_c^{(1)}, w_c^{(1)}$ can be extracted from third order Navier Stokes equations. We did this with the use of symbolic calculus (on maple) in the following way : we obtained third order Navier Stokes equations, then we carried out the φ-average and s-average of the axial and circumferential components of the third order momentum equations using the third order continuity equation to eliminate $u_c^{(2)}$. Lots of terms vanished and we found equations for $v_c^{(1)}, w_c^{(1)}$. Note that Fukumoto and Miyazaki ([4] page 378) postulated $v_c^{(1)} = 0$, kept $w_c^{(1)}$ dependent of s, and did not have equation for $w_c^{(1)}$. The following equations are obtained :

$$S^{(0)}\left(\frac{\partial v_c^{(1)}}{\partial t}-\alpha^2\left[\frac{1}{\bar{r}}(\bar{r}v_c^{(1)})_{\bar{r}}\right]_{\bar{r}}\right)-\frac{1}{2}\overset{\bullet(0)}{S}\left(\bar{r}v_c^{(1)}\right)_{\bar{r}}=\frac{\bar{r}}{2}\left(\overset{\bullet(1)}{S}-S^{(1)}\frac{\overset{\bullet(0)}{S}}{S^{(0)}}\right)\zeta^{(0)} \quad (14d)$$

where $S^{(1)}=\int_0^{2\pi}\sigma^{(1)}ds$.

$$S^{(0)}\left(\frac{\partial w_c^{(1)}}{\partial t}-\alpha^2\left[\frac{1}{\bar{r}}(\bar{r}w_{c\ \bar{r}}^{(1)})_{\bar{r}}\right]_{\bar{r}}\right)-\frac{1}{2}\overset{\bullet(0)}{S}\bar{r}^3\left(\frac{w_c^{(1)}}{\bar{r}^2}\right)_{\bar{r}}=\frac{\bar{r}^3}{2}(w^{(0)}/\bar{r}^2)_{\bar{r}}\left(\overset{\bullet(1)}{S}-S^{(1)}\frac{\overset{\bullet(0)}{S}}{S^{(0)}}\right)$$

$$+\frac{1}{4\pi}\left(\ln(\frac{S}{\varepsilon})+C-4\pi\bar{r}v^{(0)}\right)\int_0^{2\pi}K^{(0)}\vec{\mathbf{A}}_s(s,t)\bullet\vec{\mathbf{b}}^{(0)}ds-\int_0^{2\pi}\sigma^{(0)}\vec{\mathbf{A}}(s,t)\bullet\vec{\tau}^{(0)}ds \quad (14e)$$

$$-\int_0^{2\pi}\sigma^{(0)}\frac{\partial w_{cc}^{(1)}(s,t)}{\partial t}ds-\frac{\overset{\bullet(0)}{S}}{S^{(0)}}\int_0^{2\pi}\sigma^{(0)}w_{cc}^{(1)}(s,t)ds$$

The left hand side of these equations are the same for $v_c^{(1)}(\bar{r},t)$ and $w_c^{(1)}(\bar{r},t)$ than for $v^{(0)}(\bar{r},t)$ and $w^{(0)}(\bar{r},t)$. We may notice that even if initially $w_c^{(1)}(\bar{r},0)=0$, the right hand side terms will induce $w_c^{(1)}(\bar{r},0)\neq 0$ when $t\neq 0$.

These equations for $v_c{}^{(1)}, w_c{}^{(1)}$ are linked to $\vec{\mathbf{X}}^{(1)}(s,t)$, so an equation for $\vec{\mathbf{X}}^{(1)}$ is needed to have a closed system of equations for the first order solutions $v_c{}^{(1)}, w_c{}^{(1)}$ and $\vec{\mathbf{X}}^{(1)}$. The best attempt to find this equation is by Fukumoto and Miyazaki ([4] page 382), where contribution from Navier Stokes equations up to order ε^1 have been found in order to performed the asymptotic matching. We may note that in their expression the term due to first order curvature $K^{(1)}$ is missing. Moreover, their expression for $\vec{\mathbf{X}}^{(1)}$ is not complete, as local and non-local (named $\vec{\mathbf{Q}}$ in [4]) contribution from Biot and Savart integral are given only up to order ε^0 in Fukumoto and Miyazaki ([4] page 373) and Callegari and Ting ([2] page 173), while order ε^1 is obviously needed to obtain complete and correct equation for $\vec{\mathbf{X}}^{(1)}$. The complete expression, up to ε order, was performed in section 2. The matching is then done and lacking terms in equation of $\vec{\mathbf{X}}^{(1)}$ are found :

$$\overset{\bullet(1)}{\vec{\mathbf{X}}}-(\overset{\bullet(1)}{\vec{\mathbf{X}}}\bullet\vec{\tau})\vec{\tau}=\left\{\overset{*}{\mathrm{C}}_2-\frac{1}{8\pi}K^{(1)}(s,t)+\frac{1}{4\pi}K^{(1)}(s,t)\ln\varepsilon-\frac{m}{4\pi}K_s\left[3\ln\varepsilon+3+\frac{5}{6}-3\ln S\right]\right\}\vec{\mathbf{n}}$$

$$+\left\{\overset{*}{\mathrm{C}}_1+\frac{m}{4\pi}KT\left[\ln\varepsilon+\frac{5}{6}-\ln S\right]\right\}\vec{\mathbf{b}}+\vec{\mathbf{E}}_1-(\vec{\mathbf{E}}_1\bullet\vec{\tau})\vec{\tau} \quad (15a)$$

$$\overset{*}{C}_1 = \pi \int_0^\infty \xi v^{(0)}(\xi,t) Hs_{12}^{(2)}(\xi,t) d\xi \tag{15b}$$

$$\overset{*}{C}_2 = \pi \lim_{\bar{r}\to\infty} \left(\int_0^{\bar{r}} \xi v^{(0)}(\xi,t) Hs_{11}^{(2)}(\xi,t) d\xi - \frac{1}{4}\frac{K^{(1)}(s,t)}{\pi^2} \ln \bar{r} \right) \tag{15c}$$

$$Hs_{12}^{(2)} = 2\frac{\xi}{v^{(0)}(\xi,t)}\frac{\partial w^{(0)}(\xi,t)}{\partial \xi}\left(\overset{\bullet}{\vec{\tau}}^{(0)}(s,t) \bullet \vec{\mathbf{b}}^{(0)}(s,t) \right) - 2\xi \frac{w^{(0)}(\xi,t)}{\sigma^{(0)}(s,t)}\frac{\partial K^{(0)}(s,t)}{\partial s} \tag{15d}$$

$$\begin{aligned} Hs_{11}^{(2)} = {} & 2\xi K^{(0)}(s,t)\frac{\partial v_c^{(1)}(\xi,t)}{\partial \xi} + v^{(0)}(\xi,t)K^{(1)}(s,t) + 2\frac{\xi}{v^{(0)}(\xi,t)}\frac{\partial w^{(0)}(\xi,t)}{\partial \xi}\left(\overset{\bullet}{\vec{\tau}}^{(0)}(s,t) \bullet \vec{\mathbf{n}}^{(0)}(s,t) \right) \\ & + 6K^{(0)}(s,t)v_c^{(1)}(\xi,t) + 2\xi \frac{\partial v^{(0)}(\xi,t)}{\partial \xi} K^{(1)}(s,t) + 2\xi\, w^{(0)}(\xi,t)K^{(0)}(s,t)T^{(0)}(s,t) \\ & + 2\frac{\xi K^{(0)}(s,t)v_c^{(1)}(\xi,t)}{v^{(0)}(\xi,t)}\frac{\partial v^{(0)}(\xi,t)}{\partial \xi} + 2\frac{\xi K^{(0)}(s,t)w_c^{(1)}(\xi,t)}{v^{(0)}(\xi,t)}\frac{\partial w^{(0)}(\xi,t)}{\partial \xi} \\ & + 2\frac{\xi K^{(1)}(s,t)w^{(0)}(\xi,t)}{v^{(0)}(\xi,t)}\frac{\partial w^{(0)}(\xi,t)}{\partial \xi} + 2\frac{\xi K^{(0)}(s,t)w^{(0)}(\xi,t)}{v^{(0)}(\xi,t)}\frac{\partial w_c^{(1)}(\xi,t)}{\partial \xi} \end{aligned} \tag{15e}$$

This equation (15a) is the first order equation of motion of the central line. Fukumoto and Miyazaki [4] (page 373) had written this equation without $K^{(1)}$, $\vec{\mathbf{E}}_1$ and terms with the axial flux *m*.

5. Conclusion :

A closed and complete system of equations (14d,14e,15a) for the first order axisymmetric part $v_c^{(1)}, w_c^{(1)}$ of the velocity field and for the first order central line : $\vec{\mathbf{X}}^{(1)}$ of a slender *non circular* vortex ring has been given.
It would be interesting to find a simple case where a solution to these equations can be found and to compare these equations with those of a circular vortex ring [5].

6. Appendix :

In this appendix, terms that appears in formulas of section 2 are given.

$$m=\pi\int_0^\infty \omega_2 \bar{r}^{-2} d\bar{r} = 2\pi\int_0^\infty w^{(0)}\bar{r}d\bar{r} \qquad \vec{\mathbf{I}}_1 = \frac{m}{\pi}\vec{\tau} \qquad \vec{\mathbf{I}}_2 = \frac{1}{2\pi}Km\cos\varphi\vec{\tau}$$

$$\vec{\mathbf{I}}_3 = m\left[(\frac{1}{8\pi}K^2 - \frac{1}{4\pi}T^2)\vec{\tau} - \frac{1}{4\pi}\left(KT\sin\varphi + 3K_s\cos\varphi\right)\vec{\mathbf{r}} - \frac{1}{4\pi}\left(KT\cos\varphi - 3K_s\sin\varphi\right)\vec{\theta}\right]$$

$$\begin{aligned}\vec{\mathbf{I}}_4 = \frac{m}{4\pi}\Bigg\{ & \left(\left[-\frac{5}{6} + \ln S\right]KT\sin\varphi + \left[-3 - \frac{5}{6} + 3\ln S\right]K_s\cos\varphi\right)\vec{\mathbf{r}} \\ & + \left(\left[-\frac{5}{6} + \ln S\right]KT\cos\varphi + \left[4 - \frac{1}{6} - 3\ln S\right]K_s\sin\varphi\right)\vec{\theta} \\ & + (\left(\left[-\frac{1}{2}\ln(2) + \frac{5}{16} - \frac{8}{S^2} + \frac{1}{4}\cos 2\varphi\right]K^2 + \left[\ln(2) - \frac{3}{2}\right]T^2\right)\vec{\tau}\Bigg\}\end{aligned}$$

$$\begin{aligned}\vec{\mathbf{E}}_1(a) = \frac{m}{4\pi}\Bigg\{ \int_{-S/2}^{S/2} & \frac{K(a+\bar{a})\vec{\mathbf{b}}(a+\bar{a}) \times (\vec{\mathbf{X}}(a+\bar{a}) - \vec{\mathbf{X}}(a)) + 2T(a+\bar{a})}{\left|\vec{\mathbf{X}}(a+\bar{a}) - \vec{\mathbf{X}}(a)\right|^3} \\ & + 3\frac{\left[\vec{\mathbf{n}}(a+\bar{a}) \bullet (\vec{\mathbf{X}}(a+\bar{a}) - \vec{\mathbf{X}}(a))\right]\left(\vec{\mathbf{b}}(a+\bar{a}) \times (\vec{\mathbf{X}}(a+\bar{a}) - \vec{\mathbf{X}}(a))\right)}{\left|\vec{\mathbf{X}}(a+\bar{a}) - \vec{\mathbf{X}}(a)\right|^3} \\ & - 3\frac{\left[\vec{\mathbf{b}}(a+\bar{a}) \bullet (\vec{\mathbf{X}}(a+\bar{a}) - \vec{\mathbf{X}}(a))\right]\left(\vec{\mathbf{n}}(a+\bar{a}) \times (\vec{\mathbf{X}}(a+\bar{a}) - \vec{\mathbf{X}}(a))\right)}{\left|\vec{\mathbf{X}}(a+\bar{a}) - \vec{\mathbf{X}}(a)\right|^3} \\ & + \left(-2\frac{1}{\left|\bar{a}\right|^3} + \frac{K(a)^2}{4\left|\bar{a}\right|} - \frac{T(a)^2}{2\left|\bar{a}\right|}\right)\vec{\tau}(a) - \frac{K(a)T(a)}{2\left|\bar{a}\right|}\vec{\mathbf{b}}(a) - \frac{3}{2}\frac{K_a(a)}{\left|\bar{a}\right|}\vec{\mathbf{n}}(a) \quad d\bar{a}\Bigg\}\end{aligned}$$

$$\vec{\mathbf{I}}_5 = -\frac{1}{4\pi}\left[(K_s\sin\varphi - KT\cos\varphi)\vec{\tau} + \frac{3}{4}K^2\sin(2\varphi)\vec{\mathbf{r}} + \frac{3}{4}K^2\cos(2\varphi)\vec{\theta}\right]$$

$$\vec{\mathbf{I}}_6 = -\frac{1}{4\pi}\left[\begin{aligned}&(K_s\sin\varphi - KT\cos\varphi)[1 - \ln S]\vec{\tau} + \left[1 - \frac{3}{4}\ln S\right]K^2\sin(2\varphi)\vec{\mathbf{r}} \\ &+ \left[\frac{4}{S^2} - \frac{K^2}{24} + \left(\frac{5}{8} - \frac{3}{4}\ln S\right)K^2\cos(2\varphi)\right]\vec{\theta}\end{aligned}\right]$$

$$\vec{\mathbf{E}}_2(\varphi, a) = \frac{1}{4\pi}\left(\vec{\mathbf{B}}(\varphi, a) - 3\vec{\mathbf{C}}(\varphi, a)\right)$$

with :

$$\vec{\mathbf{A}}(a)=\frac{1}{4\pi}\int_{-S/2}^{+S/2}\frac{\vec{\tau}(a+\bar{a})\times(\vec{\mathbf{X}}(a)-\vec{\mathbf{X}}(a+\bar{a}))}{\left|\vec{\mathbf{X}}(a)-\vec{\mathbf{X}}(a+\bar{a})\right|^3}-\frac{K(a)\vec{\mathbf{b}}(a)}{2\left|\bar{a}\right|}d\bar{a}$$

$$\vec{\mathbf{B}}(\varphi,a)=\vec{\mathbf{r}}(\varphi,a)\times\int_{-S/2}^{+S/2}\left[-\frac{\vec{\tau}(a+\bar{a})}{\left|\vec{\mathbf{X}}(a)-\vec{\mathbf{X}}(a+\bar{a})\right|^3}+\frac{1}{\left|\bar{a}\right|^3}(\vec{\tau}(a)+K(a)\vec{\mathbf{n}}(a)\bar{a}+(K_a(a)\vec{\mathbf{n}}(a)+K(a)T(a)\vec{\mathbf{b}}(a)-\frac{3}{4}K^2(a)\vec{\tau}(a))\frac{\bar{a}^2}{2}\right]d\bar{a}$$

$$\vec{\mathbf{C}}(\varphi,a)=\int_{-S/2}^{+S/2}\left[\frac{\vec{\mathbf{r}}(\varphi,a)\bullet(\vec{\mathbf{X}}(a+\bar{a})-\vec{\mathbf{X}}(a))}{\left|\vec{\mathbf{X}}(a+\bar{a})-\vec{\mathbf{X}}(a)\right|^5}\left[\vec{\tau}(a+\bar{a})\times(\vec{\mathbf{X}}(a+\bar{a})-\vec{\mathbf{X}}(a))\right]+\frac{K^2(a)}{4}\frac{\vec{\mathbf{b}}(a)\cos\varphi}{\left|\bar{a}\right|}\right]d\bar{a}$$

where S is the length of the vortex ring.

7. References :

1. Bender C.M. and Orszag S.A. (1978) Advanced mathematical methods for scientists and engineers, *McGraw-Hill, New York,* 341-349
2. Callegari, A.J., Ting, L. (1978) Motion of a curved vortex filament with decaying vortical core and axial velocity, *SIAM J. Appl. Math.* **35** (1), 148-175
3. François, C. (1981) Les méthodes de perturbation en mécanique. ENSTA. Paris, 98-104
4. Fukumoto, Y., Miyazaki, T. (1991) Three dimensional distortions of a vortex filament with axial velocity, *J. Fluid Mech.* **222**, 369-416
5. Fukumoto, Y., Moffatt, H.K. (1997) Motion of a thin vortex ring in a viscous fluid : higher-order asymptotics , *IUTAM Symposium on dynamics of slender vortices ,RWTH Aachen*
6. Margerit, D. Mouvement et Dynamique des Filaments et des Anneaux Tourbillons de Faible Epaisseur (Dynamics and Motion of Slender Vortex Filaments and Rings) PhD Thesis INPL Nancy, France (November 1997)

SELF-INDUCED MOTION OF HELICAL VORTICES

P.A. KUIBIN, V.L. OKULOV
Institute of Thermophysics
Lavrentiev Ave 1, 630090 Novosibirsk, RUSSIA

1. Introduction

The self-induced velocity of helical vortices in an infinite space as well as in a cylindrical tube is studied. Investigation was based on an analysis of the formula for determination of the self-induced velocity of a helical vortex with a uniform vorticity distribution inside a core. A simple formula for the velocity, that permitted to fulfil its complete analysis, was derived for the first time with the help of an explicit separation of singularities in analytical solutions for velocity fields induced by an infinitely thin helical filament in an infinite space and in a cylindrical tube.

As a result the main factors are established that influence the velocity of vortices rotation and determine existence of immobile vortex structures as well as rotating in the opposite direction to the flow swirling.

2. Problem statement

The problem about self-induced motion of vortex structures for a long time attracts efforts of the numerous contributors. However even for simple structures with constants vorticity distribution in a core and with canonical form of axes the solution of a problem still far before full completion. Only for a vortex ring (Fig. 1, *a*) it is possible to consider, that it is completely investigated. In table 1 there are presented in denotations by Ricca (1994) the formulae for the self-induced velocities as well as for the asymptotic expansions of the respective velocity fields in neighbourhood of a vortex filament for a vortex ring and for a helical vortex. The velocity scale is $\Gamma\kappa/4\pi$, where Γ is the circulation; κ is the curvature of a vortex axis. The quantities $C^{(a)}$ and $C^{(s)}$ are not determined up till now.

Unlike a vortex ring for a helical vortex these quantities are functions of the dimensionless helix pitch $\tau = l/a = t/\kappa$, where $l = h/(2\pi)$, $\kappa = a/(a^2 + l^2)$, $t = l/(a^2 + l^2)$ is the torsion of a vortex axis (geometrical parameters of a helical vortex are introduced in Fig. 1, *b*). A comparative analysis was fulfilled by Ricca (1994) for the τ-dependencies of the quantity $C^{(s)}$ obtained by different authors (Kelvin, 1880, Widnall, 1972; Moore & Saffman, 1972). In particular there was shown that at large helix pitch $\tau >> 1$ a simple analytical representation for $C^{(s)}$ obtained with prescription of Kelvin's (1880) result

E. Krause and K. Gersten (eds.), IUTAM Symposium on Dynamics of Slender Vortices, 55-62.

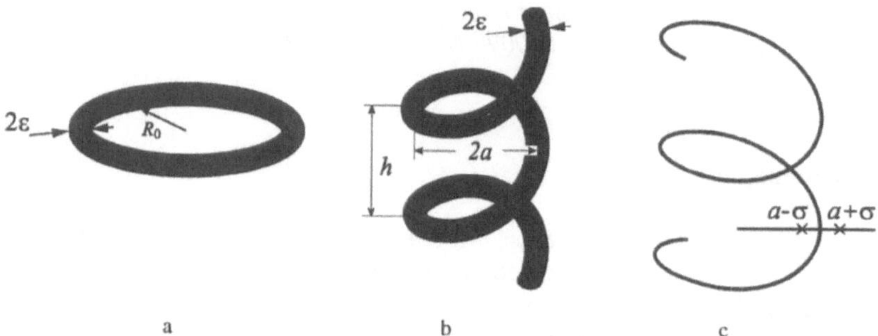

Figure 1. Vortex ring (a), helical vortex (b) and the scheme illustrating the Ricca's (1994) calculations.

$$C_K^{(s)} = -\ln\tau + 0.25 + \ln 2 - 0.5772$$

described well the calculation with an exact integral formula by Moore and Saffman (1972).

TABLE 1. A comparison of the self-induced velocity and terms in an asymptotic expansion of a velocity field responsible for the self-induced velocity

	self-induced velocity $\hat{w}_b$	asymptotic expansion
vortex ring	$\ln\dfrac{8R_0}{\varepsilon}-\dfrac{1}{4}$	$\ln\dfrac{8R_0}{\sigma}-1$
helical vortex	$\ln\dfrac{1}{\kappa\varepsilon}+C^{(s)}$	$\ln\dfrac{1}{\kappa\sigma}+C^{(a)}$

Another comparison was made by Ricca (1994) between the quantity $C^{(s)}$ calculated with formula by Moore and Saffman (1972) and the quantity $C^{(a)}$ computed on the base of the exact Hardin's (1982) solution in accordance with a formula

$$C_H^{(a)} = \frac{1}{\sqrt{\kappa a}}\left[\frac{a}{l} - \frac{l}{a+\sigma} - 2\frac{a^2}{\kappa l^2}\frac{1-\kappa\sigma}{a-\sigma}\sum_{m=1}^{\infty} mK_m'\left(\frac{ma}{l}\right)I_m\left(m\frac{a-\sigma}{l}\right)\right.$$
$$\left. - 2\frac{a^2}{\kappa l^2}\frac{1+\kappa\sigma}{a+\sigma}\sum_{m=1}^{\infty} mK_m'\left(m\frac{a+\sigma}{l}\right)I_m\left(\frac{ma}{l}\right)\right] - \ln\frac{1}{\kappa\sigma}$$

I_m and K_m are the modified Bessel's functions; the prime denotes the derivative. The difference was close to constant and was about 0.25. Due to the existence of singularities inside the logarithmic term and in square brackets the last quantity could be calculated only approximately at a finite distance σ (Fig. 1,c) from a filament.

To avoid this difficulty we propose to separate explicitly the singularities from solution for the velocity field. Such approach permits to find analytically an asymptotic representation in vicinity of a vortex filament for the binormal velocity component which is responsible for the self-induced motion.

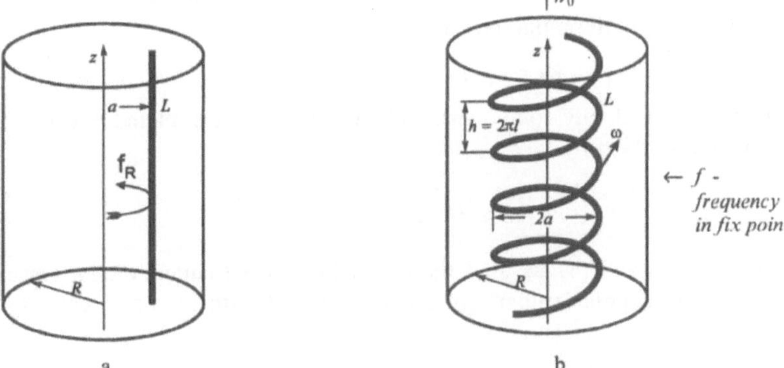

Figure 2. Rectilinear (a) and helical (b) vortices in cylindrical tube.

An other problem which can be studied at a time is the motion of a helical vortex in a cylindrical tube. In contrast to a rectilinear vortex filament (Fig. 2, a), the frequency of which rotation at specific intensity Γ is determined only by position of a vortex relative a wall (Lamb, 1940)

$$f = f_R = \frac{\Gamma}{4\pi^2(R^2 - a^2)}, \tag{1}$$

essentially more factors influence the motion of a screw vortex in a tube (Fig. 2, b). In addition to two dimensionless geometrical parameters, the relative pitch (or torsion) τ and the relative core size, ε/a, that influence a helical vortex motion in a boundless space, here we have two extra governing quantities: a relative tube radius, R/a, and dynamic parameter $\beta = w_0\, 2\pi l / \Gamma$ responsible for a uniform translation of a vortex along the tube axis with a speed w_0.

For the frequency of a vortex motion measured in a fixed point at a tube wall we can formally write

$$f = f_\kappa + f_\tau + f_R + f_{w_0} \tag{2}$$

with contributions depending on a vortex curvature, f_κ, on it's torsion, f_τ, on a tube radius, f_R, and on a value of translational speed, f_{w_0}.

One should keep in mind that the frequency of rotation of the vortex core in a fixed plane, f does not coincide with the angular velocity Ω of the vortex rotation. Such a discrepancy is explained by the existence in a general case of a translational motion. Due to the helix-like structure of a vortex, even its pure displacement along the z-axis at $\Omega = 0$ leads to the rotation of its core in a fixed plane perpendicular to the z-axis. Indeed, the angular velocity Ω of a helical vortex rotation can be presented through the dimensionless binormal velocity, $\hat{w}_b$, (Kuibin and Okulov, in press)

$$\Omega = \frac{\Gamma}{4\pi a^2}\sin^2\alpha\left(2\beta - \hat{w}_b\cos\alpha\right)$$

where $\sin^2\alpha = a^2/(a^2+l^2)$; $\cos^2\alpha = l^2/(a^2+l^2)$. The binormal velocity is correlated with the axial and circumferential ones as follows:

$$w_b = w_z\sin\alpha - w_\varphi\cos\alpha. \qquad (3)$$

In the same time the velocity components in helical vortices are related to each other as (Kuibin and Okulov, 1996)

$$w_z = w_0 - \frac{r}{l}w_\varphi \qquad (4)$$

Using Eqs. (3) and (4) we find the correlation of a frequency of a vortex core precession with an angular velocity of its rotation and with binormal velocity of its motion

$$f = \frac{1}{2\pi}\left[\frac{\Omega}{\cos^2\alpha} - \frac{w_0}{l}\right] = -\frac{\Gamma}{8\pi^2 a^2}\cdot\frac{\sin^2\alpha}{\cos\alpha}\hat{w}_b. \qquad (5)$$

Thus, to investigate the problem one needs in analytical representation for the binormal velocity component in neighbourhood of a vortex axis.

3. Formula of self-induced motion

Taking into account Eqs (3) and (4) we can consider an analytical solution for the circumferential velocity component only induced by an infinitely thin helical filament in a tube (Okulov, 1995)

$$w_\varphi = \frac{\Gamma}{2\pi r}\begin{Bmatrix}0\\1\end{Bmatrix} + \frac{\Gamma a}{\pi r l}\sum_{m=1}^{\infty} m\begin{Bmatrix} I_m\left(\frac{mr}{l}\right)Z_m\left(\frac{ma}{l}\right)\\ I'_m\left(\frac{ma}{l}\right)Z_m\left(\frac{mr}{l}\right)\end{Bmatrix}\cos m\chi, \qquad (6)$$

where $Z_m(x) = K_m(x) - a_m I_m(x)$; $a_m = K'_m(mR/l)/I'_m(mR/l)$;. $\chi = \varphi - z/l$; r, φ, z are the cylindrical coordinates. Here and further, the upper and lower lines in the brackets correspond to the cases $r < a$ and $r \geq a$ respectively.

Let's separate the singularities from the solution (6):

$$w_\varphi = \frac{\Gamma}{2\pi r}\begin{Bmatrix}0\\1\end{Bmatrix} + \frac{\Gamma a}{\pi r l}\left(S_\chi + R_\chi\right) \qquad (7)$$

where $$S_\chi = \frac{l}{2a}\sqrt{\frac{t_r}{t_a}}\left[\begin{Bmatrix}0\\-1\end{Bmatrix} + \frac{\tilde{r}^2 - \tilde{a}\tilde{r}\cos\chi}{\tilde{a}^2 + \tilde{r}^2 - 2\tilde{a}\tilde{r}\cos\chi} - \frac{\tilde{r}^2 - \tilde{a}^*\tilde{r}\cos\chi}{\tilde{a}^{*2} + \tilde{r}^2 - 2\tilde{a}^*\tilde{r}\cos\chi} + \right.$$

$$\frac{u_{1r}-v_{1a}}{2}\left(\ln\left(\tilde{a}^2+\tilde{r}^2-2\tilde{a}\tilde{r}\cos\chi\right)-\begin{Bmatrix}\ln\tilde{a}^2\\ \ln\tilde{r}^2\end{Bmatrix}\right)-\frac{u_{1r}+v_{1a}-2v_{1R}}{2}\ln\frac{\tilde{a}^{*2}+\tilde{r}^2-2\tilde{a}^*\tilde{r}\cos\chi}{\tilde{a}^{*2}}\Bigg];$$

$$R_\chi=\sum_{m=1}^{\infty}\begin{Bmatrix}mI_m\left(\frac{mr}{l}\right)Z_m\left(\frac{ma}{l}\right)+\frac{l}{2a}\sqrt{\frac{t_r}{t_a}}F_{\chi,m}(r,a)\\ mI'_m\left(\frac{ma}{l}\right)Z_m\left(\frac{mr}{l}\right)+\frac{l}{2a}\sqrt{\frac{t_r}{t_a}}F_{\chi,m}(r,a)\end{Bmatrix}\cos m\chi,$$

$$F_{r,m}(x,y)=\left(\frac{\tilde{x}}{\tilde{y}}\right)^m\left(1+\frac{v_{1x}-v_{1y}}{m}\right)-\left(\frac{\widetilde{xy}}{\tilde{R}^2}\right)^m\left(1+\frac{v_{1x}+v_{1y}-2v_{1R}}{m}\right),$$

$$F_{\chi,m}(r,a)=e^{-m\left|\ln\tilde{r}/\tilde{a}\right|}\left(\operatorname{sign}(a-r)+\frac{u_{1r}-v_{1a}}{m}\right)-\left(\frac{\widetilde{ra}}{\tilde{R}^2}\right)^m\left(1+\frac{u_{1r}+v_{1a}-2v_{1R}}{m}\right).$$

The quantities with a tilde may be interpreted as distorted radial distances $\tilde{r}=\dfrac{2rt_r}{1+t_r}\exp\left(\dfrac{1-t_r}{t_r}\right)$ ($\tilde{r}\to r$ at $l\to\infty$). A quantity $\tilde{a}^*=\tilde{R}^2/\tilde{a}$ is similar to a notion of the "reflected" vortex.

Note that at the introduced denotations (7) the velocity field is identical to the initial solution (6). At the same time separation of the singularities permits one to calculate the velocity field induced by a helical vortex filament with high precision since the singular parts are written through the elementary functions and the series quickly converge.

For the binormal velocity component, in view (3), (4) we find

$$\hat{w}_b=\hat{w}_0\sin\alpha-\frac{2}{\cos\alpha}\left(1+\frac{l^2}{ar}\right)\left[\begin{Bmatrix}0\\1\end{Bmatrix}+\frac{2a}{l}\left(S_\chi+R_\chi\right)\right] \tag{8}$$

The explicite separation of singularities in representation for the binormal velocity (8) determining the self-induced motion of a helical vortex filament allows to analyze correctly the self-induced motion in a tube of a helical vortex with the finite core size ε (Fig. 2, b). In this case it was possible to obtain a simple approximating for the solution. The full derivation of the formula for the self-induced motion is given in work by Kuibin and Okulov (in press). We write here the final result for the binormal self-induced velocity of the helical vortex with a uniform distribution of the axial component of vorticity in a core

$$\hat{w}_b^{(s)}=\ln\frac{l}{\varepsilon}+\frac{1+1.455\tau+1.723\tau^2+0.711\tau^3+0.616\tau^4}{\tau+0.486\tau^2+1.176\tau^3+\tau^4}-\frac{1}{4}$$

$$+\frac{2\left(1+\tau^2\right)^{1/2}}{\tau}\left[\beta-1-\left(1+\tau^2\right)\left(\frac{\tilde{a}^2}{\tilde{R}^2-\tilde{a}^2}-k\ln\frac{\tilde{R}^2-\tilde{a}^2}{\tilde{R}^2}\right)\right], \tag{9}$$

$$k=\frac{1}{12}(9\cos\alpha_R-7\cos^3\alpha_R-3\cos\alpha+\cos^3\alpha),\quad \mathrm{tg}\alpha_R=R/l$$

that is valid either for thin vortices with $\kappa\varepsilon << 1$ or for weakly curved thick vortices with $a << \varepsilon$. To generalise this result to the cases of non-uniform vorticity distributions in a core one should replace the third term, 1/4, with

$$\frac{1}{2}-4\int_0^1\frac{ds}{s}\left[\int_0^s\Omega(s')s'ds'\right]^2$$

in accordance with Moore and Saffman (1972), Adebiyi (1881). Here Ω is the vorticity function normalised to have mean value of 1.

4. Discussion of the results of studying of the self-induced motion of helical vortices

Substituting Eq. (9) into (5) and analysis of the obtained formula in accordance with a formal decomposition (2) yield the contributions into dimensionless frequency $\hat{f}=8\pi^2a^2f/\Gamma$ of different factors:

$$\hat{f}_\kappa=\frac{-1}{\tau\sqrt{1+\tau^2}}\ln\left[\frac{a}{\varepsilon}\left(1+\tau^2\right)\right]$$

$$\hat{f}_\tau=\frac{-1}{\tau\sqrt{1+\tau^2}}\left[\ln\left(\frac{\tau}{1+\tau^2}\right)+\frac{1+1.455\tau+1.723\tau^2+0.711\tau^3+0.616\tau^4}{\tau+0.486\tau^2+1.176\tau^3+\tau^4}-\frac{1}{4}\right]$$

$$\hat{f}_R=2\frac{1+\tau^2}{\tau^2}\left(\frac{\tilde{a}^2}{\tilde{R}^2-\tilde{a}^2}-k\ln\frac{\tilde{R}^2-\tilde{a}^2}{\tilde{R}^2}\right) \tag{10}$$

$$\hat{f}_{w_0}=\frac{-2}{\tau^2}(\beta-1)$$

In the limit $l\to\infty$ we obtain the known formula (1) for the motion frequency of a point vortex in a circular area.

In preceding our work (Kuibin and Okulov, in press) comparison of calculation with experiment showed that the contributions of all the factors influential in motion of a vortex (10) are essential. Besides the capability of existence of fixed helical vortices and vortices rotating in the direction opposite to the flow swirling was established. A problem however was not investigated how the different factors (10) influence this phenomena. Let's fill the gap. As a basic object we will consider a helical vortex with parameters

$$\tau=1,\quad \sigma=\varepsilon/a=0.3,\quad \rho=R/a=2,\quad \beta=1 \tag{11}$$

that are close to parameters of the vortex observed in the experiment (Shtork, 1994). The unit value of β means that in absence of tube the vortex induces zero velocity at infinite distance from it. In Figure 3,a the solid line presents the dependence of the vortex precession frequency on the quantity $1/\tau$ (the zero point at the abscissa corresponds to a rectilinear vortex). The dashed lines show behavior of the different components (10). Note that the frequency value at $\tau = \infty$ is 0.6661 and is close to 2/3 which corresponds to the frequency of rectilinear vortex filament calculated with Eq. (1). This means that the frequency at infinite pitch is determined by the position of the vortex relative the tube wall. The contribution from the wall decreases as the helix pitch decreases (curve 4). Predominant influence begin to exert the factors dealing with a curvature and torsion of vortex axis (curves 1 and 2, respectively). The contribution 1 is

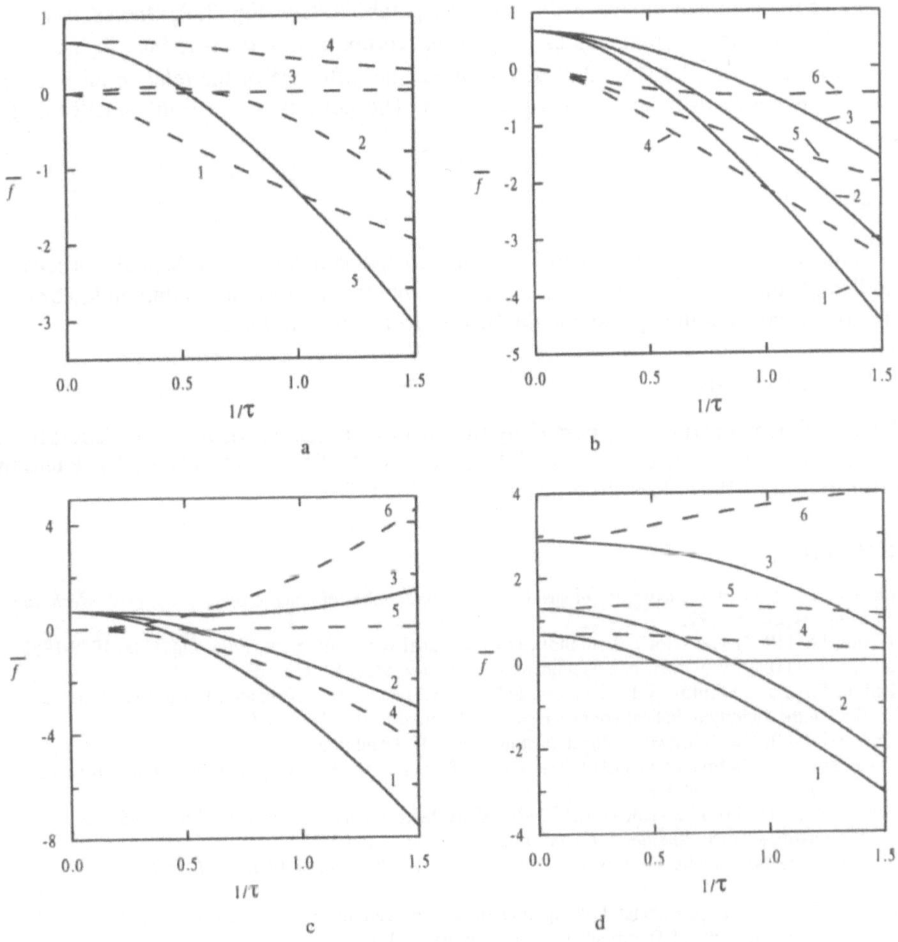

Figure 3. The influence on a precession frequency of a vortex and flow parameters.

always opposite to the contribution from a wall 4, and the contribution 2 from some values of a step also becomes negative. From a certain moment their influence becomes commensurable with influence of a wall, that corresponds to a immobile vortex, and then the vortex begins to rotate in the direction opposite to a flow swirling. The contribution from translational motion is obviously equal to zero (curve 3). It is intentionally eliminated at the given stage of research.

Additionally in Figs. 3,b - 3,d influence of each factor on the frequency is presented separately. The effect of the size of vortex core at fixed other parameters (11) on full frequency (the solid curves) and on respective component of the decomposition (2) (the dotted curves) is depicted in Fig. 3,b. The curves 1 and 4 correspond to very thin vortex whereas 3 and 6 - to the greatest possible width of a vortex permissible with remaining fixed parameters. Inside the range of the parameters variation there always exist zones where the vortices are immobile or rotate contrary the flow swirling. The action of translational motion is described by parabolic law (Fig. 3,c). One can see that at $\beta = 0$ and other parameters as in (11) the vortex at any pitch rotates in the same direction. The last plot (Fig. 3,d) demonstrates the influence of the relative tube radius on the frequency of helical vortex precession. The large is the tube the smaller is the effect.

5. Conclusion

A contribution of different factors on the self-induced motion of helical vortices is studied for the first time. The ranges of parameters that correspond to immobile vortices and to vortices rotating opposite to the flow swirling are determined.

Acknowledgements

This work was partially supported by the grant of President of Russian Federation, by the joint foundation INTAS-RFBR (grant N 95-IN-RU-1149), by Russian Foundation of Basic Research (grant N 97-05-65254).

References

Adebiyi, A. (1881) On the existence of steady helical vortex tubes of small cross-section, *Q. J. Mech. appl. Math.* **34**(2), 153-177.

Hardin, J.S. (1982) The velocity field induced by a helical vortex filament, *Phys. Fluids* **25**, 1949-1952.

Kelvin, Lord (1880) Vibrations of a columnar vortex, *Philos. Mag.* **10**, 155-168.

Kuibin, P.A., and Okulov, V.L. (in press) Self-induced motion and asymptotic Expansion of the velocity field in the vicinity of helical vortex filament, submitted to the *Phys. of Fluids*.

Lamb, H. (1940) *Hydrodynamics*, 6th ed. (Cambridge U. P., Cambridge).

Moore, D.W. and Saffman, P.G. (1972) The motion of a vortex filament with axial flow, *Phil. Trans. R. Soc. London* Ser. A **272**, 403-429.

Okulov, V.L. (1995) The velocity field induced by helical vortex filaments with cylindrical or conic supporting surface, *Russian J. Engineering Thermophys.* **5**, 63-75.

Ricca, R.L. (1994) The effect of torsion on the motion of a helical vortex filament, *J. Fluid Mech.* **273**, 241-259.

Shtork, S.I. (1994) Experimental Investigation of Vortex Structures in Tangential Chambers, Candidate dissertation, Institute of Thermophysics, Novosibirsk (in Russian).

Widnall, S.E. (1972) The stability of helical vortex filament, *J. Fluid Mech.* **54**, 641-663.

Session 2

Numerical Methods

COHERENT STRUCTURE EDUCTION IN WAVELET-FORCED TWO-DIMENSIONAL TURBULENT FLOWS

MARIE FARGE
LMD-CNRS, Ecole Normale Supérieure
24 rue Lhomond, 75231 Paris cedex 05, France

KAI SCHNEIDER
Centre de Physique Théorique, CNRS - Luminy - Case 907, 13288 Marseille cedex 09, France and ICT, Universität Karlsruhe (TH) Kaiserstraße 12, 76128 Karlsruhe, Germany

AND

NICHOLAS K.-R. KEVLAHAN
LMD-CNRS, Ecole Normale Supérieure
24 rue Lhomond, 75231 Paris cedex 05, France

Abstract. We analyze vorticity fields obtained from direct numerical simulations (DNS) of statistically stationary two-dimensional turbulence where the forcing is done in wavelet space. We introduce a new eduction method for extracting coherent structures from two-dimensional turbulent flows. Using a nonlinear wavelet technique based on an objective universal threshold we separate the vorticity field into coherent structures and background flow. Both components are multi-scale with different scaling laws, and therefore cannot be separated by Fourier filtering. We find that the coherent structures have non-Gaussian statistics while the background flow is Gaussian, and we discuss the implications of this result for turbulence modelling.

1. Introduction

Our aim is the numerical simulation of large Reynolds number flows using present computers. In this context the key question is: which quantities should be computed and which quantities can be discarded or modelled? The usual choice, based on the statistical theory of homogeneous and isotropic turbulence (in three dimensions (Kolmogorov, 1941) and two

E. Krause and K. Gersten (eds.), IUTAM Symposium on Dynamics of Slender Vortices, 65-83.

dimensions (Kraichnan, 1967)), is to compute the low-wavenumber modes, considered to be active, and to model the high-wavenumber modes, considered to be controlled by the active ones, using an *ad hoc* subgrid scale parameterization, for example an eddy-diffusivity. This wavenumber separation is assumed by most of the numerical methods presently used to compute turbulent flows, from Reynolds Averaging to Large Eddy Simulation (LES) and Nonlinear Galerkin methods. It is important to understand that this choice of a wavenumber separation is done *a priori*, supposing an hypothetical spectral gap, which has not been observed, and without reference to the first principles stated in the Navier–Stokes equations.

Statistical theories of homogeneous and isotropic turbulence rely on the hypothesis that energy (or enstrophy in two dimensions) is injected at low wavenumbers and is dissipated at high wavenumbers. This separation of scales, between energy (or enstrophy) production and dissipation, allows for the existence of an intermediate range of wavenumbers, called the inertial range, where the nonlinear dynamics is supposed to be conservative. As far as we know, the existence of such an inertial range has never been demonstrated from first principles. Moreover, according to Kolmogorov's theory, in three-dimensional turbulence energy cascades from low to high wavenumbers. Unfortunately this behaviour is only defined for ensemble averages, and in a single realization of a turbulent flow significant amounts of energy are transferred in the opposite direction, from high to low wavenumbers (Domaradzki, 1992), (Meneveau, 1991). This energy 'back-scatter' invalidates the assumption that the high wavenumber modes are passive. The back-scatter is due to the presence of organized structures, such as vortex tubes (often called 'vortex filaments' although they have nothing to do with the vortex filaments of two-dimensional turbulent flows) or horseshoe vortices, which interact and transfer energy to larger scales. Because these instabilities are local and intermittent, the Fourier representation is not able to separate them from the rest of the flow. This back-scattering problem is even worse in two-dimensional turbulence, for which Kraichnan (Kraichnan, 1967) has predicted that even ensemble averages (and not only individual realization) present an inverse energy cascade, due to the conservation of both energy and enstrophy in the inertial range.

We believe that the problem of turbulence modelling and subgrid-scale parameterization has to be reconsidered, for both two-dimensional and three-dimensional turbulent flows, and we propose that the classical division between large and small eddies be replaced by a new separation between coherent structures and background flow.

Since the pioneering paper by McWilliams (McWilliams, 1984) the emergence of coherent structures out of initially random vorticity fields has been recognized as a generic feature of two-dimensional turbulent flows. The definition of coherent structures is still rather subjective. Usually one thinks of coherent structures as localized accumulations of vorticity which concentrate most of the enstrophy, $Z = 1/2 < \omega^2 >$ (ω being the vorticity and $<>$ the space average) in a small fraction of the spatial domain and which survive on time-scales much longer than the average eddy turn-over time $\tau = \sqrt{1/Z}$. A more precise characterization of coherent structures was proposed in 1981 by Weiss (Weiss, 1992) as elliptical flow regions where rotation dominates deformation. This definition assumes that the stress tensor varies slowly, in space and time, compared to vorticity gradients.

A key question, which remains open, is the following: do coherent structures have a generic shape? The answer to this question strongly influences our analysis of vorticity fields, and in particular our interpretation in terms of scale, which is intrinsically linked to the generic shape we assume for the vortices. This hypothetical shape is not clearly defined, the notion of scale, as well as the notions of vortex size and circulation, are meaningless. This point has been a source of misunderstanding for years in the study of turbulence due to the identification of scale with the inverse of wavenumber. In fact, this traditional definition of scale makes sense only if one analyzes an homogeneous and isotropic velocity or vorticity field, e.g. a wave field, or an ensemble average of velocity or vorticity fields which is homogeneous and isotropic due to the translational and rotational invariance of the coherent structure motion. However, this definition does not make sense if one analyzes a given realization containing isolated coherent structures, which is necessarily inhomogeneous, intermittent and highly phase-correlated. In this case, to define the notion of scale one should look for the generic shape of the coherent structures observed in the vorticity field. The vorticity field should be preferred to the velocity field for dynamical reasons, because vorticity is Galilean invariant and volume preserving in the inviscid limit. It should be noted that in statistical analysis *a prioris* are as essential as hypotheses are in modelling: we should state them clearly, otherwise our results will be nonsensical. Once the reference shape has been defined we can compute the correlation between the vorticity field and this shape, which is dilated and translated in order to obtain the scale content of the vorticity field.

In this paper we analyze two-dimensional forced turbulent vorticity fields using an orthogonal wavelet basis. The mother wavelet is chosen as the generic shape for coherent structures. Such localized functions are better suited to analyze turbulent flows than the trigonometric functions used as basis elements for Fourier decomposition. This analysis allows us to perform a local spectral analysis of the flow, which is not possible with the classical Fourier transform. In selecting the analyzing functions we are limited by the Heisenberg uncertainty principle, which means that we cannot have perfect localization in both space and wavenumber. To analyze two-dimensional turbulent fields the best solution is to use functions whose shape correspond to our *a priori* model of an isolated coherent structure, namely a localized and isotropic distribution of vorticity. We can then use a self-affine set of such functions in order to generate a multi-scale analysis of the vorticity field. In order to perform a quantitative analysis, we also require isometry between the enstrophy computed in physical space and the enstrophy computed from the inner-product of the vorticity field with the analyzing functions. All these reasons have led us to use wavelets, which are well-localized in both physical and Fourier spaces, self-affine (since they are obtained by dilation from one another), and allow an isometric transform which conserves energy.

2. Wavelet–forced two–dimensional turbulence

The enstrophy dissipated by the dissipative term has to be re-injected in order to simulate a statistically stationary flow, i.e. non-decaying fully-developed turbulence. The usual technique is to inject enstrophy and energy locally in Fourier space. In this article, however, we employ a new wavelet forcing technique introduced in (Schneider and Farge, 1997). The forcing is defined in wavelet space in order to control the smoothness of the vortices thus excited. Using this method the injection of energy and enstrophy is as local as possible in both physical and spectral space. For details we refer the reader to (Schneider and Farge, 1997).

2.1. GOVERNING EQUATIONS

We consider the two-dimensional Navier–Stokes equations written in velocity–vorticity form with a forcing term $F = \nabla \times \mathbf{f}$. Furthermore we include an artificial frictional term $\lambda\Psi$, which acts as an infrared energy sink to remove the energy that accumulates at large scales due to the inverse energy cascade. The equations we solve are:

$$\partial_t \omega + \mathbf{v} \cdot \nabla\omega = \nu\nabla^2\omega + \lambda\Psi + F \quad , \qquad \nabla \cdot \mathbf{v} = 0, \tag{1}$$

where the velocity vector $\mathbf{v} = (u, v)$, the vorticity $\omega = \nabla \times \mathbf{v}$ and the stream function $\Psi = \nabla^{-2}\omega$. Further parameters are the strength of the friction term λ and the kinematic viscosity ν.

We assume periodic boundary conditions in both directions, i.e. our domain is the two-dimensional flat torus $\mathbb{T}^2$ with $\mathbb{T} = 2\pi\mathbb{R}/\mathbb{Z}$. These boundary conditions have been chosen in order to simulate turbulent flows far from walls, and thus avoid the treatment of boundary layers.

2.2. NUMERICAL METHOD

For the numerical solution of the system (1) we employ a classical Fourier pseudo–spectral method with a semi–implicit time discretization using finite differences (Euler–backwards for the viscous term and Adams–Bashforth extrapolation for the nonlinear term, both of second order). The vorticity field and the other variables are represented as Fourier series: $\omega(\mathbf{x}) = \sum_{|\mathbf{k}|\leq N/2} \widehat{\omega}(\mathbf{k})\, e^{i\mathbf{k}\cdot\mathbf{x}}$.

2.3. WAVELET FORCING

Furthermore at time step n we develop ω^n as an orthonormal wavelet series from the largest scale $l = 2^0$ to the smallest scale $l = 2^J$ using a two–dimensional multi-resolution analysis (MRA) (Daubechies, 1992) (Farge, 1992):

$$\omega^n(x,y) \;=\; c^n_{0,0,0}\, \phi_{0,0,0}(x,y) \;+\; \sum_{j=0}^{J-1} \sum_{i_x=0}^{2^j-1} \sum_{i_y=0}^{2^j-1} \sum_{\mu=1}^{3} d^{\mu,n}_{j,i_x,i_y}\, \psi^{\mu}_{j,i_x,i_y}(x,y) \quad , \tag{2}$$

with $\phi_{j,i_x,i_y}(x,y) \;=\; \phi_{j,i_x}(x)\, \phi_{j,i_y}(y)$, and

$$\psi^{\mu}_{j,i_x,i_y}(x,y) \;=\; \begin{cases} \psi_{j,i_x}(x)\, \phi_{j,i_y}(y) & ;\ \mu = 1 \quad , \\ \phi_{j,i_x}(x)\, \psi_{j,i_y}(y) & ;\ \mu = 2 \quad , \\ \psi_{j,i_x}(x)\, \psi_{j,i_y}(y) & ;\ \mu = 3 \quad , \end{cases} \tag{3}$$

where $\phi_{j,i}$ and $\psi_{j,i}$ are the 2π–periodic one–dimensional scaling function and the corresponding wavelet, respectively. Due to the orthogonality the coefficients are given by $c^n_{0,0,0} = \langle \omega^n\,,\,\phi_{0,0,0}\rangle$ and $d^{\mu,n}_{j,i_x,i_y} = \langle \omega^n\,,\,\psi^{\mu}_{j,i_x,i_y}\rangle$ where $\langle\cdot,\cdot\rangle$ denotes the inner product.

The forcing term F at time step $n+1$ is then defined as a function of ω at time step n:

$$F^{n+1}(x,y) = C \sum_{J_0<j<J_1} \sum_{k_x=0}^{2^j-1} \sum_{k_y=0}^{2^j-1} \sum_{\mu=1,2,3} \langle \omega^n\,,\,\psi^{\mu}_{j,k_x,k_y}\rangle\, \psi^{\mu}_{j,k_x,k_y}(x,y) \quad , \tag{4}$$

with $0 \leq J_0 \leq J_1 \leq J$, where J denotes the finest scale in the simulation, $C > 0$ and $|\langle \omega^n(x,y)\,,\,\psi^{\mu}_{j,k_x,k_y}\rangle| > \epsilon$. The scale parameters J_0 and J_1 define the scale range of the forcing. The restriction to wavelet coefficients above a given threshold implies that only the dynamically active part of the flow, i.e. the coherent structures (Farge, 1992), (Farge *et al.* , 1996), are forced. The constants C and λ, responsible for the strength of the forcing and the Rayleigh friction respectively, are adjusted in such a way that we obtain a statistically stationary state. For further details on the numerical simulation we refer the reader to (Schneider and Farge, 1997).

3. Coherent structure eduction

The term 'eduction' was introduced by (Hussain, 1986) to describe the extraction of coherent structures out of three-dimensional laboratory turbulent flows. We will keep the same terminology to characterize the technique we propose to separate the coherent structures from the background flow in two-dimensional turbulent flows obtained by DNS.

As we have already noted, there is at present no consensus on the precise definition of coherent structures. The only definition of a coherent structure which seems objective is that of a locally meta-stable state, such that, in the reference frame associated with the coherent structure, the nonlinearity of Navier Stokes equations becomes negligible. In consequence, a coherent structure can be characterized by a functional relation between the vorticity ω and the streamfunction ψ in the form $\omega = F(\psi)$, where F is called the coherence function (Farge and Holschneider, 1990). One possible coherent structure eduction technique would be to plot the diagram $\omega = F(\psi)$ and extract the branches which can be fitted by a function F; the points belonging to these branches would correspond to locations where the vorticity field $\omega_>$ is coherent, while the scattered points which do not belong to any branch would correspond to locations in the incoherent background flow $\omega_<$.

Other possible eduction techniques are less objective than the one described above, because they depend on a threshold value which has to be defined *a priori*. The simplest method, already proposed by several authors (McWilliams, 1984) (Babiano *et al.* , 1987), is to choose a vorticity threshold, for instance $\omega_T = \sqrt{Z}$, and retain as coherent the regions where $|\omega| > \omega_T$, while the background flow corresponds to the regions where $|\omega| \leq \omega_T$. The drawback of this method is that it does not preserve the smoothness of ω, and both fields, $\omega_>$ and $\omega_<$, will have spurious discontinuities which will affect their Fourier energy spectra. To avoid this problem we propose to replace the grid-point representation by a wavelet representation, which, on the contrary, does not introduce discontinuities and therefore conserves the spectral properties of ω.

Since the wavelet transform is invertible, it is always possible to select a subset of the coefficients and reconstruct a filtered version of the field from them. Using this property we have proposed several coherent structure eduction techniques (Farge and Philipovitch, 1993), (Farge *et al.* , 1992). The first consists of discarding all wavelet coefficients outside the influence cones, namely the spatial support of the wavelets, attached to the local maxima of the vorticity field which correspond to the centers of coherent structures (Farge and Philipovitch, 1993).

The second technique, introduced here, consists of discarding all wavelet coefficients which are smaller than a given threshold which depends only on the variance of the field $< \omega^2 >$ and on the number of samples N, such that the threshold value is: $\widetilde{\omega_T} = (2 < \omega^2 > \log_{10} N)^{1/2}$. This technique is based on the wavelet shrinkage method of Donoho (Donoho, 1992). Donoho has shown that this is the optimal denoising technique for a signal containing a Gaussian white noise of a given variance. However, we are not sure that our signal actually contains a Gaussian noise, and anyway we do not know its variance, therefore we will consider instead the variance of the total vorticity field. This is an overestimate which leads to a higher treshold value and a stronger compression rate. Anyway we can always *a posteriori* check that the discarded components correspond to a Gaussian noise (characterized by a Gaussian PDF, skewness zero and kurtosis three).

Applying the second nonlinear thresholding technique to the wavelet packet coefficients of the vorticity field, we have extracted the coherent structures from the background flow and shown that both components are multi-scale (Farge *et al.* , 1992), however they exhibit different scaling laws, the background having an energy spectra scaling in k^{-3} compatible with Kraichnan's prediction while the coherent structures scale in k^{-6} (Farge *et al.* , 1992).

We have also tried (Wickerhauser *et al.* , 1994) to use adaptive local cosines instead of wavelet packets to separate coherent structures from background flow. We showed that the local cosine representation does not compress the enstrophy as well as wavelets or wavelet packets. First, it smoothes the coherent structures and therefore loses enstrophy, and secondly it introduces spurious oscillations in the background, due to the loss of the phase information attached to the weak coefficients. These drawbacks are common to any Fourier or windowed Fourier representation, because each Fourier component contains non-local information and we need the phase information of all Fourier components to reconstruct precisely a given region of the field. Therefore no Fourier technique can properly educe coherent structures, because as the vorticity field is compressed the coherent structures disappear and become increasingly mixed up with the background flow (Wickerhauser *et al.* , 1994). In this paper we will focus on the second (variance based) nonlinear wavelet compression as the optimal solution for coherent structure eduction. We will also work with the wavelet representation rather than the wavelet packet representation.

4. Results

The wavelet forced two-dimensional Navier–Stokes equation were solved using a pseudo-spectral scheme with resolution $N = 128^2$. The dissipation is modelled by a Laplacian operator with a kinematic viscosity $\nu = 2 \times 10^{-3} m^2/s$. We have reached a statistically stationary solution (Schneider and Farge, 1997) that we have maintained for 16 000 time steps ($\Delta t = 10^{-3} s$), i.e. for 60 eddy-turn over times. We now analyze the vorticity field obtained at the final time step which corresponds to $t = 16s$.

We project the vorticity field ω onto an orthogonal spline wavelet basis of Battle-Lemarié type. Figure 1 shows the spatial and spectral support of the symmetric quintic spline wavelets used in the present analysis. In figure 2 we plot the retained enstrophy $Z_>$ as a function of the number of retained wavelet coefficients $N_>$. Note that very few wavelet coefficients retain most of the total enstrophy Z and that after $N_> = 216$ the rate of convergence abruptly changes its behaviour and becomes very slow (figure 2b). The optimal compression rate is obtained for $N_> = 216$, which contain 92.4% of the total enstrophy $Z = 9.2s^{-2}$, although they represent only 1.3% of the total number of coefficients $N = 128^2 = 16384$. The $N_< = 16168$ remaining weaker wavelet coefficients represent 98.7% of the total number of wavelet coefficients but retain only 7.6% of the total enstrophy.

Next we compute the threshold proposed by (Donoho, 1992), (Donoho *et al.* , 1995) for denoising. This criterion depends only on the total number of samples N and on the variance of the vorticity field $< \omega^2 >= 16s^{-2}$. Since we have reached a statistically stationary state $< \omega^2 >$ remains constant, therefore we compute the unique threshold value: $\widetilde{\omega_T} = (2 < \omega^2 > \log_{10} N)^{1/2} = 12s^{-1}$. The coherent structures are then extracted by projecting only the wavelet coefficients having an absolute value larger than $\widetilde{\omega_T}$ back onto grid-points, which gives the coherent vorticity field $\omega_>$. The background flow is found similarly by selecting those wavelet coefficients with absolute value smaller or equal to $\widetilde{\omega_T}$ and then reconstructing the vorticity field $\omega_<$.

In table 1 we compare the first moments up to the 6th order, the skewness and the kurtosis of the uncompressed vorticity ω, the coherent vorticity $\omega_>$ and the background vorticity $\omega_<$. Note that all fields have skewness $S = 0$ and that both the uncompressed vorticity and the coherent vorticity have kurtosis $K = 17$, characteristic of a non-Gaussian probability distribution, while the background vorticity has kurtosis $K = 3$, characteristic of a Gaussian probability distribution. These results are confirmed in figure 6 where we plot the probability distribution functions (PDF) for the three vorticity fields. The uncompressed vorticity and the coherent vorticity have nearly identical PDFs, with heavy tails (extremal values: $-37/+28$) and

TABLE 1. Statistical properties of the vorticity field at $t = 16$s.

quantity	definition	ω total	$\omega_>$ coherent	$\omega_<$ background
# of coefficients	N	16384	216	16128
% of coefficients		100 %	1.3 %	98.7 %
1st moment (mean)	$M_1 = \bar{\omega} = \frac{1}{N}\sum_{i=1}^{N}\omega_i$	0.0	0.0	0.0
2nd moment (variance)	$M_2 = <\omega^2> = \frac{1}{N}\sum_{i=1}^{N}\omega_i^2$	18.4	17.0	1.4
3rd moment	$M_3 = \frac{1}{N}\sum_{i=1}^{N}\omega_i^3$	-29.6	-31.7	-0.8
4th moment	$M_4 = \frac{1}{N}\sum_{i=1}^{N}\omega_i^4$	5763.7	4966.0	5.8
5th moment	$M_5 = \frac{1}{N}\sum_{i=1}^{N}\omega_i^5$	-55956.0	-50716.3	0.1
6th moment	$M_6 = \frac{1}{N}\sum_{i=1}^{N}\omega_i^6$	3879905.5	3258826.2	39.2
Enstrophy	$Z = \frac{M_2}{2}$	9.2	8.5	0.7
% total enstrophy		100 %	92.4 %	7.6 %
Skewness	$S = \frac{M_3}{M_2^{\frac{3}{2}}}$	-0.4	-0.4	0.0
Kurtosis	$K = \frac{M_4}{M_2^2}$	17.1	17.2	3.0

seem close to a Cauchy distribution. Such a Cauchy distribution has been predicted for the velocity gradients (a quantity similar to the vorticity we consider here) of a system of point-vortices in two and three dimensions (Min *et al.* , 1996). On the contrary, the PDF of the background flow is a parabola (in lin-log coordinates) and does not present heavy tails (extremal values: $-5/+5$), which is characteristic of a Gaussian distribution.

Figures 3 and 4 show that the coherent components retain the precise shape of each coherent structure, and therefore is inhomogeneous, while the incoherent components correspond to the homogeneous background flow.

In figure 5 we show that both components have a broad band energy spectrum, although the coherent components dominate in the low wavenumbers while the incoherent components dominate in the high wavenumbers. This is due to the fact that the spatial support of the coherent structures decreases with scale and therefore their weight in the high wavenumbers of the energy spectrum becomes negligible. However, since the background flow is homogeneous, the spatial support of the incoherent components is dense in space and therefore the background flow conditions the high wavenumber range of the energy spectrum, where we observe a k^{-4} scaling for the energy and therefore a k^{-2} (pink noise) scaling for the enstrophy. The fact that there is a break in the power-law spectra of both the coherent and incoherent components at $k = 16$ is due to the wavelet-forcing which has been limited to scales smaller than $L = 32 = 2^5$ which corresponds to $k = 16$.

In figure 7 we show that the coherence function of the coherent components $\omega_>$ is similar to that of the uncompressed vorticity field ω. The coherence function is a superposition of functions $\omega_> = F(\Psi_>)$ (corresponding to the lines drawn on figure 7), each one corresponding to a coherent structure. In contrast, the background flow does not exhibit such a correlation between $\omega_<$ and $\Psi_<$ and is therefore incoherent. This is further proof that the nonlinear wavelet thresholding technique is appropriate for coherent structures eduction in turbulent flows.

5. Conclusion

The goal for modelling or computing the evolution of turbulent flows is to take a coarse-graining point of view, namely to keep the essential information and discard the details then considered as noise. For two-dimensional turbulent flows we propose dividing the relevant dynamical field, i.e. the vorticity field ω, into its inhomogeneous, intermittent and organized components $\omega_>$, which correspond to the coherent vortices and are characterized by non-Gaussian statistics, and its homogeneous, non-intermittent and random components $\omega_<$, which correspond to the well-mixed background flow and are characterized by Gaussian statistics. Both components are multi-scale and this is why the Fourier transform, whose modulus loses the spatial information, is not the optimal functional basis to study turbulence. The grid-point representation is not suitable either, because we want to be able to detect the characteristic scaling of the two regions and therefore must avoid introducing spurious discontinuities.

By applying such an eduction technique to wavelet-forced two-dimensional turbulent flow, we have shown that the coherent vorticity is highly non-Gaussian while the background vorticity is Gaussian. This result has strong implications for two-dimensional turbulence modelling. The coherent vorticity corresponds to the dynamical components which are out of statistical equilibrium and have a very large variance characteristic of non-Gaussian behaviour. Therefore we are unable to model the coherent vorticity by a simple stochastic process. The number of degrees of freedom in the coherent vorticity is less than 2% of the total number of degrees of freedom required for a full direct numerical simulation (DNS). Therefore we suggest using a pseudo-wavelet scheme (Schneider *et al.* , 1997) to compute only these few coherent modes. The remaining degrees of freedom represent more than 98% of the total and correspond to the incoherent components having Gaussian statistics. The incoherent part may therefore be modelled by any standard statistical turbulence model, such as Reynolds averaged or $k - \epsilon$ models (Mohammadi and Pironneau, 1993).

In this paper we have proposed using orthogonal wavelets, which play the role of phase-space atoms independently defined in space and scale, to extract coherent structures from the vorticity field. This eduction method uses an objectively defined, universal threshold. The Navier–Stokes dynamics combines these "phase-space atoms" into "phase-space molecules" which correspond to the coherent structures, whose formation is observed during the flow evolution. Thus each vortex can be computed as the superposition of well-localized functions, namely wavelets, which describe its internal degrees of freedom. This method may allow us to drastically reduce the number of degrees of freedom necessary to compute the turbulent flow evolution (Fröhlich and Schneider, 1996), (Schneider *et al.* , 1997). We have already shown that this approach (Farge *et al.* , 1990) may be extended to the case of three-dimensional turbulent flows where vorticity tubes, often called vorticity filaments, play the role of coherent structures moving in an homogeneous random background flow. We believe that turbulence, both two-dimensional and three-dimensional, is a random superposition of a set of meta-stable vortices whose interactions give rise to its characteristic unpredictable behaviour, therefore the statistical tools we use in turbulence should be based on the recognition of vortices as the basis elements of turbulent flows.

Acknowledgement: We would like to acknowledge partial support through NATO contract CRG-930456 "Wavelet Methods in Computational Turbulence", and Nicholas Kevlahan gratefully acknowledges support from 'Marie Curie Fellowship' no. ERBFMBICT950365.

References

Babiano A., Basdevant C., Legras B. and Sadourny R. (1987). Vorticity and passive–scalar dynamics in two–dimansional turbulence, *J. Fluid Mech.*, **183**, 379–397.

Batchelor, G. K. (1969). Computation of the Energy Spectrum in Homogeneous Two-dimensional Turbulence, *Phys. Fluids, suppl. II*, **12**, 233–239.

Daubechies, I. (1992). *Ten Lectures on wavelets*, (SIAM, Philadelphia).

Domaradzki, J. A. (1992). Nonlocal triad interactions and the dissipation range of isotropic turbulence, *Phys. Fluids* A, **4**, 2037.

Donoho, D. (1992). Wavelet shrinkage and Wavelet -Vaguelette Decomposition: a 10–minute tour, *in "Progress in Wavelet Analysis and Applications", Proc. Int. Conf. "Wavelets and Applications", Toulouse*, 109–128.

Donoho, D., Johnstone, I., Kerkyacharian, G. and Picard, G. (1995). Wavelet shrinkage: Asymptopia? (with discussion), *J. Royal Stat. Soc., Ser. B* **57**, 301–369.

Farge, M. (1992). Wavelet Transforms and their Applications to Turbulence, *Ann. Rev. Fluid Mech.*, **24**, 395–457.

Farge, M. (1994). Wavelets and two-dimensional turbulence, *Computational Fluid Dynamics*, **1**, (John Wiley & Sons), 1–23.

Farge, M., Guezennec, Y., Ho, C. M. and Meneveau, C. (1990). Continuous wavelet analysis of coherent structures, in *Studying turbulence using numerical simulation databases - III* (ed. P. Moin, W. C. Reynolds and J. Kim), pp. 331–348. Center for Turbulence Research.

Farge, M., Goirand, E., Meyer, Y., Pascal F. and Wickerhauser, M. V. (1992). Improved predictability of two-dimensional turbulent flows using wavelet packet compression, *Fluid Dyn. Res.* **10**, 229–250.

Farge, M. and Holschneider, M. (1990). Interpretation of two-dimensional turbulence spectrum in terms of singularity in the vortex cores, *Europhys. Lett.*, **15**(7), 737–743.

Farge, M., Kevlahan, N., Perrier, V. and Goirand, E. (1996). Wavelets and Turbulence, *Proc. IEEE* **84**, 639–669.

Farge M. and Philipovitch, T. (1993). Coherent structure analysis and extraction using wavelets, in *Progress in Wavelet Analysis and Applications*, (ed. Y. Meyer and S. Roques), (Editions Frontières), 477–481.

Fröhlich, J. and Schneider, K. (1996). Numerical Simulation of Decaying Turbulence in an Adaptive Wavelet Basis, *Appl. Comput. Harm. Anal.*, **3**, 393–397.

Hussain, A. K. M. F. (1986). Coherent structures and turbulence, *J. Fluid Mech.* **173**, 303–356.

Kolmogorov, A. N. (1941). The local structure of turbulence in incompressible viscous fluid for very large Reynolds numbers, *C. R. Acad. Sci. USSR*, **30**, 301–305.

Kolmogorov, A. N. (1962). A refinement of previous hypotheses concerning the local structure of turbulence in a viscous incompressible fluid at high Reynolds number, *J. Fluid Mech.*, **13**, 82–85.

Kraichnan, R. H. (1967). Inertial ranges in two-dimensional turbulence, *Phys. Fluids*, **10**, 1417–1423.

McWilliams, J. C. (1984). The emergence of isolated coherent vortices in turbulent flows, *J. Fluid Mech.*, **146**, 21–43.

Meneveau, C. (1991). Analysis of turbulence in the orthonormal wavelet representation, *J. Fluid Mech.* **232**, 469–520.

Min, I. A., Mezić, I. and Leonard, A. (1996) Lévy stable distributions for velocity and velocity difference in systems of vortex elements, *Phys. Fluids*, **8**, 1169–1180.

Mohammadi, B. and Pironneau, O. (1993). *Analysis of the k-epsilon model.* Masson-Wiley.

Poliakov, A. M. (1993). Conformal Turbulence, *Nucl. Phys. B*, **396**, 367.

von Sachs, R. and Schneider, K. (1996). Wavelet Smoothing of evolutionary spectra by non-linear thresholding, *Appl. Comput. Harm. Anal.*, **3**, 268–282.

Schneider, K. and Farge, M. (1997). Wavelet forcing for numerical simulation of two-dimensional turbulence, *C. R. Acad. Sci. Paris Série II.* t. 325, 263–270.

Schneider, K., Kevlahan, N. and Farge, M. (1997). *Comparison of an adaptive wavelet method and nonlinearly filtered pseudo-spectral methods for the two-dimensional Navier–Stokes equations.* Preprint CPT-97/P.3494, Centre de Physique Théorique, CNRS - Luminy, Marseille, to appear in *Theoretical and Computational Fluid Dynamics.*

Weiss, J. (1992). The dynamics of enstrophy transfer in two-dimensional hydrodynamics, *Physica D*, **48**, 273.

Wickerhauser, V., Farge, M., Goirand, E., Wesfreid, E. and Cubillo, E. (1994). Efficiency comparison of wavelet packet and adapted local cosine bases for compression of a two-dimensional turbulent flow, in *Wavelets: theory, algorithms and applications* (ed. Chui *et al.*), pp. 509–531.

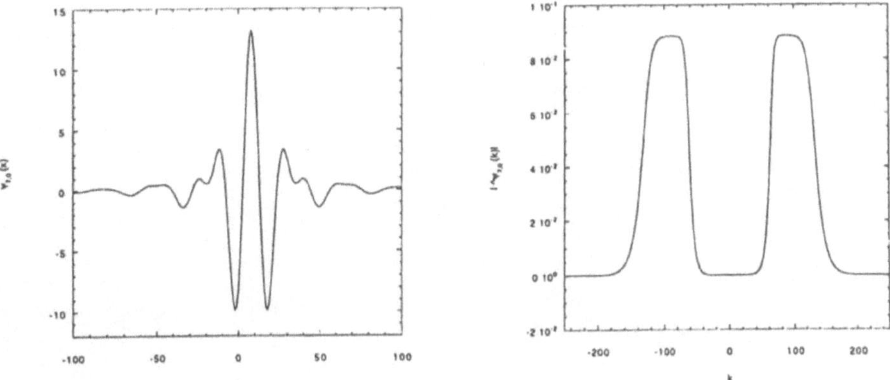

Figure 1. Quintic spline wavelet $\psi_{7,0}$ in physical and in Fourier space.

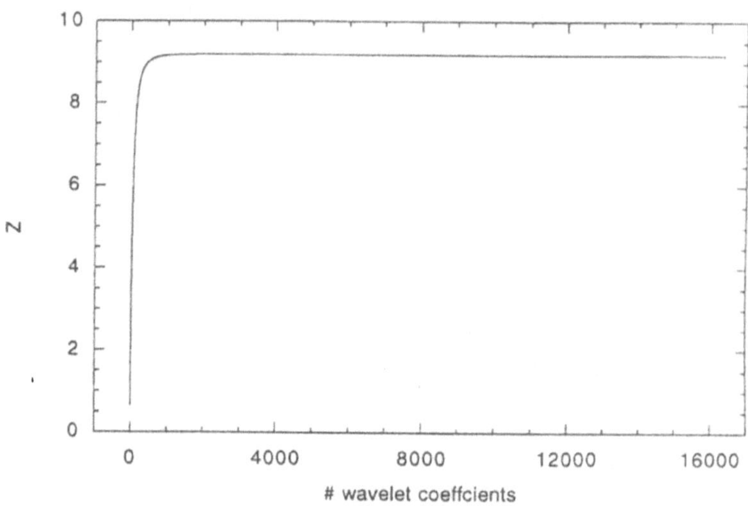

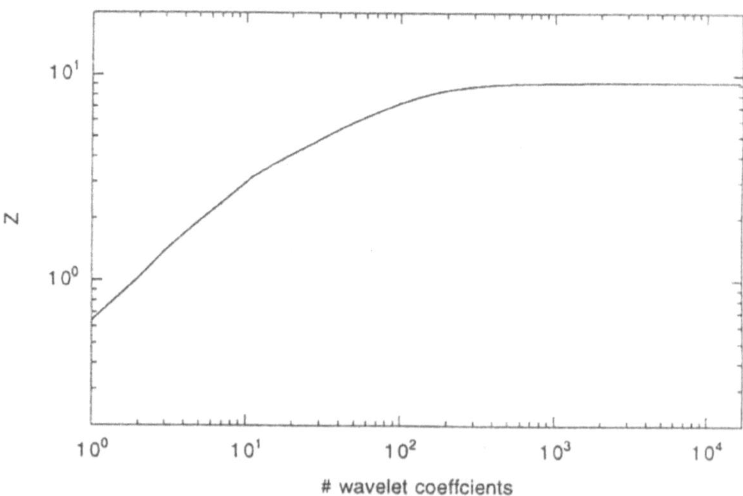

Figure 2. Compression rate: enstrophy versus the number of wavelet coefficents

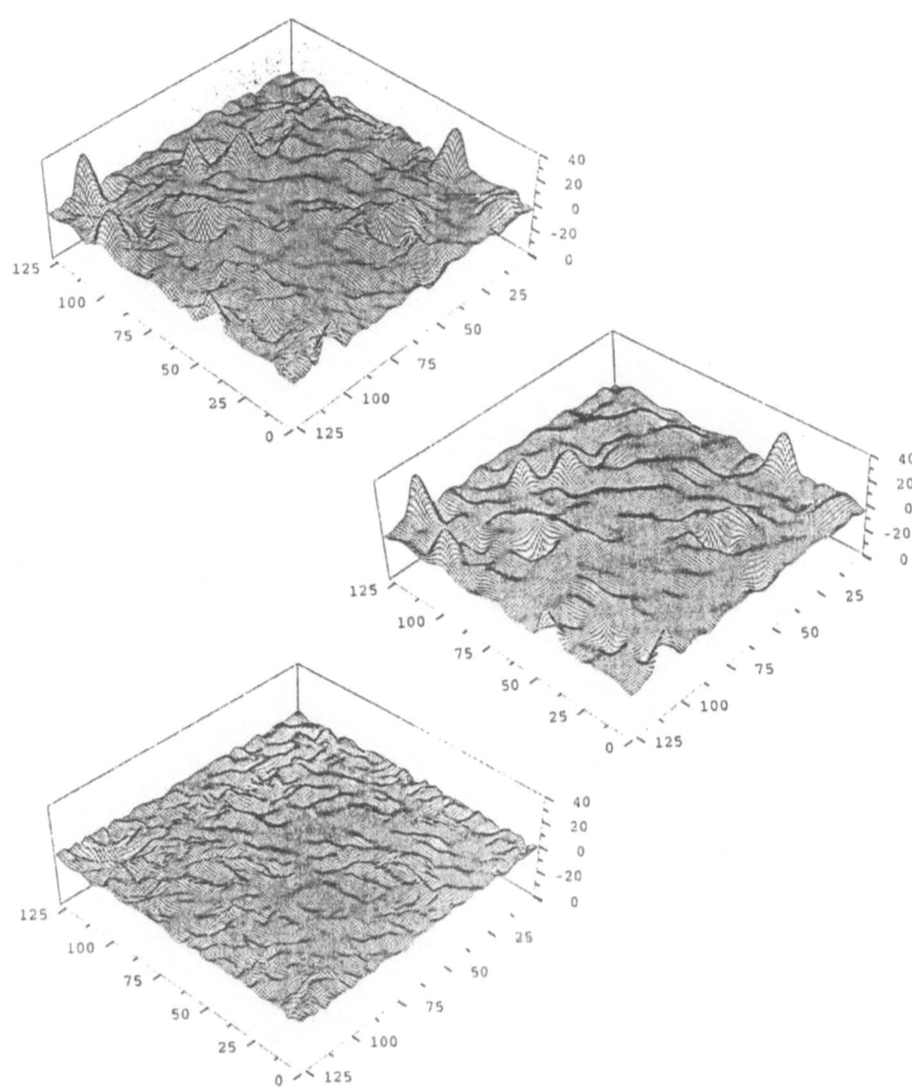

Figure 3. Surface plots of the vorticity fields ω (top), $\omega_>$ (middle) and $\omega_<$ (bottom).

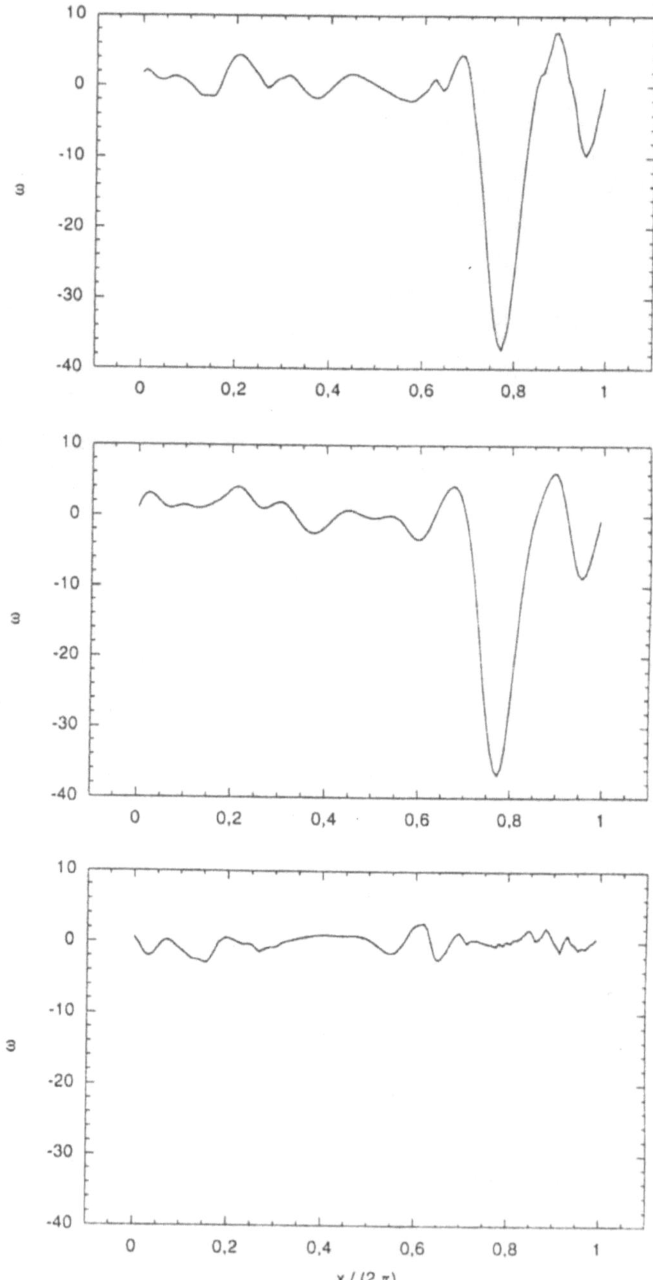

Figure 4. Cuts of the vorticity fields ω (top), $\omega_>$ (middle) and $\omega_<$ (bottom) at $y = 2\pi 77/128$.

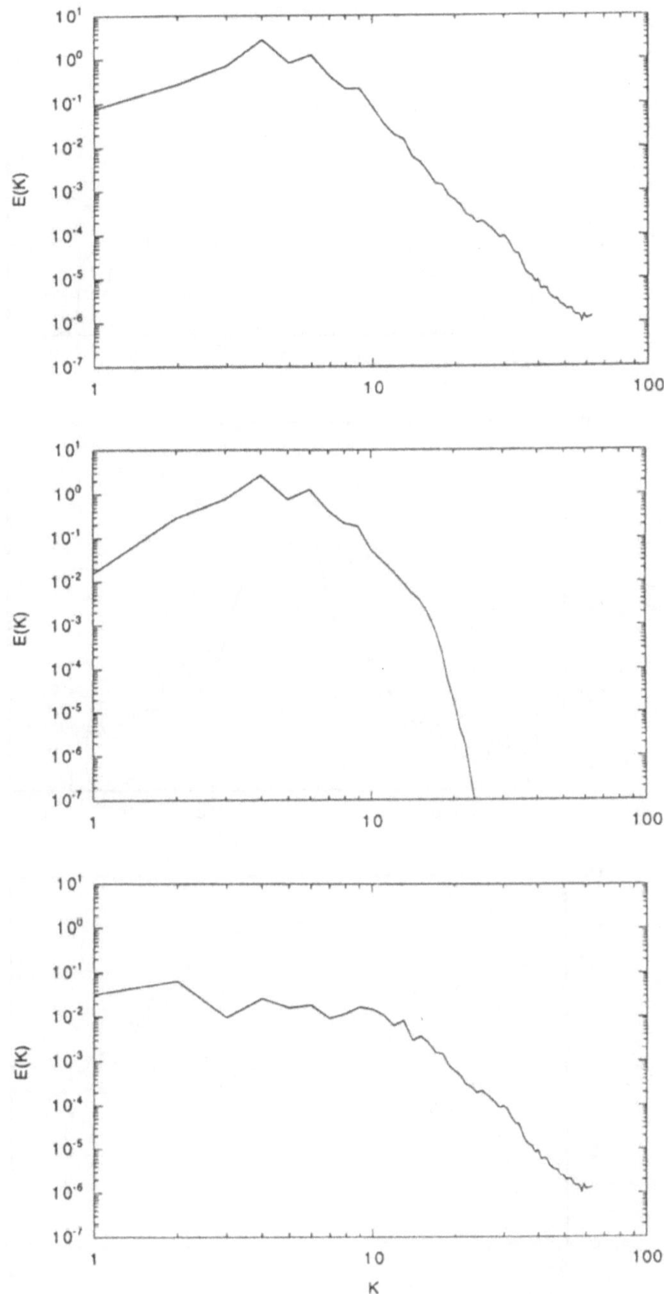

Figure 5. Energy spectra of the vorticity fields ω (top), $\omega_>$ (middle) and $\omega_<$ (bottom).

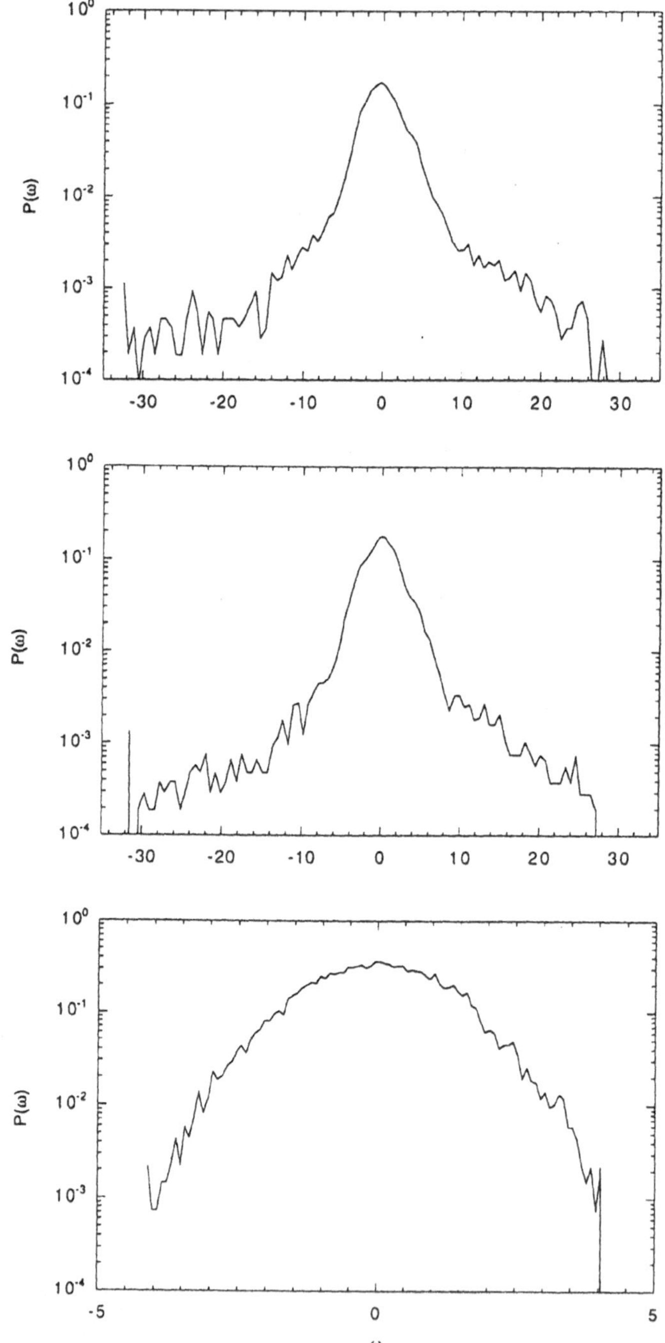

Figure 6. Probability density functions (histograms using 100 bins) of the vorticity fields ω (top), $\omega_>$ (middle) and $\omega_<$ (bottom).

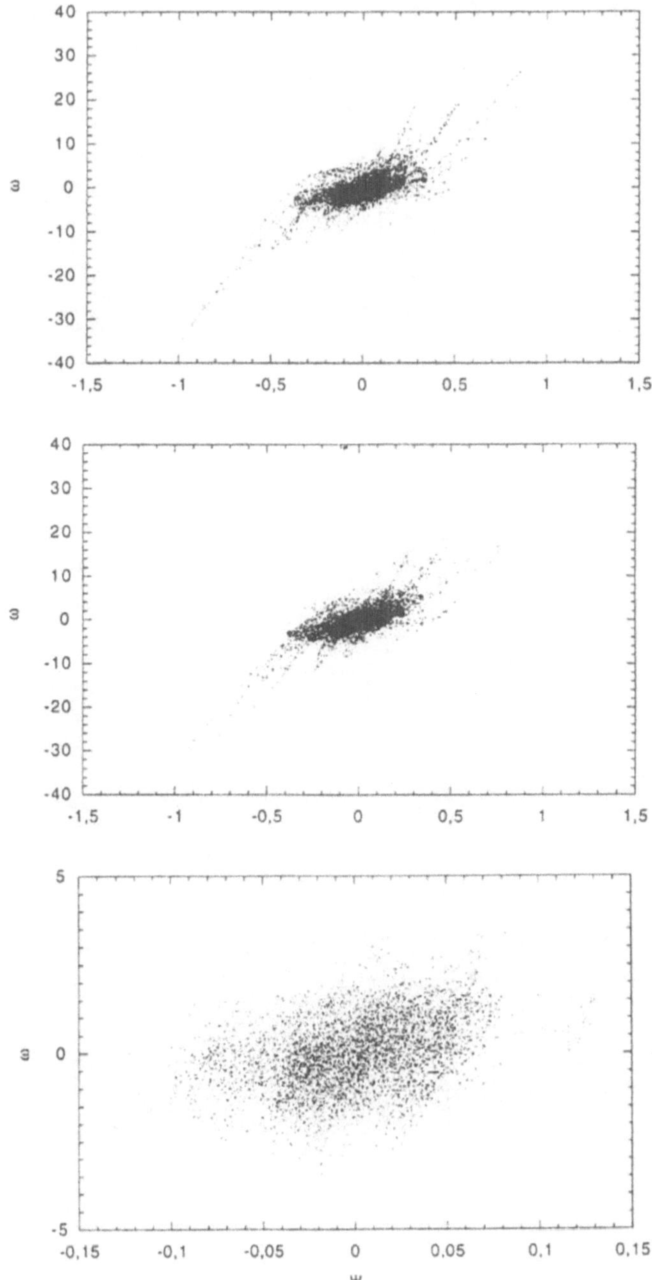

Figure 7. Scatter plots of the vorticity fields ω (top), $\omega_>$ (middle) and $\omega_<$ (bottom) versus the corresponding stream functions Ψ, $\Psi_>$ and $\Psi_<$.

OPTIMIZED VORTEX ELEMENT SCHEMES
FOR SLENDER VORTEX SIMULATIONS

Overcoming the stiffness

O.M. KNIO
Department of Mechanical Engineering, The Johns Hopkins University, 3400 N. Charles St., Baltimore, MD 21218, USA

AND

R. KLEIN
Fachbereich Sicherheitstechnik, Bergische Universität Wuppertal, Gauß-Str. 20, D-42097 Wuppertal, Germany

1. Introduction

Vortex element schemes are designed for application to flows with highly concentrated vorticity. These schemes are based on discretization of the vorticity field into spherically smoothed Lagrangian elements of overlapping cores, and transport of these elements along particle trajectories. The advantages of this approach stem from the Lagrangian discretization which naturally concentrates computational elements into regions of high vorticity, and from the Lagrangian transport procedure which minimizes numerical diffusion.

One *ad hoc* ansatz for the application of vortex element schemes to the simulation of slender filaments is to represent the filament centerline using a "chain" of regularized vortex elements (Knio & Ghoniem, 1990). When the cores of the vortex elements are overlapping, the smoothing procedure leads to a well-defined numerical vortex core structure, which is "typically" identified as the physical core vorticity distribution. However, Klein & Knio (1995) show that in the resulting thin-tube model O(1) errors appear in the prediction of the vortex filament centerline motion. Detailed asymptotic analyses of both the numerical and the physical vorticity structures in the vortex core reveal the origin of the discrepancy and naturally lead to a correction of the scheme. These ideas are extended by Klein *et al.* (1996) to include nontrivial vortex stretching as well as the effects of viscosity.

Application of thin-tube vortex element models to very thin vortices is faced with two difficulties, both due to the large scale ratio which characterizes the

E. Krause and K. Gersten (eds.), IUTAM Symposium on Dynamics of Slender Vortices, 85-94.

slender vortex. Specifically, when the ratio δ/R of the filament core size to the characteristic radius of curvature is very small, the numerical scheme suffers from very high spatial resolution requirements and a severe temporal stiffness.

The origin of the high spatial resolution requirements is that in the approaches suggested in (Klein & Knio, 1995) the numerical core radius is essentially of the same order as the physical core radius. In addition, in order to ensure that the numerical core structure is well-defined and independent of the discretization level, a high degree of overlap among the cores of neighboring elements must be maintained. Thus, when the slenderness ratio δ/R is of the order of 10^{-2} or smaller, the core overlap requirement can easily lead to an excessively large number of elements, and consequently necessitate prohibitively large CPU resources. For very small δ/R, the number of elements needed to satisfy overlap with cores of order δ may exceed by several orders of magnitude the number of elements needed for adequate representation of the filament centerline.

Another implication of having very small inter-element separation distances is a severe restriction on the integration time step. Computational experience with finely-meshed slender vortices suggests a time step restriction of the form $\Delta t \leq C\delta l/U_{\max}$ where $U_{\max}$ is the peak centerline velocity, δl is the minimum separation distance between neighboring elements, and C is a "constant" (Klein & Knio, 1995). Numerical experimentation also suggests that the constant C is generally small, and decreases with decreasing slenderness ratio. This severe temporal stiffness compounds the spatial resolution problem and substantially increases the necessary computational resources.

In this paper, we explore means to overcome the spatial and temporal stiffness of slender vortex simulations. We focus on the corrected thin-tube models proposed by Klein & Knio (1995) and Klein *et al.* (1996) and implement several approaches to enhance their performance. In order to describe these approaches, we start in the following section with a brief outline of the corrected thin tube models. Section 3 then summarizes the various approaches used to address the stiffness issues. Implementation of the optimized schemes is illustrated in section 4 in light of 3D computations of a perturbed slender ring.

2. Corrected Thin Tube Models

Governing Equations for Filament Motion

As mentioned in the introduction, thin tube vortex element models are Lagrangian methods for the simulation of a slender vortices, which by definition are characterized by a highly-localized vorticity distribution in the neighborhood of a time-dependent centerline $\mathcal{L}(t)$. The schemes "simulate" the asymptotic filament evolution equation (Callegari & Ting, 1978):

$$\dot{\boldsymbol{X}}(s,t) = \frac{\Gamma}{4\pi}\left[\ln\left(\frac{2}{\delta}\right) + C(t)\right]\kappa(s,t)\boldsymbol{b}(s,t) + \boldsymbol{Q}^f(\boldsymbol{X}(s,t)) \qquad (1)$$

where s is the arc length parameter along $\mathcal{L}$, κ and $\boldsymbol{b}$ are the curvature and unit binormal at $\boldsymbol{X}$, respectively, $C(t)$ is the time-dependent core structure coefficient, and $\boldsymbol{Q}^f$ is the so-called finite part of the line Biot-Savart integral. The core structure coefficient represents the contributions of the local swirling and axial velocities to the leading order velocity of the filament; it is expressed as:

$$C(t) = C_v(t) + C_w(t) \tag{2}$$

where

$$C_v(t) = \lim_{\bar{r}\to\infty} \left\{ \frac{4\pi^2}{\Gamma^2} \int_0^{\bar{r}} \bar{r} v^{(0)\,2} d\bar{r} - \ln \bar{r} \right\} - \frac{1}{2} \tag{3}$$

$$C_w(t) = -\frac{1}{2} \left(\frac{4\pi}{\Gamma} \right)^2 \int_0^{\infty} \bar{r} w^{(0)\,2} d\bar{r} \tag{4}$$

and $\bar{r} \equiv r/\delta$. Meanwhile, $\boldsymbol{Q}^f$ is the so-called finite part of the Biot-Savart integral, (see (Callegari & Ting, 1978) or (Klein & Knio, 1995)). It represents the nonlocal self-induction of the filament.

The core structure coefficients evolve in time according to the evolution of the leading-order axial vorticity and axial velocity distributions. As shown by Callegari & Ting (1978), the leading-order axial vorticity and velocity within the core obey inhomogeneous heat equations with a source term that depends on the stretching of the filament centerline. In the inviscid limit, simple closed-form expressions for the evolution of C_v and C_w have been obtained by Klein & Ting (1995):

$$C_v(t) = C_v(0) + \ln \sqrt{\frac{S(t)}{S(0)}} \tag{5}$$

$$C_w(t) = \left[\frac{S(0)}{S(t)} \right]^3 C_w(0) \tag{6}$$

where $S(t)$ is the total arc length of the filament. When viscous effects are present, the expressions describing the evolution of the core structure coefficient are more involved (Klein & Ting, 1995; Klein *et al.*, 1996); for brevity, they are omitted.

Numerical Simulation

Construction of thin-tube models is based on discretization of the centerline $\mathcal{L}$ into a finite number of regularized vortex elements with spherical overlapping cores. The vortex elements are described in terms of their Lagrangian position vectors, $\chi_i(t)$, $i, \ldots, N$, which are indexed consecutively such that the resulting "chain" describes the filament centerline (Knio & Ghoniem, 1990; Klein & Knio, 1995).

Based on the Lagrangian variables, a smooth representation of the *regularized* filament vorticity is obtained using the expression (Beale & Majda, 1985):

$$\omega(\boldsymbol{x},t) = \sum_{i=1}^{N} \Gamma \delta\chi_i(t)\, f_\sigma(\boldsymbol{x} - \chi_i(t)) \tag{7}$$

where f_σ is a rapidly decaying spherical core function of unit mass, $\delta\chi_i(t)$ is the arc length increment associated with the ith element, and σ is the *numerical* core radius. When inserted into the three-dimensional Biot-Savart integral, the above representation yields the following desingularized velocity field:

$$\boldsymbol{v}^{\mathrm{ttm}}(\boldsymbol{x},t) = -\frac{\Gamma}{4\pi} \sum_{i=1}^{N} \frac{(\boldsymbol{x} - \chi_i(t)) \times \delta\chi_i(t)}{|\boldsymbol{x} - \chi_i(t)|^3} \kappa_\sigma(\boldsymbol{x} - \chi_i(t)) \tag{8}$$

where $\kappa_\sigma(\boldsymbol{x})$ is the velocity smoothing kernel corresponding to f_σ. In the computations, the arc length increments $\delta\chi_i(t)$ are related to the distribution of particle positions using the procedure described by Klein & Knio (1995). It is based on a Lagrangian spectral collocation interpolation of the filament geometry onto the particle positions, and approximating the arc length based on spectral collocation derivatives of the interpolated filament centerline.

The above discretization and regularization procedure endows the thin tube with a numerical core structure that depends on the choice of the vortex element smoothing function. Unfortunately, if the numerical core vorticity distribution is identified with the physical core vorticity distribution, an O(1) velocity error in the predicted centerline velocity occurs. In order to obtain a velocity prediction that is consistent with equation (1) and the definitions of C and $\boldsymbol{Q}^f$ in (2) and (Klein & Knio, 1995), respectively, Klein & Knio (1995) propose three correction procedures for the above thin tube model. We restrict our attention to two of these approaches, which will be later used in the optimization of the numerical scheme.

In the first approach, the numerical core radius σ is taken to be equal to the physical core size δ, and an explicit velocity correction is implemented in order to remove the error associated with the thin tube model. In this case, the velocity correction:

$$\delta\boldsymbol{v} \equiv \frac{\Gamma}{4\pi}\left(C - C^{\mathrm{ttm}}\right) \kappa \boldsymbol{b} \tag{9}$$

is added to the thin-tube prediction $\boldsymbol{v}^{\mathrm{ttm}}$ in order to absorb the discrepancy in the centerline velocity. Here, C^{ttm} is the numerical core constant which corresponds to the choice of core smoothing function.

In the second approach, the numerical core radius σ is related to the physical core structure so that the regularized velocity field coincides with the theoretical prediction in (1) at the particle positions. The relationship is expressed as:

$$\sigma = \delta \exp\left(C^{\mathrm{ttm}} - C\right) \tag{10}$$

Note that an advantage of the present approach is that explicit evaluation of the curvature and binormal at the filament centerline is avoided.

Once the (corrected) centerline velocity is determined, it is used to update the positions of the vortex elements. This is achieved by integrating:

$$\frac{\partial \chi_i}{\partial t} = \boldsymbol{v}^{\mathrm{cor}}(\chi_i(t), t) \tag{11}$$

where $\boldsymbol{v}^{\mathrm{cor}}$ is the "corrected thin-tube velocity". In the computations, a second-order Adams-Bashforth scheme is used to numerically integrate the above system.

3. Optimization Techniques

In this section, we outline three methods for optimizing the corrected thin tube model outlined in the previous section. In all cases, our approach is motivated by the observation that (1) in the previous construction the number of elements required to achieve overlap is much larger than that needed to adequately describe the centerline geometry, and (2) consequently, both the temporal stiffness and high spatial resolution problems can be effectively addressed if a coarser (or generally more efficient) discretization can be adopted. This observation suggests a number of different optimization approaches, of which we discuss the following three.

Method 1: Extrapolation Technique

This approach is based on selecting a discretization level that is fine enough for the representation of the filament centerline *irrespective* of the size filament core size δ. As previously discussed, selection of numerical core size levels using either of the approaches of the previous section becomes problematic since overlap among neighboring elements may not be satisfied. In fact, for small slenderness ratios, overlap is likely to be everywhere violated.

To avoid this issue, a large value of the numerical core size is used. Let $\tilde{\sigma}$ denote the core size parameter predicted by equation (10),

$$\sigma_o = \max_{i=1,\ldots,N} |\delta \chi_i| \tag{12}$$

be the maximum inter-element separation distance, $\sigma_1 = C\sigma_o$, and $\sigma_2 = \alpha\sigma_1$. Here, C and α are taken to be real constants. Note that: (1) σ_o is time dependent, and (2) for any choice $C > 1$ and $\alpha > 1$ implementation of a thin tube model using either σ_1 and σ_2 will automatically satisfy the overlap condition.

Next, we recall that for any σ the velocity field predicted by the uncorrected thin tube model corresponds to (Klein & Knio, 1995):

$$\boldsymbol{v}^{\mathrm{ttm}} = \frac{\Gamma}{4\pi}\left[\log\left(\frac{2}{\sigma} + C^{\mathrm{ttm}}\right)\right] + \boldsymbol{Q}^f \tag{13}$$

as long as strong overlap is satisfied. Consequently, if we let $\boldsymbol{v}_1$ denote the thin tube velocity prediction based on σ_1, and $\boldsymbol{v}_2$ be the thin tube velocity prediction based on σ_2, then the *corrected* velocity field prediction can be estimated based on the following logarithmic extrapolation procedure:

$$\boldsymbol{v}^{\text{cor}} = \boldsymbol{v}_1 + (\boldsymbol{v}_1 - \boldsymbol{v}_2)\frac{\log\left(\frac{\sigma_1}{\tilde{\sigma}}\right)}{\log \alpha} \tag{14}$$

Thus, the corrected centerline velocity can be directly obtained at the cost of two coarse mesh evaluations. Note that the above procedure does not require delicate and potentially destabilizing estimates of the filament curvature and binormal (Klein & Knio, 1995). In the computations discussed below, we use $C = 3$ and $\alpha = 2$.

Method 2: Explicit Velocity Correction
The velocity correction technique is based on a single velocity evaluation using large σ and implementing an explicit correction to the corresponding velocity prediction. This approach has been first implemented in Zhou's (1996) study of Kelvin waves on a slender vortex. Here, we implement a variant of the velocity correction approach which is summarized by:

$$\boldsymbol{v}^{\text{cor}} = \boldsymbol{v}_1 + \frac{\Gamma}{4\pi}\log\left(\frac{\sigma_1}{\tilde{\sigma}}\right)\kappa \boldsymbol{b} \tag{15}$$

where σ_1 is dynamically determined during the simulation using the same definitions discussed for method 1. In the computations, the curvature and binormal are obtained using the procedure described in Klein & Knio (1995). As in the previous method, a value $C = 3$ is used.

Method 3: Virtual Local Mesh Refinement
This approach is motivated by the observation that the numerical core structure along a particular location on the axis of the thin tube is actually induced by neighboring elements only. Since the core smoothing functions decay rapidly, the numerical *vorticity* structure at a given point is determined by the fields of elements lying within a few numerical core radii of that point. This suggests the following approach to the evaluation of the corrected thin tube velocity.

As in the previous two methods, we rely on a discretization level that is fine enough to adequately represent the filament geometry. However, unlike the previous two approaches, the rescaled numerical core size predicted by Klein & Knio's original equation (10) is used. Based on this choice of parameters, a modified velocity evaluation procedure is implemented. To evaluate the velocity field at a given point, the elements are divided into two disjoint groups: a group of neighboring elements and a group of well-separated elements. A separation distance of three core radii is used as basis for this segragation procedure. The contribution of element belonging to the second group is computed directly on the coarse grid. Meanwhile, the contribution of neighboring elements is accounted for by locally remeshing the corresponding segments using a fine-enough grid for core overlap to be locally satisfied.

In the present computations, a simplified version of the local remeshing procedure is implemented. The simplified procedure consists of determining, at every time step and in a global fashion, a Lagrangian grid that is fine enough for overlap to be everywhere satisfied. The fine grid is determined by Fourier transformation of the filament geometry, extending the resulting spectrum by array padding, and then inverse transforming the extended spectrum onto physical space.

4. Results and Discussion

Implementation of the optimized schemes outlined above is described in light of three-dimensional simulations of a slender vortex ring. The initial conditions are described in terms of the normalized radius of the ring, $R = 1$, the normalized circulation, $\Gamma = 1$. In all cases, we use a re-scaled thin tube core size $\tilde{\sigma} = 0.02$ in conjunction with the core smoothing function $f(r) = \text{sech}^2(r^3)$. The velocity kernel corresponding to f is $\kappa(r) = \tanh(r^3)$ (Beale & Majda, 1985), and the numerical core structure coefficient $C^{\text{ttm}} = -0.4202$ (Klein & Knio, 1995).

In order to observe a non-trivial slender vortex evolution, and consequently contrast the predictions of the various models, the centerline of the vortex ring is perturbed using sinusoidal waves. The perturbation takes the form of an azimuthal bending wave and is given by:

$$\rho(\theta) = R\left[1 + a_1 \sin(k_1\theta + \phi_1) + a_2 \sin(k_2\theta + \phi_2)\right] \tag{16}$$

where ρ is the perturbed radius of the ring and θ is the azimuthal angle. In the computations presented below, we consider a two-mode perturbation specified by $a_1 = a_2 = 0.02$, $\phi_1 = \phi_2 = 0$, $k_1 = 2$ and $k_2 = 3$.

TABLE 1. Numerical parameters

	Original	M1	M2	M3
Δt	0.002	0.002	0.002	0.002
σ	$\tilde{\sigma}$	(σ_1, σ_2)	σ_1	$\tilde{\sigma}$
N	1024	256	256	256/1024
# steps	1500	1500	1500	1500

The evolution of the slender ring described above is computed using the rescaled numerical core radius approach developed by Klein & Knio, which we refer to as original scheme, as well as the optimized schemes based on methods 1, 2, and 3 above, which we labeled M1, M2, and M3, respectively. The parameters used in

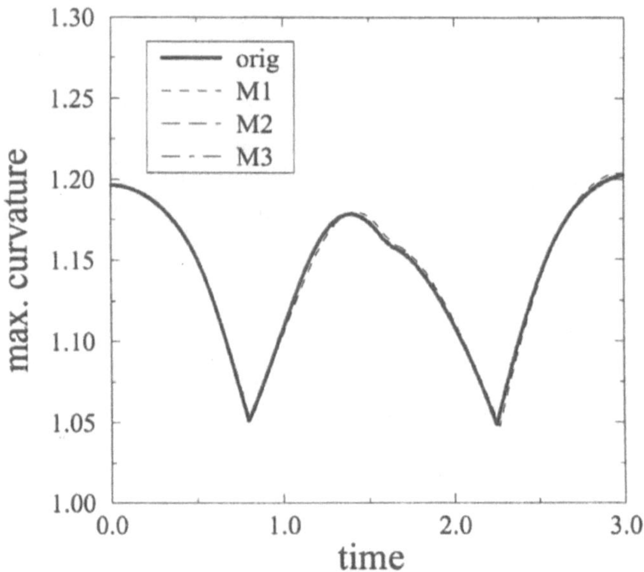

Figure 1. Comparison of the evolution of the maximum curvature obtained using the original method, and the three improved schemes M1, M2 and M3.

all of the runs are summarized in table 1. Note that core overlap using $\tilde{\sigma}$ would not be satisfied in M1, M2 and M3 at the selected grid resolution.

The predictions of the original and optimized schemes are first illustrated in figures 1–2, which respectively show the evolution of the peak centerline curvature and the peak centerline velocity throughout the computations. The results show that the peak curvature of the filament undergoes large-amplitude periodic oscillations, and that the behavior of the peak velocity follows closely that of the peak curvature. The figures show that the predictions of the optimized schemes are in close agreement with each other, and with the predictions of the original scheme.

Very close agreement between the predictions of the original and optimized schemes is also evident in more stringent tests. A sample of these tests is shown in figure 3, which depicts the spatial distribution of centerline curvature along the circumference of the ring. As observed earlier, the results of the optimized schemes are nearly identical to those obtained using the original model.

It should be emphasized that the present comparison is based on unsteady conditions during which the filament is substantially deformed, with O(1) changes in the centerline curvature. Thus, the close agreement between the results of the original and optimized schemes provides a strong support to the validity of the latter.

The performance of the original and optimized schemes is quantified in table 2. The table provides the total CPU time spent in each of the four calculations,

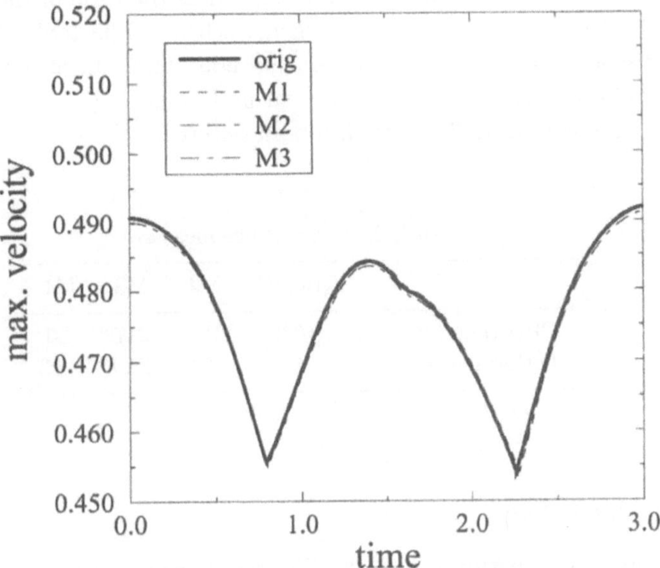

Figure 2. Comparison of the evolution of the maximum velocity obtained using the original method, and the three improved schemes M1, M2 and M3.

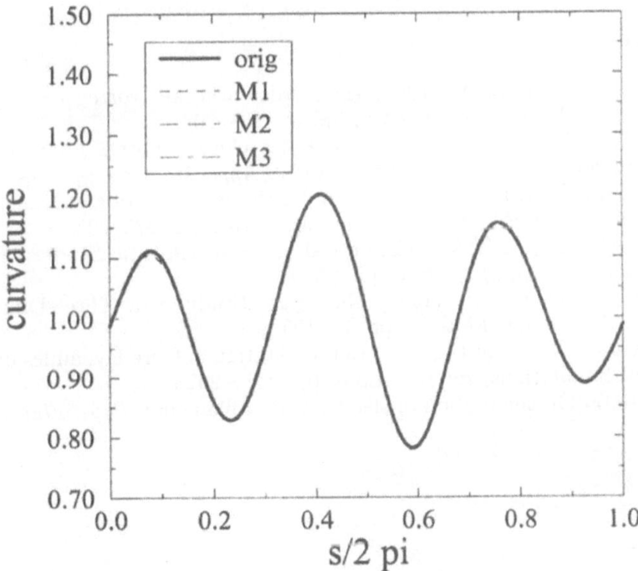

Figure 3. Spatial distribution of the curvature along the filament axis at $t = 3$. Curves are generated based on the results obtained using the original method, and the three improved schemes M1, M2 and M3.

as well as the performance gain which is defined as the ratio of the current CPU time to the CPU time spent in the original scheme. The table shows that the optimized scheme result in an order-of-magnitude enhancement in the performace of the computations. This is a substantial performance gain, especially since a relatively large core to radius ratio has been selected.

TABLE 2. Performance measures

	Original	M2	M3	M1
CPU time (s)	2792	184	217	281
Performance Gain	1	15.2	12.9	9.93

Acknowledgements

Computations were performed at the Pittsburgh Supercomputing Center. OK's research was partially supported by the National Science Foundation Grant CTS-9706701. RK's research was supported in part by Deutsche Forschungsgemeinschaft, grant KL 611/6-1.

References

Beale, J.T., and Majda, A. (1985) High order accurate vortex methods with explicit velocity kernels, *J. Comput. Phys.* **58**, pp. 188–208.

Callegari, A. and Ting, L. (1978) Motion of a Curved Vortex Filament with Decaying Vortical Core and Axial Velocity, *SIAM J. Appl. Math.***35**, pp. 148-175.

Klein, R., Knio, O.M. (1995) Asymptotic Vorticity Structure and Numerical Simulation of Slender Vortex Filaments, *J. Fluid Mech.***284**, pp. 275–293.

Klein, R., and Ting, L. (1995) Theoretical and Experimental Studies of Slender Vortex Filaments, *Appl. Math. Lett.* **8**, pp. 45–50.

Knio O.M., Ghoniem A.F. (1990) Numerical Study of a Three-Dimensional Vortex Method, *J. Comput. Phys.***86**, pp. 75–106.

Klein R., Knio O.M., Ting L. (1996) Representation of Core Dynamics in Slender Vortex Filament Simulations, *Phys. Fluids***8**, pp. 2415–2425.

Zhou, H. (1997) On the motion of slender vortex filaments, *Phys. Fluids* **9**, pp. 970–981.

A NEW APPROACH TO THE MODELING VISCOUS DIFFUSION IN VORTEX ELEMENT METHODS

B.YU. SCOBELEV AND O.A. SHMAGUNOV
Institute of Theoretical and Applied Mechanics
of the Russian Academy of Sciences
4/1 Institutskaya st., Novosibirsk, 630090, Russia

1. Introduction

Vortex methods are powerful numerical methods for computing incompressible fluid flows. Because they are grid-free and naturally adaptive, they give an opportunity to essentially simplify numerical algorithms and decrease influence of such undesirable effects as numerical diffusion. Ideal vortex elements describe sufficiently well integral characteristics of separated flows around various aircraft and large-scale turbulent structures [1,2]. To describe fine-scale turbulence, a viscosity needs to be taken into account. At present time there are various approaches to this problem [1,3,4], which use the equation of vortex viscous diffusion, in one form or another. In the present paper the totally different approach is suggested.

The main idea of the suggested method is as follows. As known, two-dimensional ideal flows have the following invariants: total vorticity, coordinates of the vorticity centre, vorticity dispersion, and a certain part of kinetic energy. For viscous flows the first two characteristics are still invariants, while the dispersion and energy vary in time according to the known laws [5,6]. As in the case of ideal flows, we will simulate continuos distribution of viscous flows vorticity by a set of circular vortices. Let us turn the vortices radius to zero and their vorticity to infinity in such a way that the vortices circulation remain finite. As this takes place, the rate of changes of the dispersion and energy infinitely increases. In order to keep the rate finite, we turn the initial viscosity to zero, so that rate of energy dissipation corresponds to a given viscosity. As a result, we will have a system of ideal point vortices with its energy dissipating in the samc way as kinetic energy of viscous flow does. To agree the energy dissipation with the equations of motion of the ideal point vortices, we postulate that the process of discretization of viscous flow vorticity field corresponds to transition into certain non-inertial coordinate system. In this system the commonly known equations of point vortices motion are correct. The numerical discretization of the motion equations is known to generate the scheme dissipation and dispersion. This process also can be viewed as a transition into another non-inertial coordinate system. Therefore, to obtain the results in the initial physical coordinate system, the inverse transformations of

E. Krause and K. Gersten (eds.), IUTAM Symposium on Dynamics of Slender Vortices, 95-104.

coordinates and time need to be performed after each step of the numerical integration. These transformations are determined by condition that dissipation of energy of a point vortices system correspond to a given viscosity.

The ideas described above are realised in a numerical algorithm for simulating two-dimensional viscous flows. Calculation of transverse viscous flow around a plate is carried out. For the large range of Reynolds numbers $50 \leq Re \leq 2000$ a good agreement with experiment in dependence of the Strouhal number for the von Karman street on the Reynolds number is obtained.

2. Viscosity Hypothesis

In two-dimensional flows of inviscid fluid which is at rest in infinity the vorticity energy W is invariant [5]

$$W = -\frac{\rho}{4\pi}\iint \omega\omega' \ln|r - r'| dS dS' \tag{1}$$

where ρ is density, $\omega \equiv \omega(x,y)$, $\omega' = \omega'(x',y')$ are vorticity distributions, $r = (x,y)$, and integrals are taken over the entire flow plane.

Let at the initial time all vorticity be concentrated in circular vortices with radius r_0, located inside some region with area $S_0 = 1$. The vorticity distribution inside a circular vortex is determined by infinitely differentiable function $\omega(p)$, which is constant inside a circle of smaller radius (p - distance from the vortex center; $\omega(p) = const$ at $p \leq r_0 - \varepsilon$). It can be shown [5] that for viscous fluid W is not conserved and varies in time according to the law

$$\frac{dW_0}{dt} = -\mu_0 \int_{S_0} \omega^2 dS, \tag{2}$$

where μ_0 is viscosity coefficient and W_0 is defined by formula (1), the integrals being taken over the area S_0.

Suppose that the quantity r_0 is small enough, and vortices are not overlapped. Then, accurate to terms $O(\varepsilon^2)$, the right-hand side of formula (2) can be presented as

$$\mu_0 \int_{S_0} \omega^2 dS = \frac{\mu_0 S_0}{\pi r_0^2} \frac{1}{S_0} \sum_{k=1}^{N} \Gamma_k^2, \tag{3}$$

where Γ_k is circulation of the k-th vortex, and N is number of vortices in region S_0.

The vorticity energy W_0 can be decomposed into three components:

$$W_0 = I_1 + I_2 + I_3, \tag{4}$$

where

$$I_1 = -\frac{\rho}{4\pi}\ln r_0 \iint_{\Omega} \omega\omega' dSdS' ; \qquad I_2 = -\frac{\rho}{4\pi}\iint_{\Omega} \omega\omega' \ln\frac{|r-r'|}{r_0} dSdS' ,$$

$$I_3 = -\frac{\rho}{4\pi}\iint_{S_0^2\setminus\Omega} \omega\omega' \ln|r-r'| dSdS' ; \ \Omega = \{(r,r'): r\in S_0, r'\in S_0, |r-r'|\le 2r_0\}$$

For comparatively small r_0 the quantity $I_1 + I_2$ is a sum of own kinetic energies of N circular vortices, I_1 being the singular part of the sum, and I_2 being its regular part. It can be shown that I_1=const and

$$I_2 = \frac{5\rho}{32\pi}\sum_{i=1}^{N}\Gamma_i^2 \ \text{ at } \varepsilon \to 0 \tag{5}$$

The integral I_3 describes the vortex interaction energy and for $\varepsilon \to 0$ it coincides with the interaction energy of point vortices.

$$I_3 = -\frac{\rho}{4\pi}\sum_{i\neq j}^{N}\Gamma_i\Gamma_j \ln|r_i - r_j| , \tag{6}$$

where $r_i = (x_i, y_i)$ are the coordinates of the i-th vortex center.

Taking into account formulas (3)-(6), equation (2) takes the following form:

$$\frac{dE}{dt} = -\frac{\mu_0 S_0}{\pi r_0^2}\frac{1}{S_0}\sum_{i=1}^{N}\Gamma_i^2 ; \quad E = \rho\left[\frac{5}{32\pi}\sum_{i=1}^{N}\Gamma_i^2 - \frac{1}{4\pi}\sum_{i\neq j}^{N}\Gamma_i\Gamma_j \ln|r_i - r_j|\right] \tag{7}$$

Let us turn $r_0 \to 0$ and require that simultaneously $\mu_0 \to 0$ following the law

$$\mu_0 = \mu\frac{\pi r_0^2}{S_0} ,$$

where μ is a given value of viscosity coefficient. Then, taking into account the equality $S_0 = 1$, we get

$$\frac{dE}{dt} = -\mu\sum_{i=1}^{N}\Gamma_i^2 . \tag{8}$$

Scaling the coordinates, each finite flow region is transformed into a region with the unit area. Therefore, the following hypothesis can be formulated.

The motion of a system of ideal point vortices simulates the viscous flow if the regular part of the vortex energy E *satisfies equation (8).*

3. Non-inertial Transformation of Coordinate System.

The equation of motion of N point vortices in the Hamiltonian form are written as

$$\Gamma_k \frac{dx_k}{dt} = \frac{\partial H}{\partial y_k}, \quad \Gamma_k \frac{dy_k}{dt} = -\frac{\partial H}{\partial x_k}; \quad k = 1,2,\dots,N \tag{9}$$

.The Hamiltonian H coincides with I_3 (6) if $\rho = 1$. It is known [5] that the vorticity center coordinates X, Y, Hamiltonian H and vorticity dispersion D^2 are the invariants of equations (9). Instead of D^2, we will consider the invariant related to it

$$L^2 = \sum_{i=1}^{N} \Gamma_i (x_i^2 + y_i^2) \tag{10}$$

(L^2, accurate to a numerical coefficient, coincides with the angular momentum). The integral analogs of the quantities X, Y, L^2 are the invariants of the inviscid flow with a continuous distribution of vorticity. It can be shown [5,6] that, given the viscosity μ_0, the values of X, Y are still retained, and the integral analog L^2 varies according to the law

$$\frac{d}{dt} \int \omega (x^2 + y^2) dS = \frac{4\mu_0}{\rho} \int \omega dS \tag{11}$$

Similar to the previous paragraph, we approximate the vorticity ω by discrete circular vortices and use the viscosity hypothesis: $\mu_0 = \mu \pi r_0^2$. Then it can be shown that the value of L^2 (10) must remain constant in the course of simulation of a viscous fluid using the point vortices.

$$\frac{dL^2}{dt} = 0 \tag{12}$$

Since the point vortices circulation and Hamiltonian H are conserved in the course of motion, $E = const$ on exact solutions of equations (9), and hence, condition (8) is not satisfied. On the other hand, the discretization of equations (9) leads to appearance of the scheme viscosity, and E and L^2 vary in an uncontrolled manner in numerical calculations. Conditions (8), (12) are not satisfied either. To meet conditions (8), (12), let us accept the following hypothesis.

Both the vorticity discretization (passage from continuous distribution to point vortices) and discretization of the motion equations (9) are equivalent to the passage to certain non-inertial coordinate systems. Therefore, obtaining the results in the initial, physical coordinate system needs the performing of inverse transformation of the coordinates and time after each step of numerical integration of the motion equations. These transformations are determined by conditions (8), (12).

Let us consider the finite-difference forms of equations (8), (12):

$$\Delta E = -\nu\Delta t\sum_{i=1}^{N}\Gamma_i^2\,; \qquad \Delta L^2 = 0\,. \tag{13}$$

where ν is the kinematic viscosity, and E denotes relation (7) at $\rho = 1$. Let us designate as $\tilde{x}_i(n+1)$, $\tilde{y}_i(n+1)$ the coordinates of the i-th vortex obtained at the $(n+1)$-th step of integration of equations (9), and as $\Gamma_i(n)$ an appropriate value of circulation (it is not changed when passing from the n-th to $(n+1)$-th time layer). To pass to the physical coordinate system, let us perform the following local transformations:

$$\begin{aligned} x_i(n+1) &= \left[1+\Delta L(n+1)\right]\tilde{x}_i(n+1) \\ y_i(n+1) &= \left[1+\Delta L(n+1)\right]\tilde{y}_i(n+1) \end{aligned} \tag{14}$$

$$\Gamma_i(n+1) = \left[1+\Delta\Gamma(n+1)\right]\Gamma_i(n)\,; \quad i = 1,2,\ldots,N \tag{15}$$

.The time step $\Delta\tilde{t}$ used for integration is transformed using the formula

$$\Delta t(n+1) = \frac{\left[1+\Delta L(n+1)\right]^2}{\left[1+\Delta\Gamma(n+1)\right]}\Delta\tilde{t} \tag{16}$$

(it follows from the definition of circulation).

The values of parameters $\Delta L(n+1)$, $\Delta\Gamma(n+1)$ are determined by conditions (13) for the increments $\Delta L^2(n+1) = L^2(n+1) - L^2(n)$ and $\Delta E(n+1) = E(n+1) - E(n)$.

$$(1+\Delta\Gamma)(1+\Delta L)^2 l^2(n+1) - L^2(n) = 0 \tag{17}$$

$$(1+\Delta\Gamma)^2 e(n+1) - \frac{1}{4\pi}(1+\Delta\Gamma)^2\ln(1+\Delta L)\times$$
$$\times\sum_{i\neq j}^{N}\Gamma_i(n)\Gamma_j(n) - E(n) = -\nu\Delta\tilde{t}\sum_{i=1}^{N}\Gamma_i^2(n) \tag{18}$$

where $l^2(n+1)$ and $e(n+1)$ are the calculated values of L^2 and E.

After transformations (14), (15) we get the values of coordinates and circulations, for which the motion of a system of ideal point vortices simulates the flow with a specified kinematic viscosity ν.

Transformations (14), (15) are convenient to be performed in two steps. The first step involves the transformation of the coordinates only. The parameter of the transformation is obtained from the second equation in (13) and has the following form

$$\Delta L_0(n+1) = \left(\frac{L^2(n)}{l^2(n+1)}\right)^{1/2} - 1\,.$$

After that the values of energy $e_0(n+1)$ and quantity $l_0^2(n+1)$ are calculated in transformed coordinates.

The second stage implies complete transformations (14), (15). By definition, $l_0^2(n+1) = L^2(n)$, therefore, we can obtain from (17) that

$$\Delta L_1 = \left(\frac{1}{1+\Delta\Gamma_1} \right)^{1/2} - 1 .$$

Substituting this expression for ΔL_1 into (18) we gain the nonlinear equation for $\Delta\Gamma_1$, which is solved by the method of iterations.

4. Simulation of Transverse Flow Around a Plate

Let us consider a transverse viscous flow around a plate (figure 1) with a constant velocity U_0. In the method of discrete vortices, the plate is substituted with a system of fixed point vortices. The points of the onset of free vortices are located at fixed distances from the plate edges. Two free vortices are generated before each step of integration of the motion equations. The circulations of fixed and free vortices are determined by two conditions: no-passing condition in the control points located between the fixed vortices and at the plate edges, and equality to zero of the total circulation. Thus, there are two types of vortices in the problem before passing to the n-th time layer: N_0 fixed vortices and $2n$ free vortices. The free vortices move in accordance with equations (9) complemented with the contribution of the fixed vortices and external flow.

The transformations of coordinates and circulations described in the previous Section are carried out at each time layer beginning with some layer n_0 before the birth of two new vortices. Thereat, it is taken into account that N free vortices which fall within the (n+1)-th time layer ($N = 2n+2$) move in external velocity field generated by the incoming flow and fixed vortices. The influence of external velocity U_0 is excluded by means of passing to a frame of reference moving with the external flow.

The velocity $U = (U_x, U_y)$ induced by the fixed vortices gives birth to additional changes in energy E and quantity L^2 [6]. Within the time interval $\Delta\tilde{t}$ these changes have the form

$$\Delta E = \Delta\tilde{t} \sum_{i=1}^{N} \Gamma_i (u_{xi} U_{yi} - u_{yi} U_{xi}); \quad \Delta L^2 = \Delta\tilde{t} \sum_{i=1}^{N} \Gamma_i (x_i U_{xi} + y_i U_{yi}) \quad (19)$$

where U_i is the value of U in the locus of the i-th vortex, and u_i is the velocity induced by all free vortices except for the i-th vortex in this point. The corresponding contributions of fixed vortices are extracted from the values of $l^2(n+1)$, $e(n+1)$, prior to determining the transformation parameters.

In the numerical simulation the plate was substituted by $N_0 = 20$ fixed vortices. The points of the onset of free vortices were located at distances $1/2N_0$ from the plate edges.

Non-simultaneity of the beginning of new vortices motion appeared to be essential for the formation of the von Karman vortex street. Moreover, the vortex with a lower circulation should start moving first. Therefore, motion equations (9) were integrated by the Euler explicit-implicit method of the following form.

$$\begin{aligned}
x_1^{(n)} &= f_1(x_1^{(n-1)}, y_1^{(n-1)}, \ldots, x_N^{(n-1)}, y_N^{(n-1)})\Delta t \\
y_1^{(n)} &= \varphi_1(x_1^{(n-1)}, y_1^{(n-1)}, \ldots, x_N^{(n-1)}, y_N^{(n-1)})\Delta t \\
x_2^{(n)} &= f_2(x_1^{(n)}, y_1^{(n)}, x_2^{(n-1)}, \ldots, x_N^{(n-1)}, y_N^{(n-1)})\Delta t \qquad (20) \\
&\cdots\cdots\cdots\cdots\cdots\cdots\cdots\cdots\cdots\cdots \\
y_N^{(n)} &= \varphi_N(x_1^{(n)}, y_1^{(n)}, \ldots y_{N-1}^{(n)}, x_N^{(n-1)}, y_N^{(n-1)})\Delta t
\end{aligned}$$

where f_i, φ_i are the right-hand sides of equations (9). Method (20) is similar to the iterative Gauss-Seidel method for solving a system of linear algebraic equations. The vortex with a smaller circulation was assigned an odd number before integration. The numerical integration was carried out with time step $\Delta t = 0.1$. For regularization of numerical simulation, the point vortices were substituted by the circular vortices with discretization radius $r_0 = 1/40$ (inside the circle, the induced velocity linearly decreased down to zero at the center). The test calculations with $r_0 = 1/80$ gave, in fact, the same results.

The problem configuration is shown in figure 1.

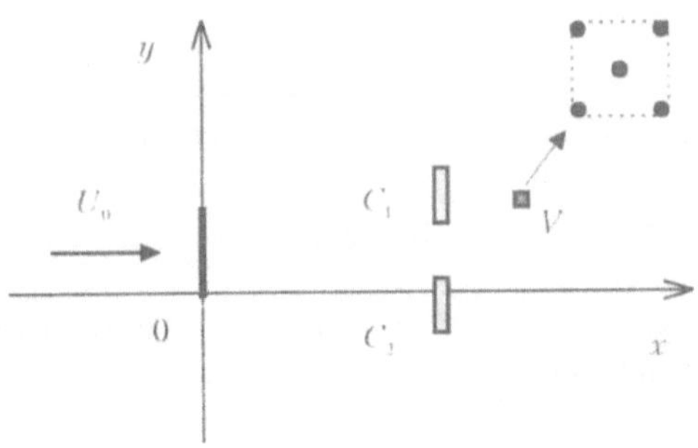

Figure 1.

Two counters of vorticity, C_1, C_2 (rectangles of size 1×0.3 summarizing circulations of all incoming vortices) at the distance of four lengths of the plate, and measuring instrument of transverse velocity at $x = 5$, $y = 1$ (the velocity fluctuations were smoothed out by averaging over 5 points) are located.

The results showed that the suggested method of viscosity consideration successfully simulates the von Karman vortex street and the dependence of the Strouhal number *Sh* on the Reynolds number *Re*. Examples of the von Karman vortex streets at the dimensionless time $t = 70$ are shown in figure 2 for $Re = 100$.

Figure 2.

Comparison of the Strouhal number *Sh* values with experimental data [7] in range $50 \le \mathrm{Re} \le 1000$ is shown in figure 3.

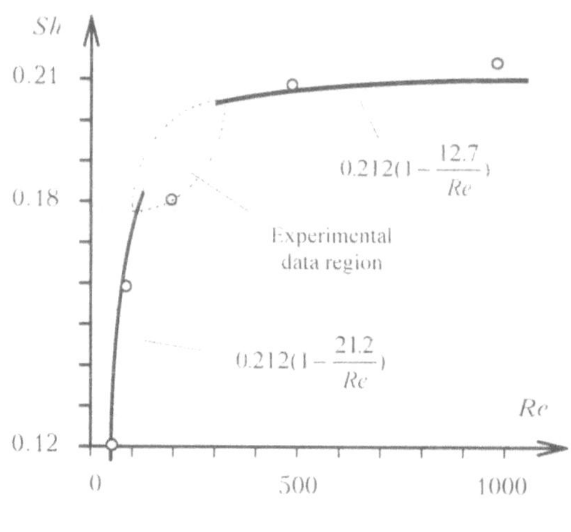

Figure 3.

The solid lines show experimental dependence of *Sh* on *Re*; the circles denote the results of the numerical simulation.

5. Simulation of Aerodynamical Characteristics of the Plate

The suggested method allows to investigate influence of viscosity on aerodynamical characteristics of the plate. Simulation of the normal force coefficient c_n and of dimensionless coordinate x_d of the pressure center at various attack angles and values of Reynolds number were carried out.

The test calculations showed that the initial stage of the plate movement $0<t<10$ can be simulated accurately enough at the time step Δt=0.04. For investigation of the aerodynamical characteristics at times $t>10$ the movement equations can be integrated with the step Δt=0.1.

The calculations made at Reynolds numbers Re=50 and Re=2000 showed that at large attack angles α=60° and α=90° (transverse flow) a form of functions c_n (t) and x_d (t) at short times $0<t<10$ virtually doesn't depend on a Reynolds number. At smaller attack angle α=30° viscosity plays more essential role.

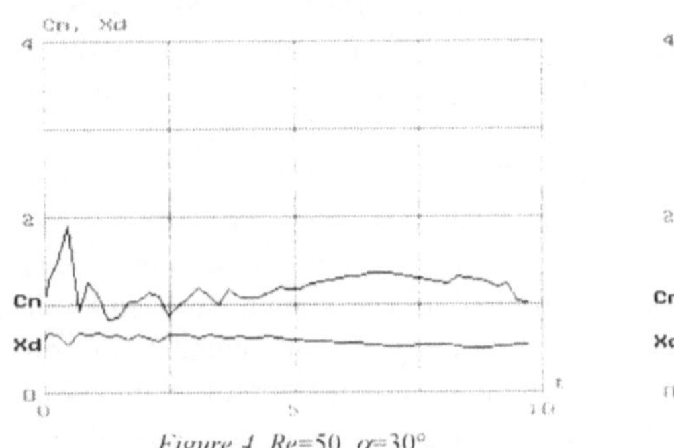

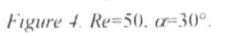

Figure 4. Re=50. α=30°.

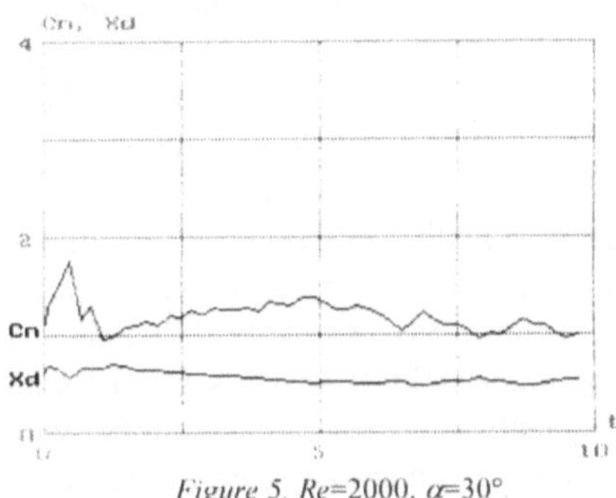

Figure 5. Re=2000. α=30°

At Re=50 and $t>5$ variation of c_n is monotonous, but at Re=5000 c_n oscillates. Average over the interval $7.5<t<10$ values of c_n are also essentially different: $\overline{c}_n$ =1.21 at Re=50; $\overline{c}_n$ =1.03 at Re=2000.

At large attack angles, in particular at α=90°, influence of viscosity becomes essential at the longer times.

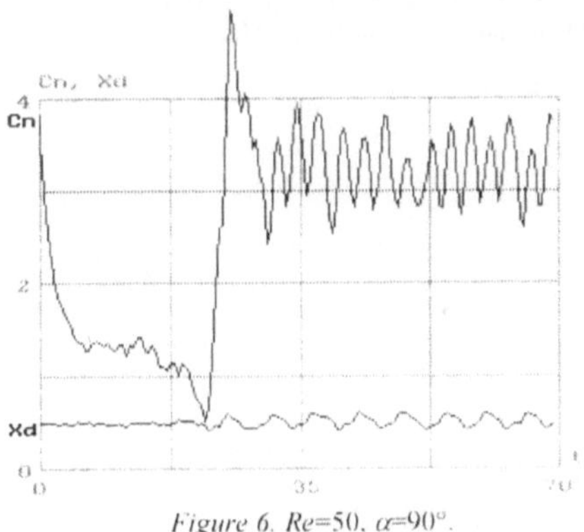

Figure 6. Re=50. α=90°.

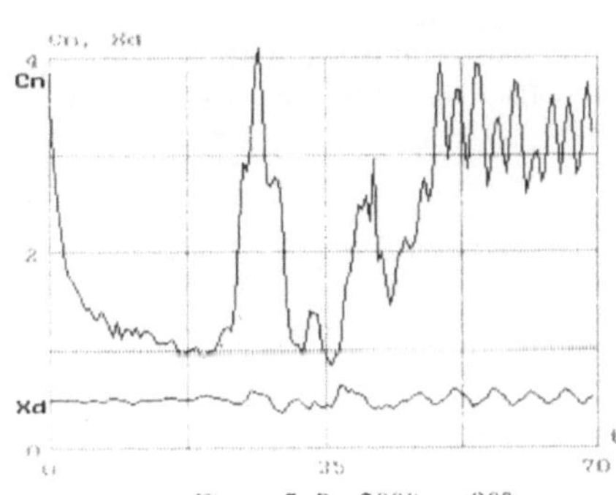

Figure 7. Re=2000. α=90°

Figures 6 and 7 show that transition regime for c_n (17<t<52) is qualitatively different at different values of Reynolds number. The periods of established at t>52 oscillatory regimes also depend on Reynolds number, although averaged over interval 52<t<70 values of c_n differ not much: $\bar{c}_n$ =3.27 at Re=50; $\bar{c}_n$ =3.2 at Re=2000.

6. Conclusion

A new concept of viscosity simulation in a point vortex system has been developed. The essential advantage of this method consists of excepting the influence of the scheme viscosity from physical processes. The process of the motion equations integration is complemented by the procedure for a local correction of coordinates and circulations of the point vortices. In result of the correction variations of energy and dispersion of vortical motion are brought to agreement with a given viscosity. The efficiency of the suggested approach is proved by the numerical simulation.

A similar approach can be developed for three-dimensional flows. The three-dimensional ideal flows have the following invariants: total vorticity, impulse, momentum, kinetic energy, and helicity. For viscous flows the first three values remain the same, but energy and helicity vary according to the known laws [5,8]. The continuos field of vorticity can be approximated by a set of short segments of vortex filaments. Each segment can be viewed as a limit of a system of closed vortex filaments. Then, an opportunity emerges to elaborate a numerical algorithm for treatment of viscosity, similar to the algorithm for two-dimensional flows.

7. References

1. Belotserkovsky S. M., Lifanov I. (1997). Method of Discrete Vortices. CRC Press, USA.
2. Leonard. (1985). *Annu. Rev. Fluid Mech.* 17, 523.
3. Chorin A. J. (1973). *J. Fluid Mech.*, 57, 785.
4. Winckelmans G. S., Leonard A. (1993). *J. Comp. Phys.*, 109, 247.
5. Batchelor G. K. (1970). An Introduction to Fluid Dynamics. Cambridge at the University Press.
6. Shavaliev M. Sh. (1989). In: Mechanics of Inhomogeneous and turbulent flows, Moscow, Nauka, p.63-69. [Russian]
7. Roshko A. (1953). NACA TN, N2913.
8. Moffat H. K. (1969). *J. Fluid Mech.*, 35, 117.

SIMULATION OF VORTEX RING INTERACTION

M. MEINKE[1], J. HOFHAUS[2] AND A. ABDELFATTAH[1]
[1] *Aerodynamisches Institut, RWTH-Aachen, Wüllnerstraße 5-7, 52062 Aachen, Germany*
[2] *BMW AG München*

Abstract. A numerical solution of the Navier-Stokes equations for incompressible flows is used to simulate the interaction of isolated vortex rings in unbounded domains. The algorithm is based on a pressure correction scheme, in which the Poisson equation for the pressure is solved with a conjugate gradient method with a preconditioning based on an incomplete lower-upper decomposition. The integration in time is carried out with an explicit Adams-Bashforth scheme on a non-staggered grid. The algorithm is efficiently implemented on vector-parallel computers. Hill's solution of the Euler equations for vortex rings is used as an initial condition of the velocity field. The time development of two vortex rings approaching each other under an angle of 40-90 degrees is simulated on a grid, moving with the propagation velocity of the vortex rings.

The generation and connection of vortex rings in a bounded domain is investigated with a solution of the Navier-Stokes equations for compressible flows in a cylinder of a piston engine. In this case an explicit time-stepping scheme with centrally discretized convective and viscous terms is applied on general curvilinear coordinates in block-structured moving grids. Two vortex rings are generated by the flow through the open intake valves, which connect to a single vortex at a crank angle of about 150°. Vortex lines are integrated to visualize the flow field of the time dependent solution.

1. Introduction

The dynamics of vortex rings has been extensively studied in the past, both in experiments and with numerical solutions, see [12, 9] for reviews of this topic. Vortex rings are very stable and relatively easy to generate, which allows a detailed study of the interaction of vortex structures. A numerical simulation with solutions of the Navier-Stokes equations is more difficult, since the flow field has to be

E. Krause and K. Gersten (eds.), IUTAM Symposium on Dynamics of Slender Vortices, 105-116.

resolved sufficiently during the propagation and the interaction of the rings. Kida, [6], applied a higher-order finite difference scheme for the numerical simulation of the oblique interaction of two rings with its merging and reseparation. Such schemes are well suited for problems in unbounded domains of simple geometry, but are more difficult to apply in block structured grids with general curvilinear coordinates. Therefore, standard second-order schemes are used here to study their applicability in such flow problems. Their advantage is the easy application also to problems in non-trivial geometries. The solution schemes considered are an explicit pressure correction scheme for incompressible and an explicit Runge-Kutta time stepping schemes for compressible flow problems. The first solution method is used for the simulation of the collision of two vortex rings in unbounded domains. The numerical simulation is carried out for different collision angles, vortex lines are integrated for the visualization of the flow.

The second algorithm is applied to simulate the flow in a 4-valve internal combustion engine. Understanding the unsteady, three-dimensional flow field in internal combustion engines during the intake and compression stroke is crucial for the development of engine design with high-performance and low emission values. The structure of the flow field, in particular the dynamics of the vortices generated during the intake stroke, determines the flame propagation rate in homogeneous charge spark-ignition engines, the fuel-air mixing and burning rates in Diesel engines. As will be shown later the development and merging of the vortex rings with each other and secondary vortex structures are the main flow phenomena during the intake and compression stroke.

2. Mathematical Model

In a dimensionless form the Navier-Stokes equations transformed in general, curvilinear coordinates ξ, η and ζ read:

$$\bar{A} \cdot \frac{\partial \vec{Q}}{\partial t} + \frac{\partial \vec{E}}{\partial \xi} + \frac{\partial \vec{F}}{\partial \eta} + \frac{\partial \vec{G}}{\partial \zeta} = 0 \qquad (1)$$

They describe the conservation of mass, momentum, and energy in unsteady, three-dimensional and viscous flow. For a gaseous compressible fluid, $\bar{A}$ is the identity matrix, and the vector of the conservative variables multiplied by the Jacobian of the coordinate transformation J is given by:

$$\vec{Q} = J\,(\rho, \rho\vec{u}, \rho e)^T$$

Here, ρ denotes the fluid's density, $\vec{u} = (u, v, w)^T$ the velocity vector, and e is the internal energy. The flux vectors $\vec{E}$, $\vec{F}$, and $\vec{G}$ are splitted in an advective and a dissipative part, e.g.: $\vec{E} = \vec{E}_A - \vec{E}_D$, with

$$\vec{E}_A = J \begin{pmatrix} \rho U \\ \rho U u + \xi_x p \\ \rho U v + \xi_y p \\ \rho U w + \xi_z p \\ U(\rho e + p) - \xi_t p \end{pmatrix} \quad \text{, and} \quad \vec{E}_D = \frac{J}{Re} \begin{pmatrix} 0 \\ \xi_x \sigma_{xx} + \xi_y \sigma_{xy} + \xi_z \sigma_{xz} \\ \xi_x \sigma_{xy} + \xi_y \sigma_{yy} + \xi_z \sigma_{yz} \\ \xi_x \sigma_{xz} + \xi_y \sigma_{yz} + \xi_z \sigma_{zz} \\ \xi_x E_{D_5} + \xi_y E_{D_5} + \xi_z E_{D_5} \end{pmatrix}$$

Herein, ξ_x, ξ_y, ξ_z are metric terms of the coordinate transformation, U the contravariant velocity $U = \xi_t + u\xi_x + v\xi_y + w\xi_z$, Re the Reynolds number and $\overline{\overline{\sigma}}$ the stress tensor. E_{D_5} is the dissipative part of the energy flux containing contributions of the stress tensor and the heat flux.

For incompressible flows with constant viscosity, the Navier-Stokes equations simplify significantly. The equation for energy conservation is decoupled from the equation for mass and momentum, and can be omitted, if the distribution of the fluid's temperature is not of interest. The vector of the conservative variables in Eq. (1) is then reduced to:

$$\vec{Q} = J\,(p, \vec{u})^T \quad .$$

The lack of a time-derivative for the pressure p in the continuity equation yields a singular matrix $\bar{A}$ and renders the integration of the governing equations more difficult. For fluids with constant density, the vectors of the advective and dissipative fluxes reduce to, e.g.:

$$\vec{E}_A = J \begin{pmatrix} U \\ Uu + \xi_x p \\ Uv + \xi_y p \\ Uw + \xi_z p \end{pmatrix} \quad \text{, and} \quad \vec{E}_D = \frac{J}{Re} \begin{pmatrix} 0 \\ g_1 u_\xi + g_2 u_\eta + g_3 u_\zeta \\ g_1 v_\xi + g_2 v_\eta + g_3 v_\zeta \\ g_1 w_\xi + g_2 w_\eta + g_3 w_\zeta \end{pmatrix} \quad ,$$

with:

$$g_1 = \xi_x^2 + \xi_y^2 + \xi_z^2 \quad , \quad g_2 = \xi_x \eta_x + \xi_y \eta_y + \xi_z \eta_z \quad , \quad g_3 = \xi_x \zeta_x + \xi_y \zeta_y + \xi_z \zeta_z \quad .$$

A block-structured moving grid is used for both the simulation of the interacting vortex rings and the flow in the piston engine. The grid movement induces a relative flow through the cell faces of the control volume, which must be taken into account with additional grid velocity terms in the governing equations. The cylinder and the valve movement requires a deformation of the grid cells. In order to retain a fully conservative formulation, the cell volume or the Jacobian of the metric terms J is recomputed within each time step with a Geometric Conservation Law (GCL), [13]

$$\frac{\partial J}{\partial t} + \frac{\partial (J\xi_t)}{\partial \xi} + \frac{\partial (J\eta_t)}{\partial \eta} + \frac{\partial (J\zeta_t)}{\partial \zeta} = 0 \quad , \tag{2}$$

Eq. (2) has the same form as the other conservation equations and the Jacobian J can therefore simply be added to the conservative variables $\boldsymbol{Q}$. The numerical solution of the GCL does not require boundary conditions, but an initial condition, which is obtained from the geometric definition of the cell volume.

3. Method of Solution

3.1. COMPRESSIBLE FLOWS

The equations for compressible flows are integrated with an explicit Runge-Kutta method. It was successfully applied to the solution of the Euler- and Navier-Stokes equations by Jameson, [5], and other authors. The coefficients in the Runge-Kutta steps are chosen as α_l=(0.25, 0.1667, 0.375, 0.5, 1), which are optimized for maximum stability for a central scheme. The accuracy in time is of the order $O(\Delta t^2)$. All spatial derivatives are discretized with central differences, artificial damping terms of fourth order were used to avoid high-frequency oscillations. The algorithm is parallelized by grid partitioning. More details can be found in [7, 1].

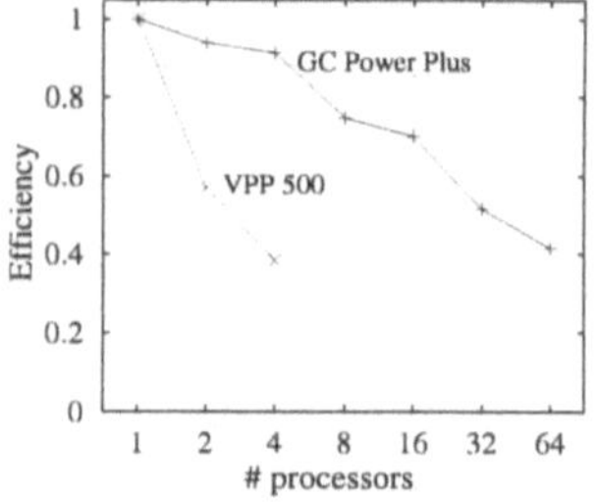

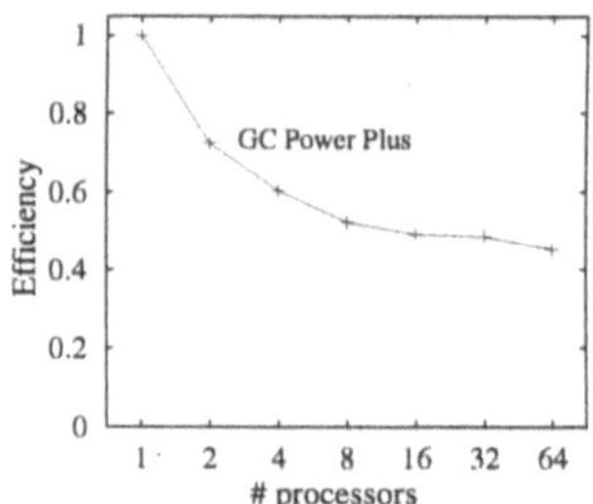

Figure 1. Efficiency of the parallelized, preconditioned BI-CGStab solver for a constant global problem size of 33×33×17 grid points (left) and a constant local problem size of 33×33×(5 × # processors) grid points.

3.2. INCOMPRESSIBLE FLOWS

An explicit Adams-Bashforth scheme is used for the integration in time for incompressible flows. The spatial derivatives are discretized with the QUICK interpolation for the convective terms and central differences of second order accuracy for all other derivatives. A divergence free velocity field is obtained in each time step by solving a Poisson equation for the pressure. The momentum interpolation of Rhie and Chow, [11], is used to avoid an odd-even decoupling of the pressure field. The Poisson equation is solved with a preconditioned Bi-CGStab method. This algorithm is also parallelized with grid partitioning. The parallel efficiency of the

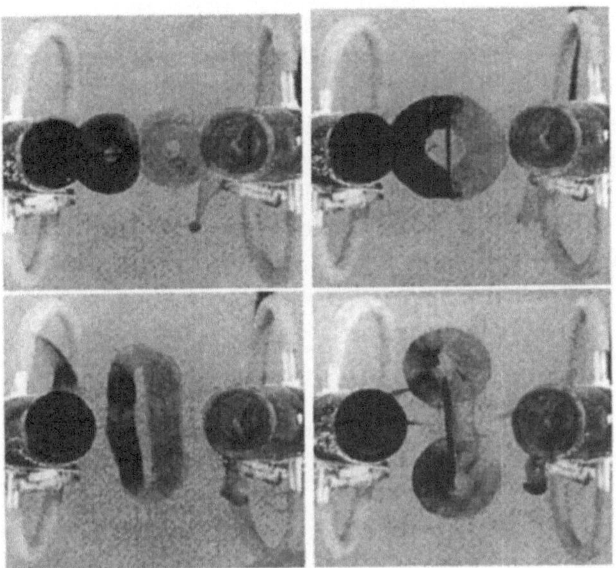

Figure 2. Experimental flow visualization taken from Lim of the interaction of two vortex rings for a Reynolds number of Re=500.

Poisson solver is shown in Fig. 1 for constant global and local problem size. More details of the algorithm and its parallelization can be found in [3, 4].

4. Results

4.1. INTERACTION OF VORTEX RINGS IN INCOMPRESSIBLE FLOW

The interaction of two colliding vortex rings is simulated in incompressible flow. The initial conditions for the velocity field for the ring vortices correspond to Hill's solution for non-viscous flows, see e. g. [8]. The initial position and orientation of the rings is chosen according to the experimental setup of Lim, [10], for the oblique collision of vortex rings. The two vortices propagate towards each other along two lines, which intersect under a certain angle Θ_c, which is varied in the range of $40^o < \Theta_c < 90^o$. The Reynolds number based on the radius of the vortex ring and its propagation velocity is 500. The numerical solution is obtained with the pressure correction scheme on a computational grid with 81×81×81 grid points on a 4 processor SNI/Fujitsu VPP300 computer. In order to keep the vortex rings within the domain of integration, the grid is moved with the propagation velocity of the interacting vortex structures.

In Fig. 2 the experimental flow visualization from [10] for an interaction angle of Θ_c=90° and in Fig. 3 the corresponding surfaces of constant pressure from the numerical solution are shown. The main characteristics of the interaction, the merging and separation, are correctly predicted by the numerical solution. Lim concluded from his experiments that helical vortex lines are generated during the merging of the rings. This is confirmed by the numerical simulation, the vortex lines integrated near the core of the vortex rings, shown in Fig. 4, also exhibit helical components. The vorticity of the initial vortex rings is not totally redistributed to the newly formed rings after the separation. From the numerical simulation it can be seen that vortex structures with closed vortex lines are formed in the leeward side of the rings, see Fig. 4 and Fig. 5.

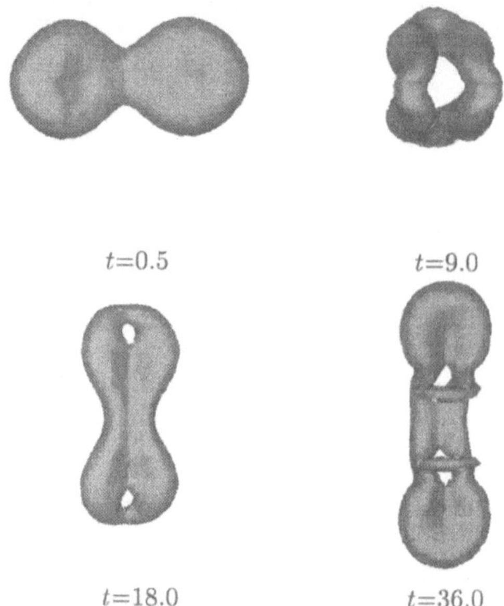

Figure 3. Surface of constant pressure from the numerical simulation of the interaction of two vortex rings for different dimensionless time levels t, a collision angle of Θ_c=90° and a Reynolds number of Re=500.

The separation of vortex rings occurs only if the collision angle is larger than a certain threshold value. In Fig. 6 vortex lines are shown after the ring interaction for different collision angles Θ_c. It can be seen that a full separation of rings occurs only if the angle is larger than about 80°. In the experiment of Fohl and Turner, [2], a reseparation of rings occurred for collision angles larger than 32°. The Reynolds number in their setup was 2000, four times higher than in the present investigation. Therefore, the critical collision angle seems to be strongly dependent on viscosity effects.

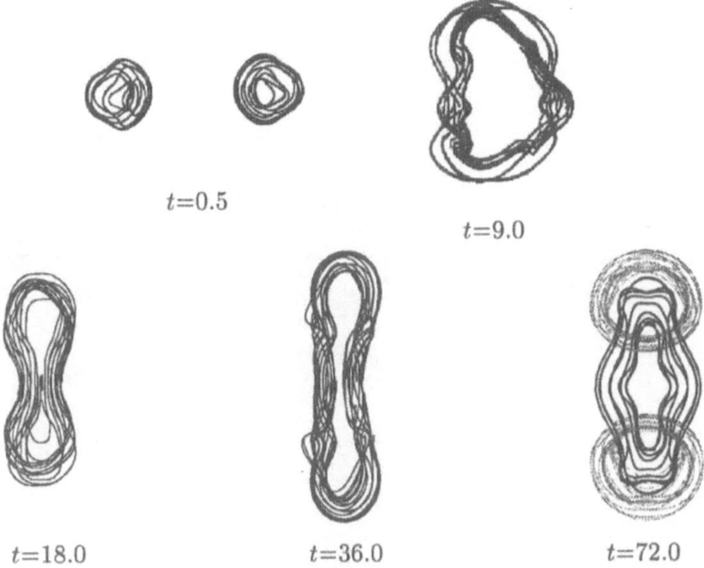

Figure 4. Vortex lines from the numerical simulation of the interaction of two vortex rings for different dimensionless time levels t, a collision angle of $\Theta_c=90^o$ and a Reynolds number of Re=500. The vortex lines are integrated near the vortex cores.

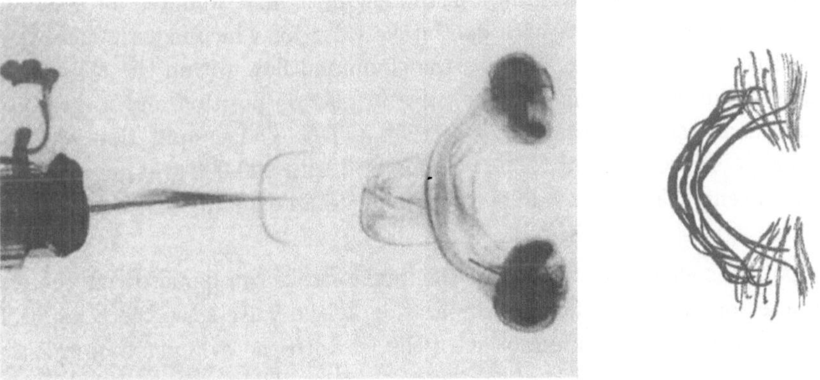

Figure 5. Experimental flow visualization taken from Lim (left) and vortex lines from the numerical simulation (right) of the interaction of two vortex rings for a collision angle of $\Theta_c=90^o$, a Reynolds number of Re=500 and a dimensionless time of t=72.0.

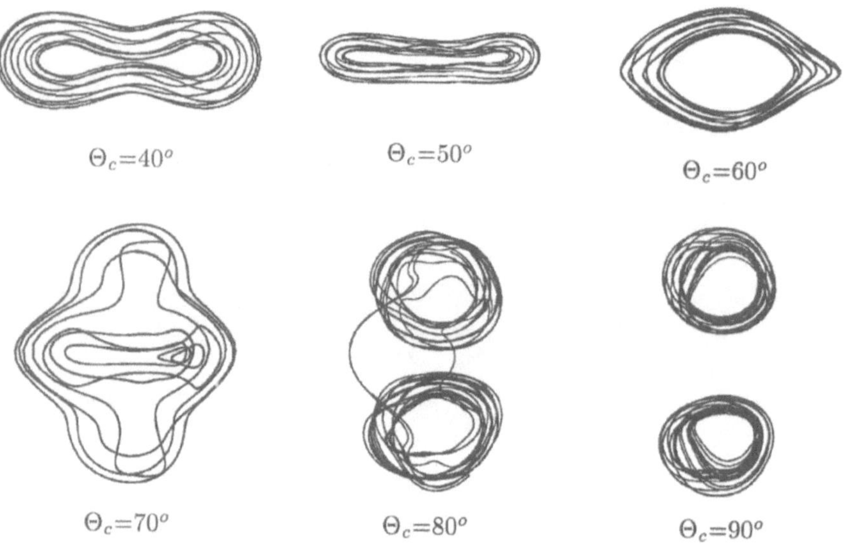

Figure 6. Vortex lines from the numerical simulation of the interaction of two vortex rings for different collision angles Θ_c at a Reynolds number of Re=500 and a dimensionless time level of t=72.0.

4.2. INTERACTION OF VORTEX RINGS IN A PISTON ENGINE

The algorithm for the simulation in a piston engine is first validated for a simplified geometry, a model engine with one intake valve for which experimental flow visualization is available. Fig. 7 shows the computed flow within the cylinder of the model engine with a single intake-valve in off-axis position and a disc-type combustion chamber at a crank angle of 125° ATDC. The computation was carried out for an engine speed of 2000 rpm on a grid with 600.000 grid points at the bottom dead center. The flow field is compared with an experimental visualization using a laser-light sheet method.

Flow patterns which are typical for the intake stroke can be observed: the jet-like character of the intake flow, its separation at the valve seat, the generation of a large scale vortex, the interaction of the jet with the cylinder wall and the moving piston and the creation of secondary vortices. The ring vortex below the intake valve is not symmetric with respect to the valve axis. It is displaced towards the center of the cylinder and its axis is inclined towards the vertical axis of the cylinder. In addition, a vortex can be observed near the piston crown in the experiment and in the numerical solution. The agreement of the flow field cannot be perfect, because of cycle to cycle variations, the main features of the flow are well predicted with the numerical solution.

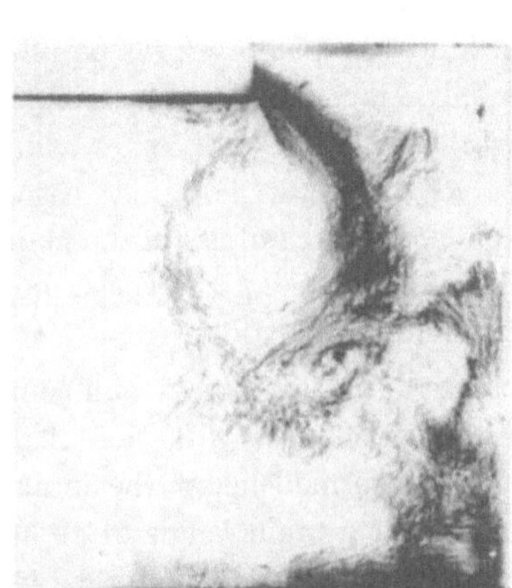

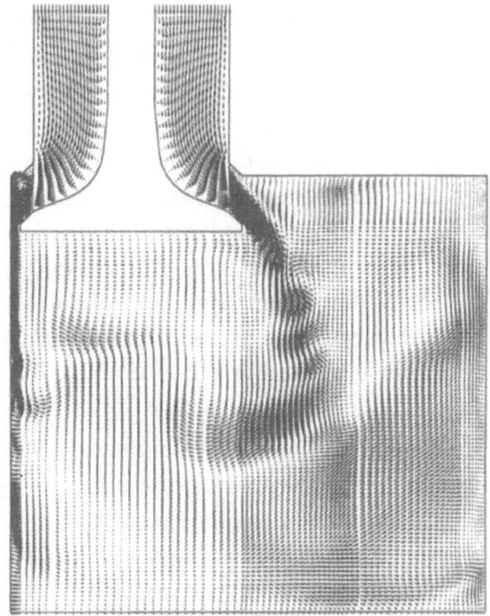

Figure 7. Comparison of the flow in a model engine with eccentric valve at a crank angle of 125° ATDC. Experimental flow visualization (left) in comparison with velocity vectors of the numerical solution (right).

For the simulation of the flow in a 4-valve engine with a pentroof combustion chamber two grids with approx. 400.000 and 2 million grid points were generated. The surface grid at the top dead center is shown in Fig. 8. For a computational grid with a maximum of 2 million grid points the CPU-time amounts to 120 hours.

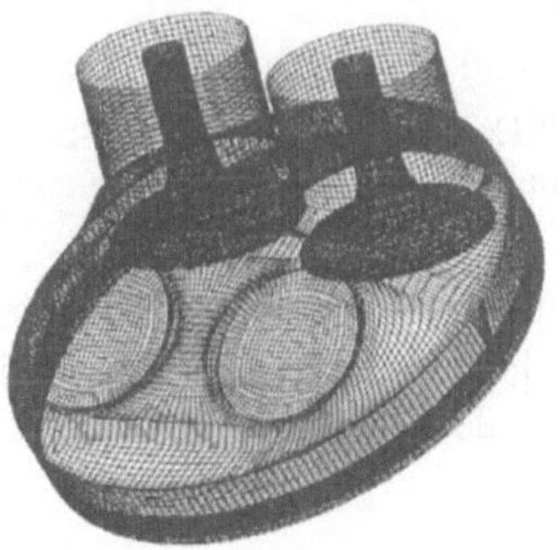

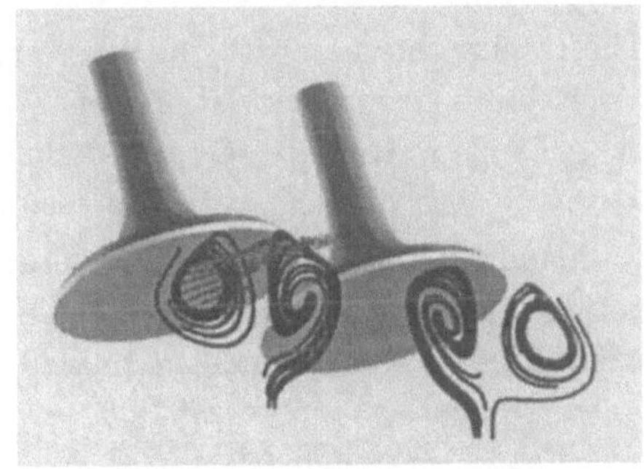

Figure 8. Computational grid (left) for the simulation of the flow in a 4-valve piston engine. Integration of vortex lines (right) are started on closed streamlines of a vortex.

The simulation is carried out for 4000 rpm, which corresponds to a Reynolds number of 220.000 based on the maximum piston speed and the diameter of the cylinder. No turbulence model is applied, since models based on the Reynolds averaged Navier-Stokes equations are not very well suited for the prediction of such flows. This topic is further discussed in citemeinke97.2. During the valve and piston movement grid lines are inserted and removed so that the resolution remained almost constant during the simulation. Since the piston moves parallel to most of all grid lines within the cylinder, a linear interpolation of variables between two grid points could be used for most of all grid blocks. For larger crank angles the flow field inside the cylinder becomes very complex. For that reason vortex lines were integrated from starting points on closed streamlines of the main vortices as indicated in Fig. 8. The number of integration steps for the vortex line were increased until the shape of the vortex lines remained constant.

The vortex lines of the dominating flow structures during the intake and compression stroke are presented in Fig. 9 for crank angles of 60°, 120°, 180° and 320° ATDC. During the intake stroke two vortex rings are formed below the intake valves. They move downward with the piston. At about a crank angle of about 100° ATDC two other vortices develop through the interaction of the intake jet with the cylinder walls and the piston. Their shape is influenced by the velocity of the intake jet, which is largest in the middle of the cylinder and decreases towards the cylinder walls. These vortices merge with the ring vortices at a crank angle of about 150° ATDC. The resulting vortex structure is a closed vortex tube that spreads out over the whole volume of the cylinder, partly stretched towards the piston and fixed by the weak intake jet near the cylinder head. It interacts with the piston crown resulting in secondary vortices in the lower part of the cylinder, not shown here. During the compression stroke the lower part of the structure is moved towards the cylinder head. The vortex structure seems to withstand during the complete compression stroke. These results clearly indicate that vortex formation, merging and bending taking place during the intake stroke, must have a marked influence on the subsequent combustion.

The results obtained with the coarse mesh and fine mesh agree qualitatively very well, but a stronger deviation from symmetry is observed for the higher resolution. All vortex structures described above could be identified in both solutions, but partially with different vortex intensities.

It should be noted that the flow visualization presented here does not contain other information than the vortex line of the major vortical structure in the cylinder. Other flow structures are present in the flow, but are omitted for clarity.

5. Summary

Solutions of vortex ring interactions have been presented for two interacting vortex rings in unbounded and bounded domains. The simulation of the interaction of

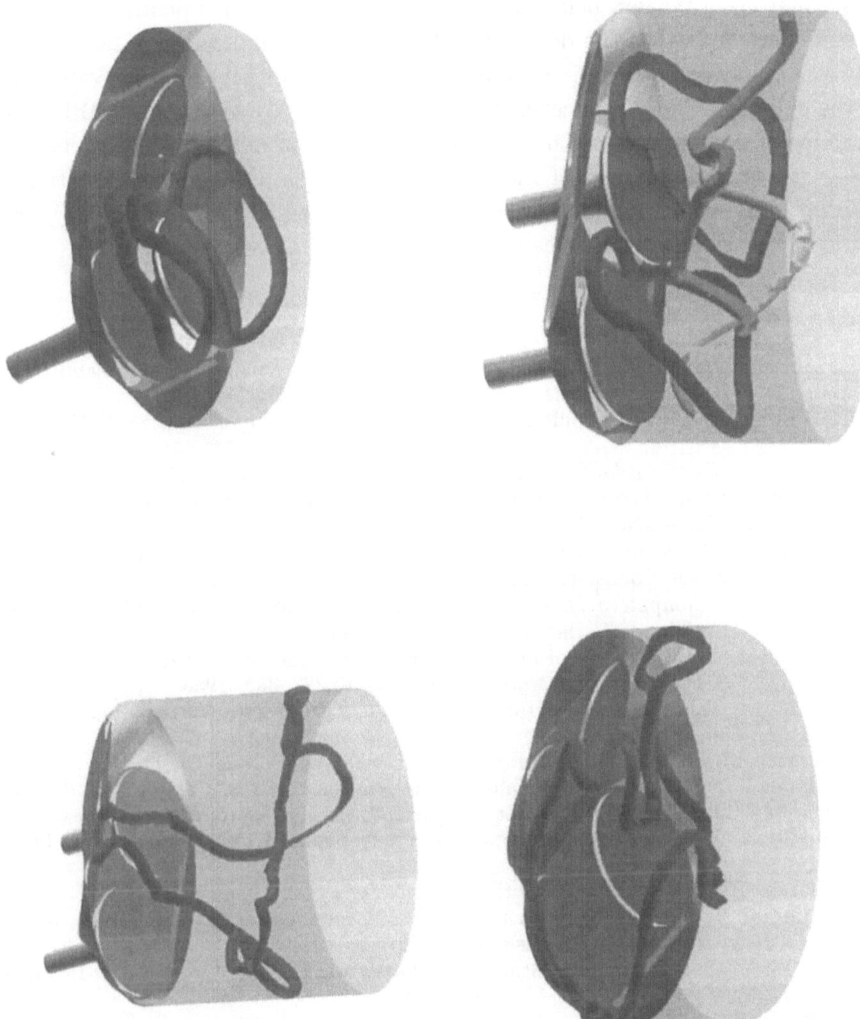

Figure 9. Vortex lines during the intake and compression stroke of a 4-valve engine. Crank angle of 60° (top left), 120° (top right), 180° (bottom left) and 320° (bottom right).

two vortex rings colliding under an angle of 90° reproduced the connection and separation of the rings in good qualitative agreement with the experimental flow visualization. A reseparation of vortex rings occurrs only for collision angles larger than 80° at a Reynolds number of 500.

The analysis of the flow field in a 4-valve piston engine with a pentroof combustion chamber shows that the dominating flow structures during the intake stroke are two ring vortices, which are generated by the jet issuing through the valve gap into the cylinder. These vortices merge with two other vortices generated by the interaction of the intake jet with the cylinder and piston walls. This flow structure remains stable until the end of the compression stroke. Grid refinement studies show that grids with 2 million points do not resolve the flow field sufficiently, but that the large scale of the flow can be predicted qualitatively correctly.

References

1. A. Abdelfattah. *Numerische Simulation von Strömungen in 2- und 4-Ventil Motoren.* Dissertation, Aerodynamisches Institut, RWTH Aachen, 1997.
2. T. Fohl and J. Turner. Colliding Vortex Rings. *Physics of Fluids*, 18(4):433–436, 1975.
3. J. Hofhaus. *Numerische Simulation reibungsbehafteter, instationärer, inkompressibler Strömungen – Vergleich zweier Lösungsansätze.* Dissertation, Aerodynamisches Institut, RWTH Aachen, 1997.
4. J. Hofhaus, M. Meinke, and E. Krause. Parallelization of Solution Schemes for the Navier-Stokes Equations. In E. Hirschel, editor, *Flow Simulation with High-Performance Computers II*, volume 52 of *Notes on Numerical Fluid Mechanics*, pages 102–116. Vieweg Verlag, Braunschweig, 1996.
5. A. Jameson. Solution of the Euler equations for two-dimensional transonic flow by a multigrid method. *Applied Math. and Comp.*, 13:327–355, 1983.
6. S. Kida, M. Takaoka, and F. Hussain. Reconnection of Two Vortex Rings. *Phys. Fluids*, A1:630–632, 1989.
7. E. Krause, M. Meinke, and J. Hofhaus. Experience with Parallel Computing in Fluid Mechanics. In S. Wagner, J. Périaux, and E. Hirschel, editors, *Computational Fluid Dynamics '94*, pages 87–95. Invited Lecture at the 2nd European Computational Fluid Dynamics Conference, Stuttgart, Germany, Sep. 5–8, John Wiley & Sons Ltd., 1994.
8. H. Lamb. *Hydrodynamics.* Cambridge University Press, 1932.
9. T. Lim and T. Nickels. Vortex rings. In S. I. Green, editor, *Fluid Vortices*, pages 95–147. Kluwer Academic Publishing, 1995.
10. T. T. Lim. An Experimental Study of a Vortex Ring Interacting with an Inclined Wall. *Experiments in Fluids*, 7:453–463, 1989.
11. C. Rhie and W. Chow. Numerical Study of the Turbulent Flow past an Airfoil with Trailing Edge Seperation. *AIAA Journal*, 21(11):1525–1532, 1983.
12. K. Shariff and A. Leonard. Vortex Rings. *Ann. Rev. Fluid. Mech.*, 24:235–279, 1992.
13. P. D. Thomas and C. K. Lombard. Geometric Conservation Law and Its Application to Flow Computations on Moving Grids. *AIAA J.*, 17(10):1030–1037, Oct. 1979.

LARGE-EDDY SIMULATIONS OF LONGITUDINAL VORTICES IN SHEAR FLOWS

PIERRE COMTE
LEGI/IMG, BP 53, F38041 Grenoble cedex 9, FRANCE
e-mail: Pierre.Comte@hmg.inpg.fr

AND

MARCEL LESIEUR
LEGI/IMG, and also Institut Universitaire de France
e-mail: Marcel.Lesieur@hmg.inpg.fr

Abstract.

The formation of three-dimensional vortices in mixing layers and boundary layers is investigated in Large-Eddy Simulation, with the aid of the subgrid-scale models presented in (Lesieur and Métais, 1996, *Ann. Rev. Fluid Mech.*, **28**, and Lesieur, 1997, *Turbulence in fluids*, third updated and revised edition). In mixing layers perturbed upstream or initially by small-amplitude random perturbations, two types of flow patterns are obtained depending on the level of three-dimensionality of the perturbations and the spanwise size of the domain. For forcing amplitudes of the order of 1% in turbulent intensity, the natural tendency is the emergence of oblique subharmonic modes yielding the formation of highly-three-dimensional vortex lattices, as in the experiments of Chandrsuda, Mehta, Weir and Bradshaw, 1978, *J. Fluid Mech.* **85**). This trend can be partially inhibitted by taking narrower domains, and sometimes just by making the perturbations more two-dimensional. In this case, a more "canonical" flow pattern is obtained, with quasi-two-dimensional billows undergoing successive pairings while stretching, in between each other, streamwise vortices that form *in a succession of stages involving local roll-up and pairing*, as conjectured by Lin and Corcos (1984, *J. Fluid Mech.*, **141**). The LES of the transition of a spatially-growing boundary layer is also shown, featuring the almost-simultaneous emergence of staggered Λ-vortices and streamwise streaks. The connection between these results and different stability analyses is discussed.

E. Krause and K. Gersten (eds.), IUTAM Symposium on Dynamics of Slender Vortices, 117-131.

1. A Rapid Overview of our Models

The subgrid-scale models used in our team derive from the Spectral-Cusp eddy-viscosity model proposed by Métais and Lesieur (1992) for isotropic turbulence (see also Lesieur and Métais, 1996, for a review). Assuming spectra $E(k) \propto k^{-m}$ for all k, the EDQNM theory (with the non-local interaction terms) yields (real) spectral eddy-viscosity and eddy-diffusivity coefficients which read

$$\nu_t(k,t) = \begin{cases} 0.31 \, \frac{5-m}{m+1} \, \sqrt{3-m} \;\; C_K^{-3/2} \; \nu_t^*(k/k_c) \sqrt{\frac{E(k_c,t)}{k_c}} & \text{for } m < 3, \\ 0 & \text{for } m \geq 3, \end{cases} \tag{1}$$

$$\kappa_t(k,t) = \nu_t(k,t)/Pr_t \qquad \text{with} \qquad Pr_t = 0.6 \quad . \tag{2}$$

C_K denotes Kolmogorov's constant, and $\nu_t^* = 1$ far away from the cut-off wavenumber ($k/k_c \leq\simeq 0.3$). It rises for higher k/k_c (cusp), a good fit of it is (at least in the case $m = 5/3$),

$$\nu_t^*(k/k_c) = 1 + 34.5 \exp[-3.03 \; k_c/k] \quad . \tag{3}$$

In the case of plane turbulent channel flows (with and without rotation) Lamballais (1995) obtained correct statistics with this model and a dynamic determination of $m(y,t)$ obtained by computing, every timestep, energy spectra in planes parallel to the wall and evaluating their slopes between k_c and $k_c/2$ by means of least-square fits. The temporal results of mixing layers shown in the next section have been obtained with this Spectral-Dynamic model.

When spectral methods cannot be used, we strive to determine eddy-viscosities out of a measure of the kinetic energy at the smallest resolved scale $\Delta = \pi/k_c$. One of these local spectra is $F_{2_\Delta}(\boldsymbol{x},t)$, the second-order structure function of the resolved velocity field, evaluated by averaging over the closest neighbours of point $\boldsymbol{x}$, either in all 3 directions of space (6-neighbour formulation) or on planes normal to the wall or mean shear (4-neighbour formulation). In the case of infinite Kolmogorov spectra, energy-conservation arguments (Leslie & Quarini, 1979) yield the *Structure-Function* model (Métais and Lesieur, 1992), defined by

$$\nu_t^{SF}(\vec{x},t) = 0.105 \, C_K^{-3/2} \, \Delta \, \sqrt{F_{2_\Delta}(\vec{x},t)} \quad , \tag{4}$$

consistent with the spectral model (2) without cusp.

This SF model appears to be slightly less dissipative than the Smagorinsky model with the constant 0.18 given by the same assumptions (infinite

Kolmogorov cascade, see e.g. Comte *et al.*, 1994). As it involves velocity increments instead of derivatives, it also has the advantage of being defined independently of the numerical scheme used. It is nevertheless not much better for transition than the Smagorinsky model: low-wavenumber velocity fluctuations corresponding to unstable modes yield ν_t's large enough to affect the growth rate of weak unstabilities like Tollmien-Schlichting waves. So far, we have found two ways of remedying this:

- apply a high-pass filter onto the resolved velocity field before computing its structure function. With a triply-iterated second-order finite-difference Laplacian filter denoted $\tilde{}$, one finds $\tilde{E}(k)/E(k) \approx 40^3\ (k/k_c)^9$ for all k, almost independently of the velocity field and resolution. With the same arguments as for the structure-function model, this yields the *Filtered Structure-Function model*, proposed by Ducros *et al.* (1996) in the form

$$\nu_t^{FSF}(\boldsymbol{x},t) = 0.0014\, C_K^{-3/2}\ \Delta\ \sqrt{\tilde{F}_{2_\Delta}(\boldsymbol{x},t)} \quad . \tag{5}$$

This model enabled them to perform the LES of a spatially-growing boundary layer, results of which will be summerized in section 2.

- switch the original structure-function model off when the flow is not three-dimensional enough in the small scales (suggested by Bartello, 1993; implemented by David, 1993). In practice, an average vorticity vector $\overline{\omega}(\boldsymbol{x},t)$ is computed over $\boldsymbol{x}$ and its (4 or 6) closest neighbours. The Structure-Function model (4) is applied only if the magnitude of the angle $\alpha = (\omega(\boldsymbol{x},t), \overline{\omega}(\boldsymbol{x},t))$ exceeds a certain threshold α_0. Simulations[1] of incompressible isotropic turbulence at resolutions ranging between 32^3 and 64^3 gave pdf's of $|\alpha|$ peaking around 20°. Having found the choice of α_0 not critical between 10 and 45°, we finally retained $\alpha_0 = 20°$. The model's constant was finally multiplied by 1.56, a least-square fit between our test-simulations yielding the average dissipation over the domain closest to the values given by the SF-model. Dispersion was found small enough to justify this in first approximation, but a lot of work has yet to be done to reduce the arbirtrariness in this model. In any case, the most surprising conclusion about the filtered and selective structure-function models (hereafter FSF and SSF, respectively) is that they can be interchanged without much difference in the results (Comte *et al.*, 1994). This comes from the fact that they both considerably shrink the support of ν_t (with respect to that of the original SF model), and that both supports are almost the same (Figure 1, middle and right plots). In any case, they do not react to Λ-vortices, whereas the SF model does (left plot in Figure 1).

[1]LES with the original Structure-Function model

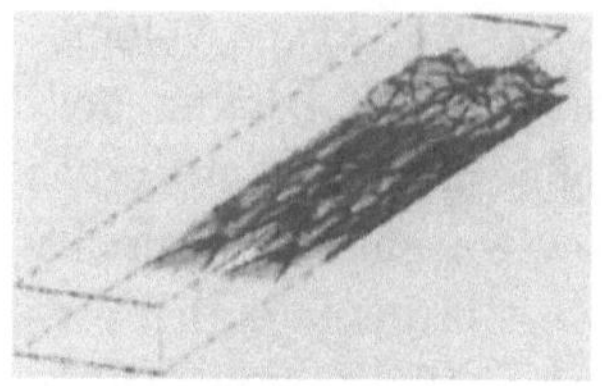
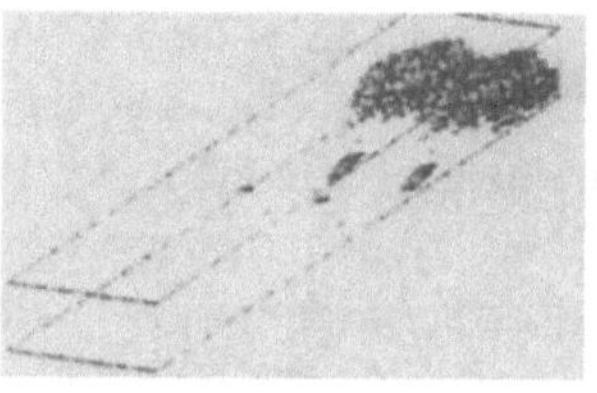
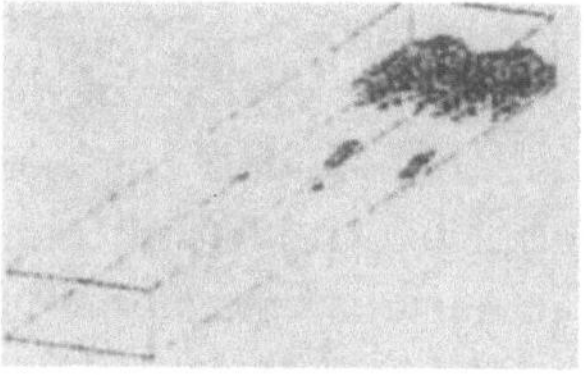

Figure 1. From left to right: isosurfaces $\nu_t = 2/3\ \nu$ given by the SF, FSF and SFS models, respectively, in the transitional portion of the spatially-growing boundary layer presented further on. The same velocity field was used for the three plots (a priori test).

2. Transitional Mixing Layers and Boundary Layers

After more than twenty years of investigations, it seems that the most commonly accepted vision of the "canonical plane mixing layer" has consisted so far of essentially spanwise Kelvin-Helmholtz billows undergoing successive quasi-two-dimensional pairings and stretching "secondary" counter-rotating streamwise vortices in between each other (Bernal & Roshko, 1986), along the principal axis of the strain field exerted by the billows which corresponds to its largest (positive) eigenvalue, as proposed almost simulteaneously by Lin and Corcos (1984) and Neu (1984). On the other hand, a completely three-dimensional vortex structure can be obtained, either "naturally",[2] as were the "helical pairings" obtained experimentally by Chandrsuda *et al.* (1978) who coined the expression, or with deterministic three-dimensional forcing as in Lasheras and Choi (1988) who used corrugated or indented splitter plates, or Nygaard and Glezer (1991; 1994) who scattered a smooth splitter plate with surface film heaters actuated by minute alternative currents. In unforced experiments, the spanwise correlation distance falls to 20% for a spanwise separation of about 6 times the local vorticity thickness δ (Browand and Ho, 1983), due to dislocations that strongly resemble helical pairings. The following quote is from Mallier and Maslowe (1994):

> *Whereas pairing was long believed to be essentially a two-dimensional process, the more recent studies, as discussed by Dallard and Browand (1993), indicate that pairing is three-dimensional. This seems to be true for both laminar and turbulent mixing layers whether or not the flow is forced.*

If we nevertheless assume that three dimensionality is a second-order effect with respect to the straight (2D) Kelvin-Helmholtz instability, the Floquet formalism can be applied to an analytical model of stready Kelvin-Helmholtz vortices like the Stuart vortices. This was done by Pierrehumbert and Widnall (1982) who proposed two candidates of comparable growth

[2] i.e. in a somewhat uncontrolled manner

rates: a couple of conjugate subharmonic modes yielding helical pairings as in Chandrsuda *et al.* (1978), and a streamwise-independent mode, referred to as *translative instability*, making the Stuart billows oscillate in phase in the spanwise direction. As a consequence of Squire's theorem in its inviscid limit, the growth rate of the oblique subharmonic instability is maximal at zero spanwise wavenumber k_z and decreases rapidly as k_z increases, a behaviour reminiscent of the instabilities subjected to the semi-circular criterion. In contrast, the growth rate of the translative instability is zero in two dimensions, reaches a maximum for $k_z = (3/2)k_x$, which matches the spanwise spacing of the secondary vortices in Bernal and Roshko (1986), and decreases very slowly for larger k_z. Pierrehumbert and Widnall (1982) also mentioned another, but less amplified, streamwise-independent mode causing bulging of the billows, and reminiscent of the Kelvin waves which grow on the "worms" of the isotropic turbulence. This mode corresponds to a case of Core Dynamics Instability, a class of instabilities proposed by Schoppa *et al.* (1995) as the main cause of the *small-scale transition* first reported by Konrad (1976). In any case, since the experimental results of Bernal and Roshko (1986), the popularity of the Floquet-type secondary instability mechanisms has extended from wall-bounded flows (Orszag and Patera, 1983, Herbert, 1988) to free-shear inflectional instabilities as well, although the assumption of a steady two-dimensional primary mode of finite (but small) amplitude is more difficult to justify.

However, other mechanisms have been proposed, in particular those in terms of non-linear triads (Kelly, 1967, Mallier and Maslowe, 1994), for which there is an evident analogy with the formalism of isotropic turbulence in the Fourier space: calling $\boldsymbol{p} = \pm k_a \boldsymbol{e_x}$ the most amplified Kelvin-Helmholz mode, the most obvious $(\boldsymbol{k}, \boldsymbol{p}, \boldsymbol{q})$ triad that might explain helical pairings is $\boldsymbol{k} = -\boldsymbol{p}/2 \pm k_b.\boldsymbol{e_z}$ and $\boldsymbol{q} = -\boldsymbol{p}/2 \mp k_b.\boldsymbol{e_z}$, in which $k_b.\boldsymbol{e_z}$ is an arbitrary spanwise wavevector. Mallier and Maslowe (1994) found resonance[3] for inclination $atan(k_b/k_a) \pm 60°$, yiedling faster than exponential growth of the oblique waves. This mechanism is analogous to the subharmonic resonant triads that Craik (1971) proposed to explain transition in boundary layers. This is also to be the case for non-resonant models (Benney and Lin, 1960, Herbert and Morkovin, 1980), that came out of fashion to the benefit of the Floquet analysis, although they provide explanations for the formation of streaks and streamwise vortices which are, at least qualitatively, compatible with the numerical observations presented below.

Coming back to mixing layers, let us finally mention the spatial analysis by Monkewitz (1988), in which slightly oblique subharmonic modes resonate,

[3] *i.e.* the same phase speed for the interacting waves, which brings the condition $c(\boldsymbol{k}) = c(\boldsymbol{p}+\boldsymbol{q}) = c(\boldsymbol{p}) + c(\boldsymbol{q})$ where $c(\boldsymbol{k}) = c(-\boldsymbol{k})$ denotes the phase speed of wavevector $\boldsymbol{k}$ given by the dispersion relation.

with preferential $k_b = 0.35(\delta_i/2)^{-1}$, where $\delta_i = \Delta U/\max(d\bar{u}/dy)$ denotes the vorticity thickness of the basic hyperbolic-tangent velocity profile $\bar{u}(y)$.

2.1. INCOMPRESSIBLE MIXING LAYERS

In three-dimensional numerical simulations of mixing layers, it has been customary since Metcalfe *et al.* (1987) to perturb such a profile with deterministic perturbations corresponding to its most unstable eigenmodes. In our team in Grenoble, we find it complementary to use stochastic perturbations and inject small-amplitude noise, as colourless as possible, expecting to see the "natural selection" of the dominant instabilities. In streamwise (and spanwise) periodic Large-eddy Simulations at zero molecular viscosity with the spectral cusp model, Comte *et al.* (1989) thus obtained the emergence, out of Gaussian isotropic random perturbations of energy $10^{-4}\ U^2$ in which U denotes half the velocity difference ΔU, of two quasi-two-dimensional Kelvin-Helmholtz billows and a set of streamwise vortices stretched in between, all that strongly resembling the "canonical mixing layer" of Bernal & Rosho (1986). The peak values of the velocity fluctuations were in acceptable agreement with the measurements of Browand & Latigo (1979). After pairing, the spectra displayed a power-law subrange near the cut-off with a slope between -2 and $-5/3$. Taking the initial vorticity thickness δ_i as unit length, the dimensions of the domain were 14 δ_i in the streamwise direction (twice the fundamental wavelength λ_0), and 7 δ_i in the other two directions, for a resolution $64 \times 32 \times 32$ collocation points. Repetition of this simulation in a domain twice as large in the spanwise direction (with twice as many collocation points) yielded the formation of four very distorted vortices undergoing helical pairings as they formed. Modal analysis confirmed the emergence, out of the initial noise, of modes $(1, \pm 1)$ instead of the expected fundamental mode $(2, 0)$ and its straight sub-harmonic $(1, 0)$. This simulation was repeated in Direct Numerical Simulation at initial Reynolds number $Re = U\delta_i/\nu = 100$, in a $(28\ \delta_i)^3$ domain at resolution 128^3. It showed the emergence of modes $(2, \pm 1)$, yielding again an oblique vortex lattice. With a quasi-two-dimensional perturbation (energy $10^{-4}\ U^2$ in two dimensions, and $10^{-5}\ U^2$ in three dimensions), a more canonical pattern was obtained. These DNS results were published in Comte *et al.* (1992), henceforth referred to as CLL92, which might be the first numerical evidence of the "natural" emergence of the three-dimensional character, not only of the pairing instability, but also of the Kelvin-Helmholtz instability, in apparent contradiction with Squire's theorem.

All these calculations have been confirmed by a systematical investigation carried out by Silvestrini (1996), in DNS and in LES with different subgrid-scale models, resolutions and molecular Reynolds numbers (up to infinity). This piece of work confirms that the three-dimensional character mentioned above emerges naturally when the initial or upstream perturbations are sufficiently three-dimensional and the domain large enough in its spanwise dimension L_z. The left plot of Figure 2, quoted from Silvestrini (1996), shows an example of vortex lattice obtained after helical pairings, in a calculation performed at $Re = U\delta_i/\nu = 2000$ with the Spectral Dynamic model, with the same initial conditions as in the *helical* case of CLL92 (Gaussian isotropic noise of energy $10^{-4}\,U^2$). The statistical data concerning velocity, rms velocity fluctuations and Reynolds stresses, are in very good agreement with the unforced experiments of Browand and Latigo (1979) The simulation with a quasi two-dimensional forcing shown in the right plot of Figure 2 is not as good from this standpoint.

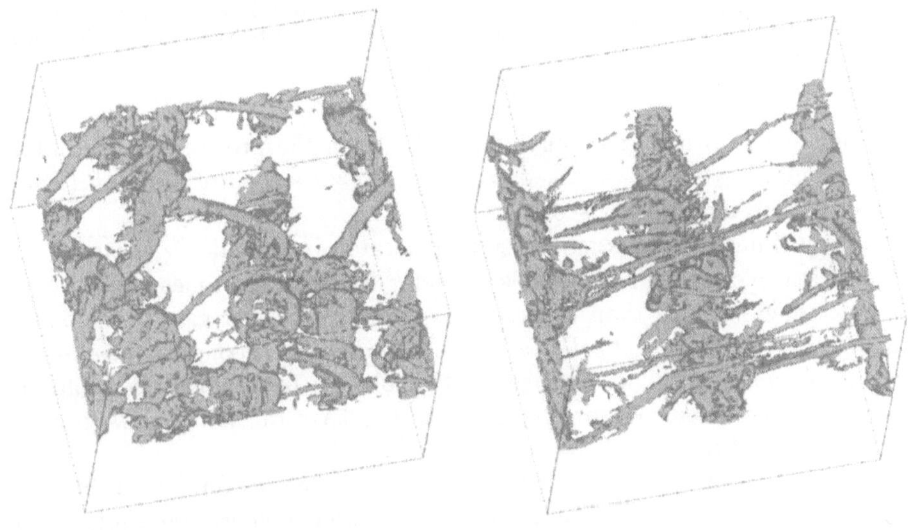

Figure 2. Vorticity magnitude in two LES of a temporal mixing layer differing only in the initial forcing: 3D on the left, quasi-2D on the right.

More realistic spatially-growing simulations have been performed by substituting pseudo-spectral schemes with compact finite-difference schemes in the streamwise direction, and prescribing an Orlansky-type outflow boundary condition, with 3rd order low-storage Runke-Kutta time stepping. Periodicity and free-slip conditions are still prescribed in the spanwise (z) and transverse (y) directions, respectively. The code, developed by Gonze (1993), has recently been parallelized, by means of slice/pencil domain decomposition and transposition. The Fourier transforms are now performed in parallel on yz slices. The domain is then reshaped and decomposed into streamwise "pencils", so that the linear systems brought about by the hermitian scheme

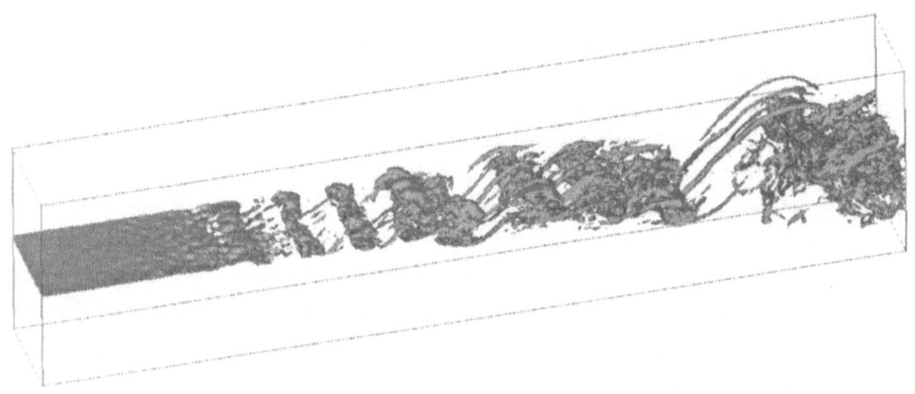

Figure 3. LES of an incompressible mixing layer forced upstream by a quasi two-dimensional random perturbation, with $L_z = 2\ \lambda_a$; the vorticity magnitude is shown at a threshold $(2/3)\Delta U/\delta_i$.

can be solved without communications between the processors. The following results have been obtained with the *filtered structure function* (FSF) model at zero molecular viscosity.

Every timestep, Gaussian noise is generated in the inlet section $x = 0$, either isotropically between modes k_y and k_z, or with ten times more energy in the modes $k_z = 0$ than in the other modes. The first case corresponds approximately to the *helical case* of CLL92, and the second to their *quasi-two-dimensional* case. In both cases, convolution by a Gaussian profile is applied to the three components, so that they have the same *r.m.s.* profile, of width δ_i and peak value $\varepsilon = \mathcal{O}(10^{-2})$ at $y = 0$.

Two values of L_z, the spanwise extension of the domain, have been considerered so far. With $L_z = 2\ \lambda_0 = 14\ \delta_i$ and a *quasi-two-dimensional* upstream forcing, intense longitudinal hairpins stretched between quasi 2D Kelvin-Helmholtz vortices are found again (Figure 3).

An interesting feature found is that longitudinal vortices of same sign may come close together and merge, contributing thus to the global self-similarity of the mixing layer. More precisely, instead of a single vortex tube going back and forth in between the rollers, as conjectured by Bernal and Roshko (1986), one can see series of co-rotating streamwise vortices which converge and merge. It seems that it is only after merging that they create visible depressions. This behaviour was predicted by Lin and Corcos (1984), who simulated the roll-up of a vortex layer strained in the direction of its vorticity. One of their conclusions was:

> *...In a layer where the sign of the vorticity alternates (in the direction along which strain is absent), each portion of the layer that contains*

vorticity of a given sign eventually contributes that vorticity to a single vortex. This may occur in a single stage if the initial layer thickness is not excessively small next to the spanwise extent of vorticity of a given sign or, otherwise, in a succession of stages involving local roll-up and pairing.

Note that low-Reynolds number DNS with the same domain dimensions and upstream forcing show hardly any roll-up of streamwise vortices, and always in a single stage. This can be taken as evidence that the effective Reynolds number in a LES performed with models like the FSF model is much larger than in a DNS at the same resolution.

Another interesting feature which appears when looking at animations of (low) pressure isosurfaces, or other quantities that reveal the large-scale vortical structure, is the chaotic nature of the vortex shedding: sometimes, the Kelvin-Helmholtz vortices appear early, with streamwise spacing close to λ_0. Sometimes, they form later, with larger spacing (more than 2 λ_0). In the first case, a high level of translative instability is generally observed, yielding early pairings. The merging of such highly distorted vortices is necessarily helical, and yields the expulsion of three-dimensional vorticity in the form of streamwise vortices of both signs. Those of the same sign converge and merge, causing the disruption of the initial billows. After evacuation at the outlet, the second case is observed, with less three-dimensional activity. The bulging mode is visible but the translative instability always dominates, which is consistent with Pierrehumbert and Widnall (1982). In both cases, streamwise vortices appear when the billows are strongly distorted. The largest values of the vorticity magnitude are $\approx 4\ \Delta U/\delta_i$, as in the temporal LES at Re = 2000 mentioned above. They are found at the tips of the streamwise vortices which connect merging billows.

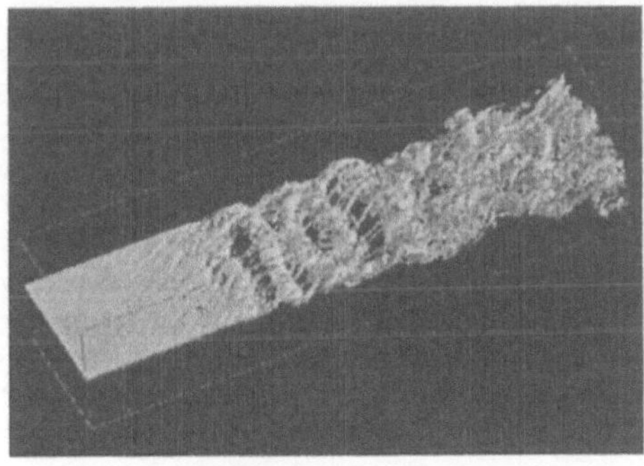

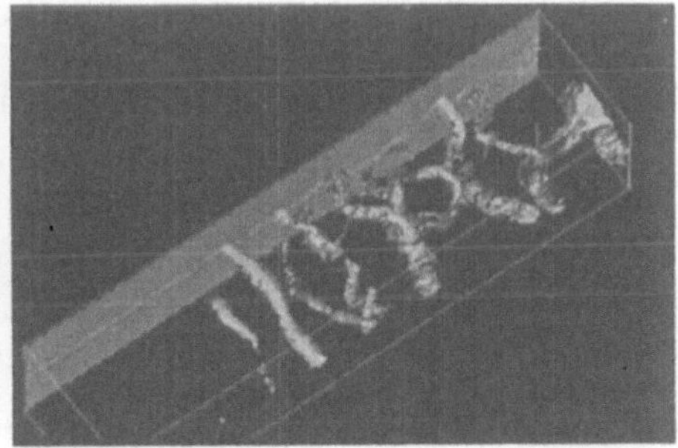

Figure 4. **Left:** counterpart of Figure 3 with $L_z = 4\ \lambda_0$. **Right:** corresponding low-pressure map.

With $L_z = 4\ \lambda_0$, the same kind of dynamics is observed, but with less translative instability and more helical pairings. Figure 4 show iso-vorticity-magnitude and low-pressure isosurfaces. With the 3D upstream forcing, the competition between these different modes persists, and a similar vortex structure is observed on average. However, none of these simulations has reached self-similarity, and calculations in longer domains are necessary: the kinetic-energy spectra in the downstream region are steeper than $k^{-5/3}$, and rms velocity fluctuations have a departure of about 20% with respect to the experiments.

2.2. LES OF A SPATIAL BOUNDARY LAYER AT MACH 0.5

A LES using the SSF model was carried out by Ducros *et al.* (1996) in a weakly-compressible case at $M_\infty = 0.5$, for an adiabatic plate. The flow upstream is the superposition of the laminar profile at this Mach number (almost indistinguishable from the Blasius solution), a two-dimensional pertubation forcing the most amplified Tollmien-Schlichting mode, and three-dimensional white noise of same amplitude. The upstream Reynolds number based on the displacement thickness is 1000. The resolution is $650 \times 32 \times 20$ in the streamwise, transverse and spanwise directions, respectively.

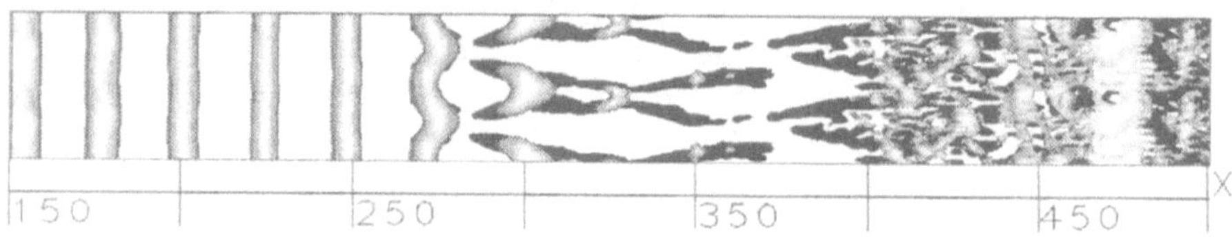

Figure 5. Spatially-developing boundary layer in LES with the Filtered Structure Function model; isosurfaces of pressure ($p = 0.999p_\infty$, grey) and longitudinal vorticity ($\omega_1 = \pm 0.1 U_\infty \delta_i$, dark) are shown

Until the beginning of transition (before the shape factor starts dropping), the TS waves travel as a whole, while the three-dimensional fluctuations grow. Transition begins as their amplitude reaches about 1%, in the form of a pair of conjugate oblique subharmonic modes (Herbert mode, 1988) mostly, growing on top of the saturated TS waves. In contrast with the mixing layers presented above, the hypothesis required for Floquet-type secondary stability analyses are satisfied here. The amplitude of the TS waves is large enough for the "Floquet system" to amplify also its second preferential mode, namely, harmonic modes (K modes, after Klebanoff *et al.*, 1962), but their amplitude remains lower.

A top view of the low pressure and longitudinal vorticity in the transitional region is shown on Figure 5. Transition begins at $x \approx 250\ \delta_i$, and the initially two-dimensional low-pressure structures (half billows) evolve into a staggered pattern which disappears shortly after (one streamwise wavelength of the subharmonic mode, *i.e.* two TS wavelengths). This disappearance corresponds to an uniformization of the pressure, during which a streamwise-independent mode develops and produces streaks, as shown in Figure 6. So far, we do not really understand the relationship between the staggered mode which appears first and the streaks which supersede them shortly farther downstream, and which were long believed to be a feature of the K-type transition only.

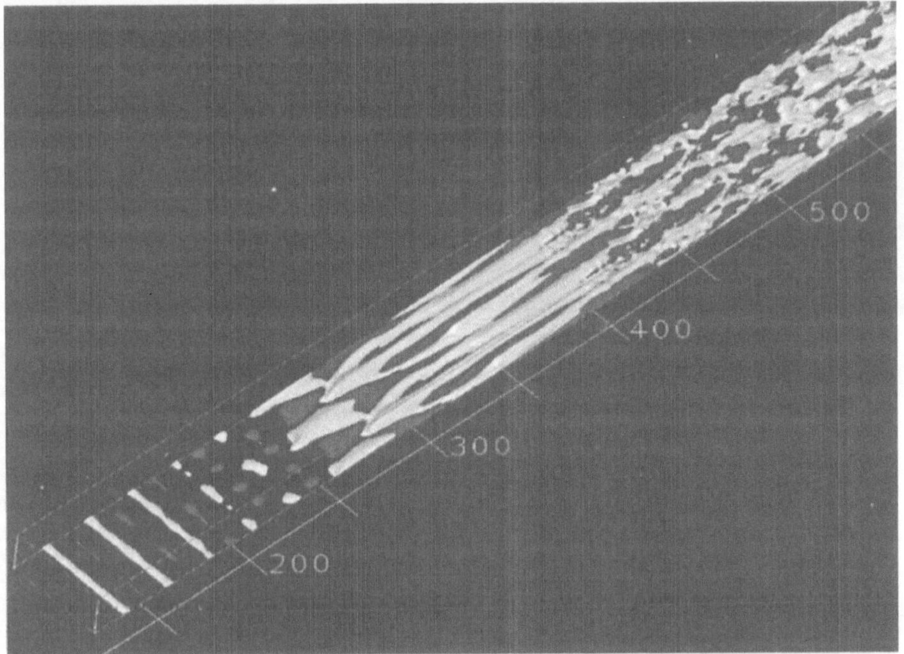

Figure 6. same calculation as Figure 5; isosurfaces of the longitudinal velocity fluctuations ($u_1' = 0.024 U_\infty$, pale grey, $u_1' = -0.024 U_\infty$ dark grey, hardly visible).

We now show in Figure 7 an enlarged view of a hairpin ejected away from the wall above a low-speed streak, just after completion of transition (when the shape factor reaches its asymptotic turbulent value $h_{12} \approx 1.4$). Such hairpins have a longitudinal vorticity which is low with respect to the spanwise vorticities attained at the wall under the high-speed streaks, where most of the drag comes from. Another remark is that we could never find in these calculations coherent alternate longitudinal vortices at the wall. On the contrary, there are several hairpins ejected above one single low-speed streak.

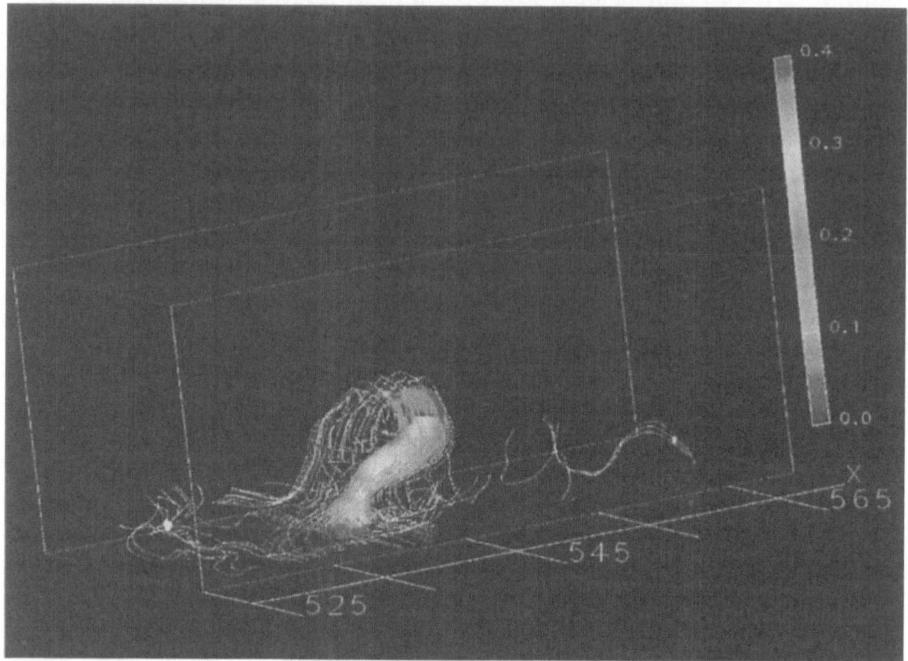

Figure 7. LES of the spatial boundary layer at Mach 0.5; vortex lines and low pressure characterizing a hairpin vortex ejected from the wall at the end of transition

Although it gives interesting informations as far as the structure of turbulence" is concerned, this LES is far from perfect, as far as statistics are concerned: in particular, it overestimates by about 15% the mean velocity in the logarithmic profile.

3. conclusion

Spatially growing Large-Eddy Simulations of mixing layers and boundary layers have been performed with the aid of subgrid scale models of the *spectral* and *structure-function* families reviewed in Lesieur and Métais (1996). In the case of mixing layers, the emergence from initial/upstream noise of an oblique sub-harmonic mode of primary instability reported in CLL92 has been confirmed, yielding *helical pairings* and, more genererally, *chain-like fence* patterns, to employ the terminology introduced by Collins *et al.* (1994). Although not presented here for want of space, similar patterns have been obtained in the case of round jets, for which straight Kelvin-Helmholz vortices correspond to the axisymetric mode and *helical pairings* to the double-helix mode (Urbin and Métais, 1997). Helical pairings have also been found over a backward-facing step (Delcayre, 1997)

Similar LES techniques have been applied to compressible flows through straightforward variable-density extension of the models, justified by a formalism proposed by Comte, David and Lesieur (1997). An example of almost-incompressible transitional boundary layer has been presented, featuring a combination of a K-type mode and a streamwise-independent mode, yielding both staggered Λ-vortices and streaks which form farther downstream and continue in the turbulent regime, where the essential features of turbulent boundary layers are qualitatively recovered.

Quantitatively correct results have been obtained in the case of temporally-growing plane turbulent channel flows with the aid of the Dynamic Spectral model. A forthcoming presentation by Dubief and Comte (1997) will also show correct results obtained with the Filtered Structure Function model in the case of a turbulent boundary layer passing over a spanwise groove. To achieve this, the resolution at the wall has been increased (first mesh line at $y^+ \approx 1$ instead of $y^+ \approx 3$), and inflow conditions generated by means of the re-injection procedure proposed by Lund *et al.* (1996).

Aknowledgements

The results presented here have been obtained by J. Silvestrini (1996) and F. Ducros (1995) during their PhD's in Grenoble. The development of the Dynamic Spectral model used for the temporal mixing layer LES of Silvestrini (1996) is due to E. Lamballais (1995) who we thank warmly. We are also indebted to P. Begou for the computational support. The supercomputing resources were allocated free by IDRIS and CGCV, the French supercomputing centres of the CNRS and CEA/CENG. One of us (M.L.) is supported by the *Institut Universitaire de France*. Our laboratory is supported by the CNRS, INPG and UJF.

References

Bartello, P., 1993, private communications.

Benney, D.J. and Lin, C.C., 1960, On the secondary motion induced by oscillations in a shear flow. *Phys. Fluids*, **3**, 656–657.

Bernal, L.P., Roshko, A., 1986, Streamwise vortex structure in plane mixing layer. *J. Fluid Mech.*, **170**, 499–525.

Browand, F.K., Ho, C.M., 1983, The mixing layer: an exemple of quasi two-dimensional turbulence, Special issue on two-dimensional turbulence, *J. Theo. and Appl. Mech.*, R. Moreau ed., 99–120.

Browand, F., Latigo, B.A., 1979, Growth of the two-dimensional mixing layer from a turbulent and non-turbulent boundary layer, *Phys. Fluids*, **22**, 1011–1019.

Chandrsuda, C., Mehta, R.D., Weir, A.D., Bradshaw, P., 1978, Effect of free-stream turbulence on large structure in turbulent mixing layers, *J. Fluid Mech.*, **85**, 693–704.

Collins, S.S., Lele, S.K., Moser, R.D., Rogers, M.M., 1994, The evolution of a plane mixing layer with spanwise nonuniform forcing, *Phys. Fluids*, **6**, 381.

Comte, P., Lesieur, M., Fouillet, Y., 1989, Coherent structures of mixing layers in large-eddy simulation, In *Topological Fluid Dynamic*, H.K. Moffatt, A. Tsinober (eds.). Cambridge University Press, 649–658.

Comte, P., Lesieur, M., Lamballais, E., 1992, Small-scale stirring of vorticity and a passive scalar in a 3D temporal mixing layer, *Phys. Fluids A*, **4**, 2761–2778.

Comte, P., Ducros, F., Silvestrini, J.H., David, E., Lamballais, E., Métais, O., Lesieur, M., 1994, Simulation des grandes échelles d'écoulements transitionnels, 74th AGARD Fluid Dynamics Panel Meeting, Crete, p. 14.

Craik, A.D.D., 1971, Nonlinear resonant instability in boundary layers. *J. Fluid Mech.*, **50**, 393–413.

Comte, P., David, E. and Lesieur, M., 1997, Un formalisme pour la simulation des grandes échelles d'écoulements compressibles, Preprint LEGI, for submission to C.R. Acad. Sci. Paris.

Dallard, T. and Browand, F.K., 1993, The growth of large scales at defect sites in the plane mixing layer, *J. Fluid Mech.*, **247**, 339–368.

David, E., 1993, *Modélisation des écoulements compressibles et hypersoniques : une approche instationnaire. Thèse* de l' Institut National Polytechnique de Grenoble.

Delcayre, F., 1997, Topology of coherent vortices in the reattachment region of a backward-facing step. In *Turbulent Shear Flows 11*, Grenoble.

Dubief, Y. and Comte, P., 1997, Large-Eddy Simulation of a boundary layer flow passing over a groove. In *Turbulent Shear Flows 11*, Grenoble.

Ducros, F., 1995, Simulations numériques directes et des grandes échelles de couches limites compressibles. *Thèse* de l' Institut National Polytechnique de Grenoble.

Ducros, F., Comte, P., Lesieur, M., 1996, Large-eddy simulation of transition to turbulence in a boundary-layer developing spatially over a flat plate, *J. Fluid Mech.*, **326**, 1–36.

Gonze, M.A., 1993, Simulation numérique des sillages en transition à la turbulence, *Thèse* de l'Institut National Polytechnique de Grenoble.

Herbert, T. and Morkovin, M., 1980, Dialogue on bridging some gaps in stability and transition research. In *Laminar-Turbulent Transition*, ed. R. Eppler, H. Fasel, 47–72, Springer Verlag, Berlin.

Herbert, T., 1988, Secondary instability of boundary layers. *Ann. Rev. Fluid Mech.*, **20**, 487–526.

Huang, L., Ho, C.M., 1990, Small scale transition in a plane mixing layer, *J. Fluid Mech.*, **210**, 475–500.

Kelly, R.E., 1967, On the stability of an inviscid shear layer which is periodic in space and time. *J. Fluid Mech.*, **27**, 657–689.

Klebanoff, P.S., Tidstrom, K.D., Sargent, L.M., 1962, The three-dimensional nature of turbulent boundary layer instability. *J. Fluid Mech.*, **12**, 1–34.

Konrad, J.H., 1976, An experimental investigation of mixing in two-dimensional turbulent shear flows with applications to diffusion-limited chemical reactions. Ph.D. Thesis, California Institute of Technology.

Lasheras, J., Choi, H., 1988, Three-dimensional instability of a plane free shear layer: an experimental study of the formation and evolution of streamwise vortices, *J. Fluid Mech.*, **189**, 53–86.

Lamballais, E., 1995, Simulations numériques de la turbulence dans un canal plan tournant. *Thèse* de l'Institut National Polytechnique de Grenoble.

Lesieur, M., 1997, *Turbulence in fluids, third edition*, Kluwer Academic Publishers.

Lesieur M., Métais O., 1996, New trends in large-eddy simulations of turbulence, *Ann. Rev. Fluid Mech.*, **28**, 45–82.

Lesieur, M., Staquet, C., Le Roy, P., Comte, P., 1988, The mixing layer and its coherence examined from the point of view of two-dimensional turbulence, *J. Fluid Mech.*, **192**, 511–534.

Leslie, D.C., Quarini, G.L., 1979, The application of turbulence theory to the formulation of subgrid modelling procedures, *J. Fluid Mech.*, **91**, 65–91.

Lin, S.J., Corcos, G.M., 1984, The mixing layer: deterministic models of a turbulent flow. Part 3. The effect of plane strain on the dynamics of streamwise vortices, *J. Fluid Mech.*, **141**, 139–178.

Lund, T. S., Wu, X., and Squires, K. D., 1996, "On the Generation of Turbulent Inflow Conditions for Boundary Layer Simulations", *Annual Research Briefs*, Center for Turbulence Research, pp. 287–295.

Mallier, R, Maslowe, S.A., 1994, Fully coupled resonant-triad interactions in a free shear layer, *J. Fluid Mech.*, **278**, 101–121.

Métais, O., Lesieur, M., 1992, Spectral large-eddy simulations of isotropic and stably-stratified turbulence, *J. Fluid Mech*, **239**, 157–194.

Metcalfe, R.W., Orszag, S.A., Brachet, M.E., Menon, S., Riley, J., 1987, Secondary instability of a temporally growing mixing layer, *J. Fluid Mech.*, **184**, 207–234.

Monkewitz, P.A., 1988, Subharmonic resonance, pairing and schredding in the mixing layer. *J. Fluid Mech.*, **188**, 223–252.

Neu, J.C., 1984, The dynamics of stretched vortices. *J. Fluid Mech.*, **143**, 253–276.

Nygaard, K., Glezer, A., 1991, Evolution of streamwise vortices and generation of small-scale motion in a plane mixing layer, *J. Fluid Mech.*, **231**, 257–301.

Nygaard, K., Glezer, A., 1994, The effect of phase variations and cross-shear on vortical structures in a plane mixing layer, *J. Fluid Mech.*, **276**, 21–59.

Orlansky, I., 1976, A simple boundary condition for unbounded hyperbolic flows, *J. Comp. Phys.*, **21** 251-269.

Orszag, S.A., Patera, A.T., 1983, Secondary instability of wall-bounded shear flows, *J. Fluid Mech.*, **128**, pp. 347-385.

Pierrehumbert, R.T., Widnall, S.E., 1982, The two- and three-dimensional instabilities of a spatially periodic shear layer, *J. Fluid Mech.*, **114**, 59–82.

Schoppa, W., Hussain, F., Metcalfe, R., 1995, A new mechanism of small-scale transition in a plane mixing layer: core dynamics of spanwise vortices, *J. Fluid Mech.*, **298**, 23–80.

Silvestrini, J., 1996, Simulation des grandes échelles des zones de mélange: application à la propulsion solide des lanceurs spatiaux. *Thèse* de l' Institut National Polytechnique de Grenoble.

Urbin, G. and Métais, O., in *Turbulent shear flows 11*, Grenoble.

NUMERICAL SIMULATION OF NONLINEAR INTERACTIONS IN SUBSONIC AND SUPERSONIC FREE SHEAR LAYERS

A.N. KUDRYAVTSEV AND D.V. KHOTYANOVSKY
Insitute of Theoretical and Applied Mechanics
Novosibirsk 630090, Russia

1. Introduction

A free shear layer (or a mixing layer) is a flow formed by mixing of two parallel flows of fluid moving with different velocities. The mechanisms leading to its instability and transition to turbulence play also an important role in other shear flows, such as jets and wakes. The dynamics of a compressible shear layer is an important part of numerous fluid dynamics problems, for example, noise generation, flow in gasdynamic lasers, astrophysical jets, and many others.

At moderate Mach numbers, the shear layer evolution is associated with intense formation of vortices. This leads to entrainment of ambient fluid into the shear layer and effective mixing of the streams. However, the mixing intensity drastically decreases if the respective velocity of the streams is supersonic. This circumstance can lead to consequences of practical importance, for example, those in operation of supersonic combustion ramjet engines designed for promising hypersonic flying vehicles [1].

Besides arising technical demands, the interest to dynamics of high-speed shear layers is also caused by attempts to understand the fundamental features of transition to turbulence at high speeds.

The linear stability of compressible shear layers has been currently well studied (see, for example [2], [3],[4],[5]. This is not true for later stages of transition to turbulence, especially for supersonic relative Mach numbers, although there is a number of investigations [6] [7],[8].

In the present study we solve two-dimensional Euler and Navier-Stokes equations to research the nonlinear development of instabilities in a high-

E. Krause and K. Gersten (eds.), IUTAM Symposium on Dynamics of Slender Vortices, 133-142.

speed shear layer. We simulate a temporal as well as a spatial evolution of the free shear layer, forced by disturbances which are superpositions of eigenfunctions of the linear stability problem. Both subsonic and supersonic disturbances are considered. Simulations of shear layer dynamics in a number of more complex (and more realistic) situations are also presented. They include the Kelvin-Helmholtz instability of the contact surface emanating from the triple point in the case of Mach reflection of an oblique shock wave, the interaction of two closely-spaced parallel shear layers and the impingement of the shock wave on the shear layer.

2. Basic Definitions

To study the nonlinear evolution of a compressible shear layer, we solve numerically the two-dimensional Navier-Stokes equations for the viscous compressible perfect gas

$$\frac{\partial Q}{\partial t} + \frac{\partial F}{\partial x} + \frac{\partial G}{\partial y} = \frac{1}{\mathrm{Re}} \left(\frac{\partial F_v}{\partial x} + \frac{\partial G_v}{\partial y} \right) \tag{1}$$

$$Q = \begin{pmatrix} \rho \\ \rho u \\ \rho v \\ e \end{pmatrix}, \quad F = \begin{pmatrix} \rho u \\ \rho u^2 + p \\ \rho u v \\ u(e+p) \end{pmatrix}, \quad G = \begin{pmatrix} \rho v \\ \rho u v \\ \rho v^2 + p \\ v(e+p) \end{pmatrix}$$

$$F_v = \begin{pmatrix} 0 \\ \frac{\partial \tau_{xx}}{\partial x} \\ \frac{\partial \tau_{xy}}{\partial y} \\ \frac{\partial r_x}{\partial x} \end{pmatrix}, \quad G_v = \begin{pmatrix} 0 \\ \frac{\partial \tau_{xy}}{\partial x} \\ \frac{\partial \tau_{yy}}{\partial y} \\ \frac{\partial r_y}{\partial y} \end{pmatrix} \qquad \begin{aligned} \tau_{xx} &= \tfrac{2}{3}\mu \left(2\tfrac{\partial u}{\partial x} - \tfrac{\partial v}{\partial y}\right) \\ \tau_{xy} &= \mu \left(\tfrac{\partial u}{\partial y} + \tfrac{\partial v}{\partial x}\right) \\ \tau_{yy} &= \tfrac{2}{3}\mu \left(2\tfrac{\partial v}{\partial y} - \tfrac{\partial u}{\partial x}\right) \end{aligned}$$

$$\begin{aligned} r_x &= u\tau_{xx} + v\tau_{xy} + \tfrac{\mu}{(\gamma-1)\mathrm{Pr}\mathrm{M}^2} \tfrac{\partial T}{\partial x} \\ r_y &= u\tau_{xy} + v\tau_{yy} + \tfrac{\mu}{(\gamma-1)\mathrm{Pr}\mathrm{M}^2} \tfrac{\partial T}{\partial y} \end{aligned} \qquad \begin{aligned} p &= (\gamma - 1)\left(e - \rho\tfrac{u^2+v^2}{2}\right) = \rho T/\gamma M^2 \\ \mu &= \mu(T) \end{aligned}$$

Here ρ, u, v, e, p, T, μ are the fluid density, velocity components along the x and y axes, the internal energy, the pressure, the temperature and the viscosity, respectively; γ is a ratio of specific heats, Pr is Prandtl number. Equations (1) are written in a dimensionless form. When simulating the evolution of a compressible shear layer between the streams 1 and 2 which move with velocities U_1 and U_2, $U_1 > U_2$, it is convenient to choose a coordinate system where the x axis is directed along the flow velocity and the y axis is directed across the shear layer towards the stream with a higher velocity. Mach number $\mathrm{M} \equiv \mathrm{M}_1$ and Reynolds number Re in (1) are determined from their values in the first stream and initial shear layer thickness.

Besides, the parameters that affect the shear layer flow are the ratio of velocities U_2/U_1 and the ratio of temperatures T_2/T_1. However, the temperature difference effects are not considered in the present paper, and it is assumed everywhere that $T_2 = T_1$. We also assume linear dependence of viscosity on temperature.

In many cases, an important parameter characterizing the influence of compressibility is a so-called convective Mach number $M_c = (U_1 - U_2)/(c_1 + c_2)$, where c_1, c_2 are speeds of sound in corresponding streams.

When simulating the spatial development of shear layer, the basic flow is determined by solving the steady Navier-Stokes equation. But when studying the linear stability and modeling the temporal evolution of the shear layer, we prescribe the profiles of the basic laminar flow. As a velocity profile, we use the known Goertler's erf(y) solution for an incompressible shear layer, which can be extended to the compressible case by introducing the Dorodnitsyn variable. The temperature profile is deduced from the Crocco integral assuming $\mathrm{Pr} = 1$.

3. Linear stability results

The linear stability characteristics determine many important features of compressible shear layer evolution in nonlinear regime. So, we briefly desribe the results of linear stability analysis.

The basic flow in the shear layer is unstable with respect to inviscid disturbances. The linear disturbances are travelling waves which have the form $e^{i(\alpha x - \omega t)}$ where α is a wavenumber and ω is a frequency. In the temporal stability problem α is real and $\omega_r + i\omega_i$ is complex and the disturbance is unstable if $\omega_i > 0$. On the contrary, in the spatial stability problem ω is assumed to be real, $\alpha = \alpha_r + i\alpha_i$ is complex and the instability arises when $\alpha_i < 0$.

Figs. 1- 2 show some results on temporal stability of compressible shear layer at $U_2/U_1 = -1$, $T_2/T_1 = 1$. At the low Mach number the flow is unstable with respect to the disturbances with $\alpha < \alpha_N$ (Fig. 1). These disturbances are well-known Kelvin-Helmholtz (KH) waves. Their phase velocity for conditions considered is $c_r = \omega_r/\alpha$. The compressibility stabilizes the KH instability and α_N decrease as M grows. Though, two new supersonic (SS) instability modes arises if $M > M_* = 0.906$ The phase velocity of the first is supersonic with respect to the first stream (slow mode) and that of the second — to the second stream (fast mode). In the case under consideration the neutral curves of both SS modes in the plane α,M coincide.

Esentially different nature of subsonic and supersonic waves become very clear if one consider the behaviour of their eigenfunctions. The eigen-

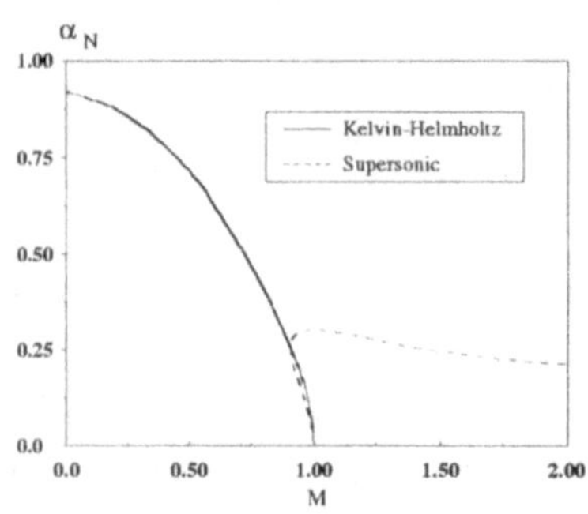

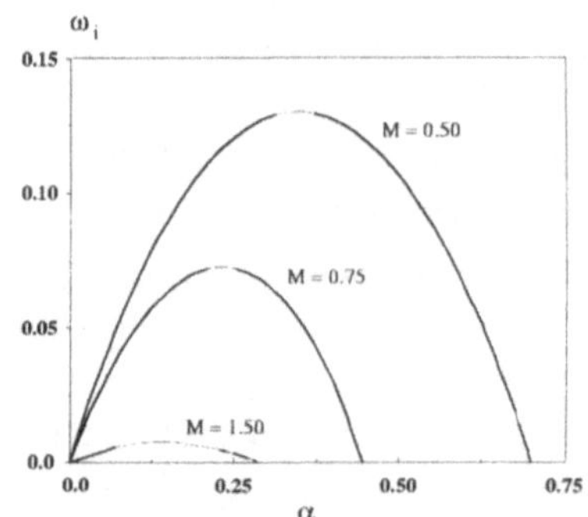

Figure 1. The neutral curves and the growth rates for compressible shear layer

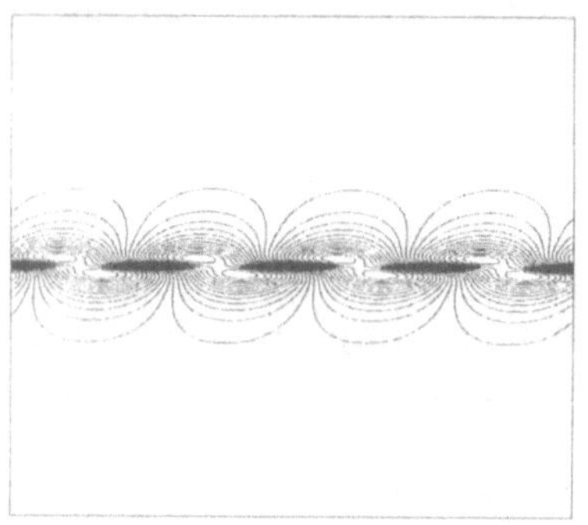

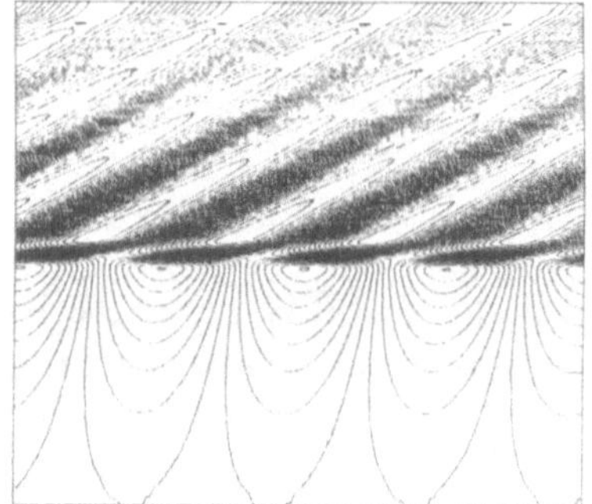

Figure 2. The linear eigenfunctions (cross-streamwise velocity) at $M = 0.5$ (the Kelvin-Helmholtz mode) and $M = 1.5$ (slow supersonic mode)

functions of the KH waves rapidly decrease out of the shear layer. On the contrary, SS disturbances in the external stream, where their velocity is supersonic, oscillate and decrease very slowly. In this stream they are oblique waves travelling away from the shear layer. Their origin is the same as that of Mach wave arising at supersonic movement of a solid body through fluid. Fig. 2 can serve as an obvious illustration.

The growth rates of SS disturbances are essentially lower than those of subsonic KH waves. It seems to be natural since significant portion of energy drawing from the basic flow emits in the ambient space. However, the situation may be different for the shear layer confined in a channel, when the outcoming waves reflect from its walls. The solution of the stability problems for 3D disturbances gives that at the subsonic Mach number larger than approximately 0.7 and the moderate supersonic Mach numbers the oblique KH disturbances have the maximim growth rate. Nevertheless, the investigation of nonlinear stages of SS instability is of interest because it may help to understand the instability mechanism unlike the well-known Kelvin-Helmholtz one. Also it may be relevant to experimentally observed Mach wave emission in the supersonic jets. Besides, as it follows from the asymptotic theory [3], namely the 2D supersonic disturbances are the most unstable at the high Mach numbers.

4. Numerical techniques

The essentially non-oscillatory (ENO) [9, 10] schemes are capable to capture the flow discontinuites without considerable oscillations and are of high order accuracy for smooth solutions. Consequenly, they are good candidate for numerical simulation of wave processes in compressible shear flows. Thus, in the present paper we apply the weighted ENO of fifth order [11] for approximation of convective terms of Eqs. (1).

In finite-difference ENO schemes [10] adaptive stencil is used to obtain the high-order polynomial interpolation of fluxes. Characteristic fluxes resulted by projection onto eigenvectors $r^\nu, \nu = 1, \ldots, 4$ of Roe averaged Jacobi matrix $\partial F/\partial Q$ are split into positive and negative parts $f^{\nu+}, f^{\nu-}$. Then, ENO interpolation is applied separately to $f^{\nu+}$,$f^{\nu-}$. For example, $f^{\nu+}$ in the point $i+1/2$ is reconstructed using one of three candidate stencils $\{i-2, i-1, i\}$,$\{i-1, i, i+1\}$, $\{i, i+1, i+2\}$ depending on the values of higher divided differences of $f^{\nu+}$. Instead weighted ENO scheme utilizes a convex weighted combination of these three fluxes in order to achieve the maximum order of accuracy (fifth in the given case) on the smooth solution. Near flow discontinuity the weight of stencils containing the discontinuity is decreased automatically by a smoothness indicator.

In this paper we use Roe splitting and Lax-Friedrichs splitting, the latter is applied only in the Mach reflection problem containing strong shock waves. If $\lambda^\nu, \nu = 1, \ldots, 4$ are the eigenvalues of Jacobi matrix then these splittings can be written as

$$f^{\nu\pm} = \frac{1}{2}\left[f^\nu \pm \mathrm{sign}(\lambda^\nu) f^\nu\right], \quad \text{for Roe spliiting,}$$

$$f^{\nu\pm} = \frac{1}{2}\left[f^\nu \pm \alpha^\nu q^\nu\right], \quad \alpha^\nu = \max_{\mathrm{all}\, i} \lambda^\nu, \quad \text{for Lax-Friedrichs spliitting}$$

In the multidimensional problems the convective terms are computed dimension by dimension. The only modification at the computation $\partial G/\partial y$ is the use of nonuniform grid along y to enhance resolution near the center of shear layer. The algebraic mappings has been used to transform equally spaced grid in computational space onto stretched grid in physical space.

The central difference formulas of 4-th order are used to approximate the viscous terms of the Navier-Stokes equations.

Time stepping was performed by the Runge-Kutta-Gill scheme of 4-th order because of its high order accuracy and low storage requirements.

5. Temporally developing shear layer

The periodic boundary condition along x is imposed when simulating temporally developing shear layer: $Q(x + L_x, y, t) = Q(x, y, t)$. This approach was widely used for direct numerical simulation of various transitional flows. The computed picture should be approximately equivalent one that an observer moving along with the instability wave can see. The periodic boundary condition saves great amount of computer time and other computer resources but it suffers from serious shortcomings. Firstly, continuous spectrum inherent in real flow is replaced by discrete one since only disturbances having wave length L_x/n, where n is an integer number, are possible. Further more, in spatially developing shear layer at subsonic velocities the feedback can exist owing to downstream influence when the distubances on later stages of evolution affect the growth of disturbances on early stages. It is impossible in temporally developing shear layer.

The simulation have been carried out in a rectangular computational domain $[0, L_x] \times [-L_y, L_y]$. The number of grid points varies from 128×200 up to 256×250. Far from shear layer, at $y = \pm L_y$ the non-reflecting boundary conditions by Thompson [12] were imposed. As an initial forcing we used a superposition of the wave with the wave number $\alpha = \alpha_m$ and the amplitude $A = A_1$ that is the most unstable linear disturbance and its subharmonic ($\alpha = \alpha_m/2$,$A = A_{1/2}$). Thus, the processes of nonlinear saturation of disturbances and interaction of disturbances with multiple wave lengths were simulated.

The run parameters were $U_2/U_1 = -1$, $T_2/T_1 = 1$, $A_1 = 0.1$, $A_1/2 = 0.01$; $M = 0.5$, 0.75, and 1.5. The Reynolds number in all runs was equal to 1000. The typical flow patterns at various Mach numbers are shown in Figs. 3. The pattern at M = 0.5 closely resembles its incompressible counterpart. However, at M = 0.75 closed regions where the velocity of fluid is supersonic with respect to large-scale structures arise and weak shock waves adjacent to vortices are formed. The computations of nonlinear development of the supersonic mode at M = 1.5 demonstrate formation of the secondary flow that is very unlike KH vortices. The oblique shock waves appear in outer region. In all cases the pairing of large scale structures occurs but for SS disturbances this process is very slow. Time when the energies of subharmonic and fundamental waves became equal was approximately $t_{1/2}$ =75, 135 and 1550 at M = 0.5, 0.75 and 1.5, respectively.

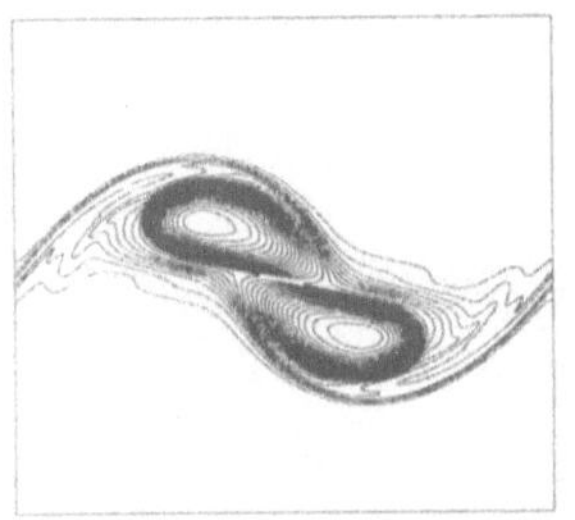
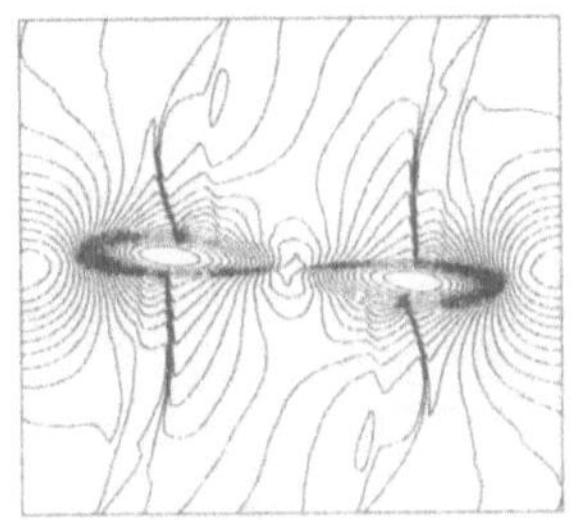
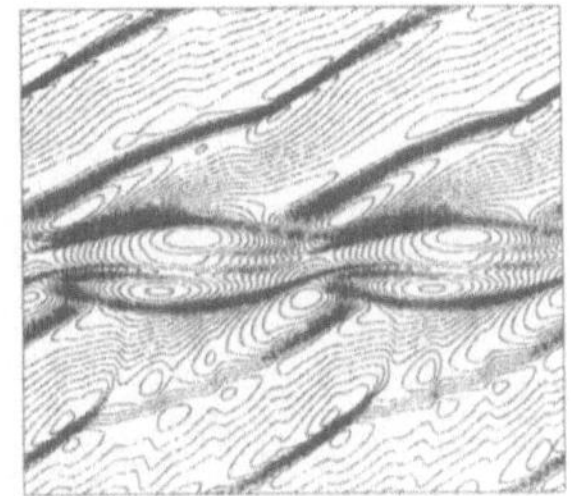

Figure 3. Typical flow patterns resulted from numerical simulations of temporal shear layer (1) M=0.5, (2) M=0.75, (3) M=1.5.

6. Spatially developing shear layer

Unlike the temporal evolution, the spatial evolution of shear layer depends on absolute values of velocities of streams, not only on their difference. The computations have been conducted with supersonic flow at the inflow boundary. The number of grid points used was 500×100. The Reynolds number at the inflow boundary was equal to 1000. The fundamenal wave and its subharmonic with $A_1 = 0.1$, $A_{1/2} = 0.01$ were introduced as a forcing at the inflow. Flowfields are presented in Fig. 4 and Fig. 5 at two different values of the convective Mach number, $M_c = 0.5$ and $M_c = 1.5$. It is evident that there are certain common features with the temporal simulations. The pairing of vortices takes place at $x = 150$ when $M_c = 0.5$ and at $x = 210$ when $M_c = 0.75$. We could not see the merging of large scale structures in the case of $M_c = 1.5$ up to the outflow boundary of the computational domain that was situated at $x = 500$.

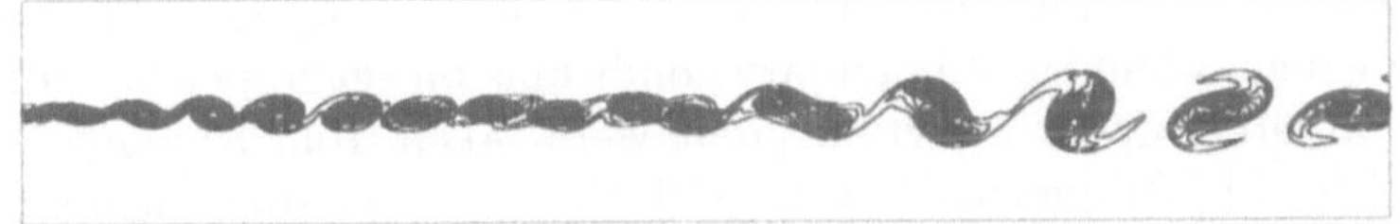

Figure 4. Vorticity contours at $M_1 = 2.5, M_2 = 1.5$

Figure 5. Density gradient flowfield at $M_1 = 4.5, M_2 = 1.5$

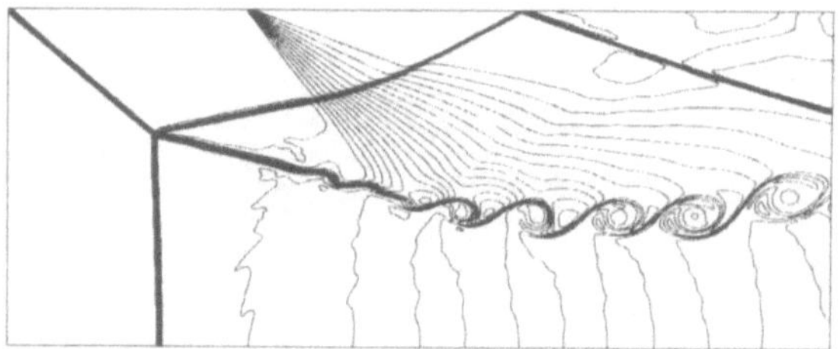

Figure 6. The density contours for steady Mach reflection

7. Examples of instability of compressible shear layer in more complex situations

In this section three examples of development of compressible shear layer instability in more complex (and more realistic) situations are given. They include (1) the KH instability of shear layer originated from triple point at the Mach reflection of shock wave, (2) interaction of two closely spaced shear layers (3) interaction of shear layer with impinging shock wave.

7.1. SHEAR LAYER ORIGINATED FROM TRIPLE POINT AT MACH REFLECTION OF SHOCK WAVE

At numerical simulation of the steady Mach reflection [13] has been found that the shear layer which originates from a triple point is unstable and rolls in a typical KH chain of vortices. This phenomenon can be illustrated by Fig. 6 which shows the flow pattern resulted from inviscid simulation of the Mach reflection of shock wave. The inflow Mach number is equal to 5 and the incident shock wave angle is 41°. The number of points in computatational domain was 480×200. The Mach configuration was obtained using the following statement of boundary conditions on the upper boundary. The flow parameters up to a certain point were taken from Rankine-Hugoniot relations for an oblique shock wave with the corresponding incidence angle. Beyond this point, solid wall boundary conditions were imposed. As a result, the expansion fan came from this point, and the incident shock wave was emanated from the left upper corner. In the shear layer downstream of the triple point, the vortices are observed that develop due to the KH instability. The convective Mach number for this case varies from approximately 0.5 near the triple point up to 0.35 at the outflow boundary, though the Mach numbers of mixing streams varies very significanly. The vortices appear during the transient process when the Mach stem grows and moves upstream, and retain during a long time after the Mach stem location does not change. An appearance of new vortices is probably connected with a feedback mechanism, when the vortices affect upstream through the sub-

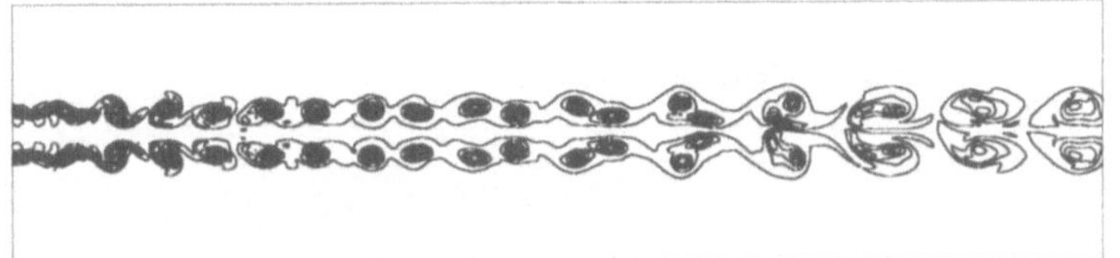

Figure 7. The vorticity contours for two closely spaced shear layers

sonic region and cause an instability of the shear layer near the triple point. This computation gives an interesting example of simultaneous action of numerical and physical factors. The instability of shear layer is physical one but the shear layer thickness and, as consequence, the size of vortices are determined in inviscid computation by numerical viscosity inherent in the weighted ENO scheme used.

7.2. INTERACTION OF TWO CLOSELY SPACED SHEAR LAYERS

Interaction of two closely spaced turbulent shear layers has been studied in experimental work [14]. This flow is of interest due to its successive transformation from two independent shear layers to wakelike flow. The simulation has been performed the following inflow parameters. The Mach number of external flow was equal to 2.5 and the Mach number of flow between shear layers was 1.5. The temperature of streams was the same. The distance between centers of shear layers $h = 7.92\delta_\omega$ and the Reynolds number based on h is 3760. The inflow forcing was taken by as superposition of fundamental wave ($A_1 = 0.1$) and its subharmonic ($A_{1/2} = 0.01$ for both shear layers. The number of grid points used was 400×100. The vorticity flowfield for this computation is shown in Fig. 7.

7.3. INTERACTION OF SHEAR LAYER WITH IMPINGING SHOCK WAVE

The interaction of shock wave with shear layer has been proposed in [1] as a mechanism to enhance the mixing at the high speeds. It has been investigated in some experimental works, for instance, in recent paper [15]. We simulate this interaction using the boundary condition statement described in Section 7.1 to generate the oblique shock wave. The shock with the angle 22° impinges on the shear layer between streams with $M_1 = 3$ (upper) and $M_2 = 2$. The shear layer at the inflow was forced by the fundamental wave with the amplitude $A_1 = 0.01$. The computed flowfielfds is shown in Fig. 8 As a result of the interaction, the shock wave refractes and the shear layer deflectes. The vortex intensity seems slightly increasing.

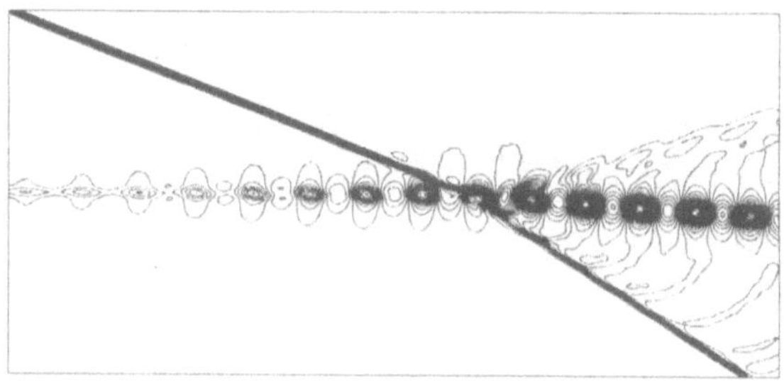

Figure 8. The density contours for interaction of shear layer with an shock wave

References

1. Kumar, A.S. Bushnell, D.M., and Hussaini, M.Y. (1987) A Mixing Augmentation Technique for Hypervelocity Scramjets, *AIAA Paper*,No. 87-1882,16 p.
2. Jackson, T.L., and Grosch, C.E. (1989) Inviscid Spatial Stability of a Compressible Mixing Layer, *J. Fluid Mech.*,**Vol. 208**,pp.609–637
3. Balsa, T.F., and Goldstein, M.E. (1990) On the Instabilities of Supersonic Mixing Layers: A High-Mach-number Asympotic Theory, *J. Fluid Mech.*,**Vol. 216**,pp.585–611
4. Kudryavtsev, A.N., and Soloviev, A.S. (1989) Shear Layer Stability in Compressible Fluid, *Prikl. Mekh. i Tekhnich. Fiz.*,**No. 6**,pp.119–127 (in Russian, English translation in *J. Appl. Mech. Tech. Phys.*)
5. Kudryavtsev, A.N., and Soloviev, A.S. (1991) Stability of Viscous Compressible Shear Layer with Temperature Difference, *Prikl. Mekh. i Tekhnich. Fiz.*,**No. 4**,pp.88–95 (in Russian, English translation in *J. Appl. Mech. Tech. Phys.*)
6. Soetrisno, M., Eberhardt, S., Riley, J.J, and McMurtry, P.A. (1989) A Study of Inviscid, Supersonic Mixing Layers Using A Second Order Total Variation Diminishing Scheme, *AIAA Journal*,**Vol. 27**,pp.1770–1778
7. Gathmann, R.J., Si-Ameur M., and Mathey, F. (1993) Numerical Simiulation of Three-dimensional Natural Transition in the Compressible Confined Shear Layer, *Phys. Fluids A*,**Vol. 5**,pp.2946–2968
8. Basset, G.M., and Woodward, P.R. (1995) Numerical Simulation of Nonlinear Kink Instabilities on Supersonic Shear Layers, *J. Fluid Mech.*,**Vol. 284**,pp.323–340
9. Harten, A., Engquist B., Osher S., and Chakravarthy S. (1987) Uniformly High Order Essentially Non-Oscillatory Schemes, III, *J. Comput. Phys.*,**Vol. 71,No.2**,pp.231–303
10. Shu, C.-W., and Osher S. (1989) Efficient Implementation of Essentially Non-oscillatory Shock-Capturing Schemes, II, *J. Comput. Phys.*,**Vol. 83,No.1**,pp.32-78
11. Jiang, G.S., and Shu, C.-W. (1996) Efficient Implementation of Weighted ENO Schemes, *J. Comput. Phys.*,**Vol. 126,No.1**,pp.202–228
12. Thompson, K.W. (1987) Time-Dependent Boundary Conditions for Hyperbolic Systems, *J. Comput. Phys.*,**Vol. 68,No. 1**,pp.1–24
13. Ivanov, M.S., Markelov, G.N., Kudryavtsev A.N., and Gimelshein, S.F. (1997) Transition between Regular and Mach Reflections od Shock Waves in Steady Flows, *AIAA Paper*,**No. 97-2511**, 13 p.
14. Clemens, N.T., Petullo, S.P., and Dolling, D.S. (1996) Large-Scale Structure Evolution in Supersonic Interacting Shear Layers, *AIAA Journal*,**Vol. 34,No.10**,pp.2062–2070
15. Ramaswamy, M., Loth, E., and Dutton, J.G. (1996) Free Shear Layer Interaction with an Expansion-Compression Wave Pair, *AIAA Journal*,**Vol. 34,No. 3**,pp.565–571

CORE DYNAMICS IN VORTEX PAIRS AND RINGS

MONIKA NITSCHE
Department of Mathematics
Tufts University, Medford, MA 02155, USA

Abstract. The roll-up of an initially flat planar and axisymmetric vortex sheet is computed numerically, using the inviscid vortex blob method. The sheets roll up into a vortex pair and a vortex ring respectively. This paper documents the onset of chaotic particle motion near the vortex core. The observed behaviour is generic for perturbed Hamiltonian dynamical systems. In the present case, the perturbation arises from self-sustained oscillations, similar to the oscillations in Kida's solutions for an elliptic vortex patch evolving in a strain field. Furthermore, the paper shows that such oscillations are also present in a viscous flow. This suggests that the core dynamics observed in the vortex sheet flow will also arise in a real viscous flow.

1. Introduction

Consider a plate immersed in an inviscid fluid which is given an impulse in direction normal to itself. The resulting potential flow is induced by a vortex sheet bound to the plate. In a reference frame fixed at infinity, this flow is shown in Figure 1(a) for the case of a flat, infinitely long rectangular plate, and in Figure 1(b) for a flat circular plate. If the plate is immediately removed or "dissolved" without disturbing the fluid, the vortex sheet remains in the fluid in place of the plate and is free to evolve under its self-induced velocity.

Klein (1910) thought of the planar and axisymmetric vortex sheets described above as generated by the impulsive motion of an oar or a coffeespoon in a fluid. He predicted that under their self-induced motion, the vortex sheets roll up into a spiral at their edges. In the planar case, the roll-up approximates a vortex pair, in the axisymmetric case, a vortex ring. Both cases are described in Saffman (1992).

This paper presents a numerical study of the evolution of the planar and the axisymmetric vortex sheets shown in Figure 1. The induced flow is highly singular near the edge of the sheet, and computations are possible only after regularizing

E. Krause and K. Gersten (eds.), IUTAM Symposium on Dynamics of Slender Vortices, 143-152.

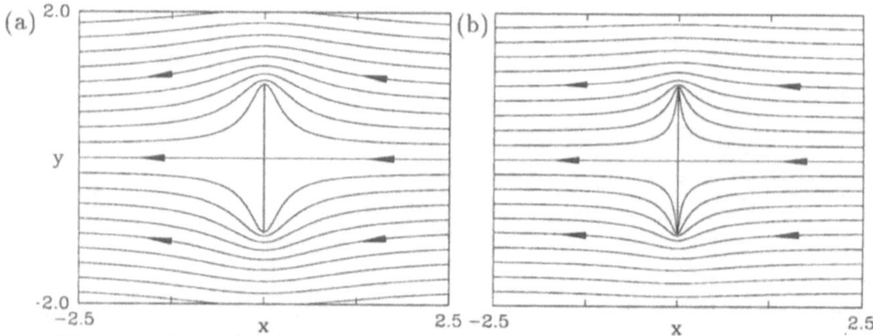

Figure 1. Potential flow past a flat rectangular (a) and circular (b) plate.

Figure 1. Potential flow past a flat rectangular (a) and circular (b) plate.

the motion. One regularization is obtained through the vortex blob method, which consists of introducing a smoothing parameter δ into the governing equations (Chorin & Bernard 1973). The computations are performed with $\delta > 0$ and information about the vortex sheet is inferred from the limit $\delta \to 0$. Another regularization is achieved by viscous diffusion in a real flow.

Section 2 presents computations of the sheet evolution using the vortex blob method. After an initial interval of self-similar flow (see Krasny & Nitsche 1997), the flow appears to settle and oscillate quasi-periodically about a steady state. Instead however, the particle motion becomes irregular near the vortex core, signaling the onset of chaos. This behaviour first develops in a thin annular region about the vortex center. It is observed both in time, using a fixed value of the smoothing parameter and at a fixed time, as the smoothing parameter decreases. In the axisymmetric case, irregular particle motion also develops near the tail of the vortex.

The observed behaviour is generic in perturbed Hamiltonian dynamical systems. In the present case, the perturbation is observed in the form of oscillations in the core vorticity, similar to the oscillations of an elliptic vortex in an external strain field (Kida 1981, Neu 1984). For both the planar and axisymmetric roll-up, the vorticity contours are indeed elliptical in shape and oscillate under a self-induced strain field. Details are found in Krasny & Nitsche (1997).

One impending question raised by the observed onset of chaotic particle motion is whether it also occurs in a real viscous fluid, a fluid which is regularized by viscosity, not by the vortex blob method. Section 3 presents preliminary results addressing this question. The Navier-Stokes equations are solved for varying fluid viscosities ν, with the initial vorticity given by the regularization with $\delta > 0$. Similar to the results of Tryggvason, Dahm & Sbeih (1991), the present results indicate that as $\nu \to 0$ and $\delta \to 0$ the vortex blob computations approximate the real viscous fluid. Moreover, the computations show that the viscous core vorticity also oscillates. It remains to be investigated whether these oscillations persist, under appropriate limits, long enough to induce the irregular particle motion observed for the vortex sheet.

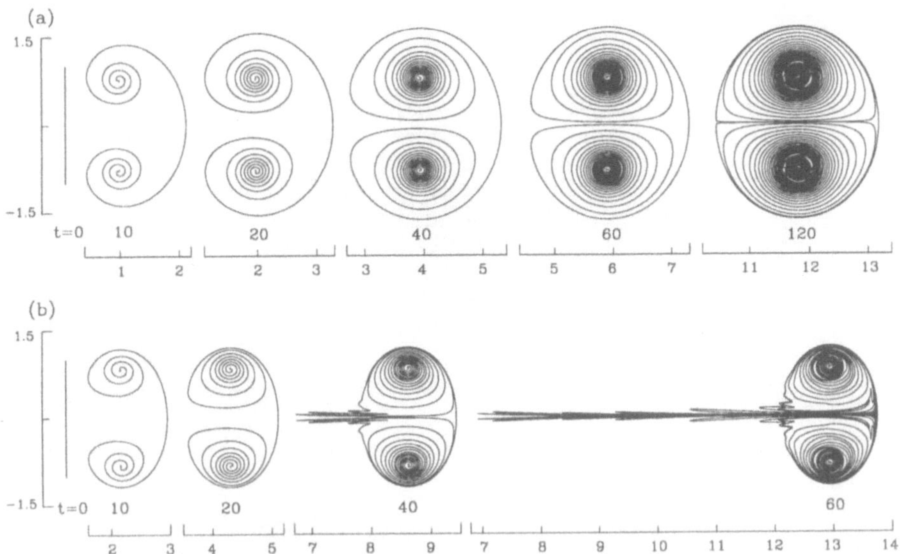

Figure 2. Computed solution at the indicated times, using the vortex blob method with $\delta = 0.2$. (a) Planar. (b) Axisymmetric.

2. Vortex Blob Computations

The initial vortex sheet shown in Figure 1 is nondimensionalized by the vortex radius and the total circulation Γ. The vorticity distribution is described by the circulation between the axis and any point y on the plate, $\Gamma(y) = \sqrt{1 - y^2}$. The initial vorticity distribution is regularized under the vortex blob method. The sheet is discretized by a finite number of regularized filaments whose evolution is solved by a system of ordinary differential equations. The system is solved using the Runge Kutta method. In addition, new filaments are inserted if the angular separation between two filaments exceeds a set parameter. For details we refer to Krasny & Nitsche (1997).

2.1. ONSET OF IRREGULAR BEHAVIOUR

Figure 2 shows the computed vortex sheet evolution with a fixed value of the smoothing parameter, $\delta = 0.2$. The planar vortex sheet rolls up into a vortex pair (Fig. 2a), the axisymmetric one forms a vortex ring (Fig. 2b).

Several large scale differences between the two cases are noticeable. The planar roll-up is larger than the axisymmetric one and more symmetric. It travels about half as fast as the axisymmetric ring. The planar flow also appears more regular than the axisymmetric one, at least near the rear of the vortex. In the axisymmetric case, some particles leave the bubble of fluid travelling with the ring and are left behind. They form a large tail, not unlike tails seen in experiments (Glezer 1988).

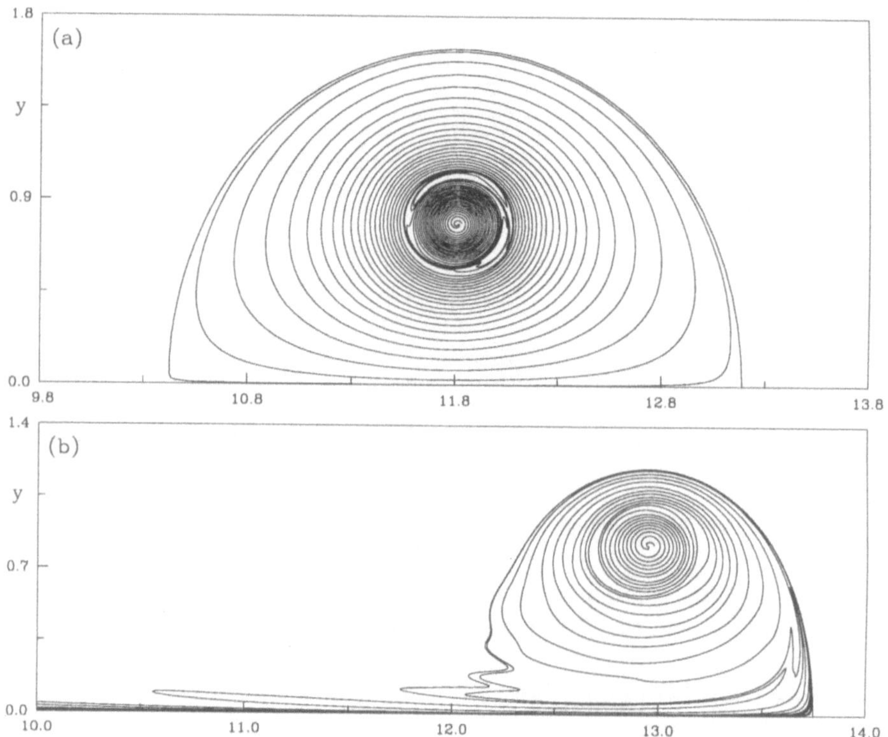

Figure 3. Closeup of the computed solution ($\delta = 0.2$). a) Planar sheet at $t = 120$. b) Axisymmetric sheet at $t = 60$.

Near the vortex core, both the planar and the axisymmetric ring develop irregular behaviour. This can be seen in a closeup of the flow at the last time in Fig. 2, shown in Fig. 3. A gap forms between two vortex sheet turns in an annular region about the vortex center. Once the gap forms, the vortex sheet folds over and the particle motion quickly becomes quite complex.

Figure 2 showed the onset of irregular behaviour in time, with a fixed value of δ. Figure 4 shows the computed solution at a fixed time, $t = 10$, for decreasing values $\delta = 0.4, 0.2, 0.1$. As $\delta \to 0$, the outer turns of the vortex sheet appear to converge. The flow near the core however develops similar structures as observed in Fig. 3. This is shown in Fig. 5, which plots a closeup of the planar flow (left column), and the axisymmetric flow (right column), at $t = 10$, using the indicated values of δ. Notice that once the irregular particle motion appears the sheet quickly folds into a complex structure.

The observations made above raise several questions. One may ask what happens to the irregular structures as time increases, or at a fixed time, as $\delta \to 0$. Do they grow and dominate the flow? First however we wish to know, what mechanisms explain the formation of these structures? Furthermore, are they also present

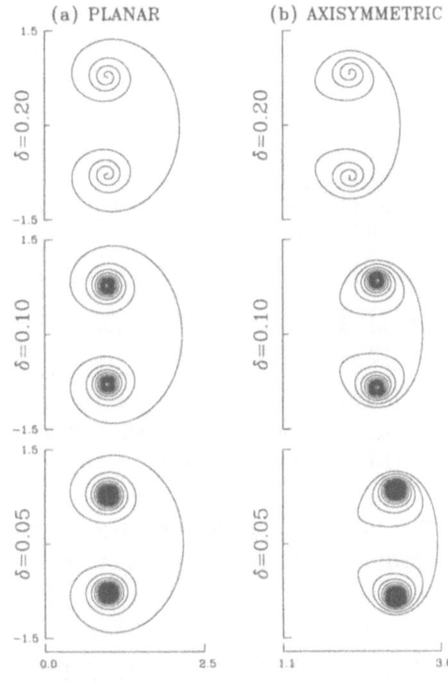

Figure 4.
Computed solution at $t = 10$, with the indicated values of δ. a) Planar sheet. b) Axisymmetric sheet.

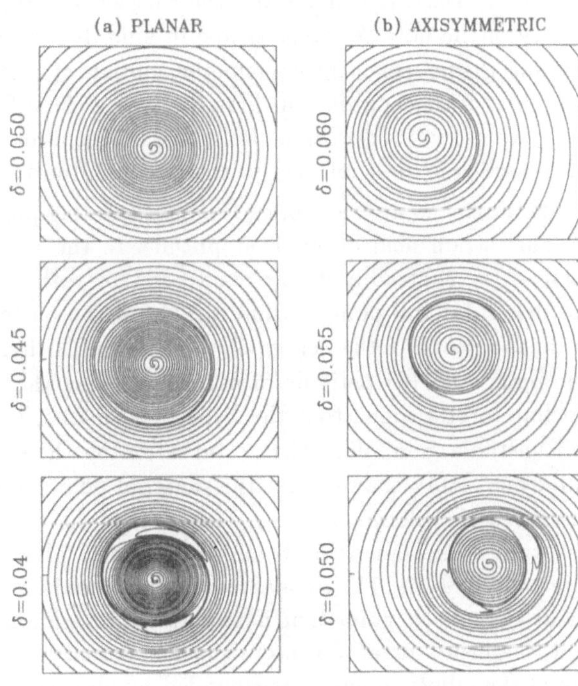

Figure 5.
Closeup of the computed solution at $t = 10$, with the indicated values of δ. a) Planar sheet. b) Axisymmetric sheet.

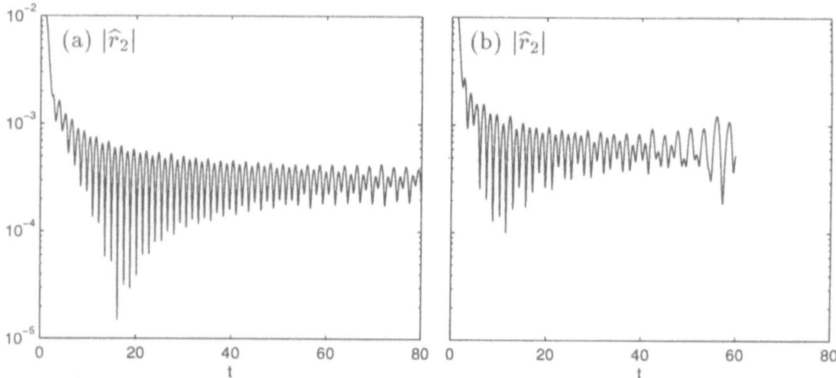

Figure 6. Magnitude of the elliptical component $\hat{r}_2$ of the vorticity contour $\omega = \omega_{max}/2$, for the vortex sheet computed with $\delta = 0.2$. (a) Planar case. (b) Axisymmetric case.

in real viscous flows? If so, this may be one of the mechanisms by which these flows become turbulent. The remainder of this paper addresses the last two of these questions.

2.2. DYNAMICS IN A PERTURBED HAMILTONIAN SYSTEM

The vortex sheet is a material curve whose shape contains information about the history of the flow. A recent theme in fluid dynamics research is that dynamical systems theory can help explain the motion of material points in a fluid flow. Examples of this approach include the work of Shariff, Leonard & Ferziger (1989). Here, we discuss the work of Rom-Kedar, Leonard & Wiggins (1992, henceforth RLW) which is most relevant to the present situation.

RLW studied a system of two counter-rotating point vortices moving in a time-periodic strain field of small amplitude. In the absence of the strain field, the vortices travel steadily downstream together with a bubble of fluid moving with the pair. Material points inside the bubble lie on four types of orbits (1) elliptic fixed points at the center of the motion (2) hyperbolic fixed points on the symmetry axis (3) periodic orbits surrounding the elliptic points, and (4) heteroclinic orbits connecting the hyperbolic points, through the associated stable and unstable manifolds. The perturbation by the strain field changes the material orbits. RLW showed that the material points are governed by a periodically perturbed Hamiltonian system. The effect of the perturbation is described by two well-known phenomena in dynamical systems. These are a *resonance band* and a *heteroclinic tangle*. The resonance band is formed by the breakup of periodic orbits about the elliptic point that have rational rotation number. Under the perturbation, the orbits break up into a sequence of elliptic and hyperbolic points. Each one of the elliptic points is surrounded by an island which forms a barrier to the

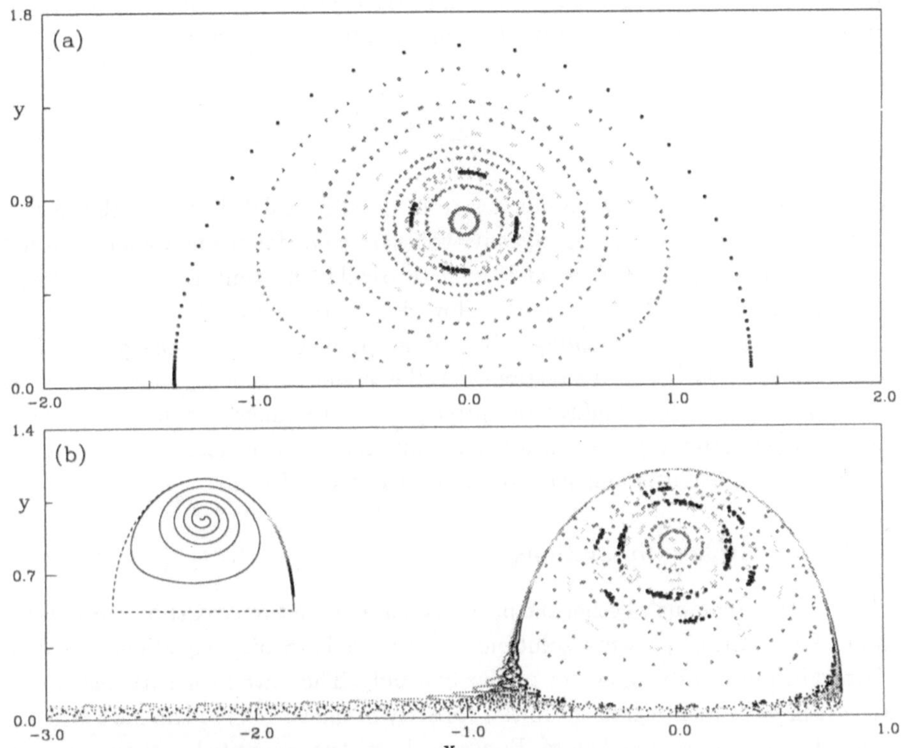

Figure 7. Poincare sections of the computed flow, using the dominant period in the core vorticity oscillations ($\delta = 0.2$). a) Planar flow. b) Axisymmetric flow.

flow. The presence of such islands is guaranteed by the KAM theory. The islands are surrounded by chaotic flow in an annular region called the resonance band. The heteroclinic tangle is a region of chaotic behaviour that forms near a hyperbolic stagnation point, if the associated stable and unstable manifolds intersect transversely. The situation is described in detail by RLW.

The flow computed here is similar to the flow studied by RLW. By analogy to their results, the irregular core dynamics and the associated gaps observed in Fig. 5 are attributed to the development of a resonance band and associated islands in the flow. The irregular behaviour near the tail in the axisymmetric case is associated to the development of a heteroclinic tangle. However, in order to confirm this, we needed to find a natural frequency in the computed flow which would allow us to plot a Poincaré section for it, and show the presence of islands directly, not by analogy to a different flow.

Kida (1982) had shown that an elliptical planar patch of vorticity immersed in a strain field oscillates periodically about a mean, under appropriate conditions. This motivated us to investigate the vorticity contours in the core, and, in particular, the elliptical component of these contours. We consider the contour $\omega = \omega_{max}/2$, where ω_{max} is the maximal vorticity. This contour may be written as

$$r(\theta) = \sum_{k=N}^{N} \hat{r}_k e^{2\pi i k \theta} .$$

The second mode in this expansion describes the elliptical distortion of the contour. Its magnitude is shown in Fig. 6, as it evolves in time, for the flow computed with $\delta = 0.2$. The elliptic component is seen to oscillate in time with a well defined dominant frequency. This frequency allowed us to compute a Poincaré section for the flow, obtained by sampling orbits at a specified value of the phase of the perturbation. The Poincaré sections for the planar and axisymmetric flow are shown in Fig. 7. They confirm the presence of a resonance band and immersed islands (black orbits), as well as a heteroclinic tangle in the axisymmetric case, in the location of the irregular patterns seen in Fig. 3.

3. Navier Stokes Simulations

The question remains whether the patterns observed in Fig. 3 exist in a real viscous fluid. This section presents solutions to the Navier-Stokes equations governing viscous incompressible flow, for planar flow only. The initial vorticity distribution was taken to be identical to the initial vorticity in the vortex blob method. The flow is nondimensionalized as before. Figure 8 shows the computed vorticity contours at $t = 10$, for $\delta = 0.4, 0.2, 0.1$, and two values of the Reynolds number, Re=1,000, 10,000. The results indicate that as Re$\rightarrow \infty$ and $\delta \rightarrow 0$ the results approximate the vortex blob results (last column).

Figure 9 shows the time evolution of the magnitude of the elliptical component of the vorticity contour $\omega = \omega_{max}$, computed for Re=100,000, $\delta = 0.2$. It shows that this component oscillates in time, as in the vortex blob computations, although its magnitude decreases in time. This decrease is attributed not to viscous diffusion - the Reynolds number in this case is quite large - but to a rearrangement of vorticity not possible in the vortex blob evolution. Thus, the mechanism to which we attribute the onset of chaotic behaviour in the vortex blob results is also present in the viscous case. It remains to be investigated whether they persist long enough, for sufficiently small viscosity ν and initial smoothing δ, to induce the irregular particle motion observed for the vortex sheet.

4. Conclusions

This paper demonstrated the onset of irregular flow patterns in the roll-up of planar and axisymmetric vortex sheets. This behaviour is caused by self-sustained

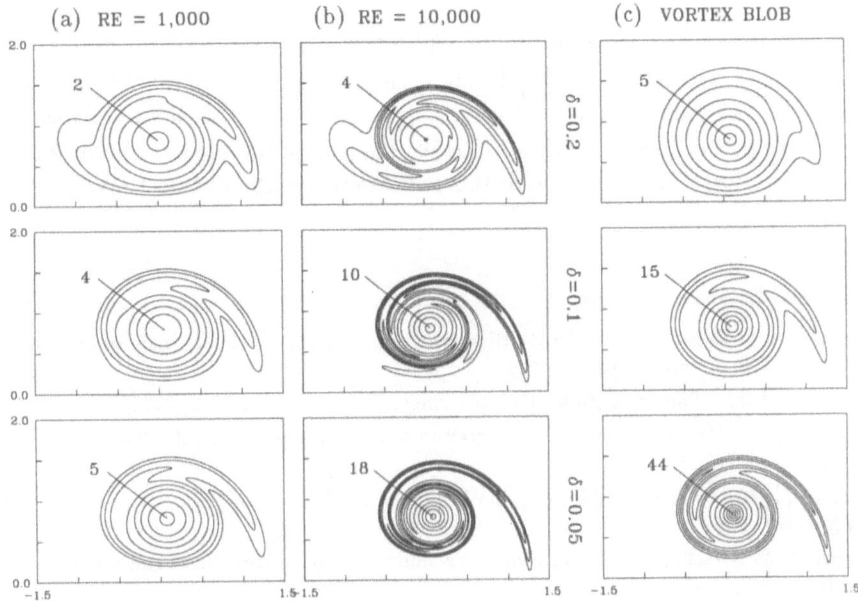

Figure 8. Vorticity contours, computed for $\delta = 0.4, 0.2, 0.1$ (as indicated). (a) Re=1,000. (b) Re=10,000. (c) Vortex blob method. The levels $\omega = 2^{-j}, j = -5, 5$ are shown and the maximum vorticity is indicated.

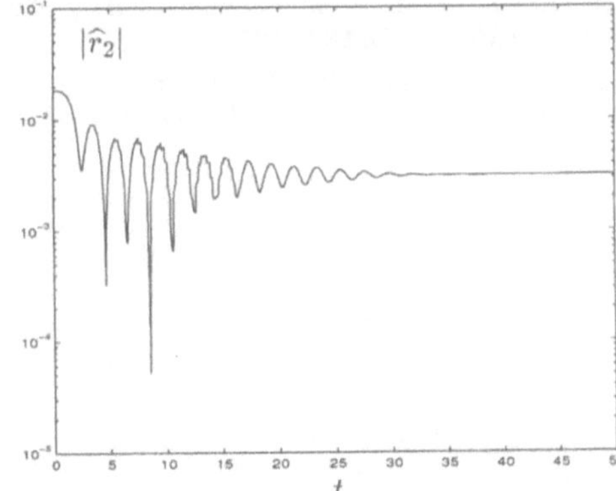

Figure 9.
Magnitude of the elliptical component $\widehat{r}_2$ of the vorticity contour $\omega = \omega_{max}/2$, for the viscous flow computed with Re=100,000, $\delta = 0.2$.

oscillations in the core vorticity. It was shown that such oscillations are also present in a real viscous flow, although the actual irregular structures near the core have not yet been found in this case.

Acknowledgements

This paper is based on extensive joint work with Robert Krasny. The viscous results were obtained with Mark Taylor during a visit last summer at NCAR. I wish to thank both for helpful comments and suggestions. I also wish to acknowledge support from NSF grant # DMS-9729158. The computations were performed at the Institute for Mathematics and its Applications, at Ohio State University and at NCAR.

References

Chorin, A.J. & Bernard, P.S. (1973) Discretization of a vortex sheet, with an example of roll-up, *J. Comput. Phys.* **13**, 423.

Glezer, A. (1988) The formation of vortex rings, *Phys. Fluids* **31**(12), 3532.

Kida, S. (1981) Motion of an elliptic vortex in a uniform shear flow, *J. Phys. Soc. Japan* **50**(10), 3517.

Klein, F. (1910) Über die Bildung von Wirbeln in reibungslosen Flüssigkeiten, *Z. Math. Phys.* **59**, 259.

Krasny, R. & Nitsche, M. (1997) Similarity and chaos in vortex sheet roll-up, *in preparation.*

Neu, J.C. (1984) The dynamics of a columnar vortex in an imposed strain, *Phys. Fluids* **27**(10), 2397.

Rom-Kedar, V., Leonard, A. & Wiggins, S. (1990) An analytical study of transport, mixing and chaos in an unsteady vortical flow, *J. Fluid Mech.* **214**, 347.

Saffman, P.G. (1992), *Dynamics of Vortex Flows, Cambridge, NY*

Shariff, K., Leonard, A. & Ferziger, J.H. (1989) Dynamics of a class of vortex rings, *NASA TM-102257* **184**, 123.

Tryggvason, G., Dahm, W.J.A., & Sbeih, K. (1991) Fine structure of vortex sheet roll-up by viscous and inviscid simulation, *J. Fluids Engin.* **113**, 31.

Session 3

Vortices in Shear Layers

DYNAMICS OF SLENDER VORTICES NEAR THE WALL IN A TURBULENT BOUNDARY LAYER

FAZLE HUSSAIN & WADE SCHOPPA
University of Houston
Department of Mechanical Engineering, Houston, TX 77204-4792

Abstract

It is now well-established that the enhanced drag and heat transfer of turbulent boundary layers are dominated by slender longitudinal vortices near the wall. Despite their immense practical significance, the geometry and dynamics of near-wall vortices are poorly understood and often controversial. To gain new insight into the near-wall dynamics, we educe *coherent structures* (CS) from a numerically simulated turbulent channel flow – using a conditional sampling scheme which extracts the entire extent of dominant vortical structures. Such structures are detected from the instantaneous flow field using our newly developed vortex definition – a region of negative λ_2, the second largest eigenvalue of the tensor $S_{ik}S_{kj} + \Omega_{ik}\Omega_{kj}$ – which accurately captures the structure details, unlike velocity, vorticity or pressure-based eduction. Extensive testing shows that λ_2 correctly captures vortical structures, even in the presence of strong shear as occurring near the wall of a boundary layer. The CS are elongated quasi-streamwise vortices, inclined 9° in (x,y) and tilted ±4° in (x,z), with vortices of alternating sign overlapping in x as a staggered array. Notably, the often heralded hairpin vortices, not to be confused with hairpin-shaped vortex line bundles, are absent both in the instantaneous and ensemble-averaged fields. Our conceptual model of the CS array reproduces nearly all experimentally observed events reported in the literature, such as VITA/VISA, Reynolds stress distributions (Q1, Q2, Q3 and Q4 and their relative contributions), wall pressure variation, elongated low-speed streaks, spanwise shear, *etc.* We also develop a new CS regeneration mechanism, in which existing CS leave behind lifted low-speed streaks, whose instability in turn initiates the formation of new streamwise vortices.

1. Introduction

Large-scale *coherent structures* (CS) have been the primary focus of turbulence research for the past four decades (*e.g.* see Kline *et al.* 1967; Crow & Champagne 1971; Cantwell 1981; Fiedler 1988; Robinson 1991). This intense research onslaught was motivated by the expectation that CS hold the key to understanding fundamental flow physics and to modeling turbulence as well as to developing control strategies for turbulence phenomena such as heat and mass transport, entrainment, mixing, combustion, and generation of drag and aerodynamic noise. In fact, we have long contended that turbulence control is possible only in the presence of CS: *i.e. no CS, no control.* While it is clear that CS must be incorporated in viable turbulence theories and models, it remains equally unclear as to

E. Krause and K. Gersten (eds.), IUTAM Symposium on Dynamics of Slender Vortices, 155-172.

how this is to be done. CS-based analytical approaches (*e.g.* Poje & Lumley 1995; Goldshtik & Hussain 1995) as well as dynamical systems modeling (*e.g.* Aubry *et al.* 1988; Broze & Hussain 1994) appear promising but fall far short of producing a viable turbulence theory. To reiterate: CS are central to turbulence in its understanding, theory, modeling, and control.

Primarily because of their pervasive technological relevance, recurring large-scale events in turbulent boundary layers have been studied extensively (mostly by flow visualization and single-sensor measurements, some by multi-sensors and more recently by DNS, LES, and flowfield quantitative study by optical methods). Yet very little is known about their geometrical features, let alone the evolutionary dynamics of CS. This understanding is crucial for modeling and realistic control of boundary layers. Numerous types of CS have been proposed to explain experimentally observed phenomena – notably, the celebrated "bursting" process – in turbulent boundary layers; see Cantwell (1981) and Robinson (1991) for detailed reviews. The various boundary layer CS proposed to date differ not only in their geometry, strength, and orientation, but also in their dynamical roles; furthermore, there is also no consensus on their evolutionary dynamics.

While several types of CS may in fact coexist in the outer region, the predominance of quasi-streamwise vortices in the buffer layer is now well accepted (*e.g.* Townsend 1956; Bakewell & Lumley 1967; Blackwelder & Eckelmann 1979; Robinson 1991). In the following, we focus exclusively on the turbulence physics in the immediate vicinity of the wall, *viz.* $y^+<60$, which is of direct relevance to turbulence production (peaking at $y^+\cong12$), and drag and heat transfer control. Although the outer layer also contains (larger) structures, our view is that their role is indirect in near-wall phenomena such as drag and heat transfer, and that the essential inner layer dynamics are autonomous (*i.e* essentially decoupled from outer layer events). Our primary objectives are to identify and delineate the geometrical and dynamical details of dominant, near-wall CS as well as to explain, by using tractable vortex dynamics arguments in the context of the educed CS, the events of significance near the wall recorded in previous conditional averaging experiments. A salient feature of our study is the eduction (*i.e.* by ensemble averaging after relative alignment in (x,z)) of the 3D field of the underlying CS from numerous 3D flow realizations using conditional sampling techniques, so that a generic model (not identical to any instantaneous event, consisting of coherent and incoherent motions) is deduced to reveal the CS and elucidate the dominant near-wall dynamics (Jeong & Hussain 1992; Hussain 1986).

The numerous boundary layer conditional averaging studies to date, typically based on Q2/Q4 (Willmarth & Lu 1972) or VITA (Blackwelder & Kaplan 1979) techniques, are rather inconclusive regarding CS. Since these stategies are based upon local single-point velocity, rather than 3D vorticity, data, they cannot directly extract vortical structures. The present study is consistent with our longstanding characterization of CS in free shear flows by instantaneous space-correlated vorticity underlying the otherwise random vorticity field of turbulence. In this regard, the widely employed method of flow visualization (*e.g.* to observe the bursting phenomenon) suffers from the limitation of being frequently confusing or even misleading (Bridges *et al.* 1990; Kida *et al.* 1991) and should be used only to supplement quantitative data. In near-wall turbulence, our prior CS definition requires modification since vortical structures cannot be detected simply by vorticity magnitude. This raises an interesting issue: how can near-wall CS and vortices in general be rigorously defined?

The concept of vortices is as old as the subject of hydrodynamics; yet, an accepted definition of a vortex is still lacking. Clearly, to utilize the CS concept, we must have effective means of identifying dynamically significant, large-scale vortices, particularly in turbulent flows, as a necessary first step, which in turn necessitates an objective definition of a vortex. We will demonstrate that popular intuitive measures, such as closed streamlines, low pressure, and vorticity magnitude are generally inappropriate as vortex identifiers. In this paper, we summarize a rigorous new definition (Jeong & Hussain 1995) which identifies vortex cores in any flow. We also discuss a general-purpose eduction scheme applicable to near-wall turbulence, which extends the algorithm developed by Hussain & Hayakawa (1987) for free shear flows.

In this paper, our objectives are twofold: (i) to develop a rigorous CS eduction scheme applicable to near-wall turbulence, based upon our new vortex definition, and (ii) to capture, for detailed analysis, all of the near-wall features classified by Kline & Robinson (1990) in a single ensemble-averaged flow field (containing the CS). Experimental acquisition of instantaneous velocity field data of evolving three-dimensional flows is currently impossible due to limitations of measurement technology, pending development of fully 3D measurement techniques such as holographic particle velocimetry (*e.g.* Meng & Hussain 1991). Thus, we use a well-resolved direct numerical simulation (DNS) database of turbulent channel flow (Kim *et al.* 1987 database) for CS eduction.

The remainder of the paper is organized as follows. In §2, we develop a general-purpose vortex definition for application to CS eduction in a turbulent channel flow (§3), including vortex dynamics-based interpretations and comparisons with previously recorded events (§4). A spatiotemporal CS regeneration scenario is outlined in §5, followed by some concluding remarks in §6.

2. Vortex Identification

Our eduction approach for free shear flows, in which CS are identified as compact regions of large-scale vorticity, is inapplicable for boundary layers. For example, vorticity is the largest at the wall, where no large-scale CS are present. The mere identification of near-wall vortices, typically embedded within the background ω_z, has proven to be a major challenge in itself. Popular free shear flow vortex identification techniques, such as $|\omega|$ isosurfaces or vortex line bundles, can be misleading and are generally ineffective near the wall (where $|\omega|$ is very large even outside CS primarily due to ω_z). As alternatives, low pressure, ω_x, and closed streamlines in (y,z) planes have all been used in other studies to identify streamwise vortices in DNS data.

2.1 Intuitive Quantities for Vortex Identification and Their Limitations

If vorticity magnitude is to be an indicator of a vortex, a subjective threshold has to be selected to identify vorticity patches, since vorticity is everywhere present in viscous flow. The presence of a high vorticity level is necessary but not sufficient for a vortex to exist; for example, a laminar shear flow has vorticity everywhere, but no vortex. As a sophistication of this intuitive idea, a vortex is often considered to be a vortex tube whose surface consists of vortex lines. However, the existence of a vortex tube does not imply the presence of a vortex; for instance, a vortex tube in a laminar pipe flow is not a vortex in any sense. Before we attempt to define a vortex, we will first clarify what is expected of an

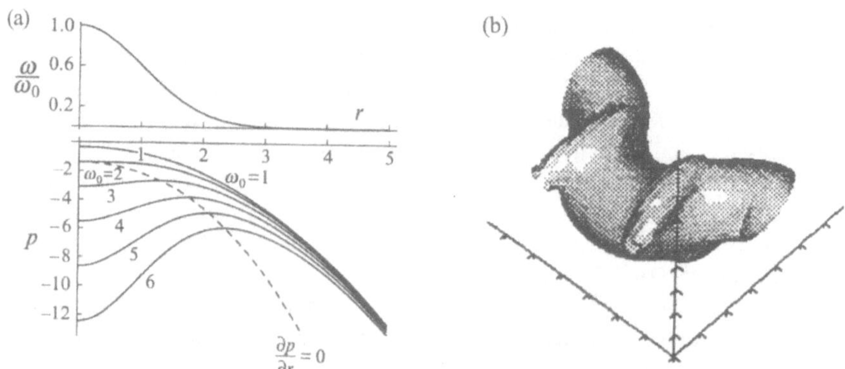

Figure 1. (a) Pressure and vorticity distributions in a Burgers vortex; ω_0 is the vorticity value at the vortex center; (b) isopressure surface in a transitional plane mixing layer.

effective definition/identification scheme.

The first necessary condition imposed is that regions identified as vortex cores must have net vorticity, *i.e.* nonzero circulation. Thus, potential flow regions are excluded from vortex cores, and a potential vortex is a line vortex. In viscous flows, we obviously expect vortices to have non-zero cross-sections. Secondly, we require the indicator to be Galilean invariant (*i.e.* identical in all inertial reference frames).

In the following, we consider three commonly used indicators of vortices: (i) low-pressure regions; (ii) regions with closed or spiral streamlines and pathlines; and (iii) iso-vorticity surfaces; for each quantity, counter-examples are presented to illustrate their inadequacy in detecting/identifying vortices in unsteady 3D flows.

Local pressure minimum. The physical reasoning for associating vortices with low pressure regions is that pressure tends to have a local minimum within regions of swirling motion when the centrifugal force is balanced by the pressure force, the so-called *cyclostrophic balance* – a simple consequence of the Euler equations for rotating flows. For instance, Robinson (1991) used a low pressure criterion to identify vortical structures in DNS data of a turbulent boundary layer. However, a cyclostrophic balance is guaranteed only when the flow is inviscid, 2D, and steady. As a counter-example, a well-defined pressure minimum can exist in an unsteady irrotational flow as well; obviously, this flow cannot contain a vortex because all contours enclose zero circulation. To illustrate, consider an *unsteady* irrotational, axisymmetric motion with a stagnation point: $u_r = -a(t)r$, $u_\theta=0$, $u_z = 2a(t)z$, where $a(t)$ is the unsteady strain rate and (r,θ,z) are the radial, azimuthal, and axial coordinates respectively. Upon integration of the Euler equation, we find that the pressure (P) is given by

$$P = (\mathrm{d}a/\mathrm{d}t - a^2)\, r^2/2 + (-\mathrm{d}a/\mathrm{d}t - a^2)\, z^2. \tag{1}$$

When $\mathrm{d}a/\mathrm{d}t - a^2 > 0$, the pressure has a local minimum in any (r,θ) plane, even though the flow has no vorticity, and hence, no vortex core. Similar problems in identifying vortices by low pressure regions arise in several other cases, including irrotational sink and source flows and Karman's viscous pump (where centrifugal forces are balanced by viscous effects). An inverse example is a 2D Stokes flow, where there is no pressure minimum even within vortical motion (explained by the maximum principle for the Poisson equation for pressure).

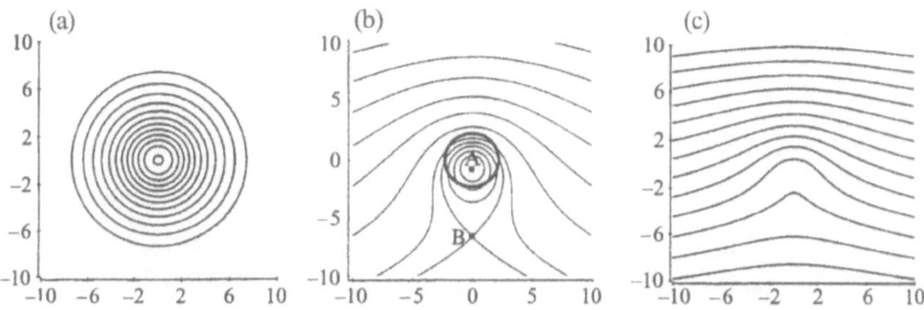

Figure 2. Streamlines of a Lamb vortex in reference frames moving (a) with the vortex center; (b) with a point within the vortex core; (c) faster than any point.

In addition to the above difficulties, the length scale of pressure may be vastly different from that of the vortex core. For the case of Burgers vortex, this scale difference is clearly illustrated in Fig. 1(a), where the lengthscale of vorticity is fixed, while the lengthscale of the pressure hump (*i.e.* the region of $\partial p/\partial r>0$) increases with the value of vorticity on the axis. This inherent scale difference between a vortex core and the associated low-pressure region (pressure, being a solution of the Poisson equation, is a nonlocal variable) may disallow a clear demarcation of weaker, yet dynamically significant vortices, especially when these are located near stronger vortices. In fact, in DNS data of a plane mixing layer, isopressure surfaces cannot capture rib (longitudinal) and roll (spanwise) vortices simultaneously (Fig. 1b); while both are dynamically dominant, the pressure level is quite different within the two vortices (see also Jeong & Hussain 1995).

Closed pathline and streamline. The use of pathlines and streamlines to qualitatively identify vortices (Lugt 1979) is fundamentally inappropriate as these are not Galilean invariant. Thus, in different reference frames, drastically different "vortices" are observed from the streamline pattern (Fig. 2). Another critical inadequacy of the closed pathline criterion is that a particle may not complete a full revolution around the vortex center (hence no closed pathline) during the lifetime of a vortex. This can occur if vortices undergo transition to turbulence due to nonlinear processes such as pairing, tearing, reconnection, core dynamics, or fine-scale transition, even though these vortices may have been dynamically significant during their lifetime. Thus, even in a swirling motion with closed streamlines, pathlines are not necessarily closed.

Vorticity magnitude surface. Vorticity magnitude is undoubtedly the most widely used quantity to represent vortices in experimental eduction and DNS data. While typically adequate for wall-free flows, this can pose a problem in wall-bounded flows. In fact, in a 2D wall-bounded flow, the maxima and minima of vorticity magnitude must occur at the wall (Lugt 1979), where a vortex obviously does not exist. In a turbulent boundary layer, the maximum vorticity magnitude occurs at the wall directly below the high-speed streaks (Jimenez *et al.* 1988). In both cases, the near-wall flows do not exhibit any swirling motion, characteristic of vortices. In addition, the use of vorticity magnitude or isovorticity surfaces to identify vortices can be inappropriate even in free shear flows. For instance, vorticity sheets or layers with locally parallel flow are not vortices. Additional counter-examples include misrepresentation of a vortex embedded in a homogeneous shear flow (Fig. 3b) and an axisymmetric vortex with axial variation of core size and vorticity (Fig. 3a). Generally speaking, nonzero vorticity is a necessary, but not sufficient condition for vortical motion. In fact, vortices with strong (*i.e.* internal) core dynamics,

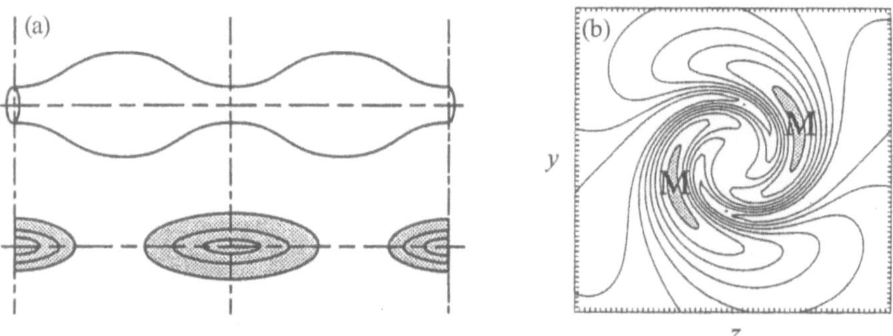

Figure 3. (a) Axisymmetric vortex with an axial variation in vorticity: top panel - vortex surface, bottom panel - vorticity contours; (b) contours of vorticity magnitude for an inviscid streamwise vortex embedded in a homogeneous shear flow.

including those involving vortex breakdown, have regions of negligible vorticity on the axis.

2.2 PRIOR DEFINITIONS

In this section, we briefly introduce two recently proposed, rather sophisticated measures for vortex identification (which satisfy the two requirements in §2.1), based on local characteristics of the velocity gradient tensor $\partial u_i/\partial x_j$.

Complex eigenvalues of velocity gradient tensor: $\Delta>0$. Dallmann (1988) and Chong *et al.* (1990) proposed to identify vortices as flow regions where the velocity gradient tensor exhibits complex eigenvalues. The physical motivation of this criterion is that in a reference frame moving with a particular fluid particle, the existence of complex eigenvalues of the velocity gradient implies a local spiral/swirling motion. Computationally, the boundary of the region with complex eigenvalues is easily obtained from the discriminant of the cubic characteristic equation

$$x^3 - Px^2 + Qx - R = 0, \tag{2}$$

where P, Q and R are the first, second, and third invariants of the velocity gradient tensor. Complex eigenvalues occur when the discriminant $\Delta = (Q/3)^3 + (R/2)^2 > 0$, so that the vortex core boundary is defined in this case by the surface $\Delta = 0$. Note that since this definition is derived from $\partial u_i/\partial x_j$, it is Galilean invariant.

Second invariant of velocity gradient tensor: $Q>0$. Hunt *et al.* (1988) defined a vortex core (eddy) as a region with positive $Q = (\Omega_{ij}\Omega_{ij} - S_{ij}S_{ij})/2$, where Ω_{ij} and S_{ij} are the antisymmetric and symmetric parts of $\partial u_i/\partial x_j$ respectively. Physically, positive Q identifies regions where the vorticity magnitude is disproportionately larger than the strain rate, thus excluding nearly parallel vorticity layers. This condition can also be expressed as $-(\lambda_1+\lambda_2+\lambda_3)/2 > 0$, where $\lambda_1>\lambda_2>\lambda_3$ are eigenvalues of the tensor $\Omega_{ik}\Omega_{kj} + S_{ik}S_{kj}$.

Similarly, Truesdell (1953) introduced the "kinematic vorticity number", defined as

$$N_k=(1+2Q/S_{ij}S_{ij})^{1/2}, \tag{3}$$

which measures the "quality" of rotation, distinguished from the local rotation rate given by $|\omega|$. Obviously, in terms of vortex core demarcation, regions with $N_k>1$ have $Q>0$ and vice-versa; thus, these two definitions are operationally equivalent. Since N_k is based on the ratio of vorticity to strain rate, it unfortunately detects as vortices regions with

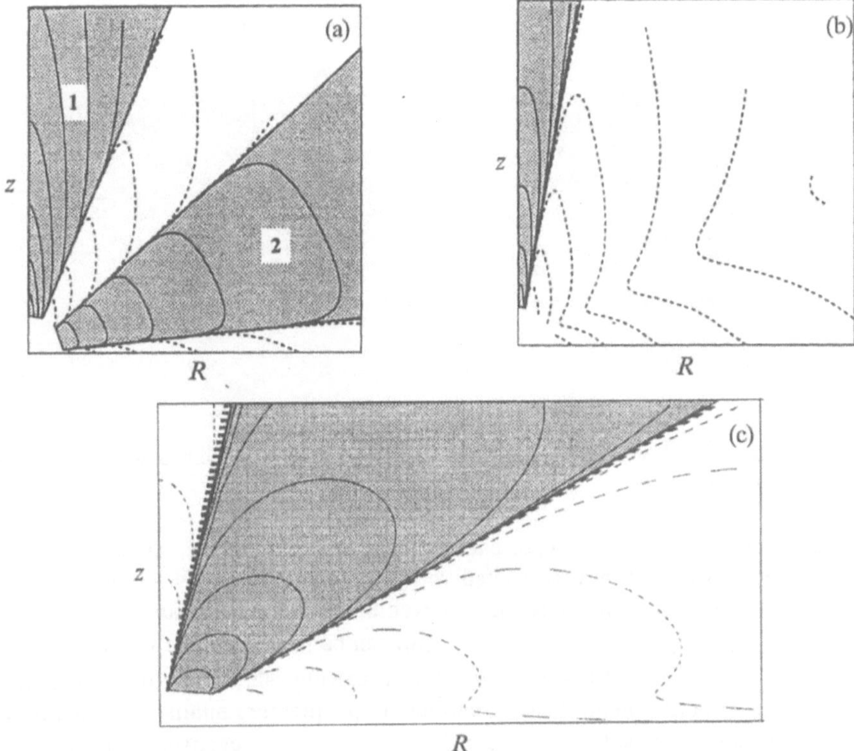

Figure 4. Vortex cores in a swirling jet with conical symmetry, depicted by shading for the (a) $\Delta>0$; (b) $\lambda_2<0$; (c) $Q>0$ vortex core identification schemes. The radius in (c) is expanded 6 times that in (a) and (b).

extremely low vorticity and negligible strain rate, even though they may not be dynamically significant.

It can be shown that both Q and Δ vanish identically at a no-slip wall, unlike vorticity magnitude, and thus may be suitable for vortex identification in boundary layer flows. Nevertheless, we will demonstrate that these definitions fail as well under certain circumstances, thus motivating our development of a new definition.

2.3 NEW VORTEX DEFINITION: $\lambda_2<0$

Although low pressure regions often poorly represent vortices as shown earlier, this concept serves as a starting point for our new definition (Jeong & Hussain 1995). In particular, we seek to isolate and eliminate consideration of non-vortical contributions to the pressure minimum, in particular, unsteady straining and viscous effects. Since information on local pressure extrema is contained in the Hessian of pressure ($P_{,ij}$), we seek an equation involving $P_{,ij}$, derived as the gradient of the Navier-Stokes equations, where the symmetric part is the tensor equation

$$DS_{ij}/Dt - \nu S_{ij,kk} + \Omega_{ik}\Omega_{kj} + S_{ik}S_{kj} = -P_{,ij}/\rho. \qquad (4)$$

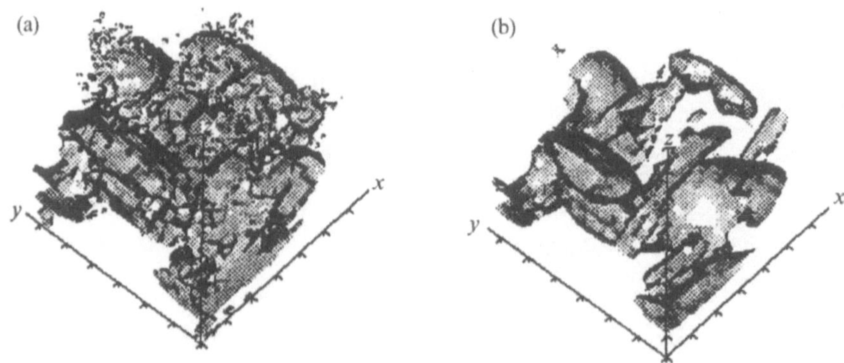

Figure 5. Vortex cores in DNS data of a transitional mixing layer, depicted as isosurfaces defined by the (a) $\Delta>0$; (b) $\lambda_2<0$ vortex identifiers.

The first term on the LHS represents unsteady straining effects, which can cause a pressure minimum without any associated vortical motion. In highly viscous flows, the second term tends to mask pressure minima even in the presence of strong vortices (*e.g.* 2D Stokes flow). In this sense, $\Omega_{ik}\Omega_{kj}$ and $S_{ik}S_{kj}$ can be isolated as the vortical contributions to pressure minima. Elementary variational calculus shows that the sum of these terms attains a local maximum (hence contributing to a pressure minimum) if and only if two eigenvalues (guaranteed to be real) of $\Omega_{ik}\Omega_{kj} + S_{ik}S_{kj}$ are negative. Equivalently, we *define a vortex core as a connected spatial region in which the* $\Omega_{ik}\Omega_{kj} + S_{ik}S_{kj}$ *has a negative second eigenvalue* λ_2.

2.4 Comparison of Definitions

In this section, we compare this λ_2-definition with the Δ- and Q-definitions discussed above, using illustrative DNS and analytical solutions, in which the vortex geometries are intuitively clear. We justify λ_2 as a more effective vortex indicator by demonstrating its success in flows where the other definitions are clearly inadequate.

Interestingly, these three definitions become identical in planar 2D flow, despite their varied physical foundations and the fact that they are not explicitly related. Thus, in simple planar flows, these three definitions are all equally effective in identifying compact vortex core regions. Furthermore, in axisymmetric flows with $v_\theta=V(r)$, $v_r=v_z=0$, all three definitions require $\mathrm{d}V^2/\mathrm{d}r > 0$, thereby excluding the surrounding, essentially potential region with decreasing azimuthal velocity in r. Nevertheless, as shown by the following counter-examples, in 3D flows (or even axisymmetric flows with swirl), our λ_2-definition is effective over a wider range of vortical flows.

Limitations of $\Delta>0$ definition. A particularly revealing test flow is our analytical model for a tornado (Shtern & Hussain 1993), consisting of a conically symmetric, swirling jet emerging into a half-space with momentum and circulation sources located at the origin. Since this flow has an axial velocity in addition to an azimuthal velocity around the vortex axis (the z axis), it exhibits helical vortical motion. In addition, due to the conical symmetry of this flow, the vortex core boundary should also be conical in shape.

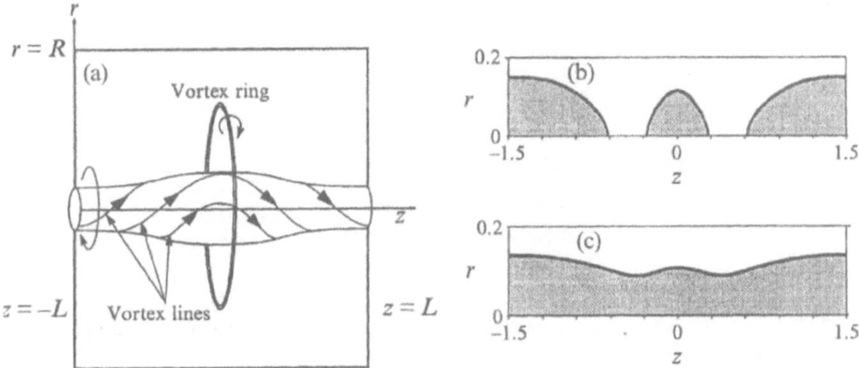

Figure 6. An axial vortex passing through the center of a vortex ring: (a) schematic; shaded vortex cores based on (b) Q>0; (c) λ_2<0 vortex definitions.

The Δ-definition shows two vortex cores marked by grey shading in Fig. 4(a): one on the axis (region 1) and another (region 2) detached from the axis. Because there is significant vortical motion only immediately near the axis, the presence of region 2 for the Δ-definition conflicts with physical intuition. In contrast, $-\lambda_2$ is maximum and positive on the axis, without a spurious detached vortex core (Fig. 4b). The effectiveness of our λ_2-definition for this idealized vortex with axial flow is especially important, noting our finding that CS generally contain helically twisted vortex lines and hence strong axial flow (*i.e.* partially Beltramized).

Additionally, in DNS data, the Δ-definition tends to produce very noisy vortex core boundaries; for example see the rib and roll vortices of a transitional plane mixing layer in Fig. 5. We have ensured that the small scales in Fig. 5(a) for Δ>0 are an artifact of the definition itself and not numerical noise. In contrast, the λ_2-definition reveals both the rib and roll vortices clearly, with smooth vortex core boundaries (Fig. 5b). Vorticity surfaces and contours are similarly smooth, confirming that the small scales in Fig. 5(a) for Δ falsely indicate the presence of fine-scale turbulence.

Limitations of Q>0 definition. In the conically symmetric, swirling jet mentioned above, the vortex core based on the Q>0 definition shows a narrow hollow core along the axis (Fig. 4c). This exclusion of the near-axis region is unphysical because fluid near the axis has the highest vorticity, undergoes nearly solid body rotation, and is clearly an integral part of the vortex. Thus, the Q-definition misrepresents the vortex geometry in this case, unlike our λ_2-definition (Fig. 4b).

As an additional test case of vortical motion with strong axial flow, we consider an inviscid, steady axisymmetric axial vortex embedded within a thin vortex ring, as illustrated in Fig. 6(a). In this flow, the axial vortex maintains its local expansion due to the induced motion of the ring. The Q-definition incorrectly shows disconnected regions (Fig. 6b) along the continuous *axisymmetric* vortex column, implying that the Q-definition becomes inadequate whenever a vortex is expanding locally due to an imposed non-uniform strain field. In contrast, the λ_2-definition correctly reveals a continuous axial vortex core (Fig. 6c).

In summary, among these potential definitions, only our λ_2-definition was found to correctly represent the topology and geometry of vortex cores for the large variety of

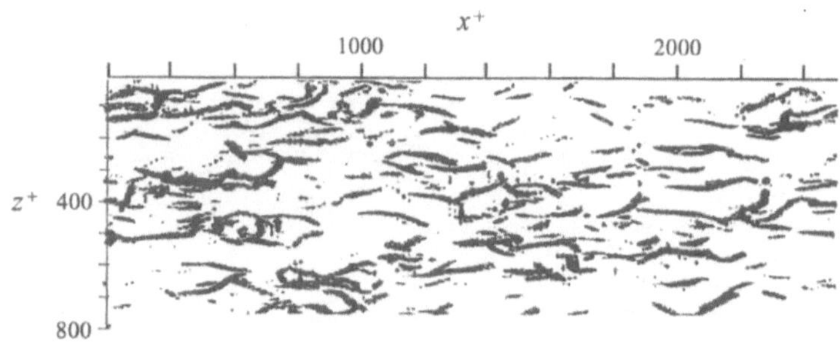

Figure 7. Top view of $\lambda_2=-\lambda_{2\,\mathrm{rms}}$ for near-wall turbulence in the range $0<y^+<60$, illustrating the predominance of quasi-streamwise vortices.

flows considered by Jeong & Hussain (1995). Thus, connected regions with $\lambda_2<0$ accurately demarcate vortex cores for studies of CS dynamics and related flow physics. Based on these results, we refine our vorticity-based definition by suggesting that a coherent structure is defined as a *connected region with phase-correlated negative λ_2 over its spatial extent.* This λ_2-definition has been used to study CS in a number of turbulent flows, of which the following discussion of near-wall turbulence is an example.

3. CS Eduction in Near-wall Turbulence

To establish the most frequently occurring vortex geometry (*i.e.* CS), we ensemble average a large number of properly aligned vortex realizations (*i.e.* CS eduction). The λ_2 definition is first used to quantify fully 3D near-wall vortices (below $y^+=60$) in DNS data (Kim *et al.* 1987), for each sign of ω_x. As shown by a representative λ_2 isosurface (Fig. 7), slender longitudinal vortices, with a streamwise extent of about 200 wall units, are predominant in instantaneous realizations of near-wall turbulence. Note the absence of individual vortices having streamwise lengths of 1000 wall units (comparable to the length of low-speed streaks) initially hypothesized by Blackwelder & Eckelman (1979). The observed vortices appear to overlap along their x extent, with chains of several overlapping vortices as long as 800 wall units (in x). In this near-wall region, no hairpin-type structures are apparent, and data for the entire boundary layer (not shown) demonstrates that z-neighboring vortices are *not* joined by spanwise-aligned heads, as would be the case if these streamwise vortices were the legs of hairpin vortices.

For eduction of these streamwise CS, only fully-formed "mature" vortices are accepted for ensemble averaging, enforced by constraints on the vortices' λ_2 magnitude ($\lambda_2<-\lambda_{2\,\mathrm{rms}}$) and minimum x extent (of $\Delta x^+=150$). To extract the entire extent of fully 3D CS, the maximum permissible deviation of the vortex axis from x is specified as 30° (appropriate for quasi-streamwise vortices). Accepted vortices of a given sign are aligned at the midplane of their x extent before ensemble averaging, which is further enhanced iteratively by relative shifting for maximum correlation between the average and instantaneous events. Our CS eduction algorithm is carefully designed to avoid: i) severe CS "smearing" inherent in velocity-based eduction, ii) false vortex detections (such as different parts of the same structure, different parts of different structures, different CS types), and iii) artificially symmetric ensemble averages (due to (x,z) homogeneity, as discussed

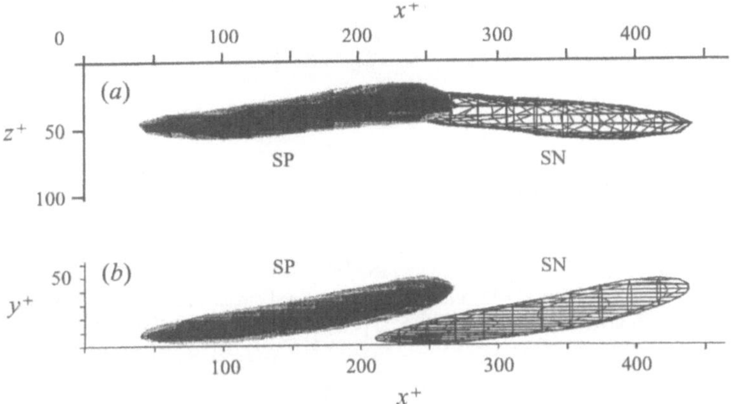

Figure 8. Isosurface plot of educed λ_2 for *SP* and *SN* structures, superimposed to show their relative spatial locations: top panel - top view, bottom panel - side view.

in Johansson *et al.* 1991). Note that once vortex realizations are identified and aligned using λ_2, ensemble-averages of all flow quantities (denoted by $\langle . \rangle$) are available. The departure of instantaneous realizations from the ensemble average of a property is that property's incoherent part. Thus, the eduction not only allows evaluation of the dynamical significance of CS, but also the phase-dependent incoherent turbulence.

4. Conceptual Model of Near-wall Structures

To account for their possible geometric differences, we educe two separate sets of streamwise vortices: those with positive ω_x (called SP) and negative ω_x (SN). As shown in Fig. 8 by $\lambda_2(\partial\langle u_i\rangle/\partial x_j)$, which represents the ensemble-averaged CS, the educed SP and SN are symmetric counterparts, both inclined 9° to the wall and tilted from x on each side by 4° in the top view. The relative spatial orientation of SP and SN is determined by aligning realizations at either end of accepted λ_2 events with $+\omega_x$ or $-\omega_x$. That is, eduction aligned at one end of SP automatically captures the adjacent end of SN and vice-versa. This CS geometry, indicating a general tendency for counter-rotating streamwise vortices to overlap in x as a staggered array, is also evident in (y,z) plane velocity vectors (Brooke & Hanratty 1993). Note that vortices neighboring in z do not survive in the ensemble average, as would otherwise be the case if hairpin vortices were significant.

From the 3D u and ω fields and their ensemble averages, we have developed a conceptual model explaining the connection of documented events, including internal shear layers and quadrant Reynolds stresses, with the near-wall CS. Both positive (+VISA) and negative (−VISA, *i.e.* internal shear layers) $\partial\langle u\rangle/\partial x$ events (Fig. 9) are direct consequences of the CS x-staggering and tilting in the (x,z) plane. The −VISA events (points E in Fig. 9) occur where high speed fluid impacts the flanks of low-speed streaks, whose waviness is generated by the induction (ejection) of the 3D CS, resulting in $-\partial\langle u\rangle/\partial x$. These −VISA events reflect inclined (in (x,y)), tongue-shaped $U(y)$ shear layers (discussed later) and are consistent with the findings of Johansson *et al.* (1991), obtained via a conditional VISA sampling technique. Additionally, +VISA events (points H in Fig. 9) occur within the

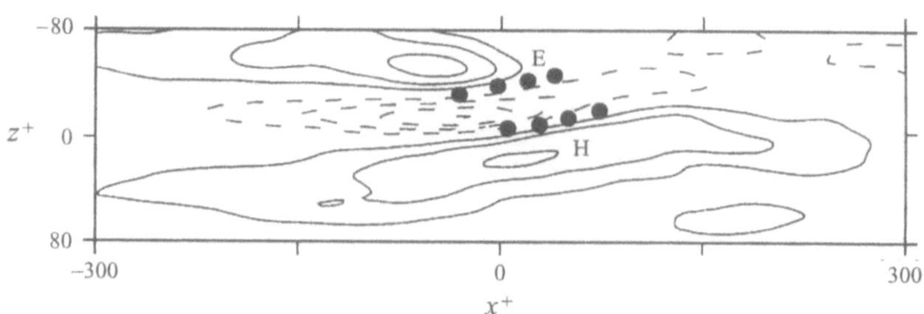

Figure 9. Contours of <u-U> at y^+=15, illustrating a low-speed streak (dashed contours) whose z-waviness produces internal shear layers with $+\partial u/\partial x$ (+VISA, at points H) and $-\partial u/\partial x$ (-VISA, at points E).

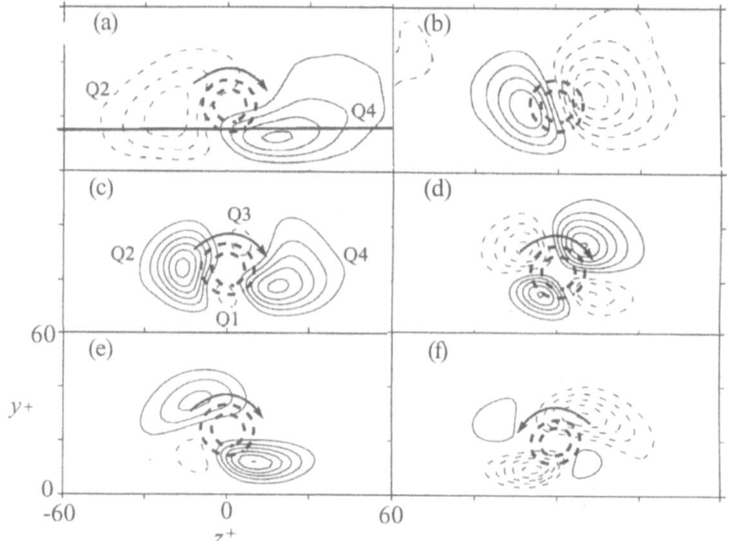

Figure 10. Coherent (*i.e.* ensemble averaged) (a-b) velocities and (c-f) Reynolds stresses at x^+=0 (the middle of SP's x extent). (a) <u-U>; (b) <v>; (c) -<u-U><v>; (d) -<v><w>; (e) -<u-U><w>; (f) -<u-U><w> for SN. Relative locations of Q1, Q2, Q3, and Q4 events with respect to the CS center are shown in (c).

cores of both SP and SN, since ejection ($-\langle u-U\rangle$) occurs upstream of sweep ($+\langle u-U\rangle$) here, due to the CS tilting in the (x,z) plane. As discussed later, vortex regeneration occurs within regions of $+\partial u/\partial x$ (generated by streak instability), via direct stretching of ω_x sheets. We observe that $-\partial\langle u\rangle/\partial x$ regions are stronger than $+\partial\langle u\rangle/\partial x$ events, due to the underlying velocity field. Namely, for $-\partial\langle u\rangle/\partial x$ regions, upstream fluid advects faster (in x) than downstream fluid, thereby progressively steepening the gradient $-\partial\langle u\rangle/\partial x$ (vice-versa for $+\partial\langle u\rangle/\partial x$). The stronger negative $\partial\langle u\rangle/\partial x$ in our educed fields is consistent with the experimental result that the frequency of occurrence of +VITA (−VISA) events is higher than that of −VITA (+VISA), for a given threshold. These results are discussed by Jeong & Hussain (1992) and in more detail by Jeong *et al.* (1997).

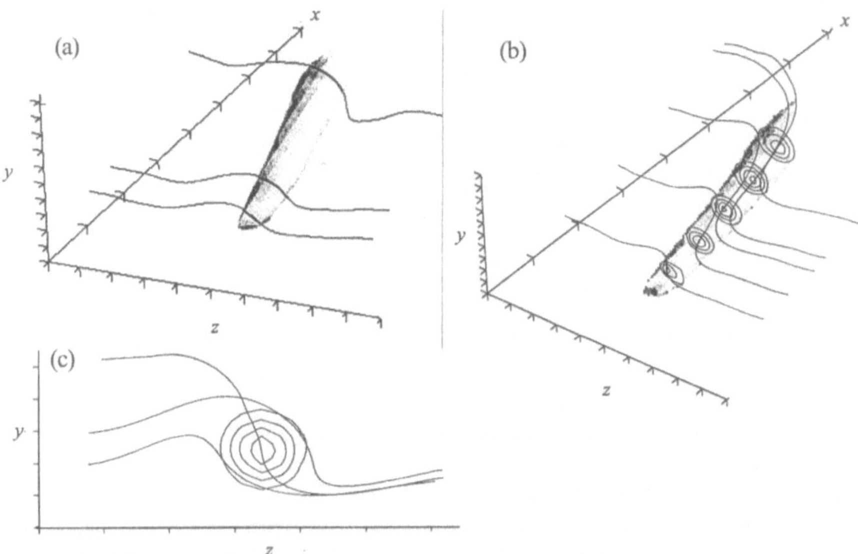

Figure 11. Vortex lines traced (a) near SP and (b) through the SP center, illustrating hairpin-shaped vortex lines despite the absence of hairpin vortices. (c) end view of vortex lines, indicating that SP resides on the $+z$ flank of the coherent low-speed streak (region of vortex line lifting); SN (not shown) appears on the $-z$ flank.

Connecting velocity fluctuations with transport of fluid across a shear region, negative u tends to accompany positive v and vice versa. Thus, in Figs. 10(a-b), a Q2 (ejection) event occurs to the left of SP, and Q4 (sweep) occurs on its right side; the opposite is true for SN. In this model, the so-called "bursting" event is simply the recurring footprints of advecting SP and SN, in which x-overlapping vortices (a few SP and SN) create a series of ejection/sweep events, but a single, long low-speed streak. This scenario is also consistent with the multiple ejections per streak noted by Bogard & Tiederman (1986). Although co-gradient Re stress transport (*i.e.* Q2, Q4) is dominant, a spatial phase difference between the $\langle u-U\rangle$ and $\langle v\rangle$ distributions produces countergradient Q1 and Q3 events below and above each CS respectively (Fig. 10c). This phase difference also causes Q4 to peak closer to the wall than Q2, consistent with experimental observations (Wallace *et al.* 1972). In Jeong *et al.* (1997), this phase difference is shown to be a direct consequence of the CS tilting (in the (x,z) plane). We find that only $-\langle u-U\rangle\langle v\rangle$ produces a nonzero spatial average. The $-\langle u-U\rangle\langle w\rangle$ (Figs. 10e,f) and $-\langle v\rangle\langle w\rangle$ (Fig. 10d) distributions exhibit large magnitudes locally, but spatial variations of sign result in their small spatial averages, when both SP and SN are accounted for. The latter, having a symmetric four-lobe distribution (Fig. 10d), has a zero average for SP and SN individually, while the former, having an asymmetric four-lobe distribution (Fig. 10e,f), vanishes when SP and SN are averaged together.

Although vortex line bundles traced in the $\langle\omega\rangle$ field show hairpin shapes, both immediately outside (Fig. 11a) and within (Fig. 11b) the SP core, these should not be confused with elongated hairpin vortices (as would be identified by λ_2), which are a rarity near the wall. Thus, vortex line tracing is not generally effective in ascertaining vortical structures within turbulent boundary layers. Note the large inclination from z of vortex lines near

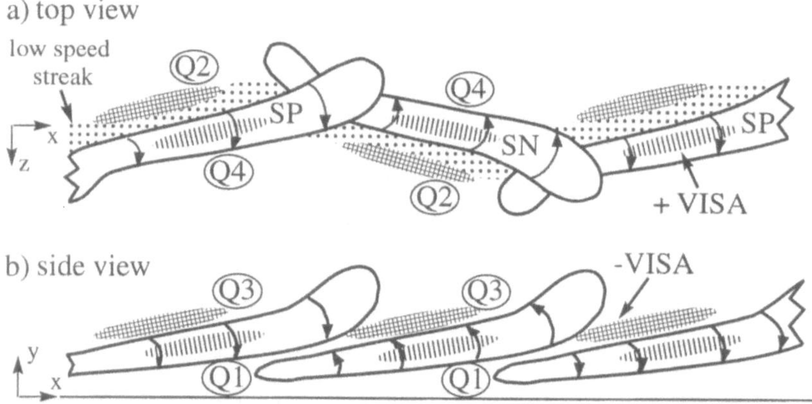

Figure 12. Summary of near-wall educed CS and associated coherent events.

SP (Fig. 11c); this illustrates the "lift-up" mechanism due to near-wall streamwise vortices, whose upward advection (ejection) lifts vortex lines to produces a low-speed streak.

In Fig. 12, we summarize the locations relative to SP and SN of identified low-speed streaks, quadrant Re stress events (Q1, Q2, Q3, Q4), and positive and negative $\partial u/\partial x$ (*i.e.* VISA events, the spatial counterpart of VITA $\partial u/\partial t$ events studied experimentally, which can be related by Taylor's hypothesis). Note that our eduction is not triggered on Reynolds stresses or internal shear layers, unlike popular event-based averaging. Thus, the prominence of these events (Q1, Q2, Q3, Q4 and ±VISA) in our ensemble-averaged flow, triggered by λ_2-based vortex detection, indicates that they are intimately linked to the CS. In other words, these events typically accompany streamwise vortices at the relative locations shown in Fig. 12. Dynamically, we should also consider whether CS are the cause or effect of these events, since a strong spatial correspondence (between these events and CS) alone does not establish the direction of causality. Clearly, quadrant Re stress events are essentially passive consequences of CS induction, but low-speed streaks appear to play a more subtle and dynamically significant role. As discussed in Schoppa & Hussain (1997a), CS and low-speed streaks are inherently coupled, in that CS create low-speed streaks via lift-up of near-wall fluid, while streak instability is responsible for the genesis of CS (and simultaneously of internal shear layers), hence regeneration.

5. CS regeneration mechanism

Having delineated the dominant CS orientation and associated near-wall dynamics, we now address the question: how are these longitudinal vortices generated? In a companion paper (Schoppa & Hussain 1997b), we demonstrate how the nonlinear instability of "vortex-less", lifted low-speed streaks leads to the genesis of streamwise vortices, internal shear layers, and arch vortices. Following up on these results, we develop here a spatio-temporal CS regeneration mechanism, explaining the formation of vortex-less streaks – whose instability generates new vortices – in the "wake" of existing streamwise vortices.

Consider streak formation (including sustenance) by pre-existing streamwise vortices (Fig. 12). First note the inherent differential advection velocity of a low-speed streak and the adjacent streamwise vortex. This can be inferred from space-time correlation data in

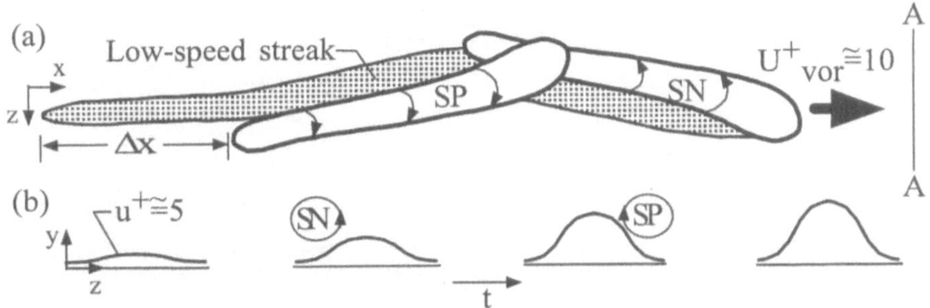

Figure 13. Schematic of streak formation by advecting streamwise vortices. The initial streak/vortex configuration is shown in (a), and the subsequent phases of streak lifting in section A-A as the vortex pair passes overhead is shown in (b) by the evolution of the u^+=5 contour.

Kim & Hussain (1993), where all velocity and vorticity fluctuations for y^+<15 are shown to propagate with velocity U^+_{vor}≅10 (the local mean velocity at y^+=15; U_{vor}=0.55U_c in outer units), *independently of y*. From this, it is easy to envision that (near-wall) streamwise vortices propagate at U^+_{vor}≅10, inducing wave-like propagation of fluctuations below y^+=15, faster than the local mean fluid velocity. Low-speed streak fluid at y^+<15, which obviously advects much slower than the local mean, is thus quickly left behind by streamwise vortices overhead.

A spatiotemporal streak sustenance scenario, which accounts for naturally occurring ends of lifted streaks, is shown schematically in Fig. 13. Here, a mature vortex pair (Fig. 13a) approaches a weak (stable) streak in section A-A. Due to the *x*-staggering of the vortices, vortex SN will encounter the low-speed fluid in section A-A first, followed by SP on the opposite side of the streak (Fig. 13b). Near-wall fluid with *u* slower than the vortex advection speed will thus experience a net upward transport (creating a lifted low-speed streak) while the vortex pair passes overhead. Consequently, the end result is a vortex-less low-speed streak (Fig. 13b), more strongly lifted than the initial state (*cf.* the initial and final *u* contours) and potentially susceptible to subsequent instability.

Having shown how streaks may be generated and sustained by advecting streamwise vortices, we now consider the subsequent streak evolution. The common viewpoint is that the streaks, once formed, are long-lived (Kline & Robinson 1990); if this were the case, each single vortex (or pair) would leave behind very long streaks (*e.g.* ~1000 wall units). Our results regarding streak diffusion/stabilization (Schoppa & Hussain 1997a) indicate that such long streaks cannot be generated by one vortex. In particular, we find that the streak decay by annihilation (due to ω_y cross-diffusion) is exponential, and streaks are stabilized on a timescale of t^+_{dif}≅ 60. Thus, streaks by themselves are far from being passive or long-lived. Simple estimates show that the length of the streak "tail" behind an advecting vortex pair (Fig. 13a) is given by $\Delta x^+=(U^+_{vor} - U^+_{stk})\, t^+_{dif}$ ~ 300 (assuming U^+_{stk}~5). Therefore, only a short (~300 wall units) streak tail would be left behind by individual streamwise vortices. However, the observed streak length is much longer (see *e.g.* Robinson 1991), with many streaks traceable downstream for over 1000 wall units (even at y^+=15), with no upstream attenuation of strength (*i.e.* magnitude of $-u$). The logical conclusion is that individual streaks are maintained by multiple new streamwise vortices.

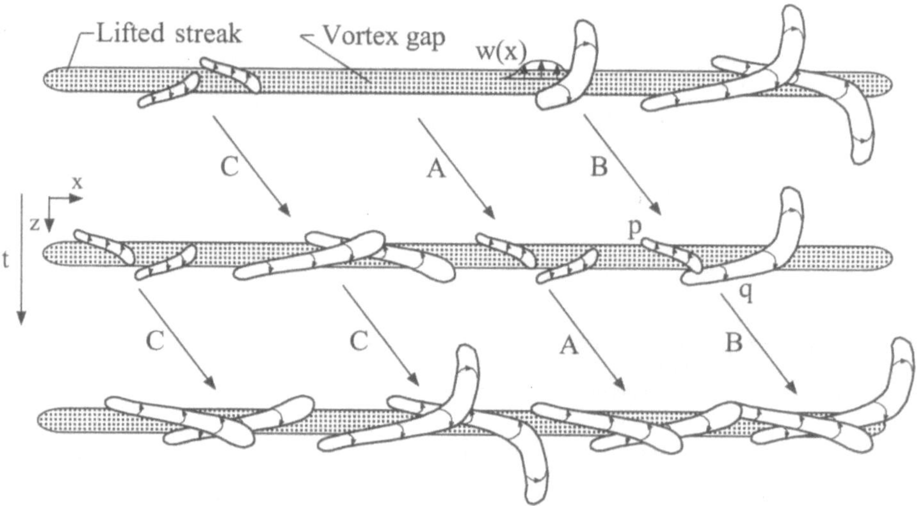

Figure 14. Spatiotemporal scenarios for vortex regeneration by instability of vortex-less streaks. Process A: regeneration within gaps between (existing) neighboring vortices. Process B: regeneration from an existing arch vortex, whose $w(x)$ profile shown excites streak instability to produce a pair of new streamwise vortices. Process C: regeneration at trailing ends of low-speed streaks.

We now consider scenarios by which streak→vortex→streak (or equivalently vortex→streak→vortex) regeneration might occur. Each case relies on the same underlying regeneration mechanism, *i.e.* vortex formation due to streak instability and streak formation by vortices, established rather rigorously here. Thus, a dominant underlying mechanism occurs although each vortex formation process superficially appears to be different.

As argued earlier, the long lifted low-speed streaks observed, coupled with the rapid streak diffusion, indicates that a given streak is sustained by strings of streamwise vortices advecting (faster) overhead. In turn, we have demonstrated that the important near-wall structures (streamwise vortices, internal shear layers, arch vortices) are generated from initially vortex-less low-speed streaks by instability. This suggests that vortex-less streaks are necessary *nucleation sites* for vortex regeneration.

The scenarios outlined in Fig. 14 indicate possible ways in which vortex-less streaks (susceptible to instability and vortex formation) can arise. For process A in Fig. 14, regeneration occurs in a naturally occurring "gap" between x-neighboring vortices along the streak. Such gaps are common in reality (Robinson 1991), indicating natural irregularities in the vortex regeneration occurring upstream. Alternatively, streaks can appear with arches overhead (process B), but without streamwise vortices (due to viscous annihilation of streamwise vortices by crowding, observed by us in minimal channel flow). In this case, the $w(x)$ induced by the tilted arch excites (finite amplitude) streak instability, generating a leg (q in Fig. 14) propagating (upstream) from the arch and also a new opposite-signed streamwise vortex (p in Fig. 14). Intuitively, we expect that the trailing ends of lifted low-speed streaks will be prominent nucleation sites for regeneration, as illustrated by process C. Due to the faster advection of vortices relative to streaks, these streak ends are "self-cleaning" in that the new vortices spawned are advected downstream. Due to their induction, the vortices sustain the streak near its trailing end, eventually leaving behind a lifted vortex-less streak (as behind the vortex pair in Fig. 13). Subsequently, incoming (incoherent) perturbations pass over the streak end, exciting streak instability and spawning a new set of streamwise vortices, and so on.

6. Concluding remarks

To summarize, these results serve as clear evidence that fully 3D CS can in fact be accurately extracted from turbulent shear flows using our general-purpose λ_2-based eduction scheme. This technique enables us to obtain an ensemble averaged flow field which not only retains the footprints of instantaneous vortices, but also exhibits their spatial arrangement with respect to well-known dynamical events. Thus, the near-wall turbulent flow is reduced to an ensemble-averaged flow unit, permitting a vortex dynamics-based analysis which reveals the fundamental turbulence physics. This illustrates that the combination of a robust eduction scheme with interpretations based on vortex dynamics is indeed a powerful approach to study fully turbulent flows, enabling development of conceptual dynamical models which can, in turn, be utilized in turbulence modeling and control.

This research was supported by ONR grant N00014-94-0510, NASA grant NCA2-317, and AFOSR grant F49620-97-1-0131. Supercomputer time was provided by the NASA Ames Research Center.

7. References

Aubry, N., Holmes, P., Lumley, J.L. & Stone, E. 1988 The dynamics of coherent structures in the wall region of a turbulent boundary layer. *J. Fluid Mech.* **192**, 115.

Bakewell, H.P. & Lumley, J.L. 1967 Viscous sublayer and adjacent wall region in turbulent pipe flow. *Phys. Fluids* **10**, 1880.

Blackwelder, R.F. & Eckelmann, H. 1979 Streamwise vortices associated with the bursting phenomenon. *J. Fluid Mech.* **94**, 577.

Blackwelder, R.F. & Kaplan, R.E. 1976 On the wall structure of the turbulent boundary layer. *J. Fluid Mech.* **76**, 89.

Bogard, D.G. & Tiederman, W.G. 1986 Burst detection with single-point velocity measurement. *J. Fluid Mech.* **162**, 389.

Bridges, J., Husain, H.S. & Hussain, F. 1990 Whither coherent structures? In *Whither Turbulence? Turbulence at the Crossroads* (ed. J.L. Lumley), p. 132. Springer.

Brooke, J.W. & Hanratty, T.J. 1993 Origin of turbulence-producing eddies in a channel flow. *Phys. Fluids A* 5, 1011.

Broze, G. & Hussain, F. 1994 Nonlinear dynamics of forced transitional jets: periodic and chaotic attractors. *J. Fluid Mech.* **263**, 93.

Cantwell, B. 1981 Organized motion in turbulent flow. *Ann. Rev. Fluid Mech.* **13**, 457.

Chong, M.S., Perry, A.E. & Cantwell, B.J. 1990 A general classification of three-dimensional flow field. *Phys. Fluids A* **2**, 765.

Crow, S.C. & Champagne, F.H. 1971 Orderly structure in jet turbulence. *J. Fluid Mech.* **48**, 547.

Dallmann, U. 1988 Three-dimensional vortex structures and vorticity topology. *Fluid Dyn. Res.* **3**, 183.

Fiedler, H. 1988 Coherent structures in turbulent flows. *Prog. Aerospace Sci.* **25**, p231.

Goldshtik, M & Hussain, F. 1995 Structural approach to the modeling of a turbulent mixing layer. *Phys. Rev. E* **52**, 2259.

Hunt, J.C.R., Wray, A.A. & Moin, P. 1988 Eddies, stream, and convergence zones in turbulent flows. *Center for Turbulence Research Rep.* CTR-S88, p.193.

Hussain, F. 1986 Coherent structures and turbulence. *J. Fluid Mech.* **173**, 303.

Hussain, F. & Hayakawa, M. 1987 Eduction of large-scale structures in a turbulent plane wake. *J. Fluid Mech.* **180**, 193.

Jeong, J. & Hussain, F. 1992 Coherent structures near the wall in a turbulent channel flow. In *Proc. of Fifth Asian Congress of Fluid Mech.*, Taejon, Korea (eds. K.S. Chang & H. Choi), p. 1262.

Jeong, J. & Hussain, F. 1995 On the identification of a vortex. *J Fluid Mech.* **285**, 69.

Jeong, J., Hussain, F., Schoppa, W. & Kim, J. 1997 Coherent structures near the wall in a turbulent channel flow. *J. Fluid Mech.* **332**, 185.

Jimenez, J., Moin, P., Moser, R. & Keefe, L. 1988 Ejection mechanisms in the sublayer of a turbulent channel. *Phys. Fluids* **31**, 1311.

Johansson, A.V., Alfredsson, P.H. & Kim, J. 1991 Evolution and dynamics of shear-layer structures in near-wall turbulence. *J. Fluid Mech.* **224**, 579.

Kida, S., Takaoka, M. & Hussain, F. 1991 Collision of two vortex rings. *J. Fluid Mech.* **230**, 583.

Kim, J. & Hussain, F. 1993 Propagation velocity of perturbations in turbulent channel flow. *Phys. Fluids A* **5**, 695.

Kim, J., Moin, P. & Moser, R.D. 1987 Turbulence statistics in fully developed channel flow at low Reynolds number. *J. Fluid Mech.* **177**, 133.

Kline, S.J., Reynolds, W.C., Schraub, F.A. & Rundstadler, P.W. 1967 The structure of turbulent boundary layers. *J. Fluid Mech.* **30**, 741.

Kline, S.J. & Robinson, S.K. 1990 Quasi-coherent structures in the turbulent boundary layer: Part 1. Status report on a community-wide survey of the data. In *Near-Wall Turbulence* (eds. S.J. Kline & N.H. Afgan). Hemisphere.

Lugt, H.J. 1979 The dilemma of defining a vortex. In *Recent Developments in Theoretical and Experimental Fluid Mechanics*, p.309. Springer.

Meng, H. & Hussain, F. 1991 Holographic particle velocimetry: a 3D measurement technique for vortex interactions, coherent structures and turbulence. *Fluid Dyn. Res.* **8**, 33.

Poje, A.C. & Lumley, J.L. 1995 A model for large-scale structures in turbulent shear flows. *J. Fluid Mech.* **285**, 349.

Robinson, S.K. 1991 Coherent motions in the turbulent boundary layer. *Ann. Rev. Fluid Mech.* **23**, 601.

Schoppa, W. & Hussain, F. 1997a Genesis and dynamics of coherent structures in near-wall turbulence. In *Self-sustaining Mechanisms of Wall Turbulence* (ed. R. Panton). Computational Mechanics Publications, p. 385.

Schoppa, W. & Hussain, F. 1997b Genesis of longitudinal vortices in near-wall turbulence. In *Proc. of IUTAM Symp. on Dynamics of Slender Vortices* (ed. E. Krause), Aachen, Germany (to appear).

Shtern, V. & Hussain, F. 1993 Hysteresis in a swirling jet as a model tornado. *Phys. Fluids A* **5**, 2183.

Townsend, A.A. 1956 *The Structure of Turbulent Shear Flows*. Cambridge University Press.

Truesdell, C. 1953 *The Kinematics of Vorticity.* Indiana University.

Wallace, J.M., Eckelmann, H. & Brodkey, R.S. 1972 The wall region in turbulent shear flow. *J Fluid Mech.* **54**, 39.

Willmarth, W.W. & Lu, S.S. 1972 Structure of the Reynolds stress near the wall. *J. Fluid Mech.* **55**, 65.

VORTICITY DYNAMICS AROUND A STRAIGHT VORTEX TUBE IN A SIMPLE SHEAR FLOW

SHIGEO KIDA
National Institute for Fusion Science
Toki 509-52, Japan

GENTA KAWAHARA
Ehime University
Matsuyama 790-77, Japan

MITSURU TANAKA
Kyoto Institute of Technology
Kyoto 606, Japan

AND

SHINICHIRO YANASE
Okayama University
Okayama 700, Japan

Abstract. The mechanism of wrap, tilt and stretch of vorticity lines around a strong thin straight tubular vortex in a simple shear flow is investigated analytically. An asymptotic expression of the vorticity field around a vortex which starts with a filament is derived at large vortex Reynolds numbers and at short times of evolution. An oblique vortex, which inclines from the streamwise direction both to the vertical and the spanwise, wraps and stretches the simple shear vorticity to form double spiral vortex layers of high azimuthal vorticity of alternating sign in adjacent spirals. These spirals induce axial shear flows of the same spiral shape and sign which in turn tilt the simple shear vorticity toward the axial direction. Vorticity lines wind helically around the vortex, which results in conversion of the simple shear vorticity to the axial direction. The effects of system rotation on this interaction are also discussed.

E. Krause and K. Gersten (eds.), IUTAM Symposium on Dynamics of Slender Vortices, 173-182.

1. Introduction

Elongated tubular vortices are ubiquitous in various kinds of turbulent flows, such as isotropic, homogeneous-shear and near-wall turbulence with and without system rotation. It has been observed that they play a significant role in the production and the dissipation of turbulence kinetic energy, the generation of high skin-friction as well as the transport processes of heat, mass and momentum. The tubular vortices are expected to be one of the key ingredients of the coherent structures to control turbulence dynamics.

In the motion of tubular vortices their interactions with background turbulence may play a crucial role. Clarification of these interactions would help to inspire a new idea for understanding and controlling the dynamics not only of tubular vortices but also of turbulence itself. In this paper, we study analytically the vorticity interaction between a strong thin straight vortex tube and a simple shear flow in the limit of large vortex Reynolds numbers. The effects of system rotation on the vorticity dynamics are also taken into consideration, which are one of the central issues in geophysical and astrophysical fluid turbulence.

2. Formulation

We consider the motion of a straight vortex tube in a simple shear flow which rotates around the spanwise axis. It is convenient to introduce a coordinate system $OX_1X_2X_3$ which rotates with the simple shear flow in such a way that the X_1-, the X_2- and the X_3-axes may align with the streamwise, the vertical and the spanwise directions, respectively. Then the velocity of the rotating simple shear flow (relative to the stationary coordinate system) is written as $\boldsymbol{U} = SX_2\widehat{\boldsymbol{X}}_1 + \boldsymbol{\Omega}^R \times \boldsymbol{X}$, where S (> 0) denotes the shear rate, $\boldsymbol{\Omega}^R = \Omega^R\widehat{\boldsymbol{X}}_3$ the angular velocity of the shear flow, $\boldsymbol{X}$ the position vector and $\widehat{\boldsymbol{X}}_i$ the unit vector in the X_i-direction $(i = 1, 2, 3)$. We assume that S and Ω^R be constant in time. The vorticity of the rotating simple shear flow and the associated pressure are written as $\nabla \times \boldsymbol{U} = (2\Omega^R - S)\widehat{\boldsymbol{X}}_3$ and $P = \frac{1}{2}\Omega^{R\,2}(X_1^2 + X_2^2) - S\Omega^R X_2^2 + \text{const.}$, respectively.

A vortex tube which is advected by the shear flow changes its direction in time. We introduce here a structural coordinate system $Ox_1x_2x_3$, of which the x_1-axis is on the central axis of the vortex tube and pointed in the direction of the vorticity (see figure 1). Hereafter, we call x_1 the axial coordinate and (x_2, x_3) the normal plane. The unit axial vectors in the rotational and the structural coordinate systems are related through

$$\widehat{\boldsymbol{X}}_i = M_{ij}\widehat{\boldsymbol{x}}_j \quad (i = 1, 2, 3), \tag{1}$$

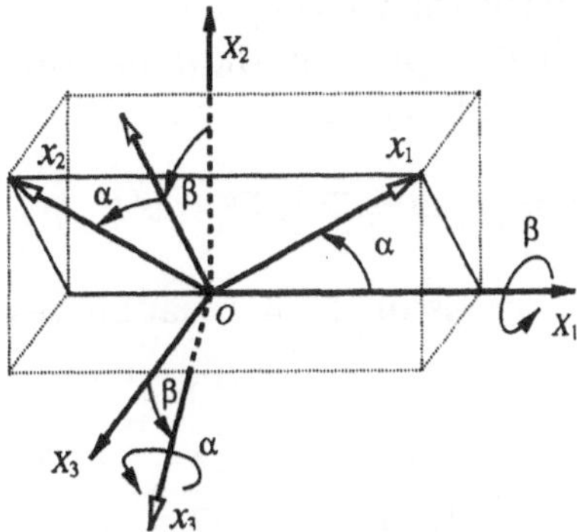

Figure 1. Structural coordinate system $Ox_1x_2x_3$.

where

$$\{M_{ij}\} = \begin{pmatrix} \cos\alpha & -\sin\alpha & 0 \\ \sin\alpha\cos\beta & \cos\alpha\cos\beta & -\sin\beta \\ \sin\alpha\sin\beta & \cos\alpha\sin\beta & \cos\beta \end{pmatrix} \quad (i,j=1,2,3). \tag{2}$$

Here, α and β denote the time-dependent angles between the X_1- and the x_1-axes, and the (X_1, X_2)- and the (x_1, X_1)- planes, respectively. Flow symmetry allows us to restrict α and β in the range of $0 \le \alpha < \pi$ and $-\frac{1}{2}\pi \le \beta \le \frac{1}{2}\pi$ without loss of generality. When the spanwise component of vorticity of a vortex tube has the same (or opposite) sign as that of the rotating simple shear flow, $2\Omega^R - S$, we call it a cyclonic (or an anti-cyclonic) vortex. The angular velocity of the structural coordinate system relative to the rotational one is expressed by

$$\boldsymbol{\Omega}^S = \mathrm{d}\beta/\mathrm{d}t\,\hat{\boldsymbol{X}}_1 + \mathrm{d}\alpha/\mathrm{d}t\,\hat{\boldsymbol{x}}_3. \tag{3}$$

The motion of an incompressible viscous fluid of uniform mass density (taken as unity) is described by the Navier–Stokes equation, or equivalently by the vorticity equation, which are respectively written in the structural coordinate system as

$$\partial_t \boldsymbol{u} + [(\boldsymbol{u} - \boldsymbol{\Omega}\times\boldsymbol{x})\cdot\nabla]\boldsymbol{u} = \boldsymbol{u}\times\boldsymbol{\Omega} - \nabla p + \nu\nabla^2\boldsymbol{u}, \tag{4}$$

$$\partial_t \boldsymbol{\omega} + [(\boldsymbol{u} - \boldsymbol{\Omega}\times\boldsymbol{x})\cdot\nabla]\boldsymbol{\omega} = \boldsymbol{\omega}\times\boldsymbol{\Omega} + (\boldsymbol{\omega}\cdot\nabla)\boldsymbol{u} + \nu\nabla^2\boldsymbol{\omega}, \tag{5}$$

where $\boldsymbol{u}(\boldsymbol{x}, t)$ is the velocity field relative to the *stationary* coordinate system, $\boldsymbol{\omega} = \nabla\times\boldsymbol{u}$ is the vorticity, p is the pressure, $\boldsymbol{\Omega} = \boldsymbol{\Omega}^R + \boldsymbol{\Omega}^S$ is the angular velocity of the structural coordinate system, and ∇ is the gradient operator. The continuity equation is written as

$$\nabla\cdot\boldsymbol{u} = 0. \tag{6}$$

We decompose the velocity, the vorticity and the pressure fields into contributions from the rotating simple shear flow and the fluctuation fields (due to the presence of a vortex tube) as

$$\boldsymbol{u} = \boldsymbol{U} + \boldsymbol{u}', \quad \boldsymbol{\omega} = \nabla \times \boldsymbol{U} + \boldsymbol{\omega}', \quad p = P + p'. \tag{7}$$

Then the temporal evolutions of the fluctuation velocity and vorticity are described by

$$\partial_t \boldsymbol{u}' + [(\boldsymbol{u}' + \boldsymbol{U} - \boldsymbol{\Omega} \times \boldsymbol{x}) \cdot \nabla]\boldsymbol{u}' = \boldsymbol{u}' \times \boldsymbol{\Omega} - (\boldsymbol{u}' \cdot \nabla)\boldsymbol{U} - \nabla p' + \nu \nabla^2 \boldsymbol{u}', \tag{8}$$

$$\begin{aligned}\partial_t \boldsymbol{\omega}' + [(\boldsymbol{u}' + \boldsymbol{U} - \boldsymbol{\Omega} \times \boldsymbol{x}) \cdot \nabla]\boldsymbol{\omega}' &= \boldsymbol{\omega}' \times \boldsymbol{\Omega} + (\boldsymbol{\omega}' \cdot \nabla)\boldsymbol{U} \\ &\quad + [(\boldsymbol{\omega}' + \nabla \times \boldsymbol{U}) \cdot \nabla]\boldsymbol{u}' + \nu \nabla^2 \boldsymbol{\omega}',\end{aligned} \tag{9}$$

$$\nabla \cdot \boldsymbol{u}' = 0, \quad \boldsymbol{\omega}' = \nabla \times \boldsymbol{u}'. \tag{10}$$

The rotating simple shear velocity and vorticity are respectively written as

$$\boldsymbol{U} = S M_{1i} M_{2j} x_j \widehat{\boldsymbol{x}}_i + \epsilon_{ijk} \Omega^R M_{3j} x_k \widehat{\boldsymbol{x}}_i, \quad \nabla \times \boldsymbol{U} = (2\Omega^R - S) M_{3i} \widehat{\boldsymbol{x}}_i, \tag{11}$$

where ϵ_{ijk} is the alternating tensor. Note that in general there appears the spatial coordinate $\boldsymbol{x}$ explicitly in (8) and (9) through $\boldsymbol{U}$ and $\boldsymbol{\Omega} \times \boldsymbol{x}$. However, if we choose the angular velocity $\boldsymbol{\Omega}^S$ as

$$\Omega_1^S = \Omega_2^S = 0, \quad \Omega_3^S = -S \sin^2 \alpha \cos \beta \ (\leq 0), \tag{12}$$

then x_1 disappears and the fluctuation fields $\boldsymbol{u}'$ and $\boldsymbol{\omega}'$ can be uniform all the time in the x_1-direction. It follows from (3) and (12) that

$$\beta = \text{const.}, \quad \cot \alpha = \cot \alpha_0 + S\, t \cos \beta, \tag{13}$$

where α_0 denotes the initial value of α. This tells us that the central axis of the vortex tube is convected passively by the rotating uniform shear flow.

Supposing that the fluctuation fields $\boldsymbol{\omega}'$, $\boldsymbol{u}'$ and p' are independent of x_1, we obtain closed equations for u_1' and ω_1' from (8) and (9) as

$$\begin{aligned}\partial_t u_1' - \frac{\partial(\psi, u_1')}{\partial(x_2, x_3)} &- S \sin \alpha \, [(\cos \alpha \cos \beta \, x_2 - \sin \beta \, x_3)\partial_2 - \cos \alpha \cos \beta] u_1' \\ &= (2\Omega^R - S)(\cos \alpha \sin \beta \, \partial_2 + \cos \beta \, \partial_3)\psi + \nu \nabla_\perp^2 u_1',\end{aligned} \tag{14}$$

$$\begin{aligned}\partial_t \omega_1' - \frac{\partial(\psi, \omega_1')}{\partial(x_2, x_3)} &- S \sin \alpha \, [(\cos \alpha \cos \beta \, x_2 - \sin \beta \, x_3)\partial_2 + \cos \alpha \cos \beta] \omega_1' \\ &= 2\Omega_3^S \partial_3 u_1' + 2\Omega^R (\cos \alpha \sin \beta \, \partial_2 + \cos \beta \, \partial_3) u_1' + \nu \nabla_\perp^2 \omega_1',\end{aligned} \tag{15}$$

where ψ is the streamfunction ($u_2' = \partial_3\psi$, $u_3' = -\partial_2\psi$), which is related to ω_1' via $\nabla_\perp^2\psi = -\omega_1'$, and $\nabla_\perp^2 = \partial_2^2 + \partial_3^2$ is a two-dimensional Laplacian operator. The first and the second terms on the right-hand side of (15) represent the production of the axial component of vorticity from the vorticity associated with the rotation of the structural coordinate system via tilting by the fluctuation velocity.

3. Asymptotic analysis at $\Gamma/\nu \gg 1$

If a straight line vortex of circulation Γ is set at an initial instant, it is diffused under the action of viscosity and the cross-section diameter increases in time as $(\nu t)^{\frac{1}{2}}$. It was shown by Kawahara *et al.* (1997) that the flow structure distant from a strong thin vortex tube (at $r \gg (\Gamma/\nu)^{\frac{1}{4}}(\nu t)^{\frac{1}{2}}$) may be analysed by replacing the vortex tube by a straight line. Separating the contribution of the vortex line from the others, we write the streamfunction as

$$\psi = -\frac{\Gamma}{2\pi}\ln r + \psi'. \tag{16}$$

Substitution into (14) and (15) leads to

$$\begin{aligned}\partial_t u_1' - \frac{\partial(\psi', u_1')}{\partial(x_2, x_3)} - S\sin\alpha\,[(\cos\alpha\cos\beta\,x_2 - \sin\beta\,x_3)\partial_2 - \cos\alpha\cos\beta]u_1' \\ = -\frac{\Gamma}{2\pi r^2}\partial_\theta u_1' - \frac{\Gamma}{2\pi r^2}(2\Omega^R - S)(\cos\alpha\sin\beta\,x_2 + \cos\beta\,x_3) \\ + (2\Omega^R - S)(\cos\alpha\sin\beta\,\partial_2 + \cos\beta\,\partial_3)\psi' + \nu\nabla_\perp^2 u_1',\end{aligned} \tag{17}$$

$$\begin{aligned}\partial_t\omega_1' - \frac{\partial(\psi', \omega_1')}{\partial(x_2, x_3)} - S\sin\alpha\,[(\cos\alpha\cos\beta\,x_2 - \sin\beta\,x_3)\partial_2 + \cos\alpha\cos\beta]\omega_1' \\ = -\frac{\Gamma}{2\pi r^2}\partial_\theta\omega_1' + 2\Omega_3^S\partial_3 u_1' + 2\Omega^R(\cos\alpha\sin\beta\,\partial_2 + \cos\beta\,\partial_3)u_1' + \nu\nabla_\perp^2\omega_1',\end{aligned} \tag{18}$$

where r and θ denote the radial and the angular coordinates, respectively ($x_2 = r\cos\theta$, $x_3 = r\sin\theta$). These equations are supplemented by the initial conditions, $u_1' = 0$ and $\omega_1' = 0$. In the following, we suppose that $\Gamma/\nu \gg 1$ and perform an asymptotic analysis of equations (17) and (18) in the region of $(\Gamma/\nu)^{\frac{1}{4}}(\nu t)^{\frac{1}{2}} \ll r \ll (\Gamma/S)^{\frac{1}{2}}$, in which the velocity induced by the line vortex dominates the rotating simple shear velocity if Ω^R is $O(S)$.

3.1. SPIRAL VORTEX LAYERS OF HIGH AZIMUTHAL VORTICITY

We first consider the evolution of the axial velocity deformed by a strong line vortex. The first term on the left-hand side and the first and the second terms on the right-hand side in (17) are dominant at $r \ll (\Gamma/S)^{\frac{1}{2}}$.

Neglecting the other terms except for the viscous one, which may become important near the vortex line, we find a solution as

$$u_1' = -(2\Omega^R - S)C(0)r\sin\left(\frac{\Gamma t}{2\pi r^2} + \varphi(0) - \theta\right)\exp\left[-\frac{8\pi\nu}{3\Gamma}\left(\frac{\Gamma t}{2\pi r^2}\right)^3\right]$$
$$+(2\Omega^R - S)C(t)r\sin(\varphi(t) - \theta), \tag{19}$$

where $C(t) = (\cos^2\alpha\sin^2\beta + \cos^2\beta)^{\frac{1}{2}}$ and $\varphi(t) = \arctan(\cos\beta/\cos\alpha\sin\beta)$. This solution is valid in region $r \gg (\Gamma/\nu)^{\frac{1}{4}}(\nu t)^{\frac{1}{2}}$ (which can exist if $St \ll (\Gamma/\nu)^{\frac{1}{2}}$). The normal components of the fluctuation vorticity are then calculated to be

$$\omega_2' = \partial_3 u_1' = -(2\Omega^R - S)\cos\alpha\sin\beta + (2\Omega^R - S)C(0)\left[\cos\left(\frac{\Gamma t}{2\pi r^2} + \varphi(0)\right)\right.$$
$$\left. + \frac{\Gamma t}{\pi r^2}\cos\left(\frac{\Gamma t}{2\pi r^2} + \varphi(0) - \theta\right)\sin\theta\right]\exp\left[-\frac{8\pi\nu}{3\Gamma}\left(\frac{\Gamma t}{2\pi r^2}\right)^3\right], \tag{20}$$

$$\omega_3' = -\partial_2 u_1' = -(2\Omega^R - S)\cos\beta + (2\Omega^R - S)C(0)\left[\sin\left(\frac{\Gamma t}{2\pi r^2} + \varphi(0)\right)\right.$$
$$\left. - \frac{\Gamma t}{\pi r^2}\cos\left(\frac{\Gamma t}{2\pi r^2} + \varphi(0) - \theta\right)\cos\theta\right]\exp\left[-\frac{8\pi\nu}{3\Gamma}\left(\frac{\Gamma t}{2\pi r^2}\right)^3\right], \tag{21}$$

which represent double spiral vortex layers, and appear as source terms in axial-vorticity equation (18).

In figure 2, we plot the spatial distribution of normal vorticity $(\omega_2^2 + \omega_3^2)^{\frac{1}{2}}$ and vorticity lines projected on the normal plane at $\Gamma/(2\pi\nu) = 1000$ for the non-rotating case ($\Omega^R = 0$). Figures (a), (b) and (c) represent the cyclonic ($\alpha_0 = \arctan\sqrt{2}$, $\beta = -\frac{1}{4}\pi$), neutral ($\alpha_0 = \frac{1}{4}\pi$, $\beta = 0$), and anti-cyclonic ($\alpha_0 = \arctan\sqrt{2}$, $\beta = \frac{1}{4}\pi$) cases, in which the vortex tube is oriented to the direction of $\widehat{\boldsymbol{X}}_1 + \widehat{\boldsymbol{X}}_2 - \widehat{\boldsymbol{X}}_3$, $\widehat{\boldsymbol{X}}_1 + \widehat{\boldsymbol{X}}_2$, and $\widehat{\boldsymbol{X}}_1 + \widehat{\boldsymbol{X}}_2 + \widehat{\boldsymbol{X}}_3$, respectively. Here, the relative magnitude of the normal vorticity is represented by colour; the red denotes the highest ($7S$) and the blue the lowest (i.e. null). It is seen that the vortex tube wraps and stretches vorticity lines around it to form two spiral vortex layers of high azimuthal vorticity oriented alternately to opposite directions. One of the most interesting features of these spirals is that the x_2-component of normal vorticity takes positive values in the outermost spiral layers of intense vorticity. It changes the sign every half turn along the spirals. An important consequence of this change of sign will be discussed in §3.2. In the near region $r \ll (\Gamma/\nu)^{\frac{1}{3}}(\nu t)^{\frac{1}{2}}$, an excessive wrapping narrows the spacing of the spiral layers and enhances the viscous diffusion to cancel out their opposite-signed vorticity. This leads to disappearance of the normal vorticity around the vortex tube (Moore 1985) and then to the stretch-and-intensification of a cyclonic vortex tube (see §3.2).

3.2. GENERATION OF AXIAL VORTICITY

In the region of $r \ll (\Gamma/S)^{\frac{1}{2}}$ and during $St \ll (\Gamma/\nu)^{\frac{1}{2}}$ the first term on the left-hand side and all the terms on the right-hand side are dominant in (18). Then a solution is obtained as

$$\begin{aligned}
\omega_1' &= (2\Omega^R - S)(\sin^2\alpha_0/\sin\alpha - \sin\alpha)\sin\beta \\
&\quad - \frac{4\pi r^2}{\Gamma}\left[1 + O\left(\frac{\nu t}{r^2}\right)\right](A(t)\cos\beta + B(t)\cos\alpha_0\sin\beta) \\
&\quad + C(0)\left\{2t(B(t)\sin\theta - A(t)\cos\theta)\cos\left(\frac{\Gamma t}{2\pi r^2} + \varphi(0) - \theta\right)\right. \\
&\quad + \frac{4\pi r^2}{\Gamma}\left[B(t)\cos\left(\frac{\Gamma t}{2\pi r^2} + \varphi(0)\right) + A(t)\sin\left(\frac{\Gamma t}{2\pi r^2} + \varphi(0)\right)\right] \\
&\quad \left. + \frac{2\pi r^2}{\Gamma}\left[B(t)\cos\left(\frac{\Gamma t}{2\pi r^2} + \varphi(0) - 2\theta\right) - A(t)\sin\left(\frac{\Gamma t}{2\pi r^2} + \varphi(0) - 2\theta\right)\right]\right\} \\
&\quad \times \exp\left[-\frac{8\pi\nu}{3\Gamma}\left(\frac{\Gamma t}{2\pi r^2}\right)^3\right] \\
&\quad - C(0)\frac{2\pi r^2}{\Gamma}\left[B(t)\cos\left(\frac{\Gamma t}{\pi r^2} + \varphi(0) - 2\theta\right) - A(t)\sin\left(\frac{\Gamma t}{\pi r^2} + \varphi(0) - 2\theta\right)\right] \\
&\quad \times \exp\left[-\frac{32\pi\nu}{3\Gamma}\left(\frac{\Gamma t}{2\pi r^2}\right)^3\right], \qquad (22)
\end{aligned}$$

where

$$A(t) = (2\Omega^R - S)(2\Omega^R\cos\beta + 2\Omega_3^S), \; B(t) = 2\Omega^R(2\Omega^R - S)\cos\alpha\sin\beta. \quad (23)$$

The first term in (22) represents the contribution from the stretching of the axial vorticity component of the rotating simple shear flow by itself. As was stated in §3.1, the normal component of the simple shear vorticity is expelled from the near field of a vortex tube and only the axial component $(\nabla \times \boldsymbol{U}) \cdot \hat{\boldsymbol{x}}_1 = (2\Omega^R - S)\sin\alpha_0\sin\beta$ is left, which may be stretched and intensified by the axial straining $\partial_1 U_1 = S\cos\alpha\sin\alpha\cos\beta$. Integration of vorticity equation $\mathrm{d}\omega_1/\mathrm{d}t = S\cos\alpha\sin\alpha\cos\beta\,\omega_1$ under initial condition $\omega_1|_{t=0} = (2\Omega^R - S)\sin\alpha_0\sin\beta$ yields the first term of (22) as a fluctuation part. As a consequence, a cyclonic vortex tube is intensified while an anti-cyclonic one is weakened (see the central regions of cyclonic and anti-cyclonic vortex tubes in figure 3). In both homogeneous shear and near-wall turbulence, streamwise vortex tubes often take a cyclonic inclination with respect to the mean shear vorticity (see Kida & Tanaka 1994; Jeong *et al.* 1997), which may be related to the above mechanism of selective intensification of a cyclonic vortex.

The remaining terms in (22) express the conversion process of the normal vorticity into the axial direction. Recall that the production of the axial vorticity is caused by the tilting by the fluctuation velocity of vorticity associated with the rotation of the structural coordinate system (see (15)). Vorticity lines are wrapped around a vortex tube by a swirling motion to form spiral layers of high azimuthal vorticity, which in turn induce axial shear flows. As a result, the rotating simple shear vorticity is tilted toward the axial direction.

The spatial distributions on the normal plane of the axial component of fluctuation vorticity (22) are drawn together with projected vorticity lines in figure 3 for the three cases corresponding to figure 2. The level of ω_1' is represented by colour; the red denotes the highest $(+S^2t)$ and the blue the lowest $(-S^2t)$. Along vorticity lines at the outermost double spirals of high azimuthal vorticity there are two crescent-shaped regions of strong negative axial vorticity, which is opposite to the vorticity of the vortex tube (cf. figure 2). Also commonly observed is relatively weak positive vorticity inside the crescent-shaped regions of negative vorticity.

In this non-rotating case only the tilting of the x_3-component of vorticity $2\Omega_3^S$ (< 0) via axial shear $\partial_3 u_1'$ $(= \omega_2')$ is active (see (15)). If it is positive (or negative), the negative (or positive) axial vorticity is generated. Since ω_2' is positive in the outermost spirals of intense azimuthal vorticity (figure 2), the crescent-shaped regions of negative axial vorticity are generated. Recently, Sendstad & Moin (1992) observed that streamwise vorticity of opposite sign appears around the near-wall streamwise vortex tube and it develops into a new streamwise vortex. The present wrapping and tilting mechanism of vorticity lines by a vortex tube is expected to express the regeneration process of streamwise vortices in near-wall turbulence.

3.3. THE EFFECTS OF SYSTEM ROTATION

The effects of the system rotation are represented by Ω^R in solutions (20)–(22). For a vortex tube nearly aligned with the streamwise direction $(\alpha \ll 1)$ we have $|\Omega_3^S| \ll |\Omega^R|$ and two functions $A(t)$ and $B(t)$ are proportional to $2\Omega^R(2\Omega^R - S)$. This means that the direction of the axial vorticity generated around a vortex tube is reversed depending upon the sign of $2\Omega^R(2\Omega^R - S)$.

In figure 4, we plot the same quantities as in figure 3 for a streamwise vortex $(\alpha = 0, \beta = 0)$ at (a) $2\Omega^R = \frac{1}{2}S$ and (b) $2\Omega^R = \frac{3}{2}S$. It is seen that

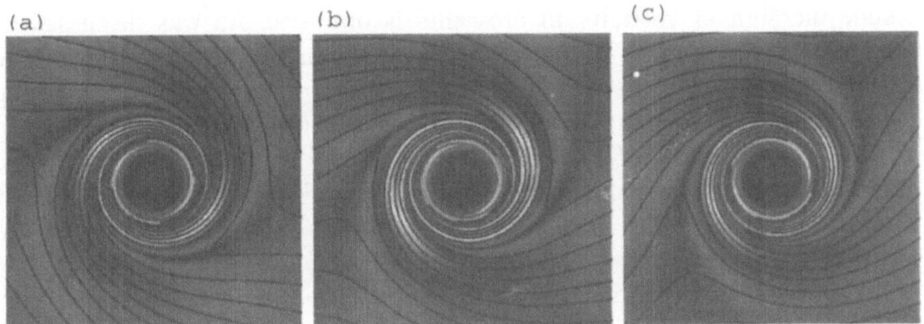

Figure 2. Spatial distribution of magnitude of normal vorticity (non-rotating case).

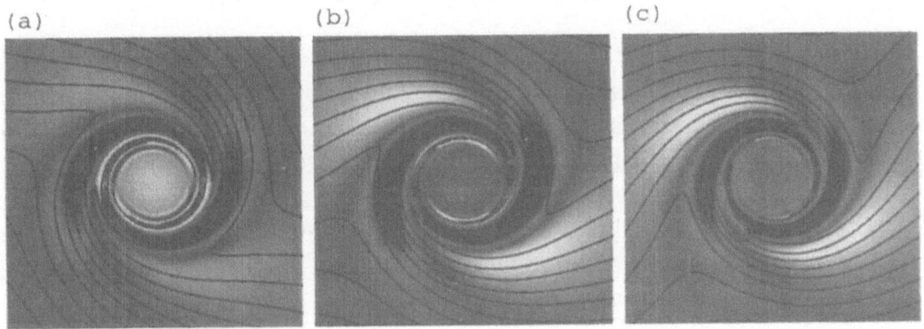

Figure 3. Spatial distribution of fluctuation axial vorticity (non-rotating case).

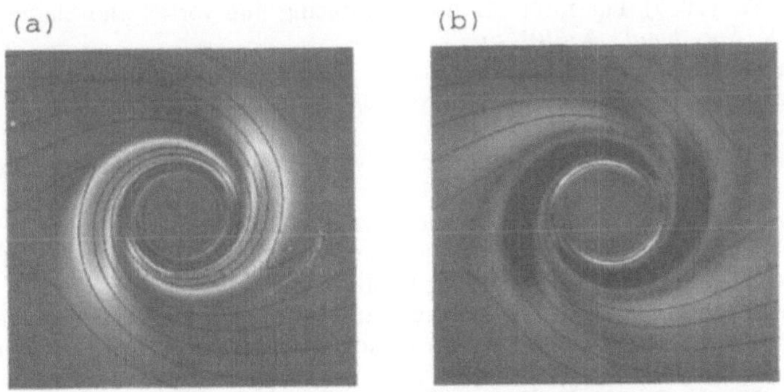

Figure 4. Spatial distribution of fluctuation axial vorticity (rotating case).

two crescent-shaped regions of negative axial vorticity appear for $2\Omega^R = \frac{3}{2}S$, which is similar to those in figure 3. For $2\Omega^R = \frac{1}{2}S$, on the other hand, the sign of vorticity in crescents is opposite. As was discussed in §3.2, the production of the axial vorticity is caused by the tilting of the normal vorticity. For a streamwise vortex ($2\Omega_3^S = 0$) only the tilting of the x_3-component of vorticity $2\Omega^R$ of the system rotation is active (see (15)). In the case of $2\Omega^R(2\Omega^R - S) < 0$ (figure 4 (a)), $2\Omega^R$ is *positive* and $(2\Omega^R - S)$ is negative. This latter condition leads to $\omega_2' = \partial_3 u_1' > 0$ in the outermost spirals, which, together with the former, results in the generation of the positive axial vorticity.

This generation of the positive axial vorticity around a streamwise vortex tube corresponds to the shear/Coriolis instability (see Yanase *et al.* 1993). Supposing the outermost spirals locally to be a planar vortex layer perpendicular to the rotation (X_3 and so x_3) axis, i.e., $\partial_1 = \partial_2 = 0$, and dropping the viscous effects, the linearized equation of (9) yields

$$\partial_t \omega_1' = 2\Omega^R \omega_2', \quad \partial_t \omega_2' = -(2\Omega^R - S)\, \omega_1', \tag{24}$$

which imply that for $2\Omega^R(2\Omega^R - S) < 0$, fluctuation vorticity whose x_1 and x_2 components have same sign can grow due to purely straining motion of a simple shear. As seen in §3.1, ω_2' is positive in the outermost spirals, and so is ω_1. This type of linear instability occurs around longitudinal vortex tubes in a rotating uniformly sheared turbulence (see Tanaka *et al.* 1997).

For a vortex tube inclined from the streamwise direction we have $|\Omega_3^S| \approx |\Omega^R|$ and thus the axial component of fluctuation vorticity may change the sign during the evolution.

References

Jeong, J., Hussain, F., Schoppa, W. & Kim, J. (1997) Coherent structures near the wall in a turbulent channel flow, *J. Fluid Mech.*, **332**, pp. 185–214.

Moore, D. W. (1985) The interaction of a diffusing line vortex and aligned shear flow, *Proc. R. Soc. Lond.*, **A 399**, pp. 367–375.

Kawahara, G., Kida, S., Tanaka, M. & Yanase, S. (1997) Wrap, tilt and stretch of vorticity lines around a strong thin straight vortex tube in a simple shear flow, *J. Fluid Mech.* (in press).

Kida, S. & Tanaka, M. (1994) Dynamics of vortical structures in a homogeneous shear flow, *J. Fluid Mech.*, **274**, pp. 43–68.

Sendstad, O. & Moin, P. (1992) The near-wall mechanics of three-dimensional boundary layers, Report No. TF-57, Thermoscience Divison, Department of Mechanical Engineering, Stanford University, Stanford, CA.

Tanaka, M., Yanase, S., Kida, S. & Kawahara, G. (1997) Uniformly sheared turbulent flow under the effect of solid-body rotation, to be presented in 11th Symposium on Turbulent Shear Flows, Grenoble.

Yanase, S., Flores, C., Métais, O. & Riley, J. J. (1993) Rotating free shear flows. Part 1: Linear stability analysis, *Phys. Fluids*, **A 5**, pp. 2725–2737.

Genesis of Longitudinal Vortices in Near-Wall Turbulence

Wade Schoppa & Fazle Hussain
University of Houston
Department of Mechanical Engineering, Houston, TX 77204-4792

Abstract

Using direct numerical simulations of turbulent channel flow, we present new insight into the formation mechanism of near-wall longitudinal vortices. Instability of lifted, vortex-free low-speed streaks is shown to generate, upon nonlinear saturation, new streamwise vortices, which dominate near-wall turbulence production, drag, and heat transfer. The instability requires sufficiently strong streaks (y circulation per unit $x > 7.6$) and is inviscid in nature, despite the proximity of the no-slip wall. Streamwise vortex formation (collapse) is dominated by stretching, rather than rollup, of instability-generated ω_x sheets. In turn, direct stretching results from the positive $\partial u/\partial x$ (*i.e.* positive VISA) associated with streak waviness in the (x,z) plane, generated upon finite-amplitude evolution of the sinuous instability mode. Significantly, the 3D features of the (instantaneous) instability-generated vortices agree well with the coherent structures educed (*i.e.* ensemble averaged) from fully turbulent flow, suggesting the prevalence of this instability mechanism. These results suggest promising new drag reduction strategies, involving large-scale (hence more durable) actuation and requiring no wall sensors or feedback logic.

1. Introduction

The boundary layers on transport vehicles and in industrial devices are invariably turbulent, with drastically increased drag and heat transfer at solid surfaces due to near-wall vortical *coherent structures* (CS). Viable control of near-wall turbulence, as yet largely unrealized in practice, has the potential to save billions of dollars per year in energy costs for engineering applications. Although massive efforts have been expended in developing drag reduction strategies, their engineering application has remained notably scarce, particularly for aircraft. The lack of success of boundary layer control to date without doubt reflects a currently limited understanding of CS initiation and evolution. In this paper, we develop a new mechanism of CS formation, well-supported by comparisons with near-wall turbulence, and briefly discuss viable large-scale control techniques.

The prominence of longitudinal vortices in near-wall turbulence is now well accepted (*e.g.* see Townsend 1956; Kline *et al.* 1967; Blackwelder & Eckelmann 1979; Robinson 1991), as is their critical role in elevating drag (Kravchenko *et al.* 1993) and heat transfer. The transport enhancing effect of near-wall vortices is easily understood. Due to their streamwise orientation, these vortices sweep near-wall fluid toward the wall on one flank

E. Krause and K. Gersten (eds.), IUTAM Symposium on Dynamics of Slender Vortices, 183-192.

and eject it away from the wall on the other. Drag and heat transfer are enhanced by the wallward motion, which steepens the wall gradients of streamwise velocity U and temperature respectively. Note that the gradient reduction on the outward motion side of vortices is relatively smaller, resulting in mean transport enhancement.

Our ensemble-averaged streamwise vortices, *i.e.* CS (Jeong & Hussain 1992; Jeong *et al.* 1997), display all previously classified near-wall features (Kline & Robinson 1990). Thus, the evolutionary dynamics of streamwise CS are the essence of near-wall turbulence. The central question addressed here is: how are streamwise vortices generated? Several widely disparate formation mechanisms have been proposed, many quite plausible and self-consistent, yet currently lacking convincing validation. Thus, a formidable challenge is to identify the correct naturally and frequently occurring dynamics.

Vortex formation must recur for turbulence to be sustained; *i.e.* existing vortices must ensure subsequent vortex regeneration. Of the numerous proposed regeneration mechanisms, most involve either: (i) the direct action (induction) of existing vortices nearby ("parent - offspring" scenarios), or (ii) local instability of a quasi-steady base flow, without directly requiring existing vortices. Note that recurring instability (ii) requires a feedback mechanism, by which previous vortices generate an unstable base flow and thus play only an indirect role.

A widely cited parent-offspring mechanism involves the generation of new vortices near existing hairpins, behind the (spanwise) arch and beside each of the (streamwise) legs (see Smith & Walker 1994 for a review). In contrast to hairpin generation, Brooke & Hanratty (1993) propose that an opposite-signed offspring vortex forms immediately underneath a parent vortex, whose downstream end has lifted from the wall. Vortex formation is also often attributed to 2D Kelvin-Helmholtz-type rollup of near-wall ω_x sheets (*e.g.* Jimenez & Orlandi 1993), with opposite sign of the streamwise vortex existing overhead, generated by the no-slip condition.

Of the numerous instability mechanisms developed to explain near-wall vortex formation, there is considerable disagreement as to the mechanisms of instability and feedback. For example, centrifugal (Brown & Thomas 1977) and Craik-Leibovich (Phillips & Wu 1994) instabilities, direct resonance of oblique modes (Jang *et al.* 1986), and local shear layer-type instabilities (Swearingen & Blackwelder 1987; Hamilton *et al.* 1995) have all been proposed. Unfortunately, physical-space, vortex-dynamics representations of these mechanisms, including comparisons with near-wall turbulence, are still not at hand.

Here we demonstrate (via DNS) that instability of streaks – without any initial (parent) vortex – directly generates new streamwise vortices, internal shear layers, and arch vortices. The instability-generated streamwise vortices are found to correspond closely with the ensemble-averaged CS educed from near-wall turbulence (Jeong *et al.* 1997), suggesting the dominance of our proposed mechanism. Physical-space, vortex dynamics-based explanations for the vortex regeneration observed here are also provided. In the following, we outline our computational approach (§2), and then demonstrate an underlying linear instability of lifted low-speed streaks (§3). The genesis of new vortices is illustrated in §4, including a brief description of the vortex dynamics involved, followed by some concluding remarks and implications for boundary layer control (§5). Additional details of these results may be found in Schoppa & Hussain (1997a).

2. Computational Approach

In the following, we address vortex regeneration using direct numerical simulations of the Navier-Stokes equations. Periodic boundary conditions are used in x and z, and the no-slip condition is applied on the two walls normal to y; see Kim *et al.* (1987) for the simulation algorithm details. The spatial discretization and Re are chosen so that all dynamically significant lengthscales are resolved (*i.e.* a finer computational grid does not markedly affect the solution); thus, no subgrid-scale turbulence model is necessary. Code validation and accuracy checks were performed by comparing the growth rates for simulated 2D and 3D Orr-Sommerfeld modes of the laminar (parabolic profile) flow with independent stability analysis (agreement within 1%).

To better isolate instability and the subsequent vortex formation, we use the minimum outer Reynolds number $Re=U_c h/\nu=2000$ (U_c is the centerline velocity of the $2h$ wide channel) and the minimum domain sizes in x and z for sustained channel flow turbulence – the so-called "minimal flow unit" of Jimenez & Moin (1991). For the simulations of isolated vortex regeneration, a constant volume flux is maintained, and 32x129x32 grid points are used in x, y, and z respectively.

3. Streak Instability

3.1 MINIMAL CHANNEL FLOW

Our own analysis of minimal channel regeneration suggests the presence of an underlying streak instability. During the quiescent phase of the regeneration cycle, when the wall shear stress is minimum, the buffer region contains only a lifted-up, long, low-speed streak, with no significant streamwise vortices or even ω_x. Shortly thereafter (t^+~40 later), a new positive streamwise vortex is created (by instability, as shown here) in the buffer region from the vorticity sheet (predominantly $+\omega_y$) flanking the streak. Thus, the observed large temporal variations in integrated wall shear stress directly reflect the vortex regeneration: the drag is minimum during the quiescent phase, when near-wall vortices are very weak, and maximum once collapsed streamwise vortices (which bring high-speed fluid toward the wall to increase drag) are generated in the buffer layer. These observations suggest that "vortex-less" low-speed streaks are unstable and serve as an agent of vortex regeneration. In full-domain flows as well, extremely long (Δx^+~1000) streaks are prevalent, and many regions along individual streaks are devoid of any streamwise vortices (Robinson 1991).

To isolate instability of vortex-less streaks, we consider a base flow of the form

$$U^+(y^+,z^+)=U_0^+(y^+)+(\Delta u_0^+/2)\cos(\beta^+z^+)(y^+/30)\exp(-\sigma y^{+2}+0.5); \qquad (1)$$
$$V^+=W^+=0$$

as a first approximation, where U_0^+ is the turbulent mean velocity profile. The streak's normal circulation per unit length Δu_0^+, spanwise wavenumber β^+ and transverse decay σ are chosen to approximate a typical U distribution, shown in Fig. 1(a) for minimal channel flow. The corresponding (y,z) distribution of (1) with $\Delta u_0^+=11.2$, $\beta^+=0.06$ (*i.e.* streak spacing $\Delta z^+=100$), and $\sigma=0.00055$ (*i.e.* maximum streak ω_y at $y^+=30$) is shown in Fig. 1(b) and closely resembles the instantaneous realization in Fig. 1(a). Note that the base flow base flow (1) *contains no* ω_x and is a steady solution of the Euler equations.

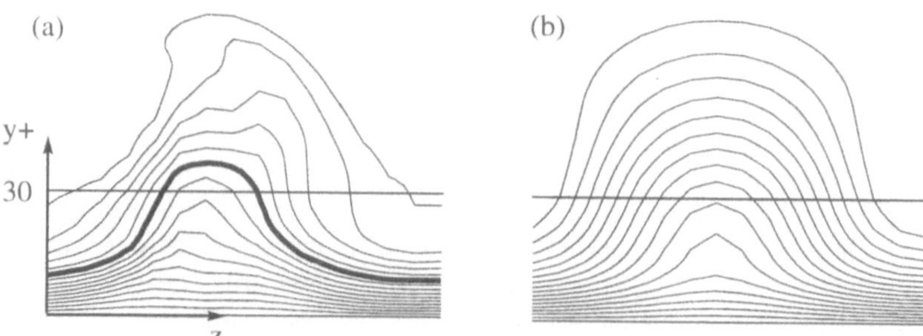

Figure 1. Lifted low-speed streak at the quiescent phase of minimal channel regeneration, illustrated by (a) a typical cross-stream distribution of U and (b) the analytical base flow (1) used for stability analysis. The bold contour shown in (a) is the $0.55\,U_c$ contour.

3.2 LINEAR STABILITY ANALYSIS

With the base flow (1) frozen and Re=2000, we find exponential growth of linear amplitude sinuous perturbations (*i.e.* streak bending in z commonly observed), indicating that lifted streaks (1) are indeed linearly unstable. The instability growth is dominated by $E_{10}(t)$ (z harmonics are also present, but less energetic), the volume-integrated energy in Fourier modes with a z-wavenumber of 0 (mean in z) and an x-wavenumber of α (x-fundamental mode). Interestingly, enhanced growth of E_{10} with increasing Re reflects an inviscid instability mechanism, found to be quite similar in nature to oblique instability modes of free shear layers. Consequently, viscous effects and the no-slip wall play no destabilizing role. This raises the question: how does viscosity, obviously crucial near the wall, enter the instability dynamics?

We find that the viscous damping of instability is quite strong for a streak spacing of Δz^+=100, the popularly accepted value. Since the peak E_{10} occurs at a linear amplitude (*i.e.* 3D perturbation amplitudes near machine accuracy; see Fig. 2b), the typical nonlinear (*i.e.* finite-amplitude) saturation is not occurring here. Instead, attenuation is due primarily to cross-diffusion (*i.e.* viscous annihilation, a kind of planar reconnection) of the opposite-signed ω_y flanking the low-speed streak. In fact, ω_y is reduced to 68% of its initial value by the E_{10} saturation time, indicating that the (exponential) streak decay rate due to cross-diffusion (Fig. 2d) is non-negligible (approximately half the instability growth rate; Fig. 2b).

We now consider the instability scaling at higher Re, keeping Δz^+=100 fixed. As shown in Figs. 2(b,d), both instability growth and streak diffusion (annihilation of normal circulation Δu^+) scale well in (inner) wall units. Although this is perhaps not surprising because of the absence of outer vortices in these flows, the possibility of autonomous inner-scaling dynamics clearly exists. These results demonstrate that streak instability grows similarly at higher Re (perhaps even at very high Re), provided that the streak velocity profile is self-similar in wall units. By considering the dimensional time evolution, one can see how this is possible. As Re is increased, the wall vorticity Ω_w (*i.e.* $U(y)$ slope) increases (according to the Blasius skin friction law), and the (dimensional) streak spacing decreases. Consequently, the streak annihilation by cross-diffusion is faster at higher Re (Fig. 2c), but the instability growth rate is also enhanced due to concomitant increased wall vorticity (Fig. 2a), their balance maintaining a nearly constant

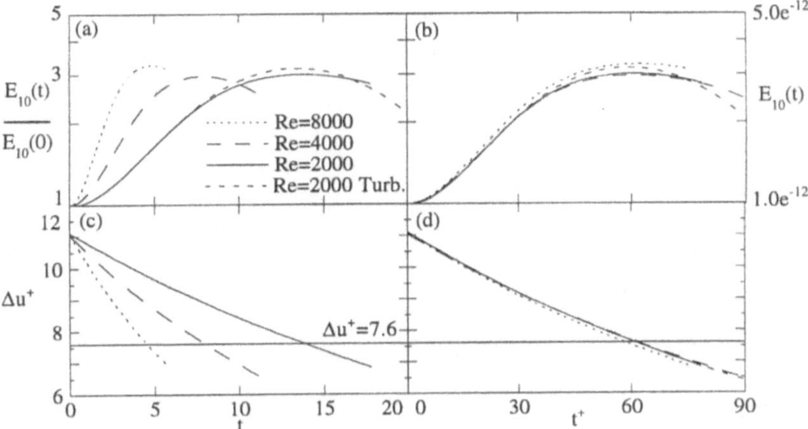

Figure 2. Temporal evolution of (a,b) E_{10} and (c,d) streak y circulation as a function of Re, in (a,c) dimensional and (b,d) wall time units, for a constant streak spacing $\Delta z^+=100$. The data collapse in (b,d) illustrate the inner scaling and balance of streak instability and viscous streak annihilation, showing Re invariance. Except for the case with a turbulent mean profile, the Reichardt relation is used for $U_0^+(y^+)$ in (1).

E_{10} amplification. Figures 2(b,d) also confirm the stabilizing role of viscous cross-diffusion across streaks; saturation occurs in each case at a critical normal circulation of $\Delta u^+=7.6$. Note that Δu^+, being a measure of the tilt angle of the vortex lines (in (y,z)) on either side of a streak, represents the extent of streak lifting (*i.e.* the crest amplitude of u contours in Fig. 1). Thus, sufficient lift-up of the low-speed streak into the buffer layer is required for instability to occur. In the following, we focus on the more computationally tractable Re=2000 case, noting that the streak instability is generic to higher Re.

4. Vortex Formation Mechanism

Having confirmed that (one-walled) streaks with sufficient y circulation (Δu^+) are indeed linearly unstable, we now consider the subsequent nonlinear evolution using DNS. We consider a "clean" flow initialization, containing only vortex-less streaks (1) and a sinusoidal w perturbation of amplitude $w'/U_c=1\%$ at $y^+=30$. Results clearly illustrate the genesis of streamwise CS, near-wall shear layers, and arch vortices, suggesting that streak instability is the dominant mechanism of vortex generation and turbulence production.

4.1 STREAMWISE VORTICES

Most significantly, as the mode grows to a nonlinear, new collapsed streamwise vortices are directly created (Fig. 3a-c). At early times, instability growth is characterized by increased circulation of flattened ω_x sheets, with the spanwise symmetry of the linear eigenmode approximately maintained. Subsequently, as nonlinear effects (described below) become prominent, $+\omega_x$ begins to concentrate on the $+z$ flank of the low-speed streak (Fig. 3b). By symmetry, the ω_x distribution at a half wavelength in x away is obtained by z reflection and sign inversion; thus, $-\omega_x$ is generated on the $-z$ flank here. As this ω_x amplification continues, collapsed (*i.e.* with compact cross-section) streamwise

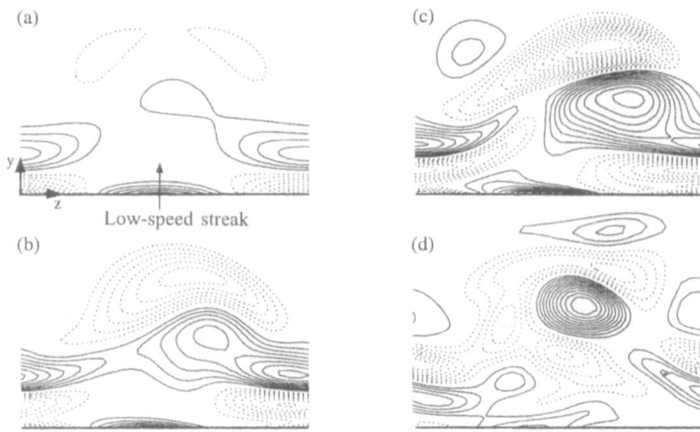

Figure 3. Streamwise vortex formation due to finite-amplitude streak instability, illustrated by cross-stream distributions of ω_x at (a) $t^+=17$, (b) $t^+=51$, (c) $t^+=103$, (d) $t^+=928$. Planes in (b) and (c) are tracked with the instability phase speed of approximately $0.6U_c$.

vortices quickly emerge (Fig. 3c). This genesis of new vortices from ω_x layers is strikingly similar to that frequently observed in minimal channel flow. Previous studies (*e.g.* Jimenez & Orlandi 1993) presumed that the layer simply rolls up due to its own (2D) advection. Our results (discussed below) indicate that the vortex formation is not in reality a (Kelvin-Helmholtz type) roll up process; the formation is inherently 3D, dominated by intense ω_x stretching. Even well past their initial formation, streamwise vortices and hence turbulence continue to be sustained (*e.g.* Fig. 3d), indicating the importance of streak instability to turbulence sustenance.

The 3D geometry of the instability-generated vortices (Figs. 4a,b) (say, the x-overlapping of tilted, opposite-signed streamwise vortices on either side of a low-speed streak) agrees well with the typical flow structure during the active phase of minimal channel regeneration. Most significantly, this vortex geometry (maintained upon evolution except for increasing overlap) is strikingly similar to that of 3D CS educed (from more than 100 vortex realizations) in full-domain turbulence (Fig. 5), which has been shown to capture all important near-wall events (Jeong *et al.* 1997). Irregularities (*e.g.* kinks) of the base flow streaks and finite-amplitude incoherent turbulence will surely occur, causing variations in vortices from one realization to another. If an underlying instability mechanism is present, it should be revealed by ensemble averaging over a large number of base flow/perturbation combinations, *i.e.* by CS eduction. The close correspondence of Figs. 4 and 5 indicates that this is in fact the case, serving as strong evidence that this vortex formation process is a dominant mechanism in fully developed near-wall turbulence.

Since the newly generated vortices are predominantly streamwise (Fig. 4a), the essential dynamics of vortex formation are those of ω_x, whose inviscid evolution is governed by

$$\frac{\partial \omega_x}{\partial t} = -u\frac{\partial \omega_x}{\partial x} \underbrace{- v\frac{\partial \omega_x}{\partial y} - w\frac{\partial \omega_x}{\partial z}}_{\text{Self-induction}} + \underbrace{\omega_x \frac{\partial u}{\partial x}}_{\text{Stretching}} + \underbrace{\frac{\partial v}{\partial x}\frac{\partial u}{\partial z} - \frac{\partial w}{\partial x}\frac{\partial u}{\partial y}}_{\text{Tilting}} . \qquad (2)$$

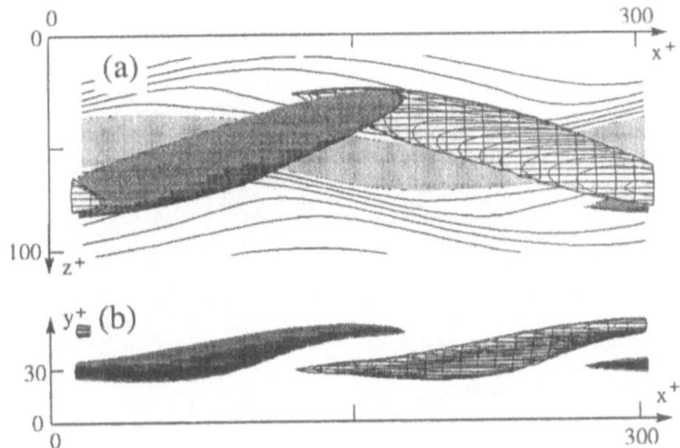

Figure 4. Streamwise vortices' (x,z) plane tilting, x-overlapping, and location relative to a low-speed streak in (a) top view, (b) side view. The 80% isosurfaces of $+\omega_x$ and $-\omega_x$ at $t^+=103$ are (dark) shaded and hatched respectively; contours of u at $y^+=20$ are overlaid in (a), with low levels of u light-shaded to demarcate the low-speed streak. Note the striking resemblance of this instantaneous realization with the ensemble-averaged CS (Fig. 5).

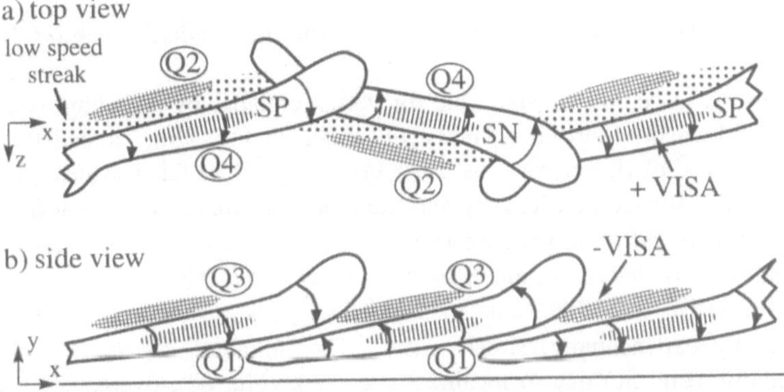

Figure 5. Near-wall educed CS and associated coherent events (adapted from Jeong *et al.* 1997); including ±VISA events ($\pm\partial u/\partial x$); quadrant *Re* stresses Q1, Q2 (ejection), Q3, and Q4 (sweep); and a kinked low-speed streak.

In Fig. 6, we observe that the circulation of the elongated near-wall ω_x layers (Fig. 6a) increases due to vortex line tilting, given by the latter production term $-(\partial w/\partial x)(\partial u/\partial y)$ (Fig. 6c), which dominates the former. Although typically largest in magnitude over all other, the $-(\partial w/\partial x)(\partial u/\partial y)$ term actually generates a flattened tail in the near-wall ω_x layer (C in Fig. 6c), *not* a vortex. Contrary to prior speculation, these layers do not roll up due to their self-advection – a purely 2D mechanism. In fact, the cross-stream transport (B in Fig. 6b) actually opposes the rollup process, due to the opposite-signed ω_x immediately overhead (SN in Fig. 6a). In reality, vortex formation is due to direct stretching of $+\omega_x$ on the $+z$ flank of the low-speed streak (also, $-\omega_x$ amplification on the $-z$ flank, at a half x wavelength away), evident from nearly circular regions of $+\omega_x\partial u/\partial x$ there (D in Fig. 6d). We find that this local ω_x stretching is sustained in time and is mainly responsible for the

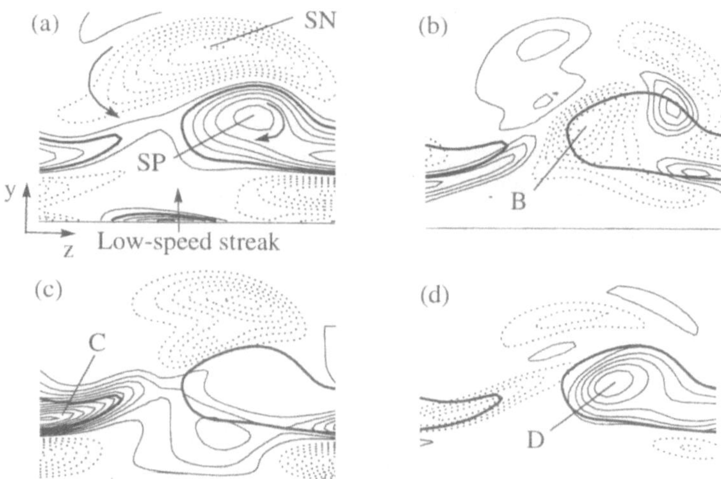

Figure 6. Distributions of (a) ω_x, and selected terms of the ω_x evolution equation: (b) self-induction (cross-stream), (c) the $-(\partial w/\partial x)\ (\partial u/\partial y)$ tilting term, and (d) direct stretching ($\omega_x \partial u/\partial x$); (a-d) are at an intermediate time during vortex formation (t^+=51). The bold line in each panel identifies the ω_x layer.

eventual vortex collapse, whose location coincides with the $+\omega_x \partial u/\partial x$ peak (*cf.* Figs. 3c, 6d).

In turn, the positive $\partial u/\partial x$ responsible for vortex collapse by stretching is a simple consequence of low-speed streak waviness, illustrated in Fig. 4(a). Recall that streak waviness is generated by (linear) sinuous streak instability. Once this waviness grows to a finite size, strong $+\partial u/\partial x$ develops downstream of the streak crests, causing direct stretching of positive (SP) and negative (SN) ω_x existing there. Since a large velocity difference exists across the streak flanks (with vorticity comparable to the mean velocity gradient at the wall), a sizable value of $+\partial u/\partial x$ is quickly generated by the rapidly growing (initially exponentially) streak wave. The initial ω_x sheets (Fig. 3a) then suddenly collapse (Fig. 3c) due to localized stretching (Fig. 6d), overcoming viscous diffusion which would otherwise cause their annihilation (on a similar timescale as the collapse). Note that these dynamics are also captured as (ensemble-averaged) +VISA events (*i.e.* $+\partial u/\partial x$) existing within the CS core (Fig. 5), indicating that this vortex generation process is indeed a dominant one.

4.2 INTERNAL SHEAR LAYERS & ARCH VORTICES

The significance of (nonlinear) streak instability is not limited to streamwise vortex formation; it also captures the genesis of new internal shear layers and spanwise arch vortices. Internal shear layers, indicated by wall-detached sheets of ω_z, form alongside (in z) the generated streamwise vortices (Figs. 7a,b). Subsequently, the downstream "end" of the internal shear layer rolls up into a new (locally spanwise) arch vortex (Figs. 7d,f). A surprising result is that the downstream tips of (newly generated) streamwise vortices tilt and propagate outward to form arches (Fig. 7c,e). Note that this process, *i.e.* streak instability → streamwise vortices → arch vortices, is contrary to the mechanism proposed by Robinson (1991), *i.e.* streak instability → arch vortices → streamwise

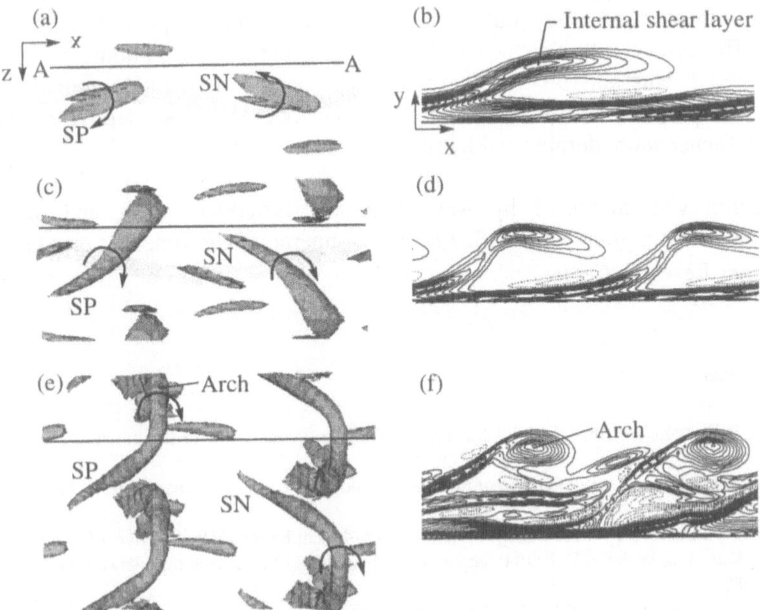

Figure 7. Genesis of internal shear layers and arch vortices due to nonlinear evolution of streak instability, illustrated by actual DNS data. (a,c,e): the evolutions of vortices SP and SN (top view) represented by λ_2 isosurfaces; (b,d,f): corresponding contours of ω_z in the section A-A (the straight line in a,c,e), indicating internal shear layer and arch vortex formation. The (periodic) z domain is expanded for clarity in (c) and (e).

vortices. Direct formation of arches through instability is unlikely, since the corresponding instability would involve varicose modes, found to be *stable* for relevant streak distributions (Schoppa & Hussain 1997a). In minimal channel flow, we find that arches without legs are commonly created, not by instability, but by viscous annihilation of a leg originally attached to an arch (like the vortices in Fig. 7e).

5. Concluding remarks

To summarize, we have shown that (nonlinear) instability of ejected low-speed streaks, initially without any vortices whatsoever, directly generates new streamwise vortices, internal shear layers, and arch vortices. The resulting 3D vortex geometry is identical to that of the dominant CS, educed from fully developed near-wall turbulence, which in turn capture all important, extensively reported near-wall events. This serves as strong evidence that vortex-less streaks are the main breeding ground for new streamwise vortices, commonly accepted as dominant in turbulence production. In turn, the geometry of the newly generated vortices constitutes a built-in mechanism which sustains ejected streaks against their otherwise rapid self-annihilation due to cross-diffusion. Vortex-less streaks, the vehicle for instability-based vortex formation, are expected to arise inherently in full-domain turbulence due to the differential advection of vortices and the streaks they generate (see Schoppa & Hussain 1997a for details).

Since vortex formation and turbulence production are critically reliant on lifted low-speed streaks, large-scale (relative to the natural streak spacing) control of streaks is a potentially effective approach to drag reduction, noting the tiny scale of near-wall structures in most engineering situations. We have found that large-scale drag reduction is in fact effective via either counterrotating vortex generators or colliding spanwise wall jets (Schoppa & Hussain 1997b). Our control approach is particularly attractive from a practical standpoint, in that no sensors are required (necessary for adaptive control) and large-scale (hence more durable and feasible) actuation is effective.

This research was supported by AFOSR grant F49620-97-1-0131 and the NASA Graduate Fellowship grant NGT-51022 of W.S. Supercomputer time was provided by the NASA Ames Research Center.

6. References

Blackwelder, R.F. & Eckelmann, H. 1979 Streamwise vortices associated with the bursting phenomenon. *J. Fluid Mech.* **94**, 577.

Brooke, J.W. & Hanratty, T.J. 1993 Origin of turbulence-producing eddies in a channel flow. *Phys. Fluids A* **5**, 1011.

Brown, G.L. & Thomas, A.S.W. 1977 Large structure in a turbulent boundary layer. *Phys. Fluids* **20**, 5243.

Hamilton, J., Kim, J. & Waleffe, F. 1995 Regeneration mechanisms of near-wall turbulence structures. *J. Fluid Mech.* **287**, 317.

Jang, P.S., Benney, D.J. & Gran, R.L. 1986 On the origin of streamwise vortices in a turbulent boundary layer. *J. Fluid Mech.* **169**, 109.

Jeong, J. & Hussain, F. 1992 Coherent structures near the wall in a turbulent channel flow. In *Proc. of Fifth Asian Congress of Fluid Mech*., Taejon, Korea (eds. K.S. Chang & H. Choi), p. 1262.

Jeong, J., Hussain, F., Schoppa, W. & Kim, J. 1997 Coherent structures near the wall in a turbulent channel flow. *J. Fluid Mech.* **332**, 185.

Jimenez, J. & Moin, P. 1991 The minimal flow unit in near-wall turbulence. *J. Fluid Mech.* **225**, 213.

Jimenez, J. & Orlandi, P. 1993 The rollup of a vortex layer near a wall. *J. Fluid Mech.* **248**, 297.

Kim, J., Moin, P. & Moser, R.D. 1987 Turbulence statistics in fully developed channel flow at low Reynolds number. *J. Fluid Mech.* **177**, 133.

Kline, S.J., Reynolds, W.C., Schraub, F.A. & Rundstadler, P.W. 1967 The structure of turbulent boundary layers. *J. Fluid Mech.* **30**, 741.

Kline, S.J. & Robinson, S.K. 1990 Quasi-coherent structures in the turbulent boundary layer: Part 1. Status report on a community-wide survey of the data. In *Near-Wall Turbulence* (eds. S.J. Kline & N.H. Afgan). Hemisphere.

Kravchenko, A.G., Choi, H. & Moin, P. 1993 On the relation of near-wall streamwise vortices to wall skin friction in turbulent boundary layers. *Phys. Fl. A* **5**, 3307.

Phillips, W.R.C. & Wu, Z. 1994 On the instability of wave-catalysed longitudinal vortices in strong shear. *J. Fluid Mech.* **272**, 235.

Robinson, S.K. 1991 Coherent motions in the turbulent boundary layer. *Ann. Rev. Fluid Mech.* **23**, 601.

Schoppa, W. & Hussain, F. 1997a Genesis and dynamics of coherent structures in near-wall turbulence. In *Self-sustaining Mechanisms of Wall Turbulence* (ed. R. Panton). Computational Mechanics Publications, p. 385.

Schoppa, W. & Hussain, F. 1997b Effective drag reduction by large-scale manipulation of streamwise vortices in near-wall turbulence. Presented at *4th AIAA Shear Flow Control Conference* (Snowmass CO), AIAA Paper No. *AIAA 97-1794*.

Swearingen, J.D. & Blackwelder, R.F. 1987 The growth and breakdown of streamwise vortices in the presence of a wall. *J. Fluid Mech.* **182**, 225.

Townsend, A.A. 1956 *The Structure of Turbulent Shear Flows*. Cambridge University Press.

THEORY OF NON-AXISYMMETRIC BURGERS VORTEX WITH ARBITRARY REYNOLDS NUMBER

K. BAJER
University of Warsaw, Institute of Geophysics
ul. Pasteura 7, 02-093 Warszawa, Poland
`www.igf.fuw.edu.pl/~kb/WWW`

AND

H.K. MOFFATT
University of Cambridge,
Department of Applied Mathematics and Theoretical Physics,
Silver Street, Cambridge, CB3 9EW, UK
`www.damtp.cam.ac.uk/user/tfd`

Abstract. We develop an asymptotic theory of the steady state of a rectilinear vortex in linear straning flow. In the special case of axisymmetric strain the solution is the familiar Burgers vortex. In the more general, non-axisymmetric situation the asymptotic theories were developed for low Reynolds number (Robinson & Saffman 1984) and for high Reynolds number (Moffatt, Kida & Ohkitani 1994). In the present paper we develop a new expansion in the parameter λ which characterises the departure from axisymmetry. Hence we obtain an expansion valid uniformly for all Reynolds numbers and thus bridgeing the gap between the low and the high Reynolds number theories. In practice the new expansion is useful when the non-axisymetric deformation of the vortex is not too large.

1. Introduction

G. I. Taylor (1938) recognised the fact that the competition between stretching and viscous diffusion of vorticity must be the mechanism controlling the dissipation of energy in turbulence. A decade later Burgers (1948) obtained exact solutions describing steady vortex tubes and layers in locally uniform straining flow where the two effects are in balance. The discovery of the

E. Krause and K. Gersten (eds.), IUTAM Symposium on Dynamics of Slender Vortices, 193-202.

exact solutions stimulated the development of the models of the dissipative scales of turbulence as random collections of vortex tubes and/or sheets.

The intermittent nature of the vorticity field was observed in experiments by taking statistical measurements which indicated the existence of the small-scale localised structures (Townsend 1951). Only recently have these structures been directly observed, first in the numerical simulations (see, for example Vincent & Meneguzzi 1991) and then in the laboratory experiments where a new visualisation technique was employed (Douady, Couder & Brachet 1991).

The Burgers vortex has axial symmetry unlikely to be found in real flows, hence the need to find solutions describing non-axisymmetric stretched vortices. Let us consider incompressible fluid with viscosity ν. We look for a steady state of a vortex having stream-function $\Psi(x, y)$, vorticity $\boldsymbol{\omega} = -(\nabla^2\Psi)\hat{\mathbf{z}}$ and total circulation Γ subjected to the ambient irrotational straining flow

$$\boldsymbol{U} = (\alpha x, \beta y, \gamma z) \quad , \qquad \alpha + \beta + \gamma = 0 \quad , \qquad \alpha < 0, \quad \gamma > 0 \tag{1}$$

characterised by the parameter

$$0 \leq \lambda = \frac{\alpha - \beta}{\alpha + \beta} \tag{2}$$

which measures the departure from axisymmetry. Taking $\sqrt{\nu/\gamma}$, γ^{-1}, $\Gamma/2\pi$, $\gamma\Gamma/2\pi\nu$ to be the units of length, time, stream-function and vorticity respectively respectively we obtain the steady state equations in polar coordinates (r, θ),

$$\frac{1}{r}\frac{\partial(\Psi, \omega)}{\partial(r, \theta)} = -R_\Gamma^{-1}\left(L_0\omega + \lambda L_1\omega\right), \tag{3}$$

$$\omega = -\nabla^2\Psi, \tag{4}$$

where $R_\Gamma = \Gamma/2\pi\nu$ is the Reynolds number of the vortex and L_0, L_1 are linear operators,

$$L_0 = 1 + \tfrac{1}{2}r\partial_r + \nabla^2 \tag{5}$$

$$L_1 = \tfrac{1}{2}\cos(2\theta)\, r\partial_r - \tfrac{1}{2}\sin(2\theta)\,\partial_\theta. \tag{6}$$

Robinson and Saffman (RS84) solved this equation numerically for $0 < R_\Gamma < 100$ and found that the vortex has a quasi-elliptical shape with the minor axis inclined at an angle $\Phi(R_\Gamma, \lambda)$ to the principal axis of strain. They found that $\Phi(R_\Gamma, \lambda) \to 0$ as $R_\Gamma \to 0$ and developed a theory for $R_\Gamma \ll 1$ involving *double* expansion in powers of both λ and R_Γ. Their computations showed that as R_Γ increases $\Phi(R_\Gamma, \lambda)$ settles to a constant value $\Phi_c \approx 45°$. The value of Φ_c was theoretically derived by Moffatt, Kida & Ohkitani (MKO94) who developed an asymptotic theory for $R_\Gamma \gg 1$ later adapted to diffusing vortices in two-dimensional strain for which much more numerical data, including details of the vortex structure, are available (Jiménez, Moffatt & Vasco 1996). Here we develop a new theory based on the expansion in powers of λ.

2. λ-expansion

We look for solutions of (3-4) in the form of the series

$$\omega = \omega_0 + \lambda\omega_1 + \lambda^2\omega_2 + \dots \quad , \tag{7}$$

$$\Psi = \Psi_0 + \lambda\Psi_1 + \lambda^2\Psi_2 + \dots \quad . \tag{8}$$

The lowest order gives an equation for the vortex in *axisymmetric* strain:

$$\frac{1}{r}\frac{\partial(\Psi_0, \omega_0)}{\partial(r, \theta)} = -R_\Gamma^{-1} L_0 \omega_0 \tag{9}$$

whose solution is the familiar Burgers vortex,

$$\omega_0 = \tfrac{1}{2}\mathrm{e}^{-\frac{1}{4}r^2}, \tag{10}$$

$$\Psi_0 = \int_0^r x^{-1}\left(\mathrm{e}^{-\frac{1}{4}x^2} - 1\right) dx. \tag{11}$$

The calculations of the $O(\lambda)$ terms become much simpler in the variables

$$w = \tfrac{1}{4}r^2 \quad , \qquad \varphi = 2\theta. \tag{12}$$

From (3-4) we obtain

$$R_\Gamma \left[\frac{d\Psi_0}{dw}\frac{\partial\omega_1}{\partial\varphi} - \frac{d\omega_0}{dw}\frac{\partial\Psi_1}{\partial\varphi}\right] + L_0\omega_1 + L_1\omega_0 = 0, \tag{13}$$

$$\nabla^2\Psi_1 = \omega_1, \tag{14}$$

where now

$$\omega_0 = \tfrac{1}{2}\mathrm{e}^{-w} \quad , \frac{d\Psi_0}{dw} = \tfrac{1}{2}w^{-1}\left(\mathrm{e}^{-w} - 1\right), \tag{15}$$

and the operators ∇^2, L_0 and L_1 now take the form:

$$\nabla^2 = \partial_w w \partial_w + w^{-1}\partial_\varphi^2 \tag{16}$$

$$L_0 = w\partial_w^2 + (w+1)\partial_w + w^{-1}\partial_\varphi^2 + 1 \tag{17}$$

$$L_1 = \cos\varphi w \partial_w - \sin\varphi \partial_\varphi. \tag{18}$$

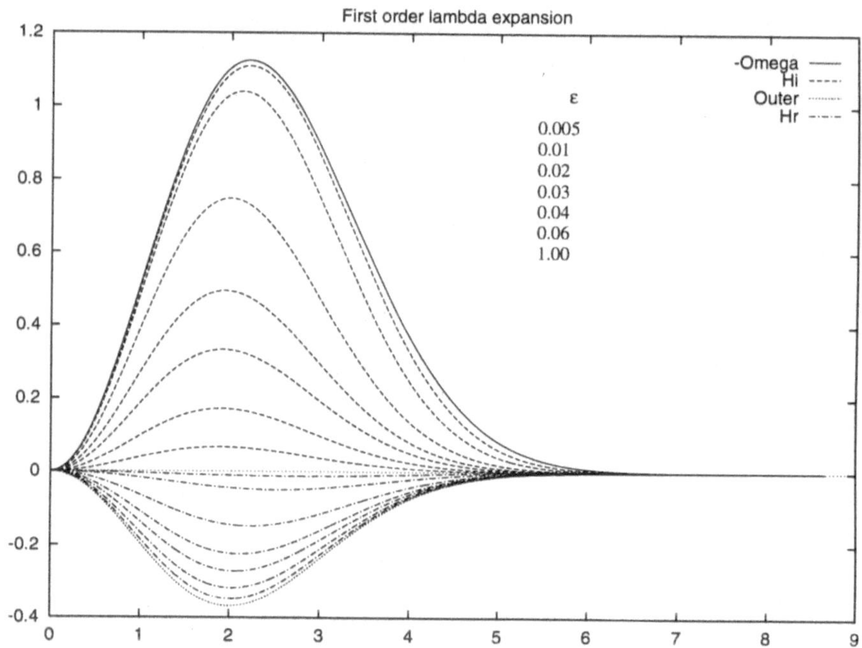

Figure 1. The functions $H_i(r)$ (positive values, dashed lines) and $H_r(r)$ (negative values, dot-dashed lines) for different values of the parameter $\epsilon = \nu/\Gamma$ listed in the figure. The peak values of H_i increase as $\epsilon \to 0$ when H_i tends to the function $-\Omega(r)$ (solid line). The peak value of $|H_r|$ increases as $\epsilon \to \infty$ ($R_\Gamma \to 0$) when H_r tend to its limiting value $-we^{-w}$.

Substituting (15) into (13) we obtain

$$\tfrac{1}{2}R_\Gamma w^{-1}\left(\mathrm{e}^{-w} - 1\right)\frac{\partial \omega_1}{\partial \varphi} + \tfrac{1}{2}R_\Gamma \mathrm{e}^{-w}\frac{\partial \Psi_1}{\partial \varphi} + L_0\omega_1 = \tfrac{1}{2}w\mathrm{e}^{-w}\cos\varphi. \tag{19}$$

This equation reveals the nature of the λ-expansion. Lower orders become *source* terms in the equations for higher orders. This is quite different from the procedure developed in MKO94 for $R_\Gamma \gg 1$. There the first order equation appeared as the integrability condition for the *second* order.

The right-hand-side of (19) consist of a single Fourier mode, so we can look for a solution in the form

$$\Psi_1 = \mathrm{Re}\left[\tfrac{1}{2}S(r)\mathrm{e}^{\mathrm{i}\varphi}\right] \quad , \qquad \omega_1 = \mathrm{Re}\left[\tfrac{1}{2}H(r)\mathrm{e}^{\mathrm{i}\varphi}\right]. \tag{20}$$

The (complex) radial functions satisfy two coupled ordinary differential

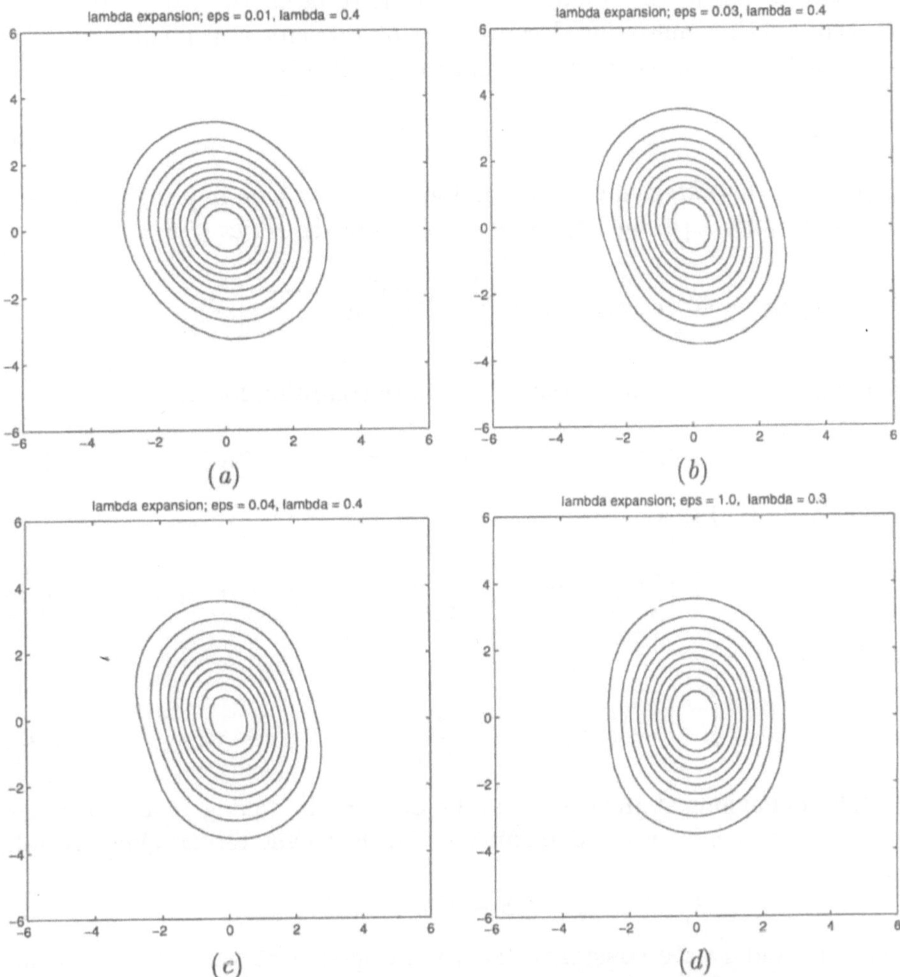

Figure 2. The level contours of $\omega_0 + \lambda\omega_1$ for $\epsilon = 0.01$, 0.03, 0.04 and 1.0. As ϵ increases the tilt of the vortex decreases from its limit value of 45° to zero while the elliptical deformation increases.

equations,

$$wH'' + (w+1)H' + \left[\mathrm{i}\tfrac{1}{2}R_\Gamma w^{-1}(\mathrm{e}^{-w} - 1) + 1 - w^{-1}\right] H + \mathrm{i}\tfrac{1}{2}R_\Gamma \mathrm{e}^{-w} S = w\mathrm{e}^{-w} \quad , \tag{21}$$

$$wS'' + S' - w^{-1}S + H = 0 \quad , \tag{22}$$

that can easily be solved numerically, but first we need to determine the appropriate boundary conditions.

Far away from the origin the nonlinear term on the left-hand side of (3) representing self-induced convection of vorticity is negligible and the *linearized* equation has a unique solution (MKO94),

$$\omega^E(w,\varphi) = \tfrac{1}{2}\mathrm{e}^{-w(1+\lambda\cos\varphi)}. \tag{23}$$

Expanding this external vorticity in powers of λ we can, in principle, obtain the asymptotic form for large r of all terms in the series (7-8),

$$\omega^E(w,\varphi) = \tfrac{1}{2}\mathrm{e}^{-w}\sum_{n=0}^{\infty}\left[\frac{(-1)^n}{n!}w^n\cos^n\varphi\right]\lambda^n. \tag{24}$$

In particular, we obtain an outer boundary condition for ω_1:

$$\omega_1 \sim -\tfrac{1}{2}w\mathrm{e}^{-w}\cos\varphi \qquad \text{as} \qquad r\to\infty, \tag{25}$$

and solving (14) gives

$$\Psi_1 \sim \tfrac{1}{2}\left[\left(1+\frac{1}{w}\right)\mathrm{e}^{-w} - \frac{1}{w}\right]\cos\varphi \qquad \text{as} \qquad r\to\infty, \tag{26}$$

which gives

$$H \sim -w\mathrm{e}^{-w}, \quad S \sim -\frac{1}{w}, \qquad \text{as} \qquad r\to\infty. \tag{27}$$

The behaviour at the origin can be deduced by taking H and S in the form of a power series and balancing the dominant terms. One possible solution is

$$H \sim Cw \quad , \qquad S \sim Dw \qquad \text{as} \qquad w\to 0, \tag{28}$$

where C and D are constants. This is in agreement with the behaviour of the asymptotic solutions of MKO94, so we take (26-27) as boundary conditions for (20-21).

The results can be verified in two limiting cases. In the limit $R_\Gamma \to 0$ the function ω_1 must be equal to $-w\mathrm{e}^{-w}$ *everywhere*, not just at large distances. In the limit $R_\Gamma \to \infty$ the asymptotic form of ω_1 can be deduced from MKO94 who derived the function Ω, such that

$$H(r) \sim -2\mathrm{i}R_\Gamma^{-1}\Omega(r) \qquad \text{as} \qquad R_\Gamma\to\infty. \tag{29}$$

In figure 1 we plot $H_i = \frac{1}{2}R_\Gamma \mathrm{Im}(H)$ (positive) and $H_r = \frac{1}{2}R_\Gamma \mathrm{Re}(H)$ (negative), as functions of r, for different values of the parameter $\epsilon = \nu/\Gamma = (2\pi R_\Gamma)^{-1}$. The functions H_i become more curved as the value of ϵ *decreases* and they tend to $-\Omega(r)$ which means that our solution, as expected, approaches that of MKO94 as $R_\Gamma \to \infty$. The functions H_r become

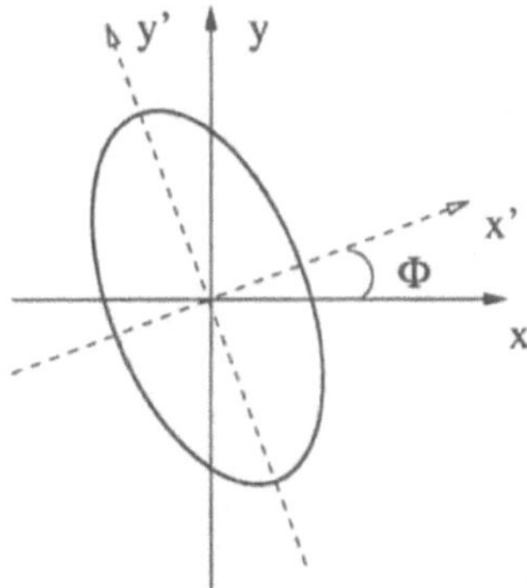

Figure 3. The frame of reference (x, y) defined by the principal axes of the ambient strain, and the frame of reference (x', y') in which the antisymmetric second moment of the vorticity distribution vanishes. The angle Φ is the tilt of the vortex.

more curved as the value of ϵ *increases* and they tend to the linearized solution $-we^{-w}$ as $R_\Gamma \to 0$.

In figure 2 we show the contour levels of vorticity in the first order solution, i.e., the contours of $\omega_0 + \lambda\omega_1$ The vortex has quasi-elliptical shape. The major axis of the ellips forms an angle Φ with the principal axis of strain (see figure 3). This tilt depends on both R_Γ (or ϵ) and λ. As $R_\Gamma \to \infty$ the angle Φ tends to 45°, as predicted by MKO94.When $R_\Gamma \to 0$ there is no self-induced rotation of the vortex, so $\Phi \to 0$. The level contours are excessively flattened along the minor axis, particularly for larger values of λ when the higher-order terms in (7) begin to play a rôle.

In order to quantify this effect the angle of inclination is defined by the rotated frame of reference,

$$x' = \cos\Phi\, x + \sin\Phi\, y \quad , \qquad y' = -\sin\Phi\, x + \cos\Phi\, y, \tag{30}$$

such that the antisymmetric second moment of the vorticity distribution vanishes in that frame (RS84),

$$\int_{-\infty}^{\infty}\int_{-\infty}^{\infty} x'y'\omega(x', y')\, dx'\, dy' = 0. \tag{31}$$

Taking $\omega = \omega_0 + \lambda\omega_1$ we obtain

$$\frac{1}{2}\int_0^{2\pi} d\theta \int_0^{\infty} dr\, r^3 \sin 2(\theta - \Phi)\left[R_\Gamma\omega_0(r) + \lambda H_r \cos 2\theta - \lambda H_i(r) \sin 2\theta\right] = 0 \tag{32}$$

Therefore, the λ-expansion yields a formula for the angle of inclination,

$$\tan 2\Phi = -\frac{\int_0^\infty dr\, r^3 H_i(r)}{\int_0^\infty dr\, r^3 H_r(r)}. \tag{33}$$

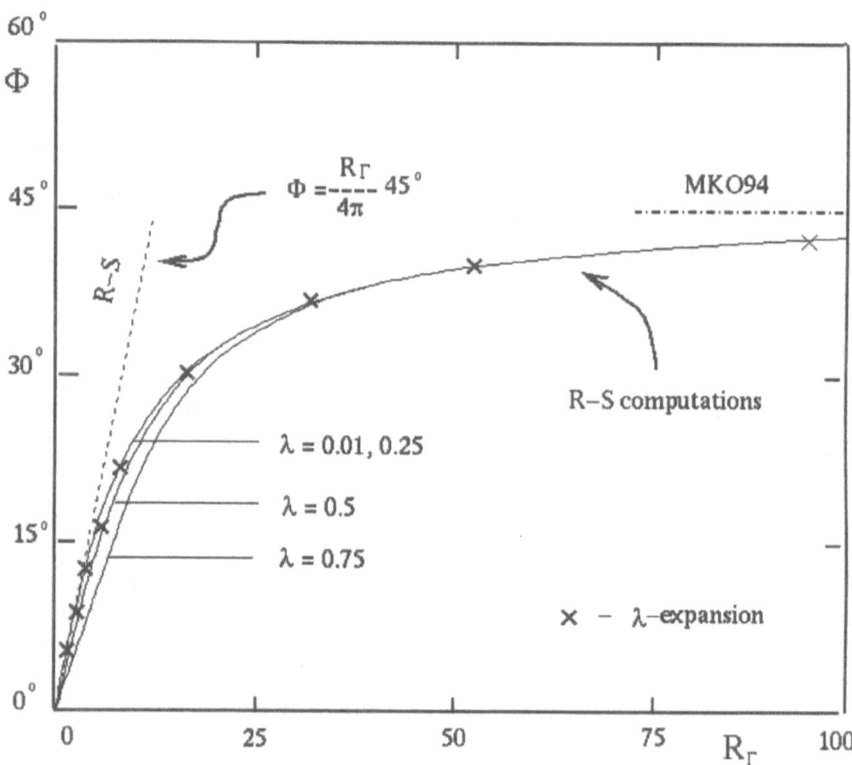

Figure 4. The tilt of the vortex computed in RS84 for four values of λ (solid lines) compared with the tilt given by (33) (crosses). The theoretical prediction of RS84 for $R_\Gamma \to 0$ (dashed line) and the asymptotic result of MKO94 for R_Γ (dot-dashed line) are also shown.

In figure 4 we show $\Phi(R_\Gamma)$ computed by Robinson and Saffman for $\lambda = 0.01$, 0.25, 0.5 and 0.75. The values calculated from (33), marked with crosses, are in good agreement, at least for $\lambda < 0.5$. The curves for $\lambda = 0.1$ and 0.25 are indistinguishable and that for $\lambda = 0.5$ differs very little. The explanation comes from (33) which shows that in the first order Φ is *independent* of λ. It means that a vortex is not easily 'tilted' by the ambient strain whose parameter λ must well exceed 0.5 to have considerable effect.

3. Conclusions

We have developed the theory describing a steady vortex in the linear irrotational ambient flow. The vortex experiences stretching along its axis and compression in the cross-sectional plane which compensates viscous diffusion of vorticity like in the classical Burgers solution which is a special axisymmetric case of this problem.

The non-linear vorticity equation is solved by expanding Ψ and ω in powers of the parameter λ measuring the departure of the ambient strain from axisymmetry. The solution is valid for any value of the Reynolds number and therefore complements previous asymptotic theories for $R_\Gamma \to 0$ and $R_\Gamma \to \infty$. For example, the tilt of the vortex can be accurately calculated for any value of R_Γ and already in the first order the results agree very well with the numerical solution of the full non-linear problem.

The first order solution is accurate only within a certain radius around the center of the vortex. Higher order corrections are needed outside this radius which depends on the value of λ and on the chosen criterion of accuracy. The first order solution is useful mainly when λ is small, say $\lambda < 0.25$, but some diagnostics, like the tilt angle, are well predicted even for larger values.

Calculating higher order terms may improve the accuracy of the result if the series is convergent, or it may make it worse if it is merely asymptotic. The convergence of the series (7) is an interesting but, due to the non-linear nature of the vorticity equation, rather difficult issue. Earlier we have developed a theory for the similar but *linear* problem of straight magnetic flux tube with a line vortex on the axis (Bajer & Moffatt 1997). The results strongly suggest that the λ-expansion is convergent in this case while the series appearing in both the low- and the high-Reynolds-nuber expansions are in fact asymptotic. The arguments used for the magnetic problem depend on its linearity, but it is plausible that the conclusion is still valid for the vortex problem.

References

BAJER, K. & MOFFATT, H. K. (1997) On the effect of a central vortex on a stretched magnetic flux tube. *J. Fluid Mech.* **339**, 121–142.

BURGERS, J. M. (1948) A mathematical model illustrating the theory of turbulence. *Adv. Appl. Mech.* **1**, 171–199.

DOUADY, S., COUDER, Y. & BRACHET, M. E. (1991) Direct observation of the intermittency of intense vorticity filaments in turbulence. *Phys. Rev. Lett.* **67**, 983–986.

JIMÉNEZ, J., MOFFATT, H. K., VASCO, C. (1996) The structure of the vortices in freely decaying two-dimensional turbulence. *J. Fluid Mech.* **313**, 209–222.

MOFFATT, H. K., KIDA, S. & OHKITANI, K. (1994) (MKO94) Stretched vortices – the sinews of turbulence; large-Reynolds-number asymptotics. *J. Fluid Mech.* **259**, 241–264.

ROBINSON, A. C. & SAFFMAN, P. G. (1984) (RS84) Stability and structure of stretched vortices. *Stud. Appl. Maths.* **70**, 163–181.

TAYLOR, G. I. (1938) Production and dissipation of vorticity in a turbulent fluid. *Proc. Roy. Soc.* A **164**, 15–23.

TOWNSEND, A. A. (1951) On the fine-scale structure of turbulence. *Proc. Roy. Soc.* A **208**, 534–542.

VINCENT, A. & MENEGUZZI, M. (1991) The spatial structure and statistical properties of homogeneous turbulence. *J. Fluid Mech.* **225**, 1–25.

Session 4

Interaction of Vortices

INTERACTION OF TWO VORTEX TUBES AND THE SINGULARITY FORMATION

A. FUKUYU
Department of Mathematical Sciences,
Tokyo Denki University,
Hatoyama, Saitama-ken 350-03 Japan

1 Introduction

Are there any three-dimensional velocity field of inviscid fluid with finite energy such that the velocity field becomes singular within a finite time T? This is a challenging problem in the mathematical theory of fluid dynamics. This problem has been investigated by many authors using phenomenological models ([1], [2]) or direct numerical simulations ([3], [4]) but yet we do not get conclusive result.

The goal of the present paper is to find a possible dynamical mechanism which may lead to finite time singularity formation in an inviscid flow. As a candidate for this mechanism, we examine the interaction of two vortex tubes having different strength. A scenario to the singularity formation, considered in this paper is as follows: Suppose that a weak curved vortex tube of intensity Γ_1 approaches to a strong tube of intensity Γ_2 ($\Gamma_1 < \Gamma_2$). If two vortex tubes are locally anti-parallel near the points of closest approach, two tubes may continue to approach and at the same time, the weak tube winds round to the strong one, and due to the induced velocity of the wound tube, the latter is notably stretched at the closest point of approach. We represent the point of the closest approach on tube 1 and tube 2 by O_1 and O_2, respectively. We call O_2 as the center of the stretching. Let the ratio of the intensity of two vortex tubes be G, that is, $G = \Gamma_1/\Gamma_2$. In the previous paper [5], each tube was replaced by a thin filament (single filament model) and examined the cases for $G = 0.1, 0.2$. The numerical results in [5] seems to show that there is a finite time T and as time t approaches to T the core radius σ of the filament decreases in a form $(T-t)^{q_\sigma}$ which means that vorticity ω increases as $(T-t)^{-2q_\sigma}$. But the numerical result also show that the inter-filament distance d decrease

E. Krause and K. Gersten (eds.), IUTAM Symposium on Dynamics of Slender Vortices, 205-214.

also in a form $(T-t)^{q_d}$ and the power q_d always greater than the power q_σ. This means that overlap of core occurs before vortex tube is stretched infinitely and singularity is formed.

In the interaction of two vortex tubes of finite core size, deformation of core has crucial importance when two tubes approach each other. In the winding process considered in this paper, when two tubes approach, for which we assume circular cross section at the initial instant, core of the tubes may become flatten and as a result approaching velocity may decrease.

To see the effect of core deformation, we replace in this paper a vortex tube of finite core size by a set of thin vortex filaments, say, tube 1 by M_1 filaments and tube 2 by M_2 filaments.

2 Numerical method

To simulate numerically the winding process considered above, we use the Biot-Savart's law for vorticity distribution

$$u(r) = -\frac{1}{4\pi}\int_\Omega \frac{(r-r')\times\omega(r')}{|r-r'|^{3/2}}dr, \tag{1}$$

where Ω represents the vorticity region. If the vorticity is concentrated in a vortex filament of intensity Γ, eq.(1) may be simplified as

$$u(r) = -\frac{\Gamma}{4\pi}\int_\Omega \frac{(r-r')\times d\tau'}{|r-r'|^3}. \tag{2}$$

In order to regularize the induced velocity, we usually introduce a core parameter σ which varies along a filament and eq.(2) may be modified as

$$u(r) = -\frac{\Gamma}{4\pi}\int_\Omega \frac{(r-r')\times d\tau'}{|(r-r')^2+\sigma^2(r')|^{3/2}}. \tag{3}$$

In this paper, we examine the three dimensional interaction of locally anti-parallel vortex tubes of intensity Γ_1 and Γ_2. We represent each tube by a set of thin vortex filaments, thus, the tube 1 by a set of M_1 filaments and tube 2 by M_2 filaments. In this case, induced velocity $u(r)$ is given by a sum of the integrals of the type (3). In the simulations below, a vortex curve, possibly infinite in length, is approximated by a finite curve, and each curved filament is replaced by a set of N segments with length δr_k and core parameter σ_k. A segment k is represented by two successive nodal points r_k and r_{k+1} on the curve. The induced velocity (3) may be modeled as

$$u(r) = -\frac{\Gamma}{4\pi}\sum_{k=1}^{N}\frac{a\times\delta r}{\{|a_k|^2+\sigma_k^2\}^{3/2}}, \tag{4}$$

where

$$a_k = \frac{1}{2}(r_{k+1}-r_k)-r, \quad \delta r_k = r_{k+1}-r_k. \tag{5}$$

The evolution of the flow is described by an evolution equation of a nodal point k of the form

$$\frac{dr_k}{dt} = u(r_k), \tag{6}$$

where $u(r_k)$ is given by eq.(4) setting $r = r_k$. In an inviscid flow, the tube volume of a filament is conserved which is a consequence of Helmholtz law. In the numerical simulations below, we adopt a core law of the form

$$\sigma^2|\delta r| = const. \tag{7}$$

Thus, if a vortex segment is stretched the core parameter of this segment decreases according to eq.(7) and since the intensity of the filament is conserved, vorticity increases in proportion to the inverse of the square of the core parameter.

For single filament model used in [5], each filament was replaced by a finite length curve and end effect of the vortex filament was ignored. But when we replace a vortex tube by a set of thin vortex filaments, it is inevitable to take into account the end effect. In our calculations below, strong tube 2 is initially straight circular tube and if we take a sufficiently long tube, both end of the tube remain almost straight during the stretching process. Thus, we add to the induced velocity given by eq.(4), contributions of half-infinite straight filaments placed at the end of each finite filament for which analytical expression can easily obtained. For curved tube 1, we assume a parabola near the point O_1 but for both end of the curve, we assume straight form. Thus, for each end point of the filament of tube 1 we also add half-infinite straight vortex filament. The equation of motion of nodal point eq.(6) should be modified as

$$\frac{dr_k}{dt} = u(r_k) + (\text{contributions from half-infinite filaments}). \tag{8}$$

As time proceeds, each vortex tube may be deformed and be stretched under the mutual interactions and two tubes may get entangled. In this situation, local length scale becomes small and to approximate a curved filaments by a set of segments, we need appropriate subdivision of segments. In this paper, we use the method of subdivision given in [5]. Then, a subdivision of a segment is taken place so as to preserve local curvature and torsion of the curve. Time integration of the evloution equation is done using the fourth order Runge-Kutta method. In a numerical simulation of entangled vortex filaments, time scale of evolution differs considerably from point to point. Thus, we use a variable time increment Δt as

$$\Delta t = \eta\, min\,\{\frac{\delta r_k}{|u(r_k)|}\}, \tag{9}$$

where *min* is taken for all segments. η is a numerical factor smaller than 1. In the simulations below, we set $\eta = 0.1 \sim 0.2$.

3 Numerical results

In this section we examine two models:

model 1 : $M_1 = 3,\ M_2 = 7$
model 2 : $M_1 = 7,\ M_2 = 7$

In all cases, initial vortex tube 2 is a straight cylindrical tube. Thus, to mimic cylindrical vortex tube by a set of 7 thin filaments, we set one filament on the center line of the cylinder (we call this filament as axial filament of the tube) and 6 filaments on the circumference of the cylinder in equal spacing (*i.e.* nodal points of same x coordinate form a hexagon, that is, the cross section is a hexagon). The center line of tube 2 is on the x axis. Initial vortex tube 1 is a parabola near the the point of the closest approach and the center line of the tube is given by $y = Ax^2$ for $-\ l \leq x \leq l$ and for $x \leq -\ l$ and $x \geq l$, this parabola continues to a straight line. This center line of tube 1 lies on the xy plane at $z = d_0$. Thus, in model 1, initially 3 filaments are set in equal distance from this center line and forming triangle. In model 2, one filament on the center line and the rest of 6 filaments are set in equal distance from the center line for which cross section is a hexagon. Initially, core parameter σ of a filament is uniform along a filament. In the simulation below, we set $G = 0.4$.

Fig.1 shows the time evolution of interaction of two tubes for model 2. In this case, the radius of two tubes are initially uniform along the tubes and taken to be 0.1 and the initial distance d_0 between tubes is 0.4. In the figure, configuration of two tubes at t=0, 0.6043, 0.9183, 1.3674 are given. The picture is drown for $x \geq 0$. To get the whole picture, a mirror image of the portion $x > 0$ should be added. We see that weak curved tube 1 winds round to tube 2.

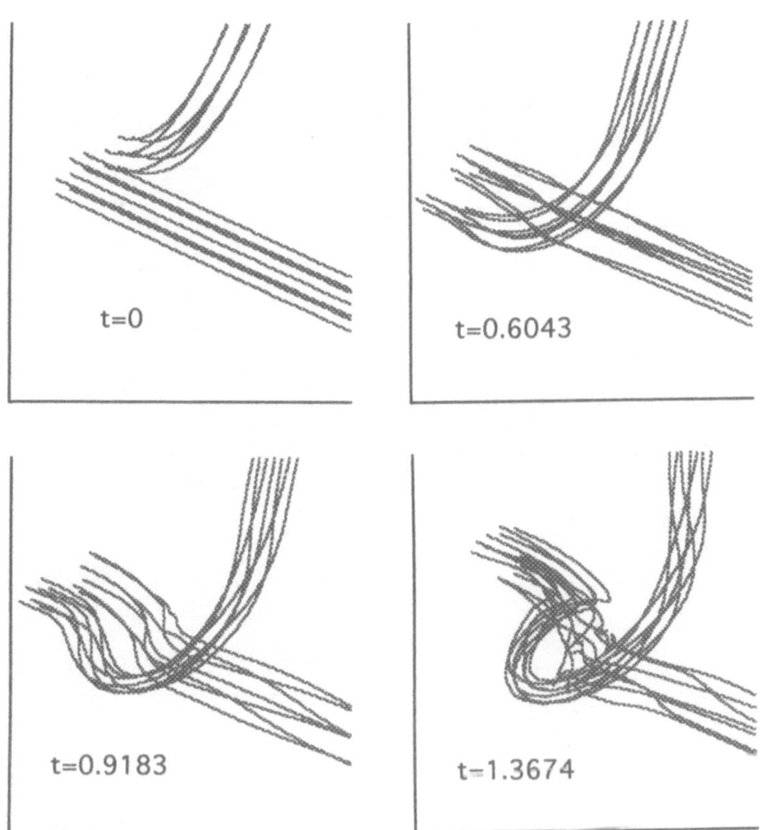

Figure 1: Evolution of interaction of two tubes. Configurations of two tubes at t=0, 0.6043, 0.9183,1.3674

In Fig.2, we show the x-y and y-z projections of two tubes at $t = 1.3674$. In this figure, solid curves correspond to tube 1 and dotted curve to tube 2. Two hexagons connecting the end points of dotted lines show the cross section at the center of the stretching (smaller hexagon) and end portion of initially straight tube 2 (larger hexagon). We see that at the end portion, the tube rotates but form of the cross section does not change essentially.

On the contrary, at the center of the tube cross section shrinks meaning the stretching of the tube.

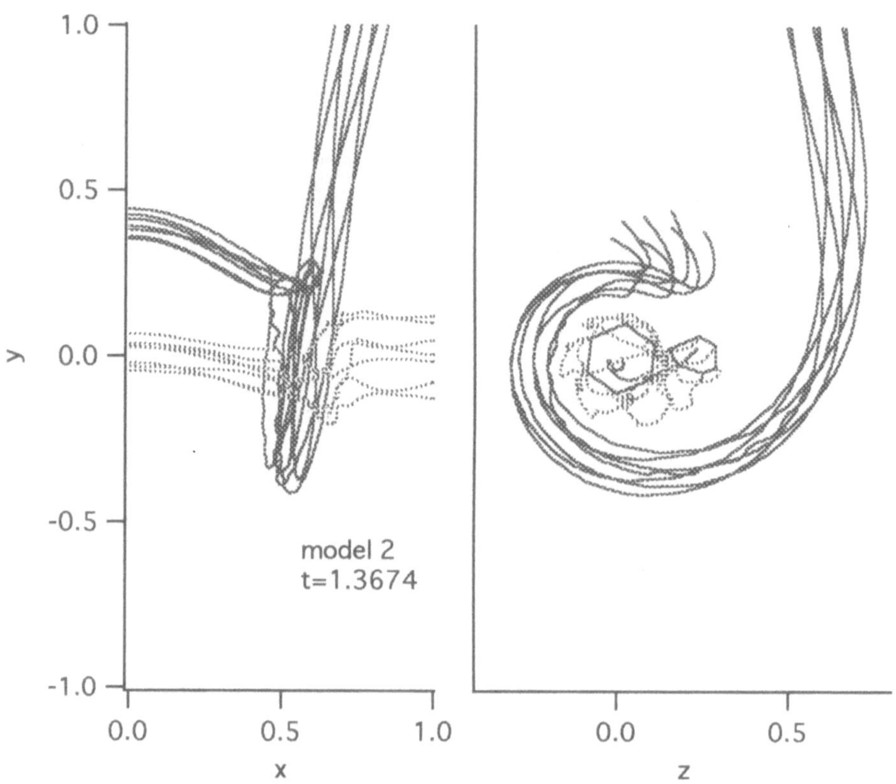

Figure 2: x-y and y-z projection of filaments at t=1.3674

Fig.3 shows the x-y and x-z projections of tube 2. We see that tube 2 is rather uniformly stretched around the center of stretching, then it continues to contracted portion and gradually to straight portion with uniform cross section. When a uniform vortex tube is shrinks at some portions of the tube, adverse pressure gradient appears which may even out the core. Thus, the stretching processes we are examining in this paper, may be seen as the competition to two effects, that is, stretching due to the induced velocity of wound tube and contraction due to the adverse pressure gradient. Fig.3 seems to show that at least at the initial stage of evolution, the stretching overcomes the contraction.

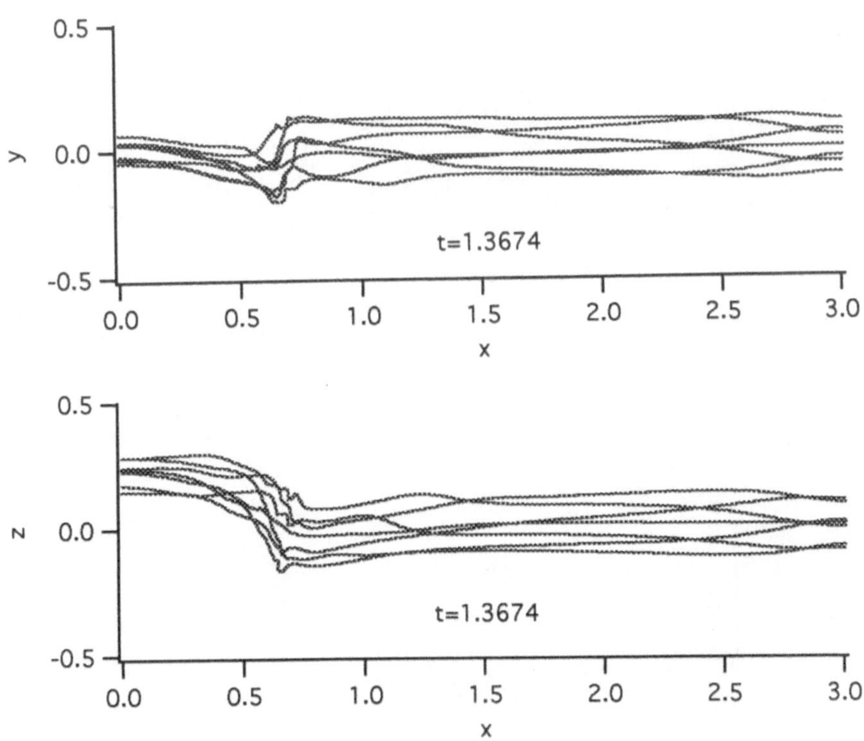

Figure 3: x-y and x-z projection of tube 2 at t=1.3674

In Fig.4, we show the time evolution of cross sections Δ_1 and Δ_2 at the center of tube 1 and tube 2 calculated from the hexagon at the center and the inverse of vorticity $1/\omega_1$ and $1/\omega_2$ at $x = 0$ of axial filaments. All values are normalized to 1 at $t = 0$. In the calculating range of time (though it may be an initial stage of evolution), $1/\omega$ and Δ decrease almost linearly in time t. We see that Δ_2 and ω_2 decrease faster than Δ_1 and ω_1. Decreasing of Δ_1 and ω_1, thus, stretching of tube 1 near the portion of O_1 may be mainly due to the differential displacement of the this portion of the tube, but decreasing of Δ_2 and ω_2 may be mainly due to the induced velocity of wound part of tube 1. This may show the possibility of singularity formation in a finite time by the winding process.

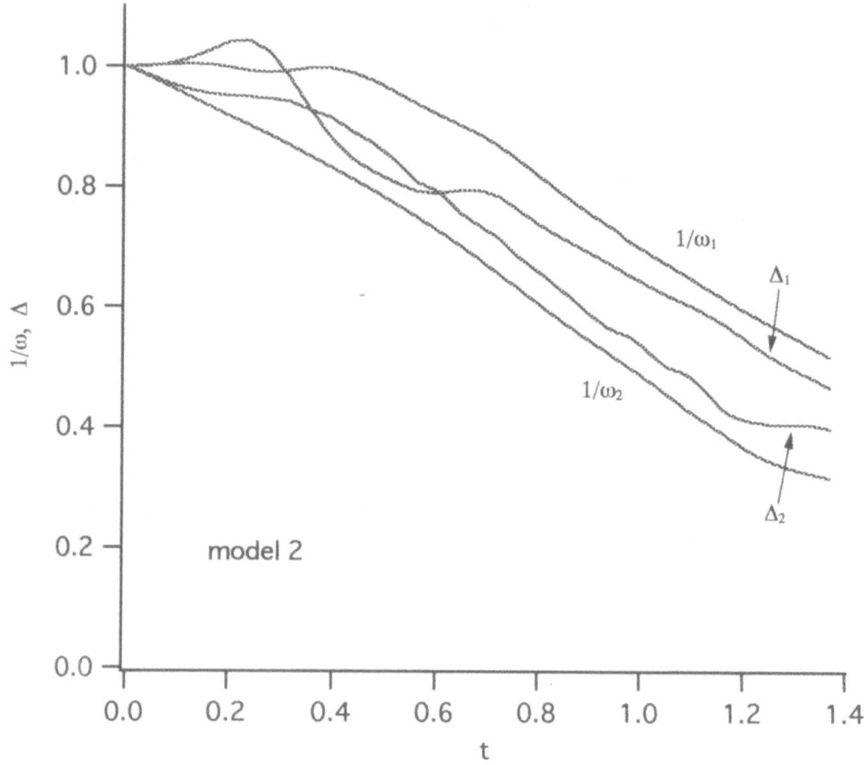

Figure 4: Evolution of the cross section and the inverse of vorticity

4 Discussions

In this paper, we examined the interaction of locally anti-parallel two vortex tubes having different intensity. The numerical scheme was based on the Biot-Savart law for vorticity distribution. To mimic a vortex tube of finite cross section, we replaced a tube by a set of M vortex filaments. From the numerical results, we see that at least at the initial stage of evolution, or more precisely, within a time duration of one winding of tube 1 around tube 2, the strong vortex tube 2 is efficiently stretched around the point

O_2. The evolution in time of the cross section of the core of tube 2 and the inverse of the vorticity seem to show that if winding process continues a singularity where vorticity becomes infinite may appear within a finite time.

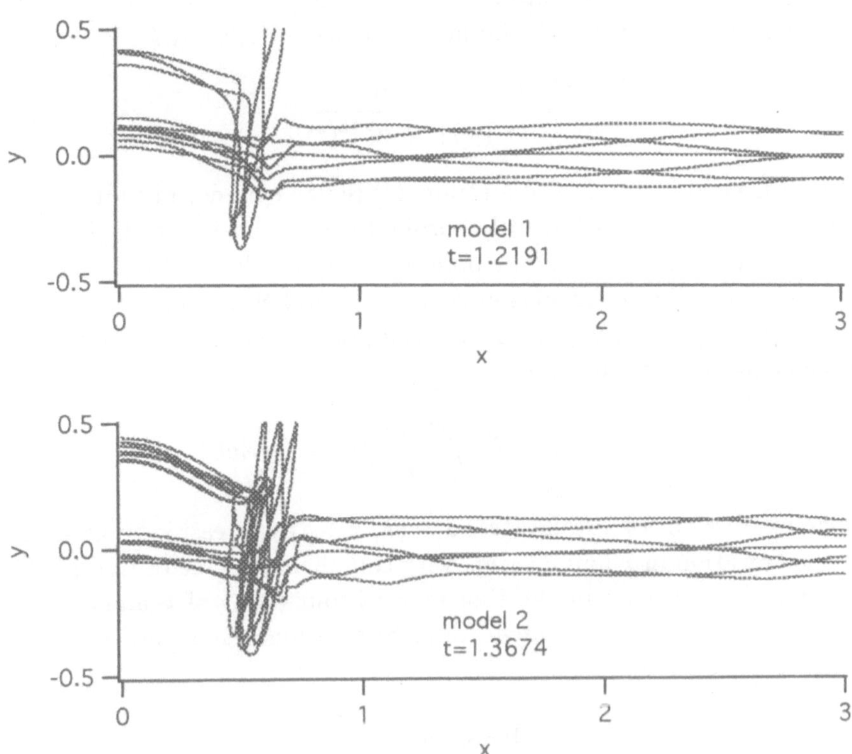

Figure 5: Comparison of model 1 and model 2

The numerical results examined in the previous section are obtained by model 2, that is, a model where two vortex tubes are represented by a set of 7 vortex filaments. We have also done numerical simulations using model 1, that is, 3 filaments for tube 1 and 7 filaments for tube 2. The numerical results seems to show that detailed profile of each filament may differ much for each model but as a whole of the tubes, numerical results do not depend much on models. In Fig.5, we compare two results calculated by model1 and model 2 both with $G = 0.4$. For model 1, tube 1 is represented by three filaments and the radius of the core is the half of that of model 2. Thus, in model1, tube 1 has the same intensity as that of model 2 but has more concentrated vorticity.

Numerical simulations done in this paper are based on the formulation (3), but this formulation is not correct when the reference point r is close from the segment. To take into account the effect of finite core size of the vortex segment, we replace the induced velocity of segment k

$$-\frac{\Gamma}{4\pi}\frac{a \times \delta r}{\{|a_k|^2 + \sigma_k^2\}^{3/2}} \tag{10}$$

by the induced velocity due to a straight circular cylinder of radius σ and length δr. We use the cylindrical coordinates $(r,\ \phi,\ z)$ in which the origin of the coordinate is on the middle point of the center line of the cylinder. If the vorticity distribution $\omega(r)$ is uniform and parallel to z axis, the induced velocity $u(r) = (u_r,\ u_p,\ u_z)$ has only ϕ component and integration in z can be done elementary resulting to

$$u(r, \phi, z) = -\frac{\omega}{4\pi}\int_0^{2\pi}\int_0^{\sigma} f(r, \theta; b) dr d\theta. \tag{11}$$

Explicit expression for the integrand $f\ (r,\ \theta\ ;\ b)$ can easily be given but farther integration in r and θ seems very complicated. One of the simplest model may be given by integrating in r at four point of θ and use four point integral formula for θ integration. Simulations using this model will be reported else where.

References

1. Siggia, E.D. (1985) *Collapse and amplification of a vortex filament*, Phys. Fluids, **28**, 794-805

2. Fukuyu, A and Arai, T. (1991) *Singularity formation in three-dimensional inviscid flow*, Fluid Dyn. Res. **7**, 229-240

3. Pumir,A. and Siggia, E.D. (1990) *Collapsing solutions to the 3-D Euler equations*, Phys. Fluids, **A2**, 220-241

4. Kerr, R.M. (1993) *Evidence for a singularity of the three-dimensional incompressible Euler equations*, Phys. Fluids, **A5**, 1725-1746

5. Fukuyu, A.(1995) *Interaction of two vortex filaments with special reference to singularity formation*, J. Phys. Soc. Japan, **64**, 2000-2011

NON-UNIQUENESS AND INSTABILITIES OF TWO-DIMENSIONAL VORTEX FLOWS IN TWO-SIDED LID-DRIVEN CAVITIES

C. BLOHM, H. KUHLMANN, M. WANSCHURA AND H. RATH
Center of Applied Space Technology and Microgravity, University of Bremen, Am Fallturm, 28359 Bremen, Germany

Abstract. The flow in rectangular cavities is investigated both numerically and experimentally. The motion is driven by two facing cavity walls which move tangentially in opposite directions. For a certain range of cavity aspect ratios the two-dimensional basic flow is not unique. At low Reynolds numbers the flow consists of separate co-rotating vortices adjacent to each of the moving walls. On an increase of the wall velocities a jump transition occurs, the two vortices partially merge, and the flow pattern appears in the form of cat's eyes. Hysteresis has been observed and investigated extensively. For high Reynolds numbers the cat's eye flow is the preferred state in the present experiment and it becomes unstable to a steady three-dimensional cellular flow that subdivides the basic stretched vortex flow into rectangular convective cells. For even higher Reynolds numbers these cells start to oscillate.

1. Introduction

The dynamics of vortices is a fundamental topic in fluid mechanics. In particular, the elliptic instability (Pierrehumbert, 1986; Bayly, 1986) seems to be responsible for the early stages of the three-dimensional evolution of plane shear flows (Bayly, Orszag & Herbert, 1988) and the flow behind bluff bodies (Williamson, 1996). Moreover, concentrated vorticity structures appear in many turbulent flows (see, e.g. She, Jackson & Orszag, 1991). Due to a symmetric strain present in the turbulent flow a vortex tube may become slightly elliptical in shape giving rise to three-dimensional instabilities. Results of basic research on the elliptic instability may find applica-

E. Krause and K. Gersten (eds.), IUTAM Symposium on Dynamics of Slender Vortices, 215-224.

tion in geophysics (Malkus, 1989) and engineering, e.g. short dwell coating (Aidun, Triantafillopoulos & Benson, 1991). Since the three-dimensional flows initiated by the elliptic instability are typically transient (Gledzer & Ponomarev, 1992) or spatially developing, it would be advantageous to study three-dimensional flow patterns caused by the elliptic instability in a stationary closed flow system. Therefore, the three-dimensional vortex flow in a two-sided lid-driven cavity is investigated here experimentally and numerically.

Previous experimental studies on one-sided lid-driven square cavities have been carried out by Koseff & Street (1984) and Aidun et al. (1991). Numerically, this problem has been treated, e.g., by Freitas, Street, Findikakis & Koseff (1985), Deville, Lê & Morchoisne (1992) and Ramanan & Homsy (1994). The latter authors investigated the linear stability of the basic two-dimensional flow.

The two-sided lid-driven cavity is a generalization of the one-sided lid-driven cavity flow problem. The geometry has been used before by Leong & Ottino (1989) and by Jana, Metcalfe & Ottino (1994) to study chaotic two-dimensional mixing in Stokes flow. Recently, Kuhlmann, Wanschura & Rath (1997) investigated the steady flow in two-sided lid-driven cavities.

2. Experimental set-up

We consider the flow in a long nearly rectangular cavity. The side walls at $x \approx \pm d/2$ are formed by two rigid circular cylinders of radii R_1 and R_2 much larger than d. They rotate independently with axes parallel to the z-axis in the plane $y = 0$. The bearings of the axes are mounted on planes $|z| > l/2$ of an outer container. The boundaries at $z = \pm l/2$ and $y = \pm h/2$ are formed by parallel Perspex plates which are rigidly attached to the outer enclosure. The flow is driven by rotating the cylinders in the same sense about their axes, such that the respective cavity walls move in opposite y-directions. A sketch of the geometry and the coordinate system is shown in Figure 1. The average width of the cavity in x-direction is $d = (56.7 \pm 0.1)$mm, the height in y-direction is $h = (29.0 \pm 0.1)$mm, the length in z-direction is $l = (190.0 \pm 0.1)$mm, and the radii of the cylinders are $R_1 = (87.55 \pm 0.03)$mm and $R_2 = (88.25 \pm 0.03)$mm, respectively. The side walls in the numerical solutions are assumed to be the planes at $x = \pm d/2$. Using h as the length scale the two aspect ratios

$$\Gamma = \frac{d}{h} = 1.96 \pm 0.05 \qquad (1)$$

and

$$\Lambda = \frac{l}{h} = 6.55 \qquad (2)$$

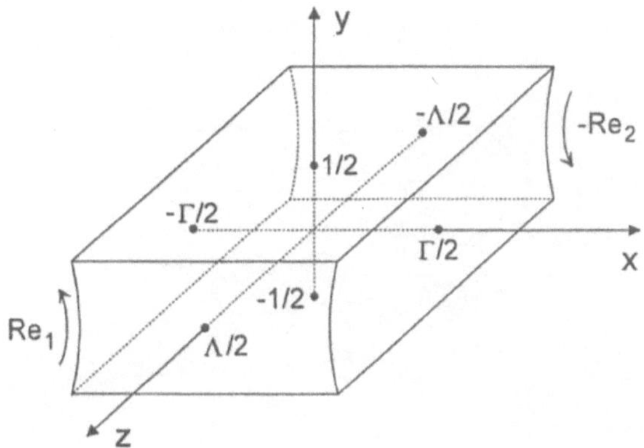

Figure 1. Sketch of the cavity and coordinate system.

arise. The Reynolds numbers are defined separately for both rotating cylinders ($i = 1, 2$)

$$Re_i = \frac{\Omega_i R_i h}{\nu}, \tag{3}$$

where $\Omega_i R_i$ is the respective wall velocity and ν the kinematic viscosity.

As a working fluid we use Bayer "Baysilone M20", a Newtonian silicone oil. The temperature in the liquid bath is permanently measured to update the two Reynolds numbers periodically. Flow visualization was accomplished by adding small amounts of aluminium particles to the liquid. To view the flow from above and from aside a light sheet was used. Hot film anemometry and laser Doppler velocimetry were applied to measure the dynamics of the vortex flow quantitatively.

3. Results

3.1. STEADY FLOWS AT SYMMETRICAL DRIVING

We first consider the case of symmetrical driving ($Re = Re_1 = Re_2$). Increasing the Reynolds number quasi-steadily we obtain a two-vortex flow as shown in Figure 2. The photography shows the streak lines in the midplane of the cavity at $Re = 230$. The left sidewall moves upwards whereas the right side-wall moves downwards. Both vortices co-rotate in clockwise direction and their strength increases with increasing Reynolds number. This two-dimensional flow is almost independent of the z-position, except for small regions close to the endwalls at $z = \pm\Lambda/2$ where the flow is three-dimensional owing to endwall effects. At the stationary top and the

Figure 2. Two-vortex flow at $z = 0$ (midplane) for $Re = 230$.

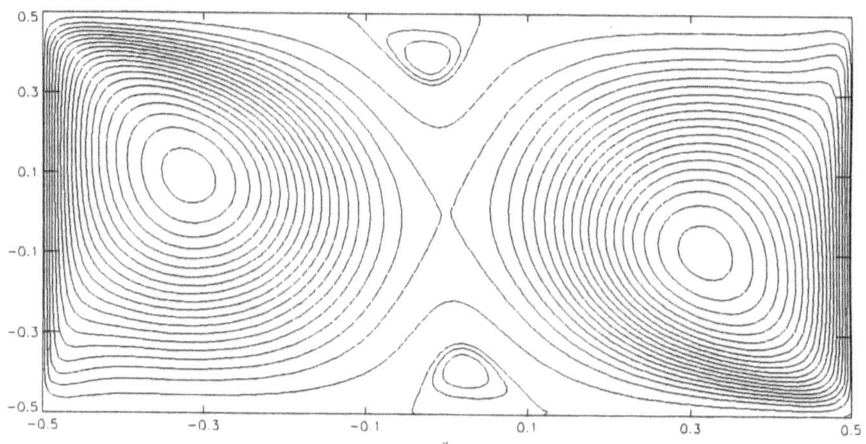

Figure 3. Numerical streamlines for $Re = 230$ and $\Gamma = 1.96$ (the streamlines have been rescaled in the separated region for better visibility); the x-axis is scaled in units of d.

bottom walls ($y = \pm 1/2$) two small regions of separated flow appear near $x = 0$. This feature is better visible in Figure 3 where the streamlines of the corresponding two-dimensional fluid flow calculated with the code of Kuhlmann et al. (1997) and Wanschura, Shevtsova, Kuhlmann & Rath (1995) are plotted.

By increasing the Reynolds number a jump transition occurs at $Re = 232$. The flow changes to the cat's eye flow shown in Figure 4. This qualitatively different flow can be explained in terms of two wall jets, one at the bottom ($y = -1/2$) and another opposite one at the top boundary ($y = 1/2$). During the transition process the two distinct vortices get in-

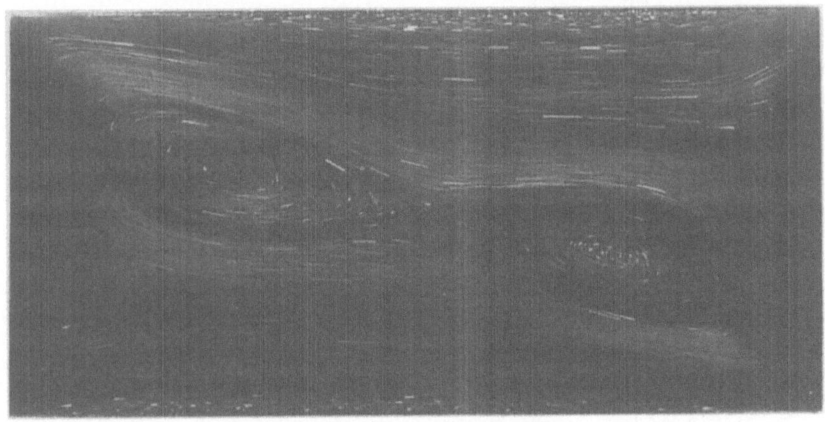

Figure 4. Cat's eye flow at $z = 0$ (midplane) for $Re = 233$.

volved and partially merge. As a result the flow in the center of the cavity takes the form of cat's eyes with two elliptic and one hyperbolic stagnation point. By decreasing the Reynolds number a transition from the cat's eye state back to the two-vortex state occurs at $Re = 224$ closing a hysteresis loop.

In Figure 5 the numerically obtained range of non-uniqueness is shown in the Γ-Re-plane. For a given aspect ratio Γ, $Re^{(0+)}$ denotes the upper existence Reynolds number for the two-vortex state and $Re^{(0-)}$ the lower boundary for the cat's eye state. Both lines represent the edges of a cusp in the solution manifold explaining the observed hysteresis loop. The location of the cusp point is indicated by an asterisk ($\Gamma = 1.87, Re = 205$). The experimental aspect ratio $\Gamma = 1.96$ is plotted as a dashed line. For this case, the calculated transition from the two-vortex flow to the cat's eye flow occurs at $Re^{(0+)} = 427.5$; the transition back to the two-vortex state is found at $Re^{(0-)} = 234.3$. The dotted line $Re^{(1)}$ is the linear stability boundary of the cat's eye flow with respect to three-dimensional perturbations. Using periodic boundary conditions in z-direction we obtain the linear stability boundary $Re^{(1)} = 257.2$ for cat's eye flow at $\Gamma = 1.96$. The instability is stationary with a critical wavenumber $k_c = 2.25$ ($\omega_c = 0$) for an infinitely long system. More details can be found in Kuhlmann et al. (1997).

In fact the transition to a steady three-dimensional fluid flow is also observed in the experiment. By increasing the Reynolds number the cat's eye flow (Figure 4) becomes three-dimensional. Kuhlmann et al. (1997) have provided strong evidence that the resulting steady flow patterns (see Figures 6 and 7) arise due to the elliptic instability (Waleffe, 1990). When Re is increased quasi-steadily, the four-cell mode in Figure 6 is preferred.

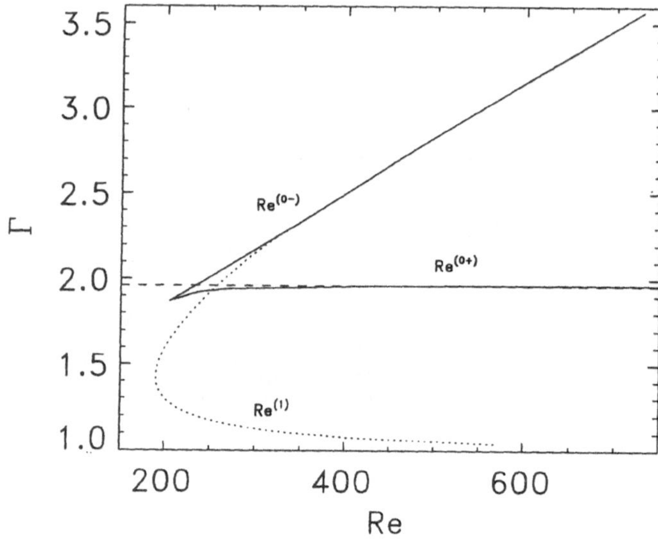

Figure 5. Region of non-uniqueness of the two-dimensional flow (full lines) bounded by $Re^{(0+)}$ (upper existence Reynolds number for the two-vortex state) and $Re^{(0-)}$ (lower existence Reynolds number for the cat's eye state) for symmetrical driving. The cusp point is indicated by an asterisk. The dotted line is the linear stability boundary $Re^{(1)}$ of the cat's eye flow, which intersects with $Re^{(0-)}$ at $\Gamma = 2.283$, $Re = 334$. The experimental aspect ratio $\Gamma = 1.96$ is shown as a dashed line.

This instability is supercritical at symmetrical driving, but subcritical at sufficiently asymmetrical driving ($Re_1 \neq Re_2$). The observable flow pattern depends on the initial conditions, e.g. on the temporal Reynolds number ramp. For instance, the five-cell flow shown in Figure 7 can be set up by a very steep Reynolds number ramp.

3.2. STEADY FLOWS AT ASYMMETRICAL DRIVING

Figure 8 shows the experimental results obtained for asymmetrical driving $Re_1 \neq Re_2$ and $\Gamma = 1.96$. The dots indicate the transition from the two-dimensional two-vortex flow to the cat's eye state for increasing Reynolds number. The transition back from this state to the two-vortex flow is shown by the open circles. The cat's eye state is nearly two-dimensional even for very different Reynolds numbers (Re_1, Re_2), whereas the vortices of the two-vortex state are highly curved near the Perspex endwalls at $z = \pm\Lambda/2$.

Once a cat's eye flow has been set up the flow pattern is stable in a finite Reynolds number range. By a quasi-steady increase of the Reynolds numbers the flow changes at the line indicated by full squares to a four-

Figure 6. Top view of the fully developed four-cell flow at $Re = 750$. The upper boundary represents the wall which moves out of the plane, whereas the wall moves into the plane at the bottom line.

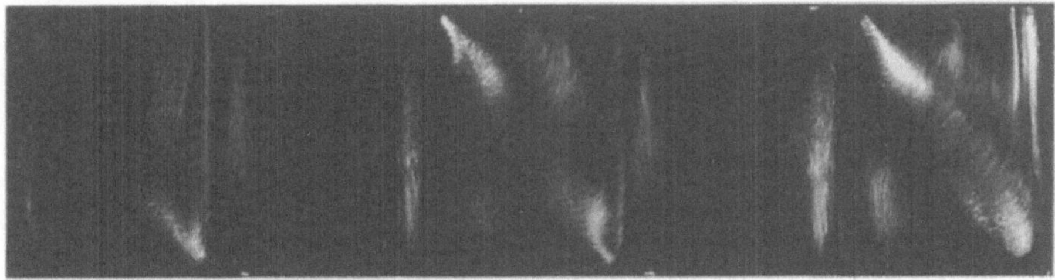

Figure 7. Top view of the fully developed five-cell flow at $Re = 500$, conditions as in Figure 6.

cell flow. For sufficiently different Reynolds numbers the cellular flow sets in with large amplitude, completely destroying the underlying cat's eye structure. The cellular flow is stable over a wide range of high Reynolds numbers. The transition back to the cat's eye state on a quasi-steady decrease of one of the Reynolds numbers is shown by the open squares. In Figure 8 the range of hysteresis can be seen very clearly. The hysteresis for the three-dimensional instability increases with increasing asymmetry of the Reynolds numbers.

At the points $(Re_1 \approx 490, Re_2 \approx 320)$ and $(Re_1 \approx 280, Re_2 \approx 430)$ the line indicated by open squares intersects with the lines of the transition between two-vortex and cat's eye flow. Due to the symmetry of the ideal system, these points should be symmetric with respect to the diagonal $Re_1 = Re_2$. The discrepancy must therefore be due to experimental imperfections. An average value would be $Re_i = 460$, $Re_j = 300$ with $i \neq j$. Thus if the larger of both Reynolds numbers is larger than its value at the intersection point ($Re_i \gtrsim 460$ in the idealized case) the cellular flow stucture is collapsing to the two-vortex flow on a decrease of the smaller Reynolds number. The latter flow, however, is already three-dimensional due to another instability (Kuhlmann et al., 1997).

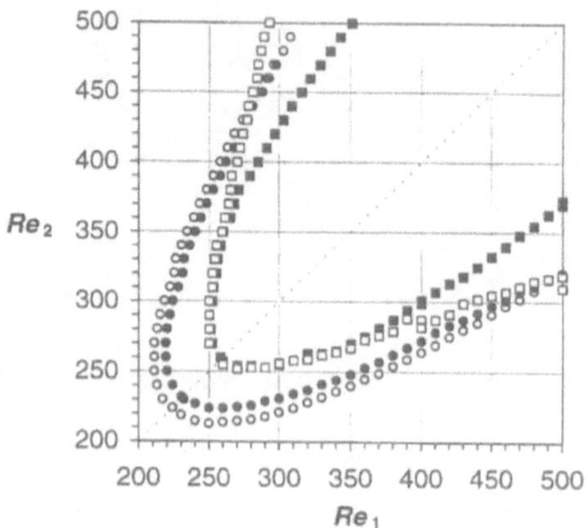

Figure 8. Experimental results for the transition boundaries for asymmetrical driving ($Re_1 \neq Re_2$) and $\Gamma = 1.96$. Full circles: two-vortex $\rightarrow$ cat's eye; open circles: cat's eye $\rightarrow$ two-vortex; full squares: cat's eye $\rightarrow$ three-dimensional cells; open squares: three-dimensional cells $\rightarrow$ cat's eye.

3.3. TIME-DEPENDENT THREE-DIMENSIONAL FLOW

When the Reynolds numbers are increased further the steady three-dimensional cellular flow becomes oscillatory with a frequency f of the order of $O(1.4\text{Hz})$. It is experimentally observed that for symmetrical driving ($Re_1 = Re_2$) the bifurcation is supercritical, whereas the flow undergoes a subcritical bifurcation for asymmetrical driving. As a preliminary result, experimental transition boundaries for the four-cell flow in the Re_1-Re_2-plane are shown in Figure 9. Crosses represent the transition to an oscillatory four-cell flow. The reverse transition to steady four-cell flow during a decrease of the Reynolds numbers is shown by dots. Owing to the very small growth rate of time-dependent perturbations, particularly for $Re_1 = Re_2$ (critical slowing down), it takes about 6h to obtain one data point. It may be noted that the characteristic momentum diffusion time over the span Λ is about $t_\Lambda = l^2/\nu \approx 28\text{min}$.

4. Conclusions and further investigations

We have shown that the two-dimensional flow for $Re_1 = Re_2$ is not unique in a finite Reynolds number range. One can expect that it is not unique for all aspect ratios $\Gamma > 2$, if Re is sufficiently high. A two-vortex flow as well as a flow with elliptic streamlines (cat's eye flow for $\Gamma = 1.96$) can be found. For the experimental aspect ratio the two-vortex flow and the cat's eye flow are topologially equivalent. It can be shown numerically

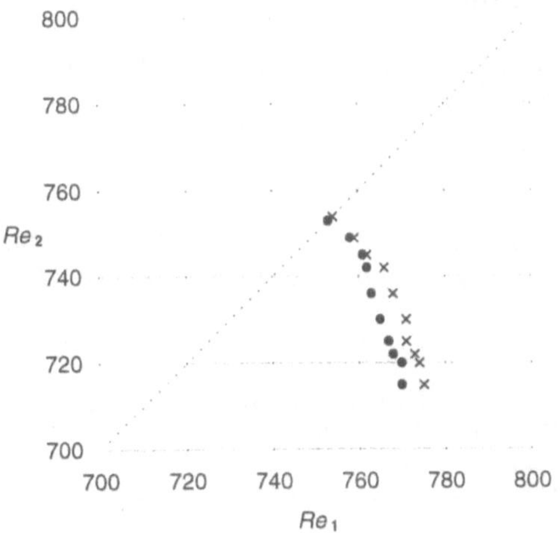

Figure 9. Preliminary experimental transition boundaries for the four-cell flow in the Re_1-Re_2-plane. Crosses: steady flow → oscillatory flow; full circles: oscillatory flow → steady flow. Due to symmetry reasons the transition boundaries must be symmetric with respect to the line $Re_1 = Re_2$.

(Kuhlmann et al., 1997) that both solution branches are smoothly connected with each other by a third unstable solution. For small aspect ratios and/or high Reynolds numbers the topology is different. The interior hyperbolic stagnation point is not present and the streamlines are essentially elliptic. It would be interesting to follow the critical line on which the hyperbolic stagnation point vanishes. For such studies, however, the aspect ratio must be varied. We note that the eddy genesis and the appearance of hyperbolic points in two-sided lid-driven cavities has recently been studied numerically by Kelmanson & Lonsdale (1996) for Stokes flow.

For asymmetrical driving and $\Gamma = 1.96$ the Reynolds number range of stable cat's eye flow is limited. Above criticality, a steady cellular flow pattern occurs as a result of the elliptic instability. This type of instability has recently been found numerically also by Sipp & Jacquin (1997) for a two-dimensional Taylor-Green vortex with aspect ratio two. Moreover, we experimentally found the onset of time-dependent cellular flow and presented preliminary transition Reynolds numbers.

Further research will concentrate on the influence of different initial conditions, other bifurcation phenomena, and routes to chaos. In addition, a new experimental set-up will allow the investigation of different aspect ratios while minimizing endwall effects.

5. Acknowledgements

This work has been supported by Deutsche Forschungsgemeinschaft under grant number Ku 896/5-1.

References

Aidun, C.K., Triantafillopoulos, N.G. and Benson J.D. (1991) Global stability of a lid-driven cavity with throughflow: flow visualization studies, *Phys. Fluids* A **3**, 2081–2091.

Bayly, B.J. (1986) Three-dimensional instability of elliptical flow, *Phys. Rev. Lett.* **57**, 2160–2163.

Bayly, B.J., Orszag, S.A. and Herbert, Th. (1988) Instability mechanism in shear-flow transition, *Ann. Rev. Fluid Mech.* **20**, 359–391.

Deville, M., Lê, T.-H. and Morchoisne, Y.(eds.) (1992) *Numerical Simulation of 3-D Incompressible Unsteady Viscous Laminar Flows.* Notes on Numerical Fluid Mechanics, vol. 36. Vieweg.

Freitas, C.J., Street, R.L., Findikakis, A.N. and Koseff, J.R. (1985) Numerical simulation of three-dimensional flow in a cavity, *Int. J. Num. Meth. Fluids* **5**, 561–575.

Gledzer, E.B. and Ponomarev, V.M. (1992) Instability of bounded flows with elliptical streamlines, *J. Fluid Mech.* **240**, 1–30.

Jana, S.C., Metcalfe, G. and Ottino, J.M. (1994) Experimental and numerical studies of mixing in complex Stokes flow: the vortex mixing flow and multicellular cavity flow, *J. Fluid Mech.* **269**, 199–246.

Kelmanson, M. A. and Lonsdale, B. (1996) Eddy genesis in the double-lid-driven cavity, *Q. J. Mech. Appl. Math.* **49**, 635-655.

Koseff, J.R. and Street, R.L. (1984) The lid-driven cavity flow: a synthesis of qualitative and quantitative observations, ASME *J. Fluids Eng.* **106**, 390–398.

Kuhlmann, H.C, Wanschura, M. and Rath, H.J. (1997) Flow in two-sided lid-driven cavities: non-uniqueness, instabilities, and cellular structures, *J. Fluid Mech.* **336**, 267–299.

Leong, C.W. and Ottino. J.M. (1989) Experiments on mixing due to chaotic advection in a cavity, *J. Fluid Mech.* **209**, 463–499.

Malkus, W.V.R. (1989) An experimental study of global instabilities due to the tidal (elliptical) distortion of a rotating elastic cylinder, *Geophys. Astrophys. Fluid Dyn.* **48**, 123–134.

Pierrehumbert, R.T. (1986) Universal short-wave instability of two-dimensional eddies in an inviscid fluid, *Phys. Rev. Lett.* **57**, 2157–2159.

Ramanan, N. and Homsy, G.M. (1994) Linear stability of lid-driven cavity flow, *Phys. Fluids* **8**, 2690–2701.

She, Z.S., Jackson, E. and Orszag, S.A. (1991) Structure and dynamics of homogeneous turbulence: models and simulation, *Proc. Roy. Soc. Lond.* A **434**, 101–124.

Sipp, D. and Jacquin, L. (1997) Elliptic instability in 2-D flattened Taylor-Green vortices (preprint).

Waleffe, F. (1990) On the three-dimensional instability of strained vortices, *Phys. Fluids* A **2**, 76–80.

Wanschura, M., Shevtsova, V.M., Kuhlmann, H.C. and Rath, H.J. (1995) Convective instability mechanisms in thermocapillary liquid bridges, *Phys. Fluids* **7**, 912–925.

Williamson, C.H.K. (1996) Three-dimensional wake transition, *J. Fluid Mech.* **328**, 345–407.

LONG-WAVELENGTH INSTABILITY AND RECONNECTION OF A VORTEX PAIR

T. LEWEKE
Institut de Recherche sur les Phénomènes Hors Équilibre
CNRS / Universités Aix-Marseille I & II, Marseille, France

AND

C.H.K. WILLIAMSON
Sibley School of Mechanical & Aerospace Engineering
Cornell University, Ithaca, NY 14853, USA

1. Introduction

The dynamics of a pair of parallel counter-rotating vortices has been the object of a large number of studies in the last three decades. The continued interest in this flow is, to a great extent, due to its relevance to the problem of aircraft trailing wakes, whose far field is primarily composed of such a pair. Due to their strength and longevity, these vortices can be dangerous to other aircraft that follow, because of the rolling moment they induce. In addition, the counter-rotating vortex pair represents one of the simplest flow configurations for the study of elementary vortex interactions, which can yield useful information for our understanding of the dynamics of more complex transitional and also turbulent flows.

A prominent feature of this flow, leading to the decay of the pair, is a long-wavelength wavy instability (see, e.g., the photographs shown in Van Dyke 1982). The first theoretical analysis of this phenomenon was made by Crow (1970), who showed that the mutual interaction of the two vortices can lead to an amplification of displacement perturbations, whose axial wavelength is typically several times the initial vortex separation distance. These sinusoidal displacements are symmetric with respect to the mid-plane between the two vortices, and they lie in planes inclined approximately 45° with respect to the line joining the vortices. The origin of this instability is linked to the balance between the stabilizing effect of self-induced rotation of the perturbations and the destabilizing influence of the strain field that each vortex induces at the location of its neighbour. A detailed physical analysis of this process can be found in Widnall, Bliss and Tsai (1974).

E. Krause and K. Gersten (eds.), IUTAM Symposium on Dynamics of Slender Vortices, 225-234.

Since Crow (1970), a great number of investigations have treated different aspects of the long-wavelength vortex pair instability: the non-linear evolution (Moore 1972, Klein and Majda 1993, Klein, Majda and Damo-daran 1995), the effects of axial flow and arbitrary vorticity distributions in the vortices (Widnall, Bliss and Zalay 1971, Moore and Saffman 1972, Klein and Knio 1995), or the influence of turbulence (Sarpkaya 1983, Liu 1992, Devenport, Zsoldos and Vogel 1997). An account on more recent results in this field can be found in the proceedings of the last NATO AGARD Wake Symposium (North Atlantic Treaty Organization, 1996).

On the experimental side, there exist surprisingly few results from controlled laboratory experiments concerning this basic flow. Although the studies of Eliason, Gartshore and Parkinson (1975), Sarpkaya (1983), and Thomas and Auerbach (1994) show qualitative agreement with theoretical predictions for the long-wavelength instability, a close comparison between theory and experiment is still lacking. Such a comparison is often not possible, either because the flow under consideration does not correspond to the one treated in theory, or because of a lack of information about the initial flow field, needed to make a meaningful comparison with the theory of Crow (1970) and Widnall *et al.* (1971). A first quantitative comparison including all the necessary data was recently made by Miller and Williamson (1997) for the trailing wake of a delta wing. Here, we use a similar approach, but for the more fundamental case of a uniform laminar vortex pair.

When the amplitude of the long-wavelength perturbation grows sufficiently large, the vortex cores touch at the location of maximum inward displacement, leading to a break-up of the pair into a series of vortex rings. This scenario is mainly known from observations of full-scale aircraft wakes (Van Dyke, 1982), and it was also observed in the recent numerical simulations of Robins and Delisi (1997). Very little is known, however, about the subsequent evolution of the flow, i.e. about the dynamics and persistence of the rings in the late stages of the Crow instability.

The transition between two wavy vortices and a series of vortex rings happens via a periodic cross-linking, or reconnection, of vorticity. This interesting phenomenon involves a change in the topology of the vortex lines and represents a fundamental interaction between concentrated vortices. So far, experimental results concerning this process are mostly qualitative (e.g. Oshima and Asaka 1977, Lim and Nickels 1995). More detailed knowledge about vortex reconnection comes from numerical simulations (see Kida and Takaoka 1994 for a recent review). Melander and Hussain (1989) and Shelley, Meiron and Orszag (1993) chose as initial conditions configurations which are very similar to a vortex pair perturbed by the symmetric mode of the Crow instability. The interacting vortices of the present study are therefore also an ideal candidate for the investigation of the reconnection phenomenon.

The rest of this paper is organized as follows. After a brief description of the experimental set-up and techniques in §2, the overall features of the long-wavelength Crow instability are presented in §3, including quantitative measurements concerning the evolution of the vortex geometry and the instability growth rate. In §4 the reconnection process is analyzed in more detail, and the conclusions follow in §5.

2. Experimental details

The vortex pair is generated in a water tank at the sharpened parallel edges of two flat plates, hinged to a common base and moved in a prescribed symmetric way by a computer-controlled step motor. The vortices are typically separated by a distance of 2.5 cm, and their length is approximately 170 cm. This high aspect ratio of the vortex pair is necessary to limit the influence of end effects that spread rapidly into the central part of the flow. Visualization is achieved using fluorescent dye, illuminated by laser light.

A vortex pair is characterized by several parameters: the circulation Γ of each vortex, the separation b between the vortex centres, and a characteristic core size a, which is the radius of the tube around the vortex centre containing most of the vorticity. These characteristics were determined from flow field measurements using Digital Particle Image Velocimetry (DPIV). These measurements also showed that the initial velocity profiles of the vortices are very well represented by the one of a Lamb-Oseen vortex with a Gaussian vorticity distribution.

Other measurements concerning the spatial structure and the growth rate of the instability were obtained from image analysis of flow visualizations recorded on video, discussed in more detail in §3.2. These measurements required a control of the phase of the long-wavelength instability in the axial direction. This was achieved by adding a low-amplitude waviness (less than 2 % of the initial vortex spacing) to the vortex-generating plate edges. The axial wavelength of these waves was chosen close to the wavelength of the naturally occurring Crow instability.

The Reynolds number based on the initial circulation ($Re = \Gamma_o/\nu$) was in the range between 1500 and 2500 in this study. In the following, time t is non-dimensionalized by the time it takes the initial vortex pair to travel one vortex spacing b, i.e. $t^* = t\ (\Gamma/2\pi b^2)$.

A detailed description of the experimental set-up and techniques is given in Leweke and Williamson (1997).

3. Characterization of the instability over long times

3.1. QUALITATIVE OVERVIEW

Figure 1 shows the overall features of the long-wavelength vortex pair instability. The initially straight and parallel vortices develop a waviness (Fig. 1*a*), which

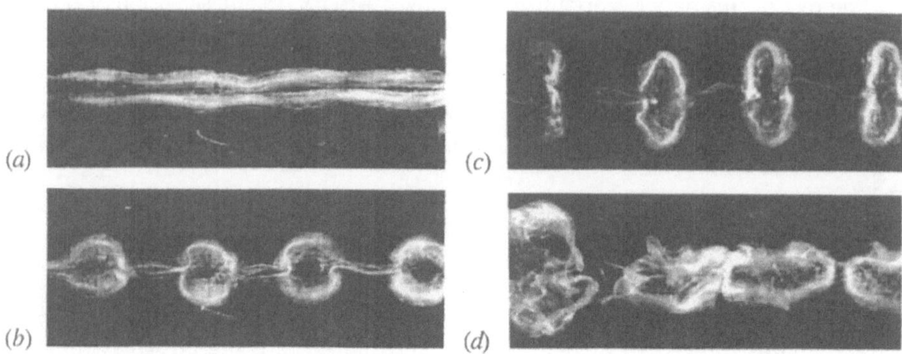

Figure 1. Visualization of the long-wavelength instability of a vortex pair for $Re = 1450$ and $a/b = 0.23$. The pair is moving towards the observer. (a) $t^* = 3.2$, (b) $t^* = 6.5$, (c) $t^* = 9.3$, (a) $t^* = 14.4$

is symmetric with respect to the plane separating the vortices, and whose axial wavelength is about 6 times the initial vortex separation. Photographs taken simultaneously from two perpendicular directions show that the plane of the waves is inclined by about 45° with respect to the plane containing the pair (see also Fig. 4b), in agreement with Crow's (1970) theoretical prediction and previous observations. This waviness is amplified in time until the vortex cores touch, break up, and reconnect to form periodic vortex rings, which initially look almost circular in the front view of Fig. 1(b). Side view visualizations show, however, that these rings are not contained in a plane; they are bent upwards in the transverse direction. Fig. 1(b) also shows that the large-scale rings are still linked by thin strands of dye. Subsequently the rings stretch out into oval vortices, which are clearly seen in Fig. 1(c). Due to their non-uniform curvature, these oval vortex rings then exhibit a well-known oscillatory behaviour (Lim and Nickels 1996), in which their principal axes exchanged periodically in time, so that at very late times (Fig. 1d), one can observe an array of oval vortex rings with their major axes in the direction of the initial vortex pair. Fig. 1(d) was taken shortly before the vortices reached the bottom of the water tank, and the whole sequence took about 35 seconds in real time.

3.2. MEASUREMENTS OF THE VORTEX GEOMETRY

In addition to these qualitative observations, a number of precise measurements concerning the time-dependent geometry of the vortex system were made using the following method. When, instead of the flood illumination used for Fig. 1, one places two light sheets perpendicular to the vortices at the locations of minimum and maximum separation, and then looks along the vortex axes, the positions of the vortex centers at these two axial locations can be determined simultaneously.

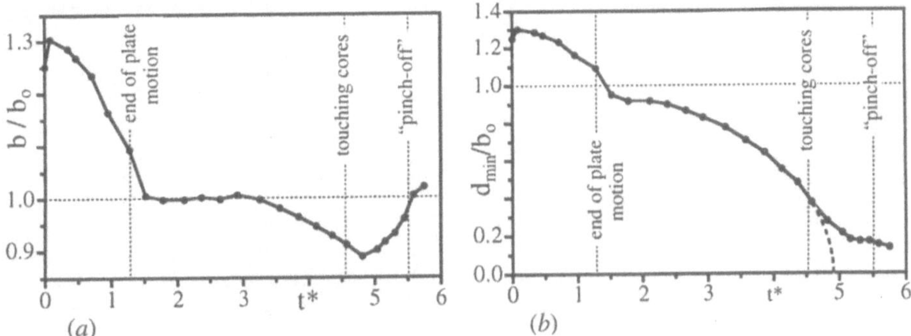

Figure 2. Evolution of the (*a*) mean and (*b*) minimum vortex spacing at $Re = 2560$. The dashed line in (*b*) shows qualitatively the prediction by Klein *et al.* (1995) for two filaments using asymptotic theory.

From video recordings of these visualizations and subsequent image analysis, time-dependent quantities like the average vertical position of the vortices, the mean and minimum separation of the vortex centers, and the inclination and amplitude of the wavy perturbations were deduced.

From these measurements it was found that the mean propagation speed of the vortices is approximately constant in time, even after the reconnection process. This is surprising, considering that the geometry and topology of the late-time vortex rings are very different from the ones of the initial vortex pair. It illustrates that, even after the cross-linking, the vortices still possess most of the initial circulation and that the energy of the flow remains in the large-scale structures.

Figure 2 shows measurements of the mean and minimum distance between the centers of the two vortices up to reconnection. After the vortex formation during the motion of the plates, the mean vortex spacing b first remains approximately constant for a certain time. As the amplitude of the wavy perturbations grows bigger, b starts to decrease slightly, only to increase again when the vortex cores start to overlap and reconnection sets in. It is known from previous calculations by Moore (1972) and Klein *et al.* (1995), that, during the second phase of the Crow instability (for $3 < t^* < 4.5$ approximately in Fig. 2*a*), the perturbed vortices no longer have a sinusoidal shape. The minimum distance between the vortex center lines (Fig. 4*b*) decreases continually, and with increasing speed, once the vortices are formed. This behaviour is very similar to the result from a calculation by Klein *et al.* (1995) with two vortex filaments using asymptotic theory. Whereas in their calculation the filaments eventually touch, the experimental observation obviously deviates from this behaviour once the separation decreases down to the order of the vortex core size.

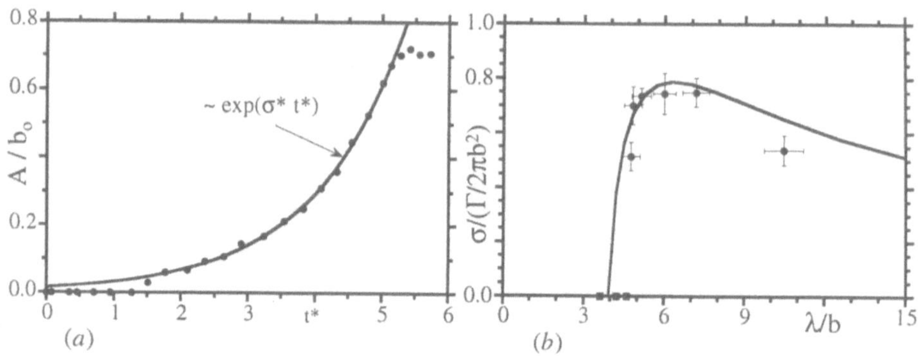

Figure 3. Determination of the growth rate of the Crow instability. (*a*) amplitue A of the wavy perturbations (line: least-squares fit to an exponential function with growth rate σ^*). (*b*) growth rate as function of axial wavelength λ (line: theoretical prediction for $a/b = 0.25$) following Crow (1970) and Widnall *et al.* (1971)).

3.3. INSTABILITY GROWTH RATE

An important aspect of this study was the experimental determination of the growth rate of the long-wavelength instability and the comparison with theoretical predictions. In Fig. 3(*a*), the time-dependent amplitude A of the wavy perturbation, as determined from visualization sequences with two light sheets, is plotted. These measurements are very closely fitted by an exponential function, allowing a straightforward determination of the non-dimensional growth rate σ^*. The growth rate was measured for a number of different axial wavelengths λ using the wavy perturbations of the vortex-generating plate edge mentioned in §2. In Fig. 3(*b*), the results are compared to the theoretical prediction, which can be derived from the works of Crow (1970) and Widnall *et al.* (1971) (see Leweke and Williamson 1997 for details). It should be pointed out that information about the vortex core size and the velocity profile is needed, in order to obtain a meaningful prediction from theory, since the dynamics of the vortex tubes depend on these quantities. In our study, they were measured with DPIV, so that Fig. 3(*b*) indeed represents a full comparison between experiment and theory, taking into account all the necessary information. To our knowledge, this is the first time that such a comparison has been achieved. The agreement between the two is found to be remarkably good.

4. Vortex reconnection

In this section, the vortex reconnection process, which marks the transition between the vortex pair and the array of vortex rings, is analyzed in more detail.

Figure 4 shows a sequence of visualizations made with a light sheet in the reconnection plane, which shows the cross-section of the pair at the location of

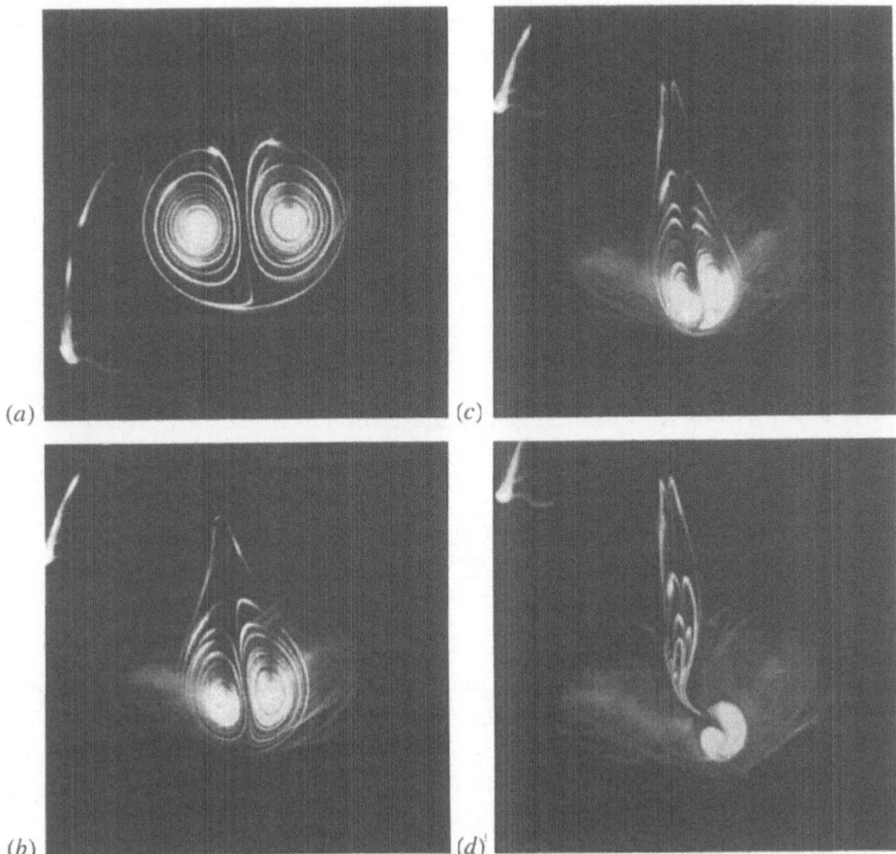

Figure 4. Visualization of the flow in the reconnection plane. (*a*) $t^* \approx 2$, (*b*) $t^* \approx 4$, (*c*) $t^* \approx 5$, (*d*) $t^* \approx 6$.

minimum separation. In Fig. 4(*a*), the pair is still symmetric, and the cores are well separated. In Fig. 4(*b*), the amplitude of the perturbation has grown quite large. The plane of the waves can be seen in the background, illuminated by scattered light from the reconnection plane. The vortex cores have come much closer and have started to elongate vertically. At the same time, a "tail" of dye is developing behind the descending pair, which is even more pronounced in Fig. 4(*c*). Towards the end of the cross-linking process, in Fig. 4(*d*), the dye has almost completely disappeared from the reconnection plane; what remains marks the thin counterrotating bridges that still link the large-scale vortex rings resulting from the reconnection (see Fig. 1(*b*). The tail appears to be cut off at this time, and it is not clear if its dye pattern still corresponds to any vorticity.

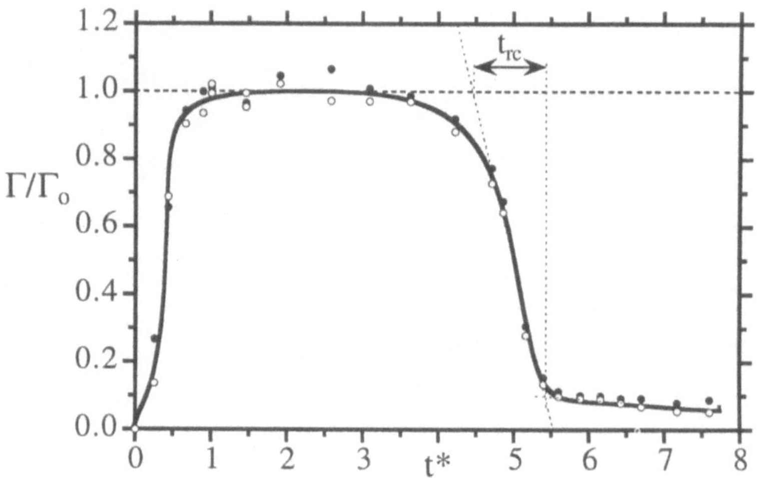

Figure 5. Evolution of the circulation in the reconnection plane for Re = 2340, and definition of the reconnection time t_c. Measurements for the two vortices are represented by different symbols.

These visualization results were complemented by measurements of the time-dependent vorticity distribution in the same plane with the help of DPIV. These measurements confirmed the qualitative observations in Fig. 4, in particular the formation of a tail of vorticity during the cross-linking. At later times, a small, slowly decaying pair of counterrotating vortices remains, corresponding to the dye threads linking the oval vortex rings. No more traces of the tail are detected at this point, which may, however, be due to the limits of the DPIV technique. These results are in good qualitative agreement with the results shown by Melander and Hussain (1989), who studied vortex reconnection in a very similar configuration by Direct Numerical Simulation.

From the vorticity measurements, the circulation of each vortex, i.e. the circulation in each half of the reconnection plane, can be calculated. The result is shown in Fig. 5. After the initial roll-up, the circulation remains constant for some time. Once the cores touch, it decreases rapidly due to the cancellation and reorientation of vorticity. At later times, it reaches a new, almost constant value of about 10% of the initial circulation Γ_o, which corresponds to the circulation of the threads between the large-scale vortex rings.

Following Melander and Hussain (1989), one can deduce from Fig. 5 a characteristic reconnection time t_c, defined by the intersections of the extensions of the different almost linear sections of the curve. From our experimental measurements we obtain a reconnection time of $t_c \approx 0.9$. If the result found in the DNS study of Melander and Hussain (1989) is transformed to the non-dimensional units used in this study, one obtains a reconnection time $t_c \approx 0.7$ for their case, which again is in reasonably good agreement with the present result.

5. Conclusions

In our experimental study of vortex pairs we have observed and characterized the initial development and the late stages of the long-wavelength Crow instability. The initially straight vortices develop a symmetric waviness before they cut and reconnect inso a system of oscillating oval vortex rings, which are found to be very energetic, even at late times.

Precise measurements of the time-dependent geometry of the vortex system were made, including quantities like the vortex separation distance and the perturbation amplitude. These results and additional measurements using Digital Particle Image Velocimetry allowed a first complete comparison, including all the necessary parameters, between experiment and theoretical predictions concerning the growth rate of the long-wavelength instability. Good agreement between the two is found.

Finally, we have obtained qualitative and quantitative information about the vortex reconnection process, including the characteristic reconnection time, which also shows good agreement with previous numerical simulations.

This work was supported by the US Office of Naval Research under Contract No. N00014-95-1-0332, and by NATO under Grant No. CRG 970259. TL acknowledges the financial support from the Deutsche Forschungsgemeinschaft, Grant No. Le 972/1-1.

References

Crow, S.C. (1970) Stability theory for a pair of trailing vortices, *AIAA J.* **8**, 2172–2179.

Devenport, W.J., Zsoldos, J.S. and Vogel, C.M. (1997) The structure and development of a counter-rotating wing-tip vortex pair. *J. Fluid Mech.* **332**, 71–104.

Eliason, B.G., Gartshore, I.S. and Parkinson, G.V. (1975) Wind tunnel investigation of Crow instability. *J. Aircraft* **12**, 985–988.

Kida, S. and Takaoka, M. (1994) Vortex reconnection, *Annu. Rev. Fluid Mech.* **26**, 168.

Klein, R. and Knio, O.M. (1995) Asymptotic vorticity structure and numerical simulation of slender vortex filaments, *J. Fluid Mech.* **284**, 275–231.

Klein, R. and Majda, A.J. (1993) An asymptotic theory for the nonlinear instability of antiparallel pairs of vortex filaments, *Phys. Fluids* A **5**, 369–379.

Klein, R., Majda, A.J. and Damodaran, K. (1995) Simplified equations for the interactions of nearly parallel vortex filaments, *J. Fluid Mech.* **288**, 201–248.

Leweke, T. and Williamson, C.H.K. (1997) Short-wavelength instability of a counter-rotating vortex pair. To appear in *J. Fluid Mech.*

Leweke, T. and Williamson, C.H.K. (1997) Experimental study of a long-wavelength vortex pair instability. Submitted to *J. Fluid Mech.*

Lim, T.T and Nickels, T.B. (1996) Vortex rings, in S.I. Green (ed.), *Fluid Vortices*, Kluwer Academic Publishers, Dordrecht, pp. 95–154.

Liu, H.-T. (1992) Effects of ambient turbulence on the decay of a trailing vortex wake. *J. Aircraft* **29**, 255.

Melander, M.V. and Hussain, F. (1989) Cross-linking of two antiparallel vortex tubes. *Phys. Fluids* A **1**, 633–636.

Miller, G.D. and Williamson, C.H.K. (1997) Dynamics in the far wake of a delta wing. Submitted to *J. Fluid Mech.*

Moore, D.W. (1972) Finite amplitude waves on aircraft trailing vortices. *Aero. Quart.* **23**, 307–314.

Moore, D.W. and Saffman, P.G. (1973) Axial flow in laminar trailing vortices. *Proc. R. Soc. Lond.* A **333**, 491-508.

North Atlantic Treaty Organization (1996) *The Characterization & Modification of Wakes from Lifting Vehicles in Fluids*, AGARD Conference Proceedings, No. 584.

Oshima, Y. and Asaka, S. (1977) Interaction of two vortex rings along parallel axes in air. *J. Phys. Soc. Japan* **42**, 708–713.

Robins, R.E. and Delisi, D.P. (1997) Numerical simulations of three-dimensional trailing vortex evolution. *AIAA J.* **35**, 1552–1555.

Sarpkaya, T. 1983 Trailing vortices in homogeneous and density-stratified media. *J. Fluid Mech.* **136**, 85–109.

Shelley, M.J., Meiron, D.I. and Orszag, S.A. (1993) Dynamical aspects of vortex reconnection of perturbed anti-parallel vortex tubes. *J. Fluid Mech.* **246**, 613–652.

Thomas, P.J. and Auerbach, D. (1994) The observation of the simultaneous development of a long- and a short-wave instability mode on a vortex pair. *J. Fluid Mech.* **265**, 289–302.

Van Dyke, M. (1982) *An Album of Fluid Motion*, Parabolic Press, Stanford.

Widnall, S.E., Bliss, D.B. and Tsai, C.-Y. (1974) The instability of short waves on a vortex ring, *J. Fluid Mech.* **66**, 33–47.

Widnall, S.E., Bliss, D.B. and Zalay, A. (1971) Theoretical and experimental study of the instability of a vortex pair, in J.H. Olsen, A. Goldberg and M. Rogers (eds.), *Aircraft Wake Turbulence and its Detection*, Plenum Press, New York, pp. 305–338.

STABILITY OF STRETCHED VORTICES IN A STRAIN FIELD

S. LE DIZÈS & C. ELOY
Institut de Recherche sur les Phénomènes Hors Équilibre,
12, avenue Général Leclerc, F-13003 Marseille, France.

Abstract. The linear stability of an arbitrarily stretched Gaussian vortex is addressed when the vortex of circulation Γ and radius δ is subjected to an additional strain field of rate s perpendicular to the vorticity axis. The resulting non-axisymmetric vortex is analysed in the limit of large Reynolds number $R_\Gamma = \Gamma/\nu$ and small strain $s \ll \Gamma/\delta^2$. The helical Kelvin wave resonance mechanism described by Moore & Saffman (1975) is shown to be active. Coupled equations for the Kelvin waves amplitudes are obtained, describing the resonance in presence of diffusion and stretching. The main effect of diffusion and stretching is shown to be stabilizing by limiting in time the resonance. The maximum gain of amplitude of the resonant Kelvin waves is computed in terms of the rescaled strain and stretching rates $s^* = s\sqrt{R_\Gamma}\delta^2/\Gamma$ and $\gamma^* = \gamma R_\Gamma \delta^2/\Gamma$. The result leads to explicit sufficient instability conditions which could explain various dynamical behaviors of vortex filaments in turbulence.

1. Introduction

The interest in vortex dynamics has been renewed by the discovery of strong vorticity filaments in turbulent flows (Cadot, Douady & Couder 1995; Vincent & Meneguzzi 1991). Vortex filaments have been generally associated with axisymmetric vortices for which the vorticity is simultaneously concentrated by axial stretching and diffused by viscosity. The simplest models are the famous Burgers vortex which is the equilibrium vortex configuration in a uniform stretching field, and the Lamb-Oseen vortex which is an unstretched diffusing vortex. More general solutions were recently obtained by Moffatt, Kida & Ohkitani (1994) and Jiménez, Moffatt & Vasco (1996) (see also Ting & Tung 1965) by submitting the above vortices to an additional plane strain field with principal axes perpendicular to the vortex axis. Non-axisymmetric corrections were calculated in the limit of large Reynolds numbers. They were shown to induce interesting modifications in

E. Krause and K. Gersten (eds.), IUTAM Symposium on Dynamics of Slender Vortices, 235-244.

the energy dissipation distribution which are in remarkably good agreement with what has been observed in numerical simulations of two and three dimensional turbulence (See Jiménez *et al.* 1996 and Moffatt *et al.* 1994 respectively). However, filaments are also known to exhibit a rich variety of dynamical behaviours which includes bursting, spliting and merging (see Arendt, Fritts & Anderssen 1997, for instance). It is then natural to address the question of the stability of the above solutions.

One of the main effect of imposing an additional external strain, perpendicular to the vortex axis, is to modify the form of the vortex core from circular to elliptical (Robinson & Saffman 1984 ; Moffatt *et al.* 1994; Jiménez *et al.* 1996). Such an effect is expected to be destabilizing since the elliptical character of the streamlines is indeed known to be a source of instability (Pierrehumbert 1986). This so-called "elliptical instability" has been put on a firm ground in the context of a pure elliptical flow by Bayly (1986) and Waleffe (1990) among others, and is now recognised as a fundamental instability which could explain the three-dimensional transition of numerous flows such as wakes (Williamson 1996), mixing layers (Landman & Saffman 1987) and vortex pairs (Leweke & Williamson 1997).

The elliptical instability is related to the short-wave instability identified by Widnall, Bliss & Tsai (1974) in vortex rings and analysed by Tsai & Widnall (1976) and Moore & Saffman (1975) in the context of two-dimensional inviscid vortices in a weak external strain field. They showed that stationary helical waves of azimuthal wavenumbers $m = -1$ and $m = 1$ are amplified for certain values of their wavenumber by the external strain field. Moore & Saffman (1975) interpreted this instability as a resonance phenomenon and gave a formal extension of Tsai & Widnall analysis for Rankine vortex to a large class of non-viscous two-dimensional vortices. In this paper we shall demonstrate that their analysis applies to a large family of stretched vortices which includes the Burgers and Lamb-Oseen vortices if the external strain field acting on the vortices is small and if the Reynolds number is large.

The paper is organised as follows. In §2, the basic flow solution is presented. In particular, the corrections to the axisymmetric stretched vortex due to the external strain are given. Section 3 starts with a simple description of the instability mechanism following Moore & Saffman (1975) analysis. The additional effects of viscosity and stretching are then discussed : they are shown to limit in time the growth of the resonant helical waves. This leads to a rough estimate for the maximum gain of the wave amplitude. In §4, an outline of the asymptotic analysis when the gain is $O(1)$ is presented. An amplitude equation describing the resonance is obtained as well as a precise value for the gain. In that section, sufficient instability conditions are given in terms of stretching and strain rates. Possible extensions of the results are discussed in §5.

2. Stretched vortices in a non-axisymmetric strain field

The Lamb-Oseen vortex is the axisymmetric solution of the Navier-Stokes equation which describes the viscous diffusion of a point vortex. It is characterised by a circulation Γ, a radius δ which evolves according to $\delta(t) = \sqrt{\nu t}$ and an axial vorticity field given by

$$\omega_z = \frac{\Gamma}{\delta^2} G\left(\frac{r}{\delta}\right), \tag{1}$$

where $G(x)$ is the normalised Gaussian function :

$$G(x) = \frac{1}{4\pi} \mathrm{e}^{-x^2/4}. \tag{2}$$

Using a transformation obtained by Lundgren (1984), this solution can be generalized to account for a arbitrary axisymmetric stretching field along the vortex axis $\mathbf{U}_\gamma = (-\frac{1}{2}\gamma x, -\frac{1}{2}\gamma y, \gamma z)$. The resulting axial vorticity is still given by (1) but the radius evolves differently as :

$$\delta(t) = \sqrt{\frac{\delta_0^2 + \nu \int_0^t S(u) du}{S(t)}}. \tag{3}$$

with

$$S(t) = \exp\left(\int_0^t \gamma(u) du\right), \tag{4}$$

where $\gamma(t)$ is the stretching rate and δ_0 is the vortex radius at $t = 0$. One recovers Burgers vortex solution for $\gamma(t) = \delta_0^2/\nu$.

When an external strain field $\mathbf{U}_s = (sx, -sy, 0)$ is added perpendicularly to the vortex axis, the axisymmetric vortex is no longer a solution of the Navier-Stokes equations. However, in the limit of large Reynolds numbers ($R_\Gamma = \Gamma/\nu$), an asymptotic study shows that the main features of the vortex are conserved near its core. Such a study was carried out in Ting & Tung (1965) and Jiménez *et al.* (1996) for the Lamb-Oseen vortex and Moffatt *et al.* (1994) for the Burgers vortex but their results extend for an arbitrarily stretched vortex. The result can be recast in the simple form :

$$\omega_z = \frac{\Gamma}{\delta^2} G\left(\frac{r}{\delta}\right) + s\eta\left(\frac{r}{\delta}\right) F\left(\frac{r}{\delta}\right) \sin 2\theta + O\left(\frac{s\gamma\delta^2}{\Gamma}, \frac{s^2\delta^2}{\Gamma}\right), \tag{5}$$

where

$$\eta(x) = \frac{x^2}{4\left(\mathrm{e}^{x^2/4} - 1\right)}, \tag{6}$$

and $F(x)$ satisfies

$$\frac{\mathrm{d}^2 F}{\mathrm{d}x^2} + \frac{1}{x}\frac{\mathrm{d}F}{\mathrm{d}x} - \frac{4}{x^2} F + \eta(x) F = 0. \tag{7}$$

The function F is subject to the boundary conditions $F(x) \sim s_0 x^2/4$ near zero and $F(x) \sim x^2/4$ near infinity, which yields, after numerical integration, $s_0 \approx 2.525$. The breakdown of approximation (5) in the far field has been discussed in Moffatt *et al.* (1994) and Jiménez *et al.* (1996). It only concerns regions where the vorticity is already exponentially small and is then expected to have a negligible influence on the instability process described below.

3. Instability mechanism

Let us assume for the moment that stretching and viscous diffusion are negligible. Expression (5) for the vorticity field derived in the previous section then describes a two-dimensional, inviscid and stationary vortex. The stability of such vortices with respect to particular three-dimensional perturbations was analysed in a general setting by Moore & Saffman (1975). The main ideas of their analysis can be applied to the present case as follows.

In the limit of small external strain ($s \ll \Gamma/\delta^2$), the vorticity is given, at leading order, by expression (1). The property of axisymmetry of the latter expression guarantees that Kelvin waves can be added, as neutral perturbations, to the basic flow. Such waves are of the form :

$$\mathbf{v} = \Phi(r)\mathrm{e}^{\mathrm{i}(kz-\omega t+m\theta)} + \text{c.c.}, \tag{8}$$

where the frequency ω is related to the axial and azimuthal wavenumbers k and m through the dispersion relation $D(\omega, k, m) = 0$. The effect of the strain field on a Kelvin wave of azimuthal wavenumber m is to generate two additional waves of azimuthal wavenumbers $m \pm 2$. As a consequence, the two waves m and $m+2$ resonate *via* the strain when they satisfy $D(\omega, k, m) = D(\omega, k, m+2) = 0$. Using the symmetry property of the dispersion relation *i.e.* $D(\omega, k, m) = D(-\omega, k, -m)$, Moore & Saffman (1975) deduced that stationary helical waves ($\omega = 0$ and $m = \pm 1$) satisfy the above resonance condition for any axial wavenumber κ such that $D(0, \kappa, 1) = 0$. They proved that this particular combination of helical modes is always amplified by the strain and gives rise to an instability. This result was simultaneously obtained by Tsai & Widnall (1976) for the Rankine vortex. This instability is characterised by a growth rate $\sigma = O(s)$ and a band $\Delta\kappa = O(s\delta^2/\Gamma)$ of unstable dimensionless wavenumbers $k\delta$ around the critical values κ.

The main effect of viscosity and stretching is to introduce a transient aspect to this instability due to the modification of the dimensionless wavenumber $k\delta$ in time. As seen on (3), the time-variation of δ is due to both viscosity and stretching which tends to respectively increase and diminish the vortex radius. By contrast, the time evolution of k is only due to stretching. Indeed, if we consider an arbitrary perturbation of axial wavenumber k in the stretching field, the phase factor $\mathrm{e}^{\mathrm{i}kz}$ must be a quantity conserved in the flow to prevent the apparition of a singularity at infinity along the z axis. This leads to the following evolution for k (Rossi & Le

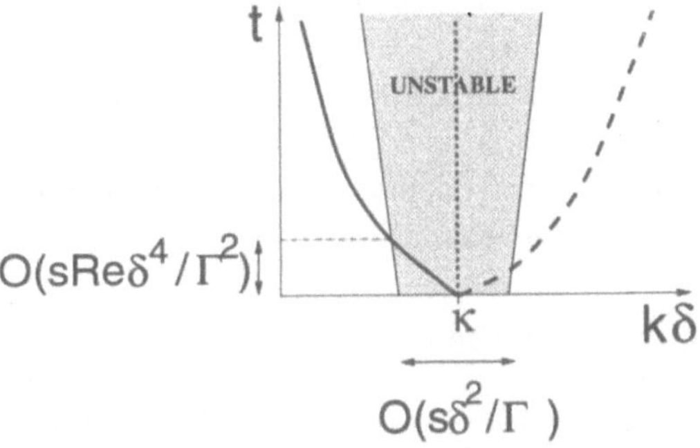

Figure 1. Time evolution of the product $k\delta$ for various stretching rate $\gamma(t)$. Solid line: $\gamma(t) = \nu/\delta_0^2$ (Burgers vortex), dashed line: $\gamma(t) = 0$ (Lamb-Oseen vortex), dotted line: $\gamma(t) = 1/(2t + 3\delta_0^2/\nu)$.

Dizès, 1997) :

$$k(t) = \frac{k_0}{S(t)}, \tag{9}$$

where k_0 is the initial wavenumber and $S(t)$ is given in (4). Using (3) and (9), one then obtains a general expression for the dimensionless wavenumber $k(t)\delta(t)$

$$k(t)\delta(t) = \frac{k_0}{(S(t))^{\frac{3}{2}}} \left(\delta_0^2 + \nu \int_0^t S(u) du \right)^{\frac{1}{2}}. \tag{10}$$

Except for the particular stretching rate $\gamma(t) = 1/(2t + 3\delta_0^2/\nu)$ which corresponds to $k(t)\delta(t) = k_0\delta_0 = Cst$, the dimensionless wavenumber $k(t)\delta(t)$ is then a function of time. In particular, note that it increases for Lamb-Oseen vortex but decreases for Burgers vortex.

It follows that in general if we start from an unstable configuration of helical waves, after a finite period of time, $k\delta$ leaves the unstable band around a critical wavenumber κ and the helical waves are no longer unstable (see sketch on figure 1). In other words, the growth of the unstable helical waves is only transient. The maximum gain of amplitude $G_{\max}$ of these waves across the unstable domain can be roughly estimated as $G_{\max} \propto \exp(\tau\sigma)$ where τ is the time spent in the unstable band and σ the mean growth rate of the instability. Here $\tau = O\left(s\frac{\delta_0^2}{\Gamma}\frac{(k\delta)(0)}{\partial_t(k\delta)(0)}\right) = O\left(\frac{s\delta_0^2}{\Gamma|\gamma_0 - \Gamma/(3R_\Gamma\delta_0^2)|}\right)$ and $\sigma = O(s)$. As a result, the instability is expected to governed by a single parameter

$$p = s^*/\sqrt{|\gamma^* - 1/3|} \tag{11}$$

where s^* and γ^* are rescaled strain and stretching rates at resonance :

$$s^* = \frac{s\delta_0^2\sqrt{R_\Gamma}}{\Gamma}, \tag{12}$$

$$\gamma^* = \frac{\gamma_0 \delta_0^2 R_\Gamma}{\Gamma}. \tag{13}$$

In particular, if $p \gg 1$, the maximum gain of amplitude of the helical waves is exponentially large : the vortex is then unstable.

In the next section, the exact value of the maximum gain is computed for the critical scaling $p = O(1)$ in the generic case.

4. Asymptotic analysis for the critical scaling

In this section, an asymptotic analysis is carried out with respect to the small parameter $\varepsilon = 1/\sqrt{R_\Gamma}$ for the particular scaling $s^* = O(1)$ and $\gamma^* - \frac{1}{3} = O(1)$. With this scaling, the parameter p is of order unity. All quantities are non-dimensionalised with respect to the characteristic time and space scales δ^2/Γ and δ respectively. According to the above discussion, the helical modes are resonant during a finite period of time which is of order $1/\varepsilon$. This time scale is slow compared to the $O(1)$ vorticity time scale but fast compared to the $O(1/\varepsilon^2)$ evolution scale of k. The description of the resonance phenomenon in the limit of small ε then requires the introduction of a new intermediate time variable

$$\overline{T} = \frac{T}{\varepsilon} = \varepsilon t. \tag{14}$$

Following classical asymptotic methods (see for instance Van Dyke 1975), perturbations are expanded in terms of ε as

$$\mathbf{v} = \mathbf{v}_0(\overline{T}, \theta, r, z) + \varepsilon \mathbf{v}_1(\overline{T}, \theta, r, z) + \varepsilon^2 \mathbf{v}_2 + \cdots, \tag{15}$$

where the leading order term $\mathbf{v}_0$ is a combination of the resonant Kelvin waves multiplied by a slowly varying amplitude :

$$\mathbf{v}_0 = A^+(\overline{T}) \mathrm{e}^{\mathrm{i}\theta + \mathrm{i}kz} \Phi^+(r) + A^-(\overline{T}) \mathrm{e}^{-\mathrm{i}\theta + \mathrm{i}kz} \Phi^-(r). \tag{16}$$

Here, the (dimensionless) axial wavenumber $k\delta$ evolves according to (10) with the initial condition that $(k\delta)(0) = \kappa$. Canceling terms at the order ε leads to a non-homogeneous equation for $\mathbf{v}_1$ which admits non-degenerate solutions only if the amplitudes A^+ and A^- satisfy the solvability conditions:

$$\frac{\partial A^+}{\partial \overline{T}} - \mathrm{i}q\left(\gamma^* - \frac{1}{3}\right)\overline{T} A^+ - s^* n A^- = 0, \tag{17}$$

$$\frac{\partial A^-}{\partial \overline{T}} + \mathrm{i}q\left(\gamma^* - \frac{1}{3}\right)\overline{T} A^- - s^* n A^+ = 0. \tag{18}$$

where[1] $q \approx 0.02$ and $n \approx 1.38$. This system of coupled equations can be easily solved in terms of special functions as shown in Le Dizès *et al.* (1996). The typical behavior of a solution is exemplified on figure 2.

[1]There is a slight variation (less than 5%) of these coefficients with respect to the resonant wavenumber κ.

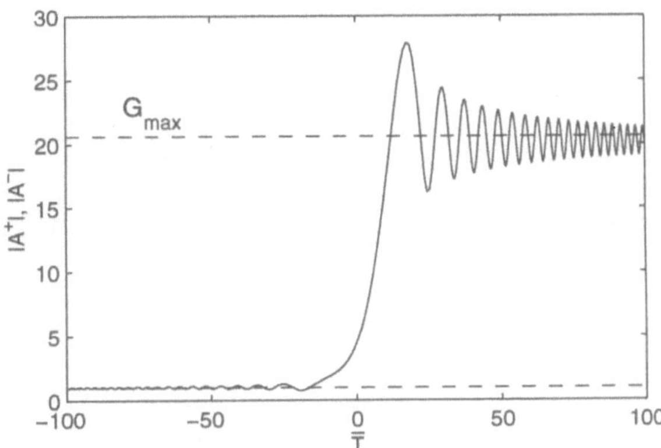

Figure 2. Typical temporal behaviour of the perturbation amplitude $|A^{\pm}|$.

If one defines the gain of amplitude across the region of resonance by the quantity

$$G = \lim_{\overline{T}\to+\infty} \left(\frac{|A^+(+\overline{T})|^2 + |A^-(+\overline{T})|^2}{|A^+(-\overline{T})|^2 + |A^-(-\overline{T})|^2} \right)^{\frac{1}{2}}, \qquad (19)$$

an exact expression for the maximum gain can be calculated as a function of $p = s^*/\sqrt{|\gamma^* - 1/3|}$. That expression reduces in the limit of large p to

$$G_{\max} \sim 2e^{Kp^2}, \qquad (20)$$

where $K = \pi n^2/(2q) \approx 150$. Figure 3 shows that (20) is a good estimate for $G_{\max}$ as soon as $p \geq 0.1$.

On figure 4 are diplayed the level curves of the maximum gain in the γ^*–s^* plane. This figure provides a sufficient condition of instability. If one assumes for instance that $G_{\max} > 10^4$ is sufficient, any vortex which corresponds to a point above the bold curve in figure 4 is unstable. This condition reduces to $s^* \geq 0.4$ for Burgers vortex, and $s^* \geq 0.3$ for Lamb-Oseen vortex. For the particular stretching parameter $\gamma^* = \frac{1}{3}$, vortices are unstable for any s^* of order unity. This stretching rate corresponds to a configuration for which the slope of $k(t)\delta(t)$ is zero at resonance [see expression (10)]. The time spend in the resonance region is then much longer than the estimate used in section 3. In that case, a specific analysis with new scaling should be carried out to determine the exact gain. Figure 4 also illustrates the non-trivial role of stretching : increasing the stretching rate is destabilising for the Lamb-Oseen vortex but stabilising for the Burgers vortex. Moreover, it is worth noting that for given radius and circulation, the Lamb-Oseen vortex is more unstable than the Burgers vortex.

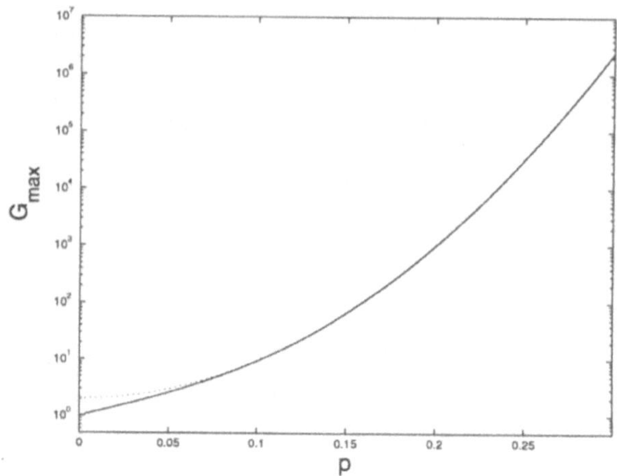

Figure 3. Plot of the maximum gain $G_{\max}$ as a function of p. Solid line : exact value, dotted line : expression (20).

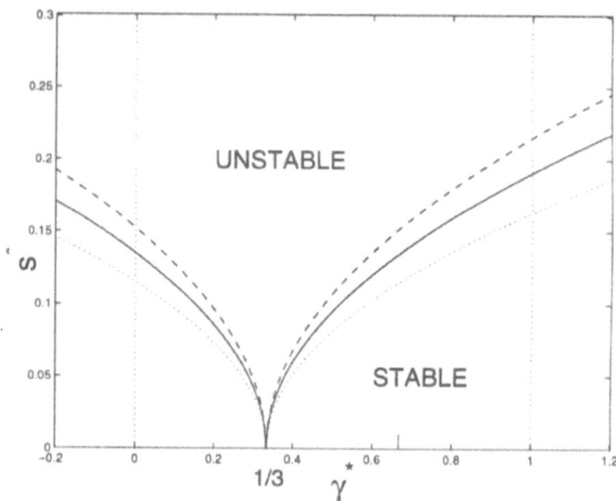

Figure 4. Level curves of the maximum gain $G_{\max}$ in the γ^*–s^* plane. Burgers and Lamb-Oseen vortices correspond to $\gamma^* = 1$ and $\gamma^* = 0$ respectively. Dotted line: $G_{\max} = 10^3$, solid line: $G_{\max} = 10^4$, dashed line: $G_{\max} = 10^5$.

5. Discussion

In this paper, we have considered a particular perturbation composed of helical modes with azimuthal wavenumber $m = \pm 1$ which leads to a sinuous deformation of the vortex. Such a deformation is well-known in the mecanism leading to the breakup of vortex rings (Widnall *et al.* 1974) or vortex pairs (Leweke & Williamson, 1997).

The present study can be easily applied to other combinations of modes. Basically any combination of two waves having same frequency, same axial wavenumber, and azimuthal wavenumbers differing by 2 constitutes a configuration which resonates with the strain field. For a bounded elliptical vortex with uniform vorticity, Waleffe (1989) showed that the maximum growth rate of such combinations depends slightly on their azimuthal wavenumber. This growth rate similarity could explain the rich dynamical behaviours of vortices in experiments and numerical simulations of high-Reynolds-number flows (Cadot *et al.* 1995, Arendt *et al.* 1997). Indeed, a combination of modes $m = 0$ and $m = 2$ could account for the spliting of vortices while modes $m = 1$ and $m = 3$ could be responsible for the formation of strands. The undulation of vortex filaments, or eventually the spliting of such vortices could then spontaneously occur if the turbulent background flow modifies the parameters s^* and γ^* in such a way that the instability criterion becomes satisfied.

References

ARENDT, S., FRITTS, D. C. & ANDREASSEN, Ø. 1997 Kelvin twist waves in the transition to turbulence. *Submitted to Eur. J. Mech.*

BAYLY, B. J. 1986 Three-dimensional Instability of Elliptical Flow. *Phys. Rev. Lett.*, **57**, 2160–63.

CADOT, O., DOUADY, S. & COUDER, Y. 1995 Characterization of the low pressure filaments in three-dimensional turbulent shear flow. *Phys. Fluids*, **7**(3), 630–646.

JIMÉNEZ, J., MOFFATT, H. K. & VASCO, C. 1996 The structure of the vortices in freely decaying two dimensional turbulence. *J. Fluid Mech.*, **313**, 209–222.

LANDMAN, M. J. & SAFFMAN, P. G. 1987 The three-dimensional instability of strained vortices in a viscous fluid. *Phys. Fluids*, **30**(8), 2339–2342.

LE DIZÈS, S., ROSSI, M. & MOFFATT, H. K. 1996 On the three-dimensional instability of elliptical vortex subjected to stretching. *Phys. Fluids*, **8**(8), 2084–2090.

LEWEKE, T. & WILLIAMSON, C. H. K. 1997 Short wavelength instability of a counter-rotating vortex pair. *Submitted to J. Fluid Mech.*

MOFFATT, H. K., KIDA, S. & OHKITANI, K. 1994 Stretched vortices -the sinews of turbulence; large-Reynolds-number asymptotics. *J. Fluid Mech.*, **259**, 241–264.

MOORE, D. W. & SAFFMAN, P. G. 1975 The instability of a straight vortex filament in a strain field. *Proc. R. Soc. Lond. A.*, **346**, 413–425.

PIERREHUMBERT, R. T. 1986 Universal short-wave instability of two-dimensional eddies in an inviscid fluid. *Phys. Rev. Lett.*, **57**, 2157–60.

ROBINSON, A. C. & SAFFMAN, P. G. 1984 Stability and structure of stretched vortices. *Stud. Appl. Math.*, **70**, 163–181.

ROSSI, M. & LE DIZÈS, S. 1997 Three-dimensional stability spectrum of stretched vortices. *Phys. Rev. Lett.*, **78**, 2567–2569.

TING, L. & TUNG, C. 1965 Motion and decay of a vortex in a nonuniform stream. *Phys. Fluids*, **8**(6), 1039–1051.

TSAI, C.-Y. & WIDNALL, S. E. 1976 The stability of short waves on a straight vortex filament in a weak externally imposed strain field. *J. Fluid Mech.*, **73**(4), 721–733.

VAN DYKE, M. 1975 *Perturbation method in fluid mechanics.* Stanford: The Parabolic

Press.

WALEFFE, F. 1990 On the three-dimensional instability of strained vortices. *Phys. Fluids* A, **2**(1), 76–80.

WIDNALL, S. E., BLISS, D. & TSAI, C.-Y. 1974 The instability of short waves on a vortex ring. *J. Fluid Mech.*, **66**(1), 35–47.

WILLIAMSON, C. H. K. 1996 Three-dimensional wake transition. *J. Fluid Mech.*, **328**, 345–407.

THE INFLUENCE OF THE SWIRL VELOCITY RATIO ON THE NATURE OF THE DOMINANT HELICAL STRUCTURES IN A SWIRLING JET MODEL

J.E. MARTIN
Department of Mathematics
Christopher Newport University
Newport News, VA 23606-2998

AND

E. MEIBURG
Department of Aerospace Engineering
University of Southern California
Los Angeles, CA 90089-1191

Abstract.

The nonlinear evolution of helical perturbations is investigated in a swirling jet model, consisting of a line vortex along the jet centerline surrounded by a jet shear layer with both azimuthal and streamwise vorticity. Inviscid Lagrangian vortex dynamics simulations demonstrate the nonlinear competition between a centrifugal Rayleigh instability and a Kelvin-Helmholtz instability feeding on both components of the base flow vorticity. The interaction of these two instabilities allows for very different flow behaviors to emerge, depending on (1) the swirl velocity ratio and (2) whether the helical perturbation wave and the vortex lines of the jet shear layer wind around the jet axis in the same or in opposite directions. For very high levels of swirl, large scale vortex helices evolve that can contain azimuthal vorticity either of the same or of opposite sign to that initially present in the jet shear layer. These different evolutions are triggered by the differences in the direction of the strain field set up by the dominant vortex helix. The effect of lowering the level of swirl is most prominent when perturbation occurs in a direction opposite to that of the vortex lines. In this case, both Kelvin-Helmholtz and centrifugal instabilities grow at similar rates leading to pairs of counterrotating helical vortices.

E. Krause and K. Gersten (eds.), IUTAM Symposium on Dynamics of Slender Vortices, 245-254.

1. Introduction

The present investigation continues our research into the stability and nonlinear evolution of swirling jets. Our primary objectives are (1) to obtain a better understanding of the dynamical interaction of shear and centrifugal instabilities which take place in free swirling jets subject to axisymmetric and helical perturbation and (2) to determine and analyze the resulting organized structures in the near field of such jets. For these purposes we have been considering a particular model of a swirling jet which will be described below. Studies on the nonlinear evolution of axisymmetric perturbations in this swirling jet model were reported by Martin and Meiburg (1994a,1996). In this paper, we describe the evolution of a nominally axisymmetric swirling jet subject to a helical perturbation. In these helically symmetric simulations, we investigate the impact that perturbation wavenumber and swirl velocity ratio have on the large scale vortices which form.

In spite of a long history of research on swirling jets (cf. the references in Martin and Meiburg 1996), researchers have begun only recently to pay attention to the dominant role played by the underlying vortical flow structures and their dynamical evolution, e.g. Panda and McLaughlin (1994). The recent results obtained by Krause and colleagues for vortex breakdown (reviewed by Althaus et al. 1993), and the three-dimensional vortex dynamics simulations by the present authors for simple models of swirling flows suggest that computations can provide fundamental insights into the flow physics of swirling jets. Very recent investigations by Huerre and coworkers (Delbende et al. 1996, Billant et al. 1997, Loiseleux et al. 1997) analyze vortex breakdown in swirling jets in the context of absolute and convective instabilities.

We consider a simplified model that is an extension to earlier models proposed by Batchelor and Gill (1962), Rotunno (1978), and Caflisch, Li and Shelley (1993). The model consists of an axial centerline vortex, which is surrounded by a nominally axisymmetric vortex sheet containing both streamwise and circumferential vorticity. While this model has obvious limitations when it comes to reproducing the detailed features of experimentally generated, and often geometry dependent velocity profiles, its simplicity offers several advantages. First of all, it allows for some analytical progress (Martin and Meiburg 1994b) in terms of a straightforward linear stability analysis, which illuminates the competition of centrifugal and Kelvin-Helmholtz instability waves. Secondly, the model enables us to study the nonlinear interaction and competition of the various instability mechanisms involved, by means of fully nonlinear Lagrangian vortex dynamics calculations.

In Martin and Meiburg (1994a), it was shown that under axisymmetric perturbation *counterrotating* vortex rings emerge in the braid regions between the primary vortex rings generated by the Kelvin-Helmholtz instability of the axisymmetric shear layer. These counterrotating vortex rings produce a pinch off effect in the jet shear layer leading to a dramatic decrease in the local jet diameter. The circulation strength of the swirling rings becomes time-dependent, in contrast to the dominant vortex structures found in nonswirling jets. Martin and Meiburg (1996) extended these earlier investigations to the three-dimensional evolution of axisymmetric waves under azimuthal perturbations. They observe that the evolution of the flow depends strongly on the initial ratio of the axisymmetric and azimuthal perturbation amplitudes. The long term dynamics of the jet can be dominated by counterrotating vortex rings connected by braid vortices, by like-signed rings and streamwise braid vortices, or by wavy streamwise vortices alone.

After a brief review of the flow model in section 2, we will investigate the nonlinear helically symmetric evolution of the above swirling jet model in more detail in section 3. We will compare the resulting helical structures for two different levels of the swirl velocity ratio. For each value, helical perturbation waves with both positive and negative azimuthal wavenumbers will be considered. Both vorticity contours and vortex line configurations will be analyzed in order to obtain a more complete picture of the nature of the dominant helical structures.

2. Flow Model

The present flow model of an axial line vortex surrounded by a nominally axisymmetric cylindrical shear layer containing streamwise and circumferential vorticity represents an extension of earlier ones investigated by several researchers, dating back to the analyses by Batchelor and Gill (1962) as well as Rotunno (1978) of the stability of an axisymmetric layer of circular or helical vortex lines. More recently, Caflisch, Li and Shelley (1993) introduced the effect of swirl by placing the additional line vortex at the center of the axisymmetric layer. However, their unperturbed vortex sheet had an axial vorticity component only, so that a jet-like velocity component was absent. In the present investigation, we employ a slightly more complicated model (fig. 1), consisting of a line vortex of strength Γ_o at radius r=0, surrounded by a cylindrical vortex sheet at $r = R$. The unperturbed axisymmetric vortex sheet contains both azimuthal vorticity (corresponding to a jump ΔU_x in the axial velocity) and streamwise vorticity (representing a jump ΔU_θ in the circumferential velocity). The strength of the vortex sheet is taken to be equal and opposite to that of the line vortex. The vortex lines in the sheet hence are of helical shape, with their pitch angle ψ being

$$\psi = \tan^{-1}\left(\frac{\Delta U_\theta}{\Delta U_x}\right) \qquad (1)$$

In the literature on swirling flows, it is common to quantify the effect of swirl in terms of a swirl number S. However, as discussed in Martin and Meiburg (1996), the standard definition of a swirl number does not take into account the presence of streamwise vorticity in the jet shear layer, and so is not very meaningful in the present model. Therefore, we instead characterize the level of swirl by the swirl velocity ratio, $\Delta U_\theta/\Delta U_x$.

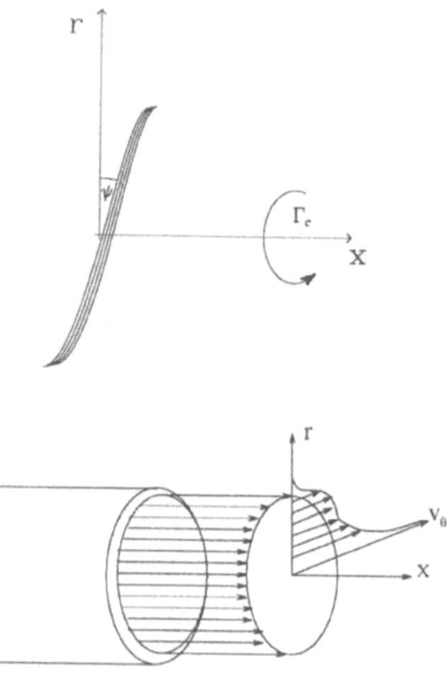

Figure 1. Simplified model of a swirling jet flow. The centerline vortex of strength Γ_c is surrounded by a nominally axisymmetric jet shear layer containing helical vortex lines of pitch ψ. The azimuthal vorticity component is related to the top hat axial velocity profile, whereas the streamwise vorticity component results in the centrifugally unstable stratification. Also shown are the streamwise and azimuthal base flow velocity profiles associated with our vortex filament simulations.

The particular features of our model were chosen on the basis of the following considerations. While an axisymmetric cylindrical layer of *circular* vortex lines represents a unidirectional flow with a top hat like profile shape, *helical* vortex lines result in an additional azimuthal velocity component, which jumps at the location of the vortex layer. If there is no streamwise circulation present at radii smaller than that of the cylindrical layer, this azimuthal velocity component vanishes inside the cylinder and exhibits a $1/r$-dependence on the outside. Consequently, since the magnitude of the circulation increases across the vortical layer, this flow is centrifugally stable on the basis of Rayleigh's circulation theorem. However, if some streamwise circulation is contained inside the cylinder, and if this circulation is of opposite sign to the streamwise circulation of the layer itself, then the magnitude of the circulation can decrease across the vortical layer, so that we obtain a centrifugally unstable flow. The line vortex at the center of the jet is introduced exactly for this purpose. Its strength is taken to be equal and opposite to that of the streamwise circulation contained in the vortical layer, in particular $\Gamma_c = -2\pi R \Delta U_\theta$. Thus the azimuthal velocity component of the base flow vanishes outside the jet. In this way, our simplified model mimics a swirling jet entering fluid at rest.

By introducing both axial and azimuthal vorticity along with the central line vortex, this model allows for the investigation of competing Kelvin-Helmholtz and centrifugal instabilities, which can be expected to lead to interesting nonlinear dynamical behavior. For the nonswirling top hat jet velocity profile it is known that axisymmetric and helical perturbations will result in the formation of vortex rings or helices, respectively, all of the same sign (Martin and Meiburg 1991, 1992). For purely swirling flow, on the other hand, Caflisch et al. (1993) demonstrated that axisymmetric perturbations lead to the emergence of counterrotating vortex rings. By superimposing a top hat streamwise velocity profile upon the purely swirling flow, a breaking of the symmetry exhibited by the purely swirling flow alone occurs. In the present investigation, our goal is to simulate the nonlinear growth of a helical perturbation, characterized by a given streamwise wavelength and an assigned azimuthal wavenumber m.

In order to compute the nonlinear evolution of the flow in response to certain imposed perturbations, we employ a vortex filament technique that is essentially identical to the one used in earlier investigations of plane shear layers and jets (Ashurst and Meiburg 1988, Martin and Meiburg 1991, 1992, 1994a, 1996, 1997). It is based on the theorems of Kelvin and Helmholtz and follows the general concepts reviewed by Leonard (1985) and Meiburg (1995). Using the vortex filament technique, the vorticity layer now has a finite thickness. A detailed description of the general technique is provided in these earlier references. For the particular discretization employed in the present simulations, cf. the preliminary account given by Martin and Meiburg (1997). In each of the simulations to be described, we include an initial helical perturbation which displaces the vortex filament centerlines in the radial direction, with an amplitude of five percent of the jet radius. The perturbed location r of the filaments' centerlines is therefore specified as a function of the circumferential coordinate ϕ by

$$r = R\left(1 + .05\cos\left(x + m\phi\right)\right). \tag{2}$$

The streamwise wavelength of the perturbation, i.e., the length of the control volume, is fixed at 2π to match the wavelength of maximum growth given by Michalke and Hermann (1982). A sketch of the fluid velocity profile associated with our filament simulations is included in fig. 1.

3. Results

3.1. $\Delta U_\theta/\Delta U_X = 8.2$

This case was described in detail by Martin and Meiburg (1997). Here the jump in the azimuthal velocity component across the jet shear layer is much larger than that of the axial component, so that, for the unperturbed flow, the vortex lines are predominantly oriented in the streamwise direction. For an m=-2 perturbation,

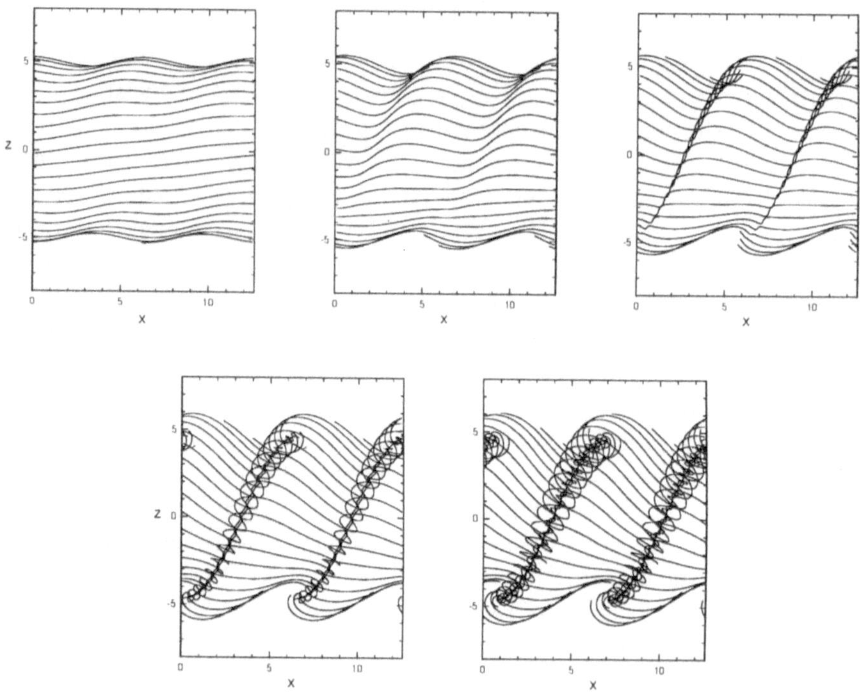

Figure 2. The evolution of a swirling jet with $\Delta U_\theta/\Delta U_x = 8.2$ subject to a helical perturbation with $m = -2$. Shown are side views of the vortex filaments at times 0.039, 0.430, 0.625, 0.820, and 0.947. Only every fifth vortex filament is drawn. For clarity, two streamwise wavelengths are shown. Note the roll-up of the jet shear layer into a large-scale vortex helix, as well as the reorientation of the braid vortex line segments into the opposite azimuthal direction.

the helical wave winds around the jet centerline in the same direction as the helical vortex lines. Side views shown in figure 2 demonstrate the roll-up of the vorticity layer into a large-scale helix, whose azimuthal vorticity is of the same sign as that initially present in the jet shear layer. The sign of the azimuthal vorticity component in the braid region is reversed, so that the azimuthal braid vorticity points in the opposite direction to the azimuthal vorticity making up the large scale helix. This appearance of azimuthal braid vorticity of the opposite sign reflects the influence of the centrifugal Rayleigh instability and its tendency to generate pairs of counterrotating vortical structures.

Figure 3 contrasts the above evolution of the swirling jet model with that obtained for m=2, i.e., a helical perturbation of opposite azimuthal wavenumber. The strain field generated by the evolving helix now *amplifies* the azimuthal braid vorticity from the start, rather than reversing it, as for m=-2. Now it is the primary helical vortex itself whose sign is reversed. As a result, the vortex helix now is made up of azimuthal vorticity of the opposite sign to that initially present in the jet shear layer.

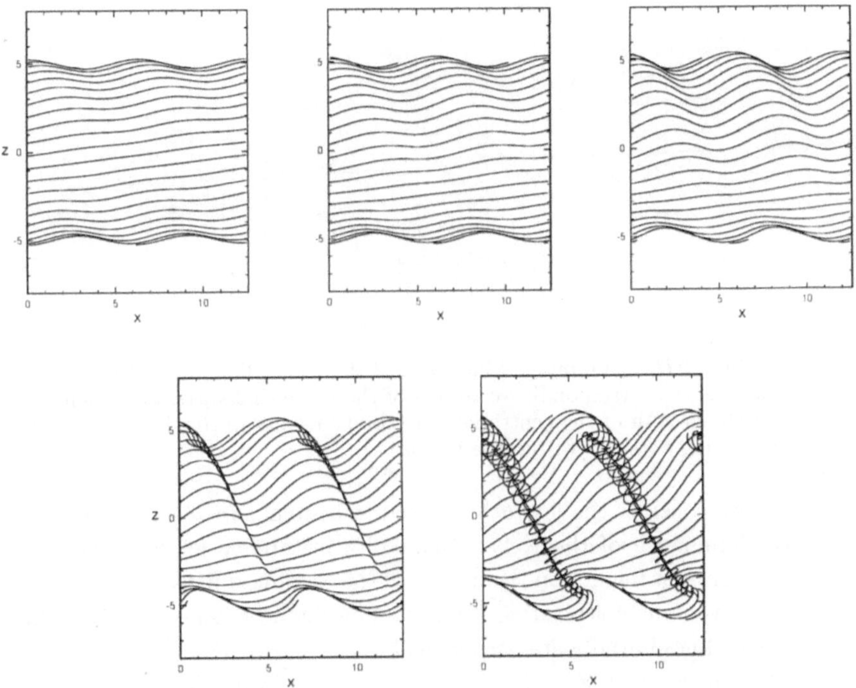

Figure 3. The evolution of a swirling jet with $\Delta U_\theta/\Delta U_x = 8.2$ subject to a helical perturbation with m=2. Shown are side views of the vortex filaments at times 0.039, 0.234, 0.430, 0.684, and 0.908. The helical perturbation wave and the vortex lines of the jet shear layer wind around the jet axis in opposite directions. In contrast to the m=-2 case, the vortex lines now reverse their azimuthal direction in the core region, rather than in the braid region.

3.2. $\Delta U_\theta/\Delta U_X = 1.0$

We now consider the flow evolution for the situation in which the shear layer contains equal amounts of azimuthal and streamwise vorticity, so that $\Delta U_\theta/\Delta U_x = 1.0$. Figure 4 shows side views and cross cuts of the vortex lines for a $m = -2$ perturbation. Most notable in the side views of the filaments is again the roll-up of the vorticity into a large scale helical vortex of the same sense of rotation as in nonswirling jets. The azimuthal vorticity contours indicate only a small amount of counterrotating vorticity occurring within the braids. This implies that growth of the centrifugal instability has been slowed down by the reduction of the amount of swirl. As a result, the Kelvin-Helmholtz helices are allowed to more strongly dominate the resulting flow.

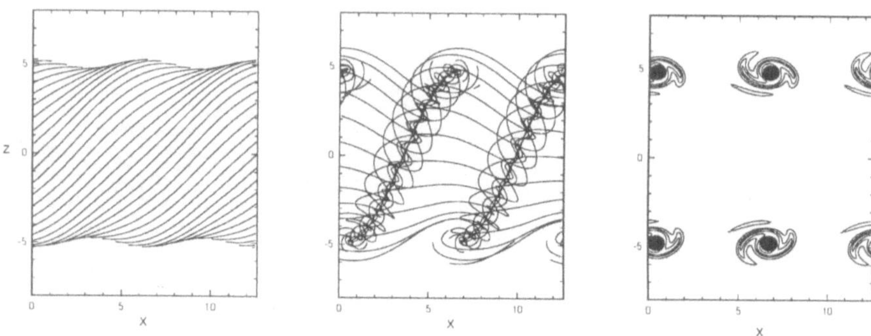

Figure 4. $\Delta U_\theta/\Delta U_x = 1.0$,$m$=-2: Side views of the vortex filaments for times 0.312 and 6.367, as well as the corresponding contours of the azimuthal vorticity for time 6.367. A decrease in the growth of the centrifugal instability results in the decreased production of counterrotating vorticity within the braids.

Lastly, we consider the jet with reduced swirl under $m = 2$ perturbation. The increased dominance of the Kelvin-Helmholtz instability for low levels of swirl has an even greater impact on the nature of the dominant flow structure when the azimuthal wavenumber is positive. Figure 5 shows side views and cross cuts for $m = 2$. The vortex lines initially are perpendicular to the perturbation wave. Unlike our findings for wavenumber $m = 2$ in the jet with strong swirl, the azimuthal vorticity of the emerging large-scale helix now has the same sign as the azimuthal vorticity initially present in the jet shear layer, which indicates that, at this reduced level of swirl, the Kelvin-Helmholtz instability is allowed to dominate over the centrifugal instability and its tendency to create dominant vortical structures with azimuthal vorticity of opposite sign. However, in time, the centrifugal instability begins to generate some counterrotating vorticity within the braids. By the final time shown in figure 5, the centrifugal instability has gained influence to the point that a counterrotating helical vortex of near equal strength has emerged within the braids.

4. Summary and Conclusions

The present investigation represents an attempt to study the main mechanisms resulting in the complex nonlinear dynamical evolution of swirling jets subject to helical perturbations, by performing Lagrangian vortex dynamics simulations for a simplified model that nevertheless contains the essential features characterizing such flows. We observe that the nonlinear evolution resulting from the interaction of the dominant instabilities allows for very different flow behaviors to emerge,

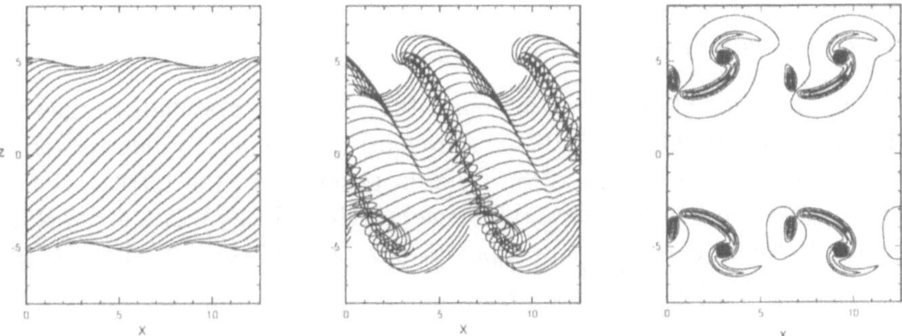

Figure 5. $\Delta U_\theta/\Delta U_x = 1.0$,$m$=2: Side views of the vortex filaments and the corresponding contours of the azimuthal vorticity for times 0.312 and 7.187, along with the contours of the azimuthal vorticity for this latter time. In contrast to the $m = 2$ results for high velocity ratio, the initial vortical structure which emerges contains azimuthal vorticity of the same sign as is initially present in the jet shear layer. By decreasing the level of swirl, the Kelvin-Helmholtz and centrifugal instabilities each grow at similar rates, ultimately producing a pair of counterrotating, helical vortices.

depending on the swirl velocity ratio and the helical perturbation wavenumber. For very high levels of swirl, a perturbation with wavenumber $m = -2$ results in a large-scale vortex helix which contains azimuthal vorticity of the same sign as that initially present in the jet shear layer. The generation of opposite signed vorticity is fueled by the centrifugal instability. We observe that by decreasing the level of swirl for this perturbation wavenumber, we diminish the centrifugal instability and correspondingly dampen the generation of a paired counterrotating structure in the braids.

For the case of positive wavenumber, if a high level of swirl is used, the centrifugal instability is again allowed to generate both signs of azimuthal vorticity. However, in this case, it is the emerging large-scale vortex helix which reverses its sign of azimuthal vorticity. Thus for high levels of swirl the flow can be dominated by a helical vortex with azimuthal vorticity of opposite sign to that initially present in the jet shear layer. With positive wavenumber, the effect of lowering the swirl velocity is to allow the Kelvin-Helmholtz instability to gain in prominence in relation to the centrifugal instability. By lowering the swirl level in the positive wavenumber case, the Kelvin-Helmholtz instability generates a dominant helix with vorticity of the same sign as the jet shear layer. Ultimately, the centrifugal instability gains in strength to create counterrotating vorticity within the braids. This situation of positive wavenumber and equal levels of axial and azimuthal velocity ultimately results in a flow with paired counterrotating vortices of similar strengths. As a next step, we plan to perform an investigation in which the helical symmetry of the flow will be broken, for example, by an additional, azimuthal perturbation.

Acknowledgements

Support by the National Science Foundation to EM is gratefully acknowledged. JEM was supported by the National Aeronautics and Space Administration under NASA Contract No. NAS1-19480 while in residence at the Institute for Computer Applications in Science and Engineering (ICASE), NASA Langley Research Center. Computing resources were provided by the NSF-supported San Diego Supercomputer Center.

References

Althaus, W., Brucker, C., and Weimer, M. (1995) Breakdown of slender vortices, in: *Fluid Vortices*, ed. S.I. Green, Kluwer, pp. 373-426.

Ashurst, W.T. and Meiburg, E. (1988) Three-dimensional shear layers via vortex dynamics, J. Fluid Mech. **189**, pp. 87-116.

Batchelor, G.K. and Gill, A.E. (1962) Analysis of the stability of axisymmetric jets, J. Fluid Mech. **14**, pp. 529-551.

Billant, P., Chomaz, J.-M., and Huerre, P. (1997) Experimental study of vortex breakdown in swirling jets, submitted to J. Fluid Mech.

Caflisch, R.E., Li, X., and Shelley, M.J. (1993) The collapse of an axi-symmetric swirling vortex sheet, Nonlinearity **6**, pp. 843-867.

Delbende, I., Chomaz, J.-M., and Huerre, P. (1996) Absolute/convective instabilities in the Batchelor vortex: a numerical study of the linear impulse response, submitted to J. Fluid Mech.

Leonard, A. (1985) Computing three-dimensional flows with vortex elements, Ann. Rev. Fluid Mechanics **17**, pp. 523-559.

Loiseleux, T., Chomaz, J.-M., and Huerre, P. (1996) The effect of swirl on jets and wakes: Linear instability of the Rankine vortex with axial flow, submitted to Phys. Fluids.

Martin, J.E. and Meiburg, E. (1991) Numerical investigation of three-dimensionally evolving jets subject to axisymmetric and azimuthal perturbations, J. Fluid Mech. **230**, pp. 271-318.

Martin, J.E. and Meiburg, E. (1992) Numerical investigation of three-dimensionally evolving jets under helical perturbations, J. Fluid Mech. **243**, pp. 457-487.

Martin, J.E. and Meiburg, E. (1994a) The nonlinear evolution of swirling jets, Meccanica **29**, pp. 331-341.

Martin, J.E. and Meiburg, E. (1994b) On the stability of the swirling jet shear layer, Phys. Fluids **6**, pp. 424-426.

Martin, J.E. and Meiburg, E. (1996) Nonlinear axisymmetric and three-dimensional vorticity dynamics in a swirling jet model, Phys. Fluids **8**, pp. 1917-1928.

Martin, J.E. and Meiburg, E. (1997) The growth and nonlinear evolution of helical perturbations in a swirling jet model, Submitted to the European Journal of Mechanics B/Fluids.

Meiburg, E. (1995) Three-dimensional vortex dynamics simulations, in: *Fluid Vortices*, ed. S.I. Green, Kluwer, pp. 651-685.

Michalke, A. and Hermann, G. (1982) On the inviscid instability of a circular jet with external flow, J. Fluid Mech. **114**, pp. 343-359.

Panda, J. and McLaughlin, D.K. (1994) Experiments on the instabilities of a swirling jet, Phys. Fluids **6**, pp. 263-276.

Rotunno, R. (1978) A note on the stability of a cylindrical vortex sheet, J. Fluid Mech. **87**, pp. 761-771.

THEORY OF HELICAL VORTICES

S.V. ALEKSEENKO, P.A. KUIBIN, V.L. OKULOV, S.I. SHTORK
Institute of Thermophysics
Lavrentiev Ave 1, 630090 Novosibirsk, RUSSIA

1. Introduction

The numerous facts confirm wide spreading of the concentrated vortices of the shape, canonical helical or close to it, in swirl flows. The assumption about a helical symmetry in theoretical models results in essential simplification of a problem. Attempts earlier were undertaken to construct simple models for flows with a helical symmetry (see, for example, Dritschel (1991)). However, as additional simplification the Beltrami condition as a rule was used. The class of such flows assumes identical coincidence of fields of velocity and vorticity to within a constant factor in the whole area of flow. The concentrated vortices do not concern to the given class. In our work another class of flows covering the concentrated helical vortices is investigated. Within the framework of the model, constructed by us, the new elementary solutions permitting to explain numerous features of swirl flows are obtained. The cases of change of the helical symmetry observed in experiments are described.

2. Basic suppositions of model

Let's consider motion of inviscid incompressible fluid in the infinite cylinder or boundless space. As well as in case of helical Beltrami flows we shall assume, that the flow has a helical symmetry on spatial coordinates with a constants pitch $2\pi l$. It means, that one can to pass from three spatial variables (r, φ, z) of cylindrical coordinate system to two ones $(r, \chi = \varphi - z/l)$, and the operator $\boldsymbol{h}\cdot\nabla$ applied to any scalar function from r,χ and time t is equal to zero, where $\boldsymbol{h}$ is the Beltrami vector (Dritschel, 1991). The further simplification of problems can be made with the help of additional assumption.

Dritschel (1991) requires coincidence of fields of velocity $\boldsymbol{w}$ and vorticity rot($\boldsymbol{w}$) to within a constant factor in the whole area of flow, i.e. $\mathrm{rot}(\boldsymbol{w}) = -\alpha\cdot\boldsymbol{w}$ (Beltrami condition), that allows to reduce a problem to determination of one scalar function.

For the class of flows under consideration the second simplifying supposition will be formulated alternatively. With this purpose we shall rewrite the Euler equations in variables r,χ :

E. Krause and K. Gersten (eds.), IUTAM Symposium on Dynamics of Slender Vortices, 255-264.

$$\frac{Dw_r}{Dt} - \frac{w_\varphi^2}{r} = -\frac{1}{\rho}\frac{\partial p}{\partial r};$$

$$\frac{D}{Dt}\left(w_z + \frac{r}{l}w_\varphi\right) = 0; \tag{1}$$

$$\frac{D}{Dt}\left(w_\varphi - \frac{r}{l}w_z\right) + \left(w_\varphi + \frac{r}{l}w_z\right)\frac{w_r}{r} = -\frac{1}{\rho}\frac{l^2 + r^2}{l^2}\frac{\partial p}{r\partial\chi}.$$

Further we shall limit by the class of flows, that may be called two-dimensional ones, in which the second equation of a system (1) is satisfied identically. The obvious solution of this equation $w_z + w_\varphi\, r/l = \text{const}$ is taken as the second simplifying condition. The constant corresponds to value of axial velocity w_0 on the axis oz determining a uniform flow, imposed on a basic one i.e.

$$w_z = w_0 - \frac{r}{l}w_\varphi \quad \text{or} \quad \frac{w_z - w_0}{w_\varphi} = -\frac{r}{l}. \tag{2}$$

The condition (2) may be presented in an equivalent form in terms of the vorticity vector components rot($\boldsymbol{w}$) = $(\omega_r, \omega_\varphi, \omega_z)$:

$$\omega_r = 0, \quad \omega_\varphi/\omega_z = r/l. \tag{3}$$

The condition in the form (3) requires that vorticity vector should be directed on tangent to helixes, along which the characteristics of flow are conserved pursuant to the first condition. Thus given class of flows is possible to call as two-dimensional flows with a helical symmetry, or flows with helical vortices.

It is impossible to describe the whole area of flow by model with a condition (2) or (3) at constant helix pitch. However, in swirl flows there are rather extended areas, where the structure of flow varies weakly. In these local areas the hypothesis (2) was checked up in fixed cross sections by comparison of the measured axial velocity with values calculated through the measured value of circumferential velocity according to formula (2). The check of fulfillment of a ratio (2) in actual swirl flows was conducted by us for different types of swirl generators, flow regimes and ways of diagnostics of flows. There was found that the helical symmetry across a flow is realized practically in the whole area of flow, except for area near to walls of a pipe (Alekseenko *et al.*, in press). The evaluation of a helical symmetry along the flow axis or check of a condition (3) can be executed by an evaluation of quality of the helical form at vortex structures formed in swirl flows. The idea of such check is based on the fact that the projection of a helix to a plane is a sine wave. For determination of a projection of actual helical structures Cherny (1997) developed a computer code for processing of the instantaneous frontal videoimages of an air cavity, visualizing an axis of fixed vortex. The comparison of sine with obtained axes for various flow regimes allows to conclude that a helical symmetry in a swirl flow exists for significant stretch in a longitudinal direction. The listed data indicate existence of a helical symmetry in swirl flows for all types of swirl generators and capability of effective application for their description of model of perfect fluid and even of its simpler two-dimensional helical form practically in the whole area of flow, except for a neighbourhood of walls of tubes and end faces of working sites.

It is important to note that at zero uniform flow ($w_0 = 0$) and zero radial velocity ($w_r = 0$) Eqs. (2) and (3) yield that fields of velocity and vorticity are orthogonal to each other as opposed to the helical Beltrami flows, where they are parallel.

For the considered here class of flows with helical vortices, similarly to helical Beltrami flows (Dritschel, 1991), the problem can also be simplified up to the solution of one equation concerning one unknown function. Actually, the continuity allows to introduce a stream function $\Psi(r,\chi)$:

$$w_r = \frac{1}{r}\frac{\partial \Psi}{\partial \chi}, \qquad w_\chi \equiv w_\varphi - \frac{r}{l} w_z = -\frac{\partial \Psi}{\partial r}. \tag{4}$$

Equations (4) together with the definition of the vorticity field result to a single equation for the stream function (Alekseenko *et al.*, in press)

$$\frac{\partial^2 \Psi}{\partial r^2} + \frac{l^2 - r^2}{l^2 + r^2}\frac{1}{r}\frac{\partial \Psi}{\partial r} + \frac{r^2 + l^2}{l^2}\frac{\partial^2 \Psi}{r^2 \partial \chi^2} = \frac{2 l w_0}{r^2 + l^2} - \omega_z \frac{(r^2 + l^2)}{l^2}. \tag{5}$$

Unlike the Biot-Savart approach the way with solution of the equation (5) allows to find solution of the problem in a bounded space - in a cylindrical tube. If the axial component of the vorticity field ω_z is given the problem may be considered as fully stated. Nonetheless the vorticity field has to satisfy the Helmholtz equations. In our case they are reduced to a single equation

$$D\,\omega_z / Dt = 0$$

It means, that ω_z does not vary along trajectory of a liquid particle, and in the steady-state case ω_z is arbitrary function of the stream function Ψ. Note that in a general case it is very hard to solve the equation (5) for a given ω_z. So below there will be considered three special classes: 1) flows induced by an infinitely thin vortex filament (see Fig. 1a; solution in this case represents a fundamental solution of the equation (5)); 2) steady-state flows with axisymmetric vortices (Fig. 1b; in this case we have $\Psi = \Psi(r)$ and $\omega_z = \omega_z(r)$ and the Helmholtz equation is satisfied); 3) flows in frames of a model of a helical vortex with the finite size of a core with a uniform distribution of the axial vorticity component (see Fig. 1c).

3. Elementary solutions for flows with helical vortices

As the simplest example of the flow with helical vortex we shall consider flow with an infinitely thin vortex line, twisted on a cylinder of radius a. This elementary vortex structure is a fundamental object in the theory of vortex flows similarly to well investigated rectilinear filament and vortex ring. Analytical representation of the solution for a helical filament in boundless space was obtained by Hardin (1982). He has transformed integral representation of the solution in the form of the Biot-Savart law in

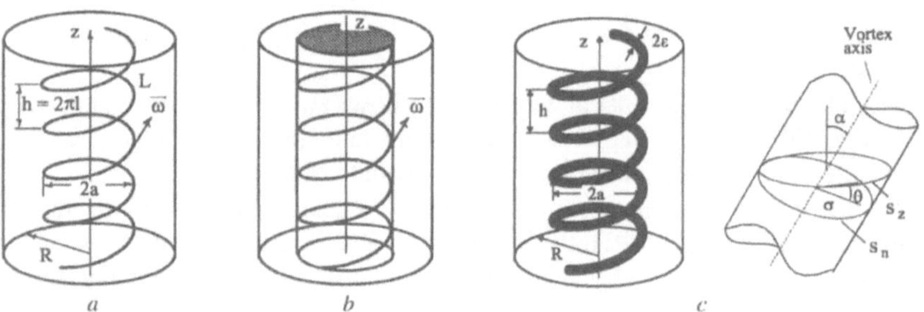

Figure 1. Elementary distributions of vorticity in swirl flows: an infinitely thin filament (a); a columnar-like vortex (b); vortex with a helical axis (c).

view of the helical symmetry of a filament and used the integral definition of the cylindrical functions. Similar representation of the solution, but already as the solution of the equation (5) for a helical filament in bounded space (in a tube of radius R, Fig. 1, a) was derived by Okulov (1995):

$$\Psi = \frac{w_0 r^2}{2l} - \frac{\Gamma}{4\pi}\left\{\begin{matrix} \dfrac{a^2}{l^2} + \ln a^2 \\ \dfrac{r^2}{l^2} + \ln r^2 \end{matrix}\right\} - \frac{\Gamma a r}{\pi l^2}\sum_{m=1}^{\infty}\left\{\begin{matrix} I'_m\left(\dfrac{mr}{l}\right) Z'_m\left(\dfrac{ma}{l}\right) \\ I'_m\left(\dfrac{ma}{l}\right) Z'_m\left(\dfrac{mr}{l}\right) \end{matrix}\right\}\cos m\chi\,, \tag{6}$$

where $Z_m(x) = K_m(x) - a_m I_m(x)$; $a_m(x) = K'_m(mR/l)/I'_m(mR/l)$; $I_m(x)$, $K_m(x)$ - are the modified cylindrical functions. The components of the velocity vector are determined under the formulas:

$$w_r = \frac{1}{r}\frac{\partial\Psi}{\partial\chi},\quad w_\varphi = \frac{l}{r^2 + l^2}\left(r w_0 - l\frac{\partial\Psi}{\partial r}\right),\quad w_z = \frac{l}{r^2 + l^2}\left(l w_0 + r\frac{\partial\Psi}{\partial r}\right). \tag{7}$$

Note first, that at $R \to \infty$ we deal with the case of a helical filament in boundless space. Then $a_m \to 0$ and the solution (7) for the velocity components w_r and w_φ completely coincides with the result by Hardin (1982). The value w_z differs only by the constant w_0. Secondly, the solution (6) is represented through the modified cylindrical functions unlike the case of the helical Beltrami flows (Dritschel, 1991) where one deals with the ordinary cylindrical functions.

We start to consider this class of helical vortices with the simplest particular case of axisymmetrical or columnar vortices. The elementary models of such vortices have long been known. These are the vortices by Rankine, Lamb, Burgers etc. (Hopfinger and van Heijst, 1993). However, the above models yield only the radial distribution of the circumferential component of a velocity vector. The axial component does not depend on radius that is inconsistent with the numerous experimental data. This disadvantage is a consequence of the transition to axial symmetry without taking into account the helical structure of the vortex core (Fig. 1, b). Analysis of the governing equations shows that the condition of the axial symmetry allows to design strictly an enough general model of axisymmetrical helical vortices. In the case under consideration one have $\partial/\partial\chi \equiv 0$,

$w_r \equiv 0$. Therefore arbitrary radial distribution of the vertical component of the vorticity ω_z satisfies the Helmholtz equation. Let's consider three important examples of the vorticity distributions: a step-function; Cache and Gaussian distributions:

$$\omega_z = \frac{\Gamma}{\pi\varepsilon^2}\begin{cases}1, & r<\varepsilon,\\ 0, & r\geq\varepsilon,\end{cases} \quad \omega_z = \frac{\Gamma}{\pi\varepsilon^2}\left(1+r^2/\varepsilon^2\right)^{-2}, \quad \omega_z = \frac{\Gamma}{\pi\varepsilon^2}\exp\left(-r^2/\varepsilon^2\right),$$

where ε is the length scale of the vortex. Integrating yields the respective velocity fields

$$w_\varphi = \frac{\Gamma}{2\pi r}\begin{cases}r^2/\varepsilon^2, & r\leq\varepsilon\\ 1, & r>\varepsilon\end{cases}, \quad w_z = w_0 + \frac{\Gamma}{2\pi r}\begin{cases}r^2/\varepsilon^2, & r\leq\varepsilon\\ 1, & r>\varepsilon\end{cases} \tag{8}$$

$$w_\varphi = \frac{\Gamma}{2\pi}\frac{r}{r^2+\varepsilon^2}, \quad w_z = w_0 - \frac{\Gamma}{2\pi l}\frac{r^2}{r^2+\varepsilon^2} \tag{9}$$

$$w_\varphi = \frac{\Gamma}{2\pi r}\left[1-\exp(-r^2/\varepsilon^2)\right], \quad w_z = w_0 - \frac{\Gamma}{2\pi l}\left[1-\exp(-r^2/\varepsilon^2)\right]. \tag{10}$$

At $l \to \infty$ solution (8) describes the Rankine vortex, (9) - Scully (1975) vortex and (10) - Lamb vortex (Hopfinger and van Heijst, 1993).

Now let's construct a model of a vortex having a helix-like core with a round cross-section of radius ε in a plane perpendicular to the vortex axis (Fig. 1,c). We consider the simplest distribution of vorticity to satisfy the Helmholtz equation i.e. ω_z = const within a core. The velocity field induced by such a vortex may be easily represented through solution (7) assuming that the vortex is the superposition of infinitely thin vortex filaments uniformly distributed within the vortex core

$$\vec{w}_\varepsilon(r,\varphi,z) = \frac{1}{lS_z}\int_0^\varepsilon\int_0^{2\pi}\sqrt{l^2+r'^2}\cdot\vec{w}(r,\varphi,z;r',\varphi',z')\sigma d\sigma\, d\theta. \tag{11}$$

The integration is made over the circle of radius ε with a center at the vortex axis $(a,\varphi_0,0)$; σ, θ are the local polar coordinates.

The simple representation for the solution (11) manages to be received only for velocity profiles averaged over angular coordinate:

$$\langle\vec{w}_\varepsilon\rangle = \frac{1}{2\pi}\int_0^{2\pi}\vec{w}_\varepsilon\, d\varphi \tag{12}$$

Substitution of (11) into (12) and change the sequence of integration yield

$$\langle w_{r\varepsilon}\rangle \equiv 0, \quad \langle w_{\varphi\varepsilon}\rangle = \frac{\Gamma}{2\pi r}F(r), \quad \langle w_{z\varepsilon}\rangle = w_0 - \frac{\Gamma}{2\pi l}F(r), \quad F(r) = S^{(0)}/\pi\varepsilon^2, \tag{13}$$

where $S^{(0)}$ is the area of intersection of a circle $\sigma = \varepsilon$ with an ellipse specified by formula

$$(a+\sigma\cos\theta)^2/l^2 + (\sigma\sin\theta)^2/(a^2+l^2) = \sigma^2.$$

4. Results of the description of swirl flows by two-dimensional model

We shall put some results of the preliminary analysis of actual flows with the help of the constructed elementary solutions. At the same flow swirling (Γ is positive) the existence two solutions (7) with a with the right of and left-hand helical symmetry is possible. In experiment we also observed vortices with different helical symmetry (Figs. 2, a and 2, b). The right-hand helical filament initiates intensive near-axis flow, and the left-hand one is characterized by braking of a flow near the axis. Computed flow patterns for the left-hand and the right-hand filaments at the same flow rate through cross-section of a pipe are shown in Fig. 2.

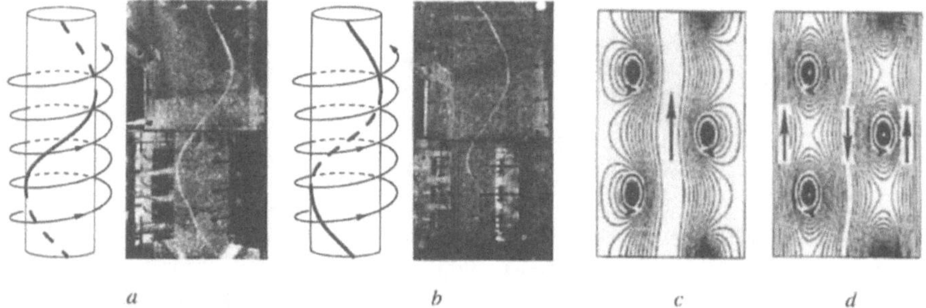

Figure 2. Different types of a helical symmetry in swirl flows: scheme and visualization of the right-hand helical vortex (a) and the left-hand one (b); calculation of thev flow patterns with an intensive stream along the tube axis for right-hand vortex (c) and the reverse flow for the left-hand one (d).

It should be noted that the origin in a swirl flow of vortices with a left-hand helical symmetry explains appearance of a reverse flow at some regimes of flow.

Other interesting fact is that the solution (10) on the structure coincides with the empirical formulas used for the description of flow before and after the vortex breakdown (Garg and Leibovich, 1979):

$$w_\varphi = \frac{K}{r}\left(1-\exp\left(-\alpha r^2\right)\right); \tag{14}$$

$$w_z = W_1 + W_2 \exp\left(-\alpha r^2\right) = W_1 + W_2 - W_2\left(1-\exp\left(-\alpha r^2\right)\right),$$

where K, W_1, W_2, α are empirical constants. The direct comparison to the exact solution (10) gives:

$$\Gamma = 2\pi K; \quad l = K/W_2; \quad w_0 = W_1 + W_2 \ , \ \ \varepsilon = 1/\sqrt{\alpha}\ .$$

The coincidence of the solution (10) with for a long time used relations not only allows to clarify the physical sense of the empirical constants, but also explains their good matching with experiment. In Figure 3, a and 3, b comparisons of the elementary models of flow (8) - (10) among themselves and with experimental data by Garg and Leibovich

(1979) are shown. Let's mark, that here a good matching with models is obtained not only for the circumferential velocity but also for the axial one, that wasn't described by known models by Rankine, Lamb and Burgers earlier.

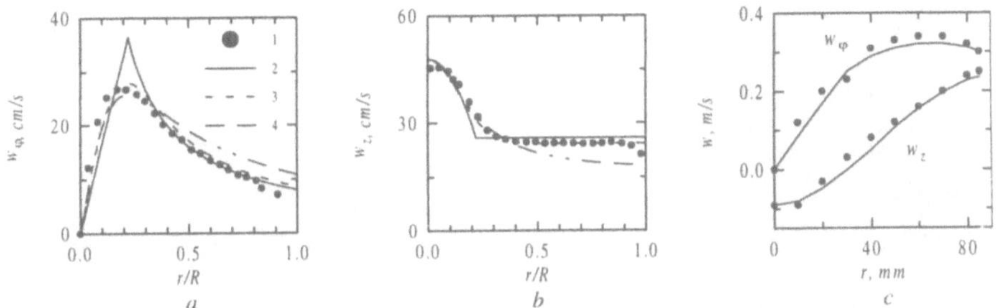

Figure 3. Comparison of the elementary solutions with the experimental data. (a), (b) points correspond to data by Garg and Leibovich (1979), numbers 2 - 4 correspond to Eqs. (8) - (10); (c) points correspond to data by Alekseenko et al. (in press), lines - to calculation according to formulae (13).

To check up a capability of the description of actual flows by approximate model (13) let's compare it with experimental data for a regime with a precesing helical vortex (Alekseenko *et al.*, in press). Due to rotation of a vortex, averaging on angular coordinate φ made in (13) corresponds to averaging on time in experiment. Figure 3, c presents calculation of the averaged velocity profiles well matched with experiment.

The conclusion follows from performed comparisons that the proposed model of two-dimensional flows with a helical symmetry of the vorticity field describes the experiment well for wide class of swirl flows and explains origin in them of a reverse flow as a consequence of generation of vortex structures with a left-hand helical symmetry.

5. Change of a helical symmetry in swirl flows

It is necessary to mark, that in actual flows the condition of a helical symmetry for the vorticity field is true not for the whole area of flow. The most obvious example here is the interaction of a swirl flow with face surfaces. Our numerous observations (Alekseenko *et al.*, in press) show that the axis of a helical vortex in a neighbourhood of a face wall is distorted, loses its helical form and mates a plane of an end face perpendicularly. In separate experiments we observed a conical shape of a flow pattern (Fig. 4, a). In this case, the supposed above models are unsuitable. Here it is necessary to use or generalization on conical case of the solution (6) - (7) (Okulov, 1995), or model considered in work by Shtern, Borissov and Hussain (1997). Other example of violation of a helical structure of a vortex axis was observed by Alekseenko and Shtork (1997) at interception of a vortex filament by a thin rod. The vortex axis failed, a structure visually similar to axisymmetrical vortex breakdown was formed which moved upstream with a constant speed (Fig. 4, b)

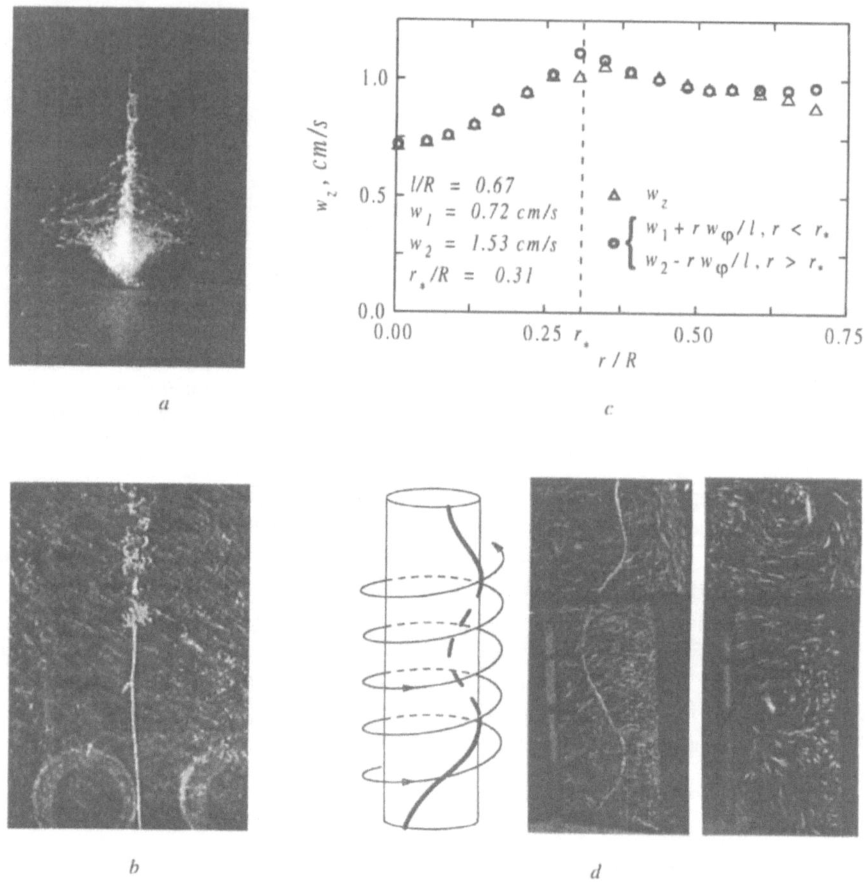

Figure 4. Violation of a helical symmetry in actual swirl flows: a conical vortex near the bottom (a); a running vortex breakdown (b); vortex with a changing sign of the helical pitch in transversal direction (c); vortex with changing helical symmetry along the tube axis.

The change of a helical symmetry is observed not only at active control of a swirl flow by end faces or bodies, disturbing a flow. In a number of experiments (Alekseenko *et al.*, in press; Escudier *et al*, 1980; Brücker and Althaus, 1992), the vortex structure with a depression on profiles of the axial velocity and profiles of the tangential velocity essentially deviating from empirical profiles (14) (or from the exact solution (10)) were observed. A question arises does the helical symmetry exist in flows of such kind? Let's consider this problem with an example of the velocity profiles measured by Brücker and Althaus (1992). Split the area of flow into two zones: a zone 1 - from an axis up to a maximum of axial velocity ($r_* = 0.31\ R$) and zone 2 - annular domain from r_* up to a periphery (wall of the chamber). Parameters of a helical symmetry may be defined for each zone of flow separately. We have for a zone 1:

$l = -0.67\,R$, $w_0 = 0.72$ cm/s, and for a zone 2:

$l = 0.67\,R$, $w_0 = 1.53$ cm/s.

As seen, the pitch of a helical symmetry has the same value but the different sign (i.e. the helical symmetry for the given flow varies step-wise from a left-hand to a right-hand in a radial direction). Satisfaction of a condition (2) with account for sign of pitch in each zone (Fig. 4, c) is amazing!

An example of change of a helical symmetry in an axial direction of a swirl flow is the phenomenon of a vortex breakdown (Okulov, 1996). The existence of the symmetry up to a zone of breakdown and after it follows from the fact that the measured velocity profiles (Garg and Leibovich, 1979) are well described by relations (14), which obey the symmetry condition (2). Unlike the practically step-like change of a symmetry in the transversal direction (Fig. 4, c), at the vortex breakdown a zone of transition from the right-hand vortex to the left-hand one (the zone of breakdown) is more extended and can have different forms (bubble, helical etc.).

The answer to a question whether is possible smooth transition from one symmetry to other was obtained in our experiments. We observed a regime of flow with a smooth transition from a right-handed vortex to a left-handed one (Fig. 4, d). In a zone of transition an air cavity, visualizing the axis of a vortex, is curved smoothly in such a manner that in a point of change of a symmetry tangent to an axis of a vortex is orthogonal to the chamber cross-section. The flow has mirror symmetry relative the given cross-section.

6. Conclusion

The supposition about two-dimensionality of swirl flows with a helical symmetry of a vorticity field has allowed to construct simple mathematical models which generalize the known models of vortices by Rankine, Lamb, Burgers etc. They describe swirl flows well for regimes without change of a helical symmetry. With their help it was possible to analyze also complex phenomena arising in swirl flows and related to transition from a right-hand helical symmetry to a left-hand one.

Acknowledgments

This work was partially supported by the grant of President of Russian Federation, by the joint foundation INTAS-RFBR (grant N 95-IN-RU-1149), by Russian Foundation of Basic Research (grants N 96-01-01667 and N 97-05-65254).

References

Alekseenko, S.V., Kuibin, P.A., Okulov, V.L., and Shtork, S.I. (in press) Helical vortices in swirl flow, submitted to the *J. Fluid Mech.*

Alekseenko, S.V. and Shtork, S.I. (in press) Running vortex breakdown, submitted to the *Tech. Phys. Lett.*

Brücker, Ch. and Althaus, W. (1992) Study of vortex breakdown by particle tracking velocimetry (PTV). Part 1: Bubble-type vortex breakdown, *Exps Fluids* **13**, 339-349.

Cherny, I.S. (1997) Study of swirl flows structure by the computational processing of videoimages, in N.V. Medvetskaya and R.S. Gromadskaya (eds.), *The Physics of Heat Transfer in Boiling and Condensation*, Institute for High Temperature, Russian Academy of Sciences, Moscow, pp. 515-517.

Dritschel, D. G. (1991) Generalized helical Beltrami flows in hydrodynamics and magnetohydrodynamics, *J. Fluid Mech.* **222**, 525-541.

Escudier, M.P., Bornstein, J., and Zehnder, N. (1980) Observations and LDA measurements of confined turbulent vortex flow, *J. Fluid Mech.* **98**, 49-63.

Garg, A.K. and Leibovich, S. (1979) Spectral characteristics of vortex breakdown flowfields, *Phys. Fluids* **22**, 2053-2064.

Hardin, J.S. (1982) The velocity field induced by a helical vortex filament, *Phys. Fluids* **25**, 1949-1952.

Hopfinger, E.J. and van Heijst, G.J.F. (1993) Vortices in rotating fluids, *Ann. Rev. Fluid Mech.* **25**, 241-289.

Okulov, V.L. (1995) The velocity field induced by helical vortex filaments with cylindrical or conic supporting surface, *Russian J. Engineering Thermophys.* **5**, 63-75.

Okulov, V.L. (1996) Transition from right to left helical symmetry during vortex breakdown, *Tech. Phys. Lett.* **22** (10), 798-800.

Scully, M.P. (1975) Computation of helicopter rotor wake geometry and its influence on rotor harmonic airloads, *Massachusets Inst. of Technology Pub.* ARSL TR 178-1, Cambridge.

Shtern, V., Borissov, A., and Hussain, F. (1997) Vortex sinks with axial flow: Solution and applications, *Phys. Fluids* **9** (10), 2591-2609.

Session 5

Vortex Breakdown

INSTABILITIES AND VORTEX BREAKDOWN IN SWIRLING JETS AND WAKES

Paul BILLANT, Jean-Marc CHOMAZ, Ivan DELBENDE,
Patrick HUERRE, Thomas LOISELEUX, Cornel OLENDRARU,
Maurice ROSSI and Antoine SELLIER
*LadHyX - Laboratoire d'Hydrodynamique, École Polytechnique,
F-91128 Palaiseau Cedex, France*

Abstract. The convective/absolute nature of the instability is determined for two distinct families of swirling wake/jet flows: the Batchelor vortex with external axial flow and the Rankine vortex with plug-flow velocity profile. It is demonstrated that, even in the absence of external axial counterflow, wakes and jets may exhibit a transition to absolute instability for sufficiently large swirl. Distinct transitional helical modes are found for wakes and jets and the transitional swirl value for the Rankine vortex model is consistent with observed values for vortex breakdown onset. An experimental study of vortex breakdown in swirling jets reveals the existence of four distinct breakdown states (bubble, cone, spiral bubble, spiral cone) above a critical swirl value $S_c \approx 1.3 - 1.4$. This threshold can be predicted from a simple inviscid criterion based on the existence of a stagnation point within the flow. Extensive hysteretic behaviour prevails close to the critical swirl value for breakdown onset.

1. Introduction

Swirling jets or wakes may be viewed as primary examples of configurations in which transverse shear and streamwise rotation are intimately coupled to give rise to a variety of flow phenomena: Kelvin-Helmholtz instabilities, Kelvin inertial waves, vortex breakdown, centrifugal instabilities, etc. The objective of the present paper is to give a synthetic account of recent theoretical [1-3] and experimental [4] investigations of instabilities and vortex breakdown in jet/wake flows.

On the theoretical side, we focus on the determination of the absolute/convective nature of the instability, as defined for instance in [5], for two families of basic flows: the Batchelor or q-vortex [1, 2] that is typical of fully developed swirling

E. Krause and K. Gersten (eds.), IUTAM Symposium on Dynamics of Slender Vortices, 267-286.

flows such as trailing vortices, and the Rankine vortex [3], which crudely models near field swirling jets or wakes. These notions can be viewed as a generalisation of the subcritical/supercritical flow concepts introduced by Benjamin [6] in the context of neutral nondispersive Kelvin waves, in order to account for the vortex breakdown phenomenon in swirling flows [7-10]. We briefly recall that in the framework of swirling flows an absolutely unstable (subcritical) basic state is capable of propagating information in both upstream and downstream directions via instability waves (respectively neutral Kelvin waves). By contrast a convectively unstable (supercritical) basic state only propagates information downstream via instability waves (respectively neutral Kelvin waves). According to Benjamin [6], vortex breakdown may be viewed by analogy with hydraulic jumps, as the transition region between an incoming supercritical flow and an outgoing subcritical flow. We wish to adopt an alternate point of view and investigate the possibility of a transition between a convectively unstable regime and an absolutely unstable regime as the swirl parameter is being varied.

On the experimental side, a study of vortex breakdown onset [4] in swirling liquid jets is presented, which documents the various states that are observed as a function of swirl parameter and Reynolds number. These experimental observations have been compared to a simple vortex breakdown criterion based on the existence of a stagnation point within the flow.

2. Absolute/convective instabilities in the Batchelor vortex

The Batchelor vortex [11] is frequently used as a convenient basic flow representation for trailing line vortices past airplanes. It is a typical example of a fully developed velocity profile with the following nondimensional axial, radial and azimuthal velocity components

$$U(r) = a + \mathrm{e}^{-r^2}, \quad V(r) = 0, \quad W(r) = q\,\frac{1-\mathrm{e}^{-r^2}}{r}, \tag{1}$$

where the control parameters

$$a \equiv \frac{U_\infty}{\Delta U}, \qquad q \equiv \frac{\Omega_c R}{\Delta U} \tag{2}$$

measure the relative magnitudes of the imposed axial external flow and swirl respectively. In (2), U_∞ is the free-stream axial velocity, ΔU is the axial velocity difference between the core flow and the external stream, Ω_c is the rotation rate prevailing on the axis $r = 0$ and R is a measure of the core size. Typical axial and azimuthal velocity distributions are displayed on Figure 1. The effect of varying the external flow parameter a is illustrated on Figures 1(a,b,c). Co-flowing jets correspond to $a > 0$ (Fig. 1a), counter-flowing jets or wakes to $-1 < a < 0$ (Fig. 1b), co-flowing wakes to $a < -1$ (Fig. 1c). As outlined below, two distinct methods have been used in order to determine the absolute/convective nature of the instability supported by the Batchelor vortex [11].

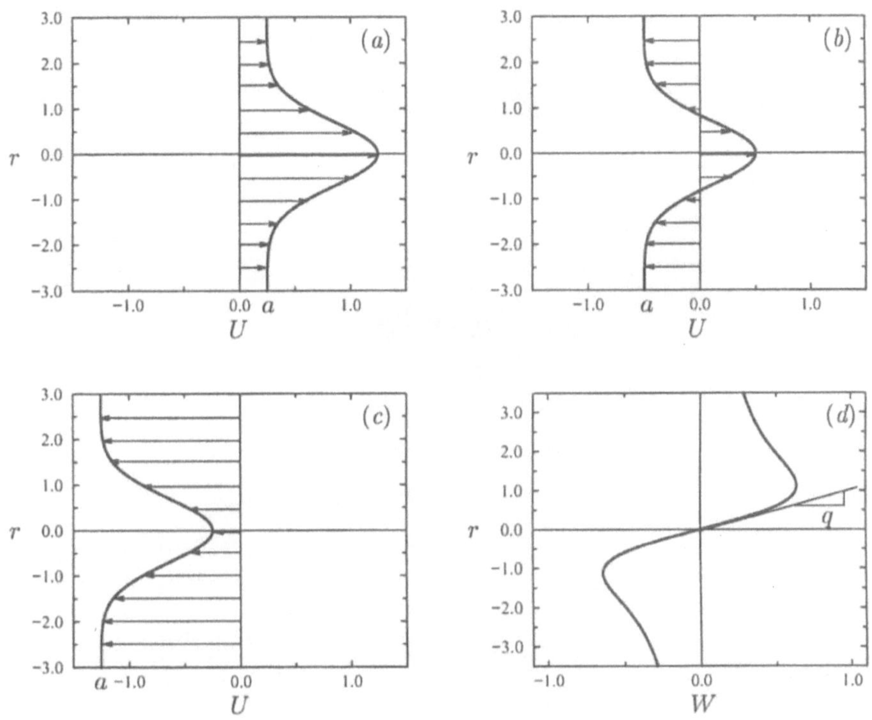

Figure 1. Axial velocity profiles: (a) $a > 0$ co-flowing jet, (b) $-1 < a < 0$ counter-flowing wake/jet, (c) $a < -1$ co-flowing wake. Azimuthal velocity profile (d).

2.1. NUMERICAL SIMULATION OF THE LINEAR IMPULSE RESPONSE.

In the first approach [1], the Navier-Stokes equations linearized around the basic flow (1), have been integrated numerically, subject to a spatially localized initial condition centered in the jet shear layer at $r_0 = 0.75$. Such a numerical simulation is tantamount to the study of the impulse response of the Batchelor vortex, provided that the initial perturbation is sufficiently limited in spatial extent. Typical calculated impulse responses are displayed in figures 2 and 3 for a non-rotating jet ($q = 0$) and a rotating jet ($q \neq 0$) respectively.

In the non-rotating case (Fig. 2), the jet response is dominated by counter-rotating helical modes $m = \pm 1$ that are of equal intensity, this feature being consistent with the fact that these are the only unstable modes. In the swirling case (Fig. 3), the impulse response is composed of a complex pattern of helical waves which rotate in a direction opposite to the basic flow. All the observable

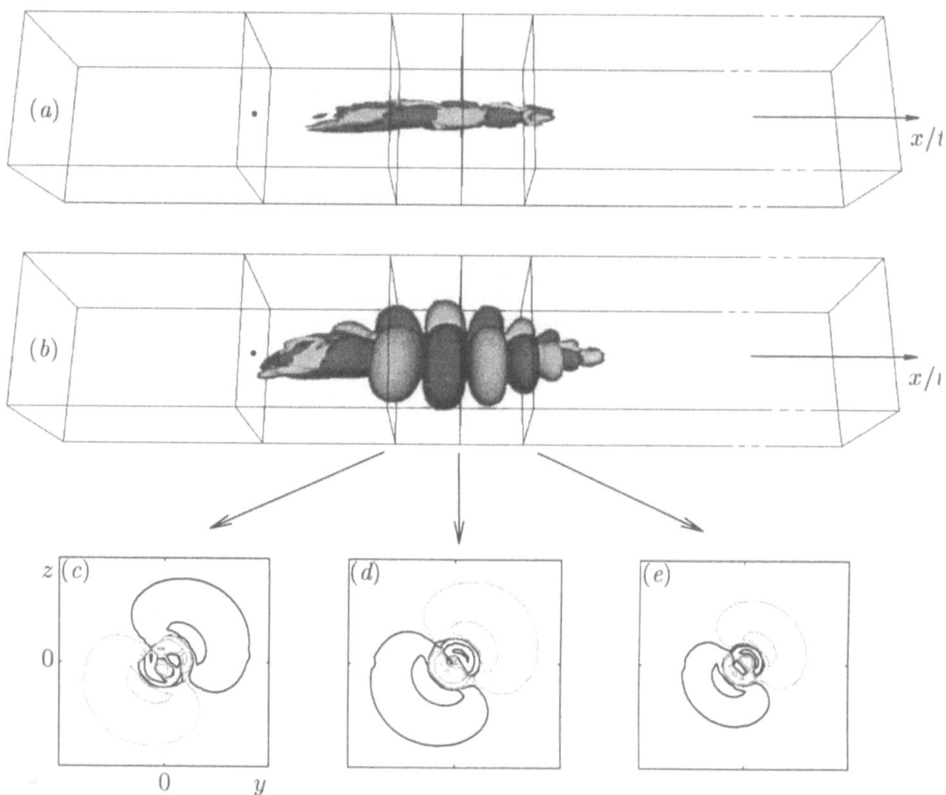

Figure 2. Wavepacket spatial distribution at $t = 32$ for a non rotating jet at $a = 0$, $q = 0$, $Re = 667$ [1]. Isosurfaces of axial perturbation velocity component u: (*a*) $u = \pm u^{\max}/20$ and (*b*) $u = \pm u^{\max}/1000$, where $u^{\max}$ is the maximum value of u in the entire domain. Isocontours of axial perturbation velocity component u at various cross-sections corresponding to distinct x/t stations: (*c*) at the wavepacket trailing-edge $x/t = 0.4$ for $u = \pm 10^{-4,-5,-6}$; (*d*) at the wavepacket maximum $x/t = 0.6$ for $u = \pm 10^{-3,-4,-5,-6}$; (*e*) at the wavepacket leading-edge $x/t = 0.8$ for $u = \pm 10^{-4,-5,-6}$. In these plots, dark regions or lines correspond to positive values of u, grey regions or lines to negative values of u.

characteristics of the linear dispersion relation

$$D[k, \omega; q, a, Re] = 0, \tag{3}$$

between axial wavenumber k and frequency ω may be retrieved from the computed impulse response at given values of the swirl q, external flow a and Reynolds number $Re = \Delta U\, R/\nu$, where ν is the kinematic viscosity. An appropriate Fourier decomposition procedure in the azimuthal and axial directions is applied to the

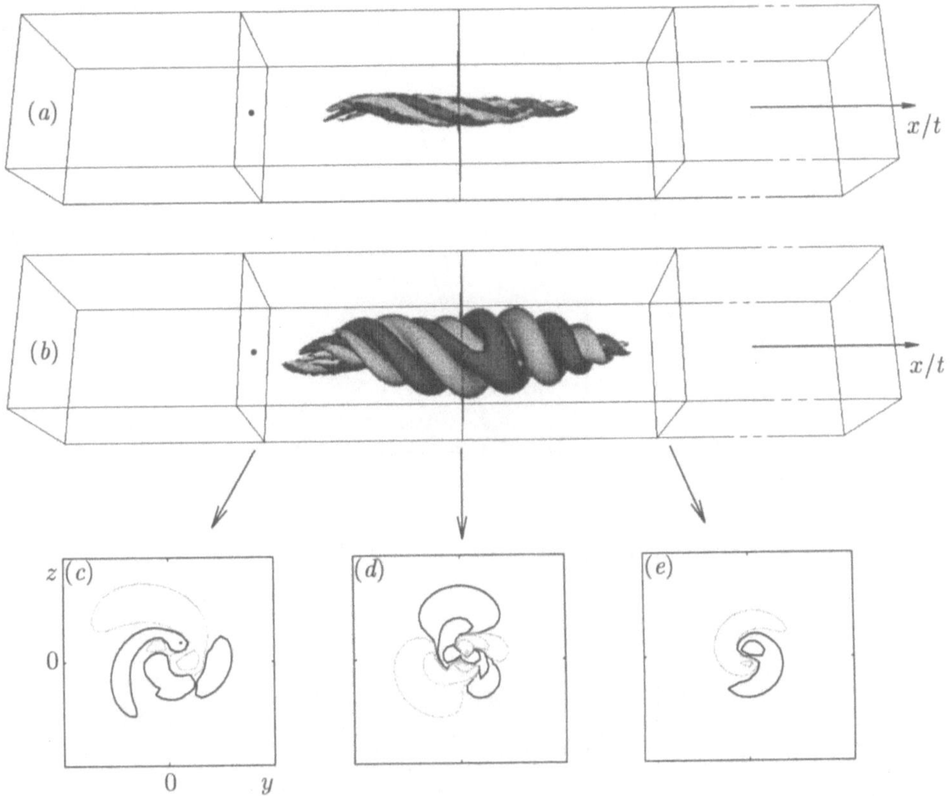

Figure 3. Wavepacket spatial distribution at $t = 32$ for a rotating jet at $a = 0$, $q = 0.4$, $Re = 667$ [1]. Isosurfaces of axial perturbation velocity component u: (a) $u = \pm u^{\max}/20$ and (b) $u = \pm u^{\max}/1000$, where $u^{\max}$ is the maximum value of u in the entire domain. Isocontours of axial perturbation velocity component u at various cross-sections corresponding to distinct x/t stations: (c) at the wavepacket trailing-edge near $x/t = 0$ for $u = \pm 10^{-5,-6}$; (d) at the wavepacket maximum $x/t = 0.6$ for $u = \pm 10^{-1,-2,-3}$; (e) at the wavepacket leading-edge $x/t = 0.12$ for $u = \pm 10^{-5,-6}$. In these plots, dark regions or lines correspond to positive values of u, grey regions or lines to negative values of u.

impulse response of Figs. 2 or 3, so as to determine the complex axial wavenumber $k_m(v_g)$ and complex frequency $\omega_m(v_g)$ observed along each spatio-temporal ray $x/t = v_g$ as $t \to \infty$. A typical amplitude distribution of the various helical modes $A_m(x,t)$ is illustrated on Figure 4(a) for the case of no external flow ($a = 0$) at $q = 0.8$ and $Re = 667$. By comparing such amplitude plots at different times, it is possible to determine the growth rate

$$\sigma_m(v_g; a) = \omega_{m,i}(v_g; a) - k_{m,i}(v_g; a) v_g, \tag{4}$$

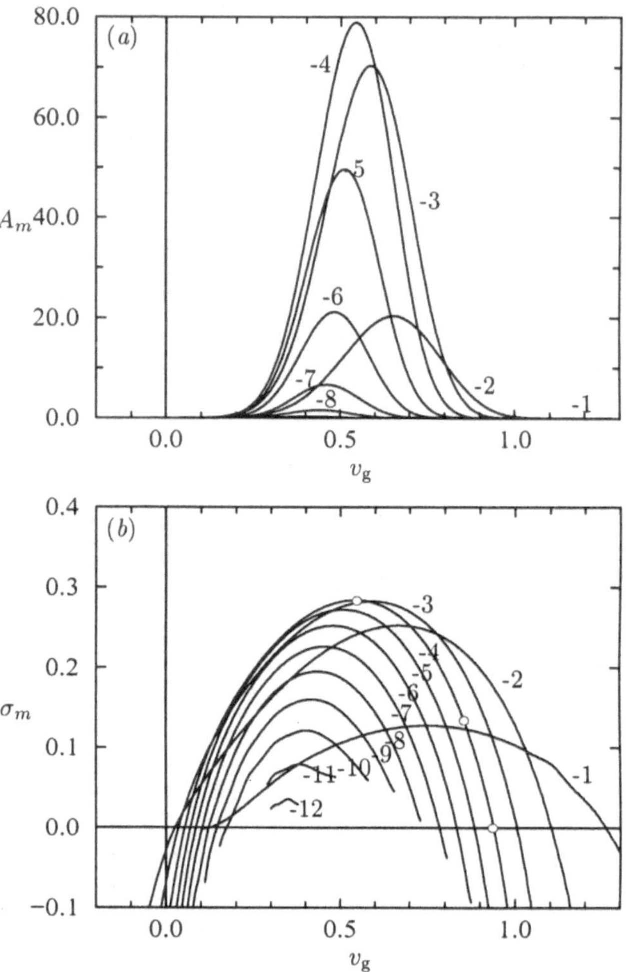

Figure 4. (*a*) Amplitude distribution $A_m(x,t)$ of dominant azimuthal modes $-8 \leq m \leq -1$ at $t = 32$ for $a = 0$, $q = 0.8$ and $Re = 667$ [1]. Linear scale. (*b*) Corresponding growth rate $\sigma_m(v_g)$ "observed" along each spatio-temporal ray $x/t = v_g$ for $-12 \leq m \leq -1$.

"observed" along a particular spatio-temporal ray $x/t = v_g$ for azimuthal mode m (The subscript "i" denotes the imaginary part). The growth rate curves $\sigma_m(v_g)$ displayed on Figure 4(*b*) contain all the essential information necessary to characterize the spatio-temporal behaviour of each azimuthal mode m. The axial extent of each wavepacket is limited by rays $v^+_{g,m}$ and $v^-_{g,m}$ such that $\sigma_m(v_g) = 0$. More importantly, to each helical mode m, one may associate an absolute growth rate $\omega_{0,m,\mathrm{i}} = \sigma_m(0)$ observed along the particular ray $x/t = v_{\mathrm{g}} = 0$, i.e. in the laboratory frame. From such plots, one may determine the maximum absolute growth rate $\omega_{0,\mathrm{i}}$ over all m. If $\omega_{0,\mathrm{i}} > 0$ ($\omega_{0,\mathrm{i}} < 0$), the flow is absolutely (convectively) unstable. For the parameter setting $a = 0$, $q = 0.8$ and $Re = 667$ of Fig. 4(b), the leading- and trailing-edges of the full wavepacket are seen to be associated with helical modes $m = -1$ and $m = -2$ respectively and they propagate with positive velocities v_{g}^{+} and v_{g}^{-}. The overall absolute growth rate $\omega_{0,\mathrm{i}}$ is negative and the flow is therefore convectively unstable.

A convenient feature of the method lies in its ability to conclude about the absolute/convective nature of the instability for arbitrary values of the external flow parameter a, solely from the knowledge of the results for zero external flow ($a = 0$): The growth rate $\sigma_m(v_{\mathrm{g}}; a)$ for *finite* external flow is in fact deduced from the previously computed growth rate $\sigma_m(v_{\mathrm{g}}; 0)$ for *zero* external flow by performing a simple translation of amount a along the v_{g}-axis of Fig. 4(b), according to the relation

$$\sigma_m(v_{\mathrm{g}}; a) = \sigma_m(v_{\mathrm{g}} - a; 0)\,. \tag{5}$$

This property greatly reduces the computations which only need to be carried out at $a = 0$ for different values of q.

Final results are represented in the a–q parameter plane on Figure 5 at $Re = 667$. The various absolute/convective transition curves $\omega_{0,m,\mathrm{i}} = 0$ have been represented for helical modes $m = 1$ and $m = -1, \cdots, -7$. The global AI/CI transition boundary is given by the outermost envelope of the individual transition curves (thick curve on Fig. 5).

In the no-swirl case ($q = 0$), AI is first triggered by the bending modes $m = \pm 1$ and it prevails in the external flow range $-0.80 < a < -0.39$, for wake/jet configurations with finite counterflow (Fig. 1b). The application of swirl q is seen to widen considerably the extent of the AI range towards $a = 0$ (zero-counterflow jet) and $a = -1$ (zero-counterflow wake).

On the wake side ($a < -0.5$), the transitional mode leading to AI remains $m = -1$ at all q. For swirl values as low as $q = 0.13$, wakes may undergo a transition to AI for $a < -1$, i.e. without the presence of axial counterflow (Fig. 1c).

On the jet side ($a > -0.5$), the transitional mode leading to AI varies in discrete steps as q increases: $m = -1$ to $m = -2, -3, -4, -5, -4, -3, -2, -1$. AI may be triggered for a counterflow as small as $a = -0.015$ (1.5% of the centerline axial velocity) when swirl reaches the value $q = 0.55 \pm 0.05$. Over an extended range of swirl $0.2 < q < 1$, only a slight amount of counterflow is necessary to induce AI onset.

Finally, note that large swirl values gradually dampen all helical modes until complete stabilization of the flow for $q \approx 1.5$.

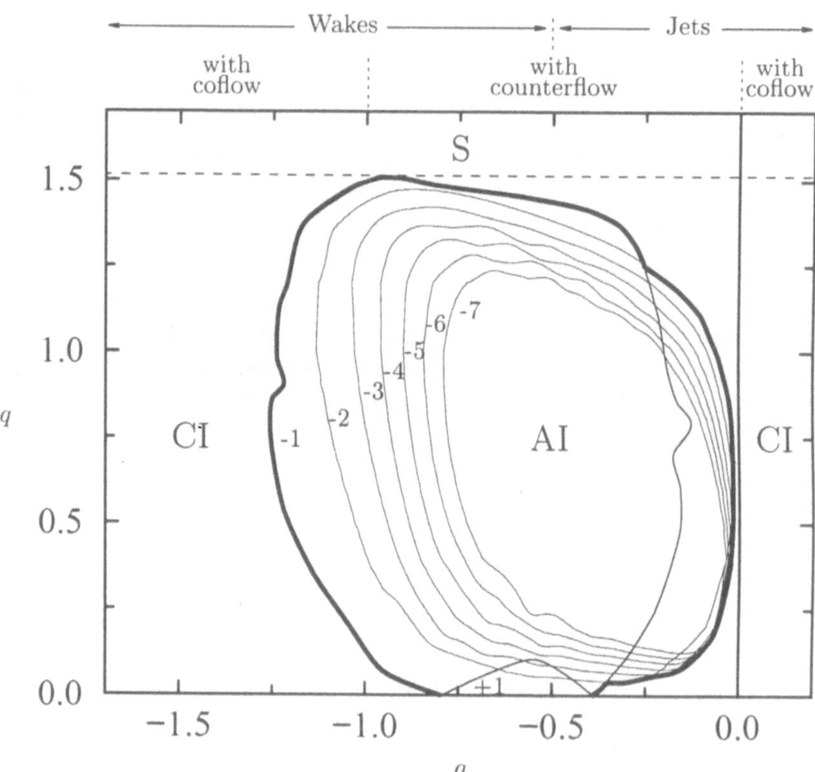

Figure 5. Regions of stability (S), convective instability (CI) and absolute instability (AI) in a–q parameter plane for the Batchelor vortex at $Re = 667$ [1]. Thin lines indicate AI/CI transition curves for each azimuthal mode $m = +1$ and $m = -1, \cdots, -7$. Bold line denotes outermost boundary of AI region.

2.2. ZERO-GROUP VELOCITY CRITERION

The AI/CI nature of the instability has also been approached directly [2] by investigating the properties of the inviscid dispersion relation determined from the Howard-Gupta [12] equation subject to appropriate boundary conditions on the axis $r = 0$ and at $r = \infty$. The numerical solution of this eigenvalue problem leads to relation (3) at $Re = \infty$, to which one may apply the Briggs-Bers [13, 14] zero-group velocity criterion

$$\frac{\partial \omega}{\partial k}(k_{0,m}; m, q, a) = 0\,, \qquad \omega_{0,m} = \omega(k_{0,m}; m, q, a)\,, \tag{6}$$

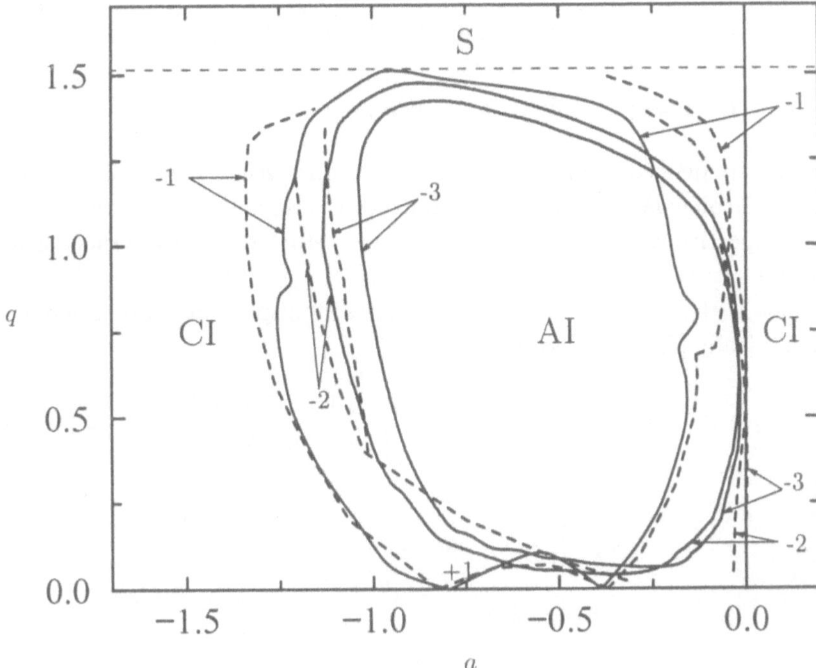

Figure 6. AI/CI transition curves in a–q parameter plane for azimuthal modes $m = \pm 1, -2, -3$. Solid lines: study of Delbende *et al.* [1] at $Re = 667$. Dashed lines: results of Olendraru *et al.* [2] obtained by application of the zero-group velocity criterion to the inviscid dispersion relation of the Batchelor vortex.

in order to calculate the absolute wavenumber $k_{0,m}$ and frequency $\omega_{0,m}$. AI/CI transition curves obtained via each method are overlaid on Figure 6. Satisfactory agreement prevails over a wide domain in the a–q plane. The direct numerical simulation procedure of the previous section yields a slightly smaller domain of absolute instability. The inviscid dispersion relation is seen to predict the occurence of AI even on the co-flowing jet side $a > 0$. However viscous effects push the AI/CI transition boundary towards the counter-flow side $a < 0$.

3. Absolute/convective instabilities in the Rankine vortex with axial flow

An idealized representation of the flow conditions prevailing in the very near field of swirling jets is given by the Rankine vortex with superimposed plug-like axial velocity profile. In this case, the axial, radial and azimuthal velocity components read, in nondimensional variables:

$$
\begin{aligned}
U(r) &= a+1, \quad V(r)=0, \quad W(r)=qr \quad \text{if } r<1,\\
U(r) &= a, \qquad\; V(r)=0, \quad W(r)=\frac{q}{r} \quad \text{if } r>1,
\end{aligned}
\tag{7}
$$

where a and q are defined as in (2). Note that the above velocity distributions for $r<1$ and $r>1$ coincide with those of the Batchelor vortex (1) in the limits $r \ll 1$ and $r \gg 1$, respectively. The effect of varying the external flow a is the same as in the case of the Batchelor vortex, see Figs. 1(a,b,c).

In contrast with the Batchelor vortex, the inviscid dispersion relation may be derived analytically [3] in the form:

$$
D[k,\omega;q,a,m] \equiv (\omega_j+k)^2[-2mq+\omega_j\beta\frac{J'_m(\beta)}{J_m(\beta)}] + \frac{\beta^2\omega_j^3}{sk}\frac{K'_m(sk)}{K_m(sk)} = 0\,, \tag{8}
$$

where

$$
\omega_j \equiv \omega - mq - (1+a)k\,, \tag{9}
$$

$$
\beta \equiv k\sqrt{\frac{4q^2-\omega_j^2}{\omega_j^2}}\,, \tag{10}
$$

$s \equiv sgn(k_r)$, and J_m,K_m denote the Bessel functions of the first kind and modified Bessel function of the second kind respectively.

In the absence of external flow ($a=0$), one recovers the result of Krishnamoorthy[15]. If in addition $q=0$, the dispersion relation for plug-flow jets[16] is obtained.

The absolute/convective nature of the instability has been determined in the a–q plane by numerical implementation of the Briggs-Bers criterion stated in (6). Overall results are displayed in Figure 7. As in the case of the Batchelor vortex, only negative helical modes are involved in defining the AI/CI transition boundary. The magnitude of the transitional mode index increases continuously with swirl q on both jet ($a>-0.5$) and wake ($a<-0.5$) sides. As in Fig. 5 for the Batchelor vortex, the a-interval of AI widens very significantly as q increases.

In the no-swirl case ($q=0$), AI is first triggered by the bending mode $m=\pm1$ on the wake side and by the axisymmetric mode $m=0$ on the jet side. Note the difference with the fully developed jet in the absence of swirl (see Fig. 5 for $q=0$) which cannot sustain an unstable axisymmetric mode $m=0$ and where the transitional mode is necessarily $m=\pm1$.

On the wake side ($a<-0.5$), higher-order negative helical modes gradually take over to define the transition curve for large q. AI takes place for co-flowing wakes ($a<-1$), as soon as swirl exceeds the value $q=0.47$.

On the jet side ($a>-0.5$), the AI/CI boundary takes the shape of overlapping tongues pertaining to distinct negative helical modes. As q increases, the

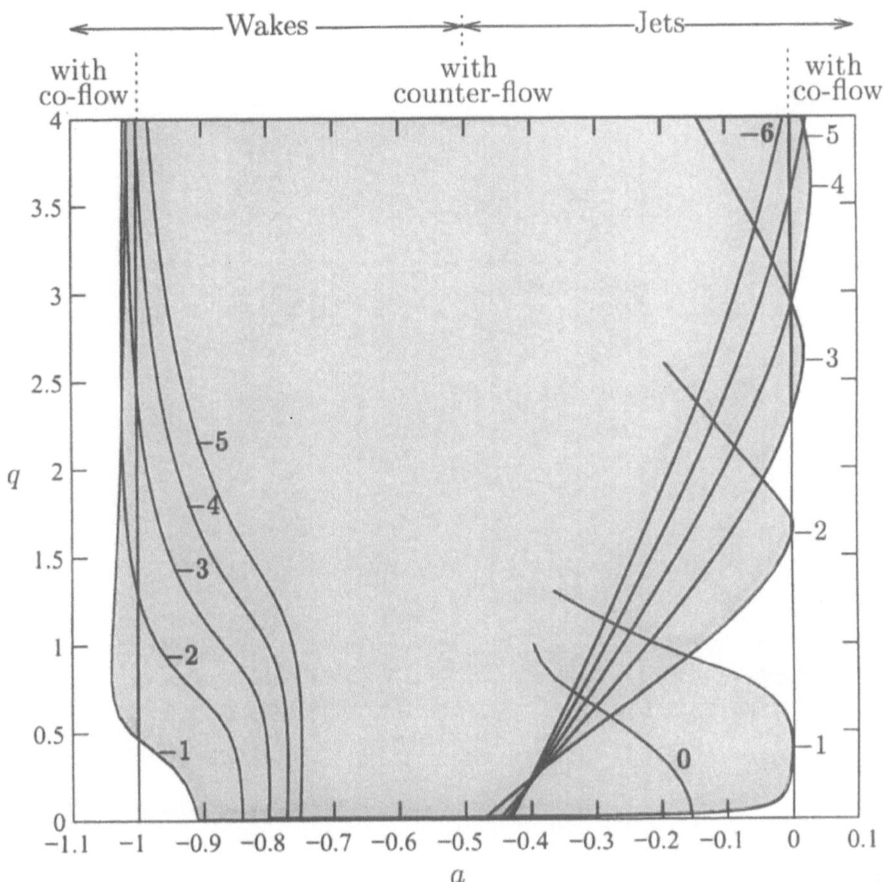

Figure 7. Absolute/convective instability domains in $a-q$ parameter plane for jets and wakes [3]. Shaded (clear) areas denote absolute (convective) instability.

transitional mode $m = 0$ is almost immediately superseded by the bending mode $m = -1$. At $q = 0.43$, the tip of the $m = -1$ tongue barely fails to touch the zero external flow axis. The $m = -2$ tongue penetrates into the co-flowing jet-side $a > 0$ in the range $1.61 < q < 1.71$, thereby first leading to AI. Above $q = 2.30$, weak co-flow jets may exhibit AI for successively higher order helical modes $m = -3$, -4, etc.

In contrast with the Batchelor vortex, the Rankine vortex remains unstable for large swirl parameter values and the AI region stays open on the high-q side. In this context, it is important to bear in mind that the plug flow axial velocity

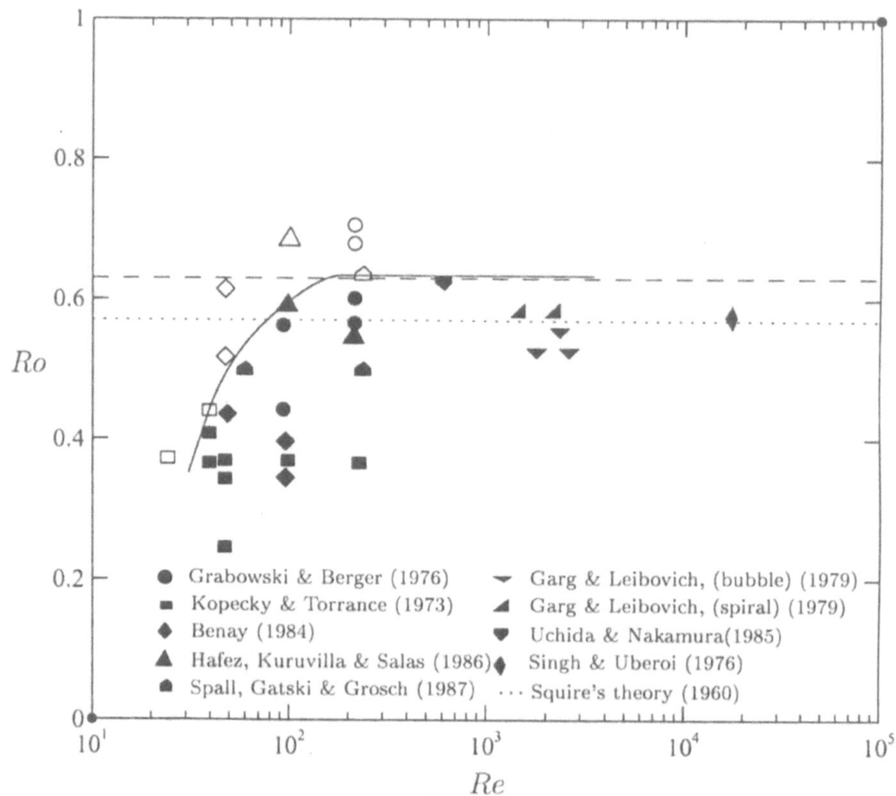

Figure 8. Observed states in Reynolds number - Rossby number plane for various numerical and experimental studies of swirling flows and trailing wing-tip vortices. The open symbols denote no breakdown and the solid symbols breakdown. After Spall, Gatski & Grosch[17]. Solid line indicates border between breakdown and no breakdown. Dashed line (- - -) corresponds to CI/AI transition for Rankine vortex with plug flow and zero external axial velocity ($a = 0$) predicted in [3].

profile is unstable at arbitrarily high wavenumbers. Causality cannot therefore be enforced. As a consequence, the present results are only relevant as long as the length scales of interest are much larger than the typical thickness of the shear layer separating the core axial flow from the external flow.

The analysis predicts that zero axial external flow jets ($a = 0$) exhibit a transition to AI of the $m = -2$ helical mode as soon as swirl reaches the value $q = 1.61$ (see Fig. 7). Various experimental and numerical results on vortex breakdown onset, reproduced from [17], have been gathered on a state diagram in the Reynolds number (Re) - Rossby number (Ro) plane of Figure 8. The critical CI/AI transi-

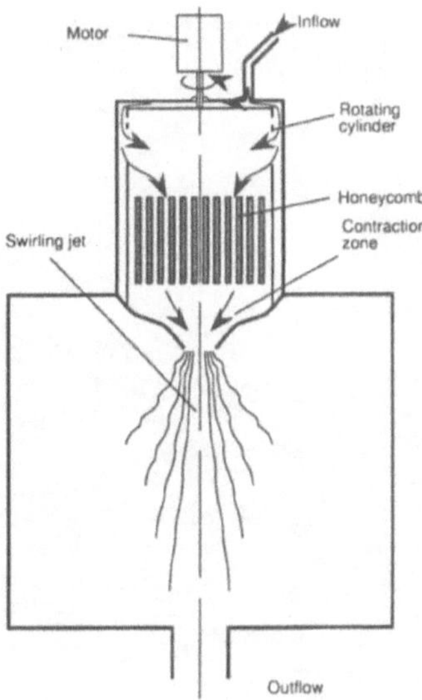

Figure 9. Sketch of experimental apparatus.

tion curve $Ro \equiv q^{-1} = 0.62$ corresponding to $q = 1.61$ is indicated by a dashed horizontal line. In spite of the relative simplicity of the Rankine vortex model, the AI/CI "criterion" is seen to provide a reasonable estimate of vortex breakdown onset that is frequently assumed to occur above the empirically determined value $Ro = 0.65$.

4. Experimental study of vortex breakdown in swirling jets

Concurrently with the preceding theoretical analyses, the various breakdown states sustained by a swirling water jet have been investigated as the swirl ratio S and Reynolds number Re are varied [4]. The experimental set–up is sketched in Figure 9: a gravity-driven water jet discharges into a large tank, swirl being imparted by means of a motor and two concentric cylinders. The inner cylinder, which is 40 cm long and 18.5 cm in diameter, is set into rotational motion by a motor. The flow is kept into a state of solid body rotation by inserting an 18 cm long honeycomb into the inner rotating cylinder. The swirling jet flow discharges from a smooth converging nozzle attached to the outer cylinder into a large tank of 120×40×40 cm^3 square cross section. Two nozzle exit diameters have been used: $D_1 = 40$ mm and $D_2 = 25$ mm. The swirl parameter and Reynolds number are defined as

$$S \equiv \frac{2W(R/2,x_0)}{U(0,x_0)}, \qquad Re \equiv \frac{2R\bar{U}(x_0)}{\nu}, \tag{11}$$

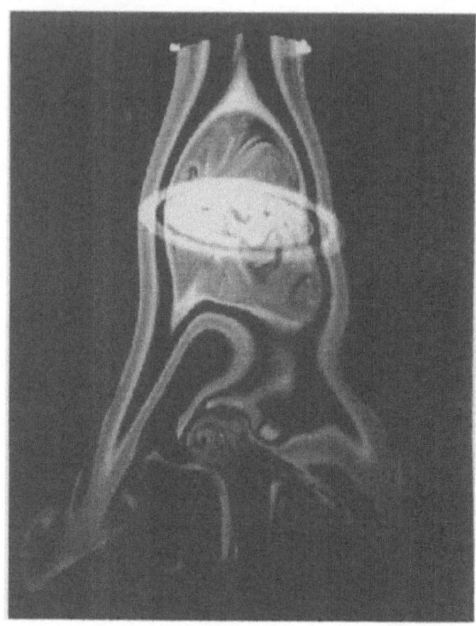

Figure 10. Flow visualisations of bubble for $Re = 606$, $S = 1.42$, D_2. Two laser sheets are simultaneously produced in a meridional plane and in a slanted plane from the horizontal [4].

where R is the nozzle exit radius, x_0 is a short axial distance from the nozzle exit plane and $\bar{U}(x_0)$ is the mean axial velocity in the jet. The swirl parameter S is approximately the same as the q parameter introduced in (2), provided that there is no external axial flow ($a = 0$). The experiments are conducted by varying the swirl ratio S while maintaining the Reynolds number Re fixed in the range $300 < Re < 1200$. Breakdown, characterized by the presence of a stagnation point within the flow, is observed to take place when S reaches a well defined threshold $Sc \approx 1.3 - 1.4$, which is independent of Re and nozzle diameter used (Figure 12).

4.1. BREAKDOWN STATES

Four distinct breakdown states have been identified. The two basic steady configurations are the bubble (Figure 10) and the cone (Figure 11). The bubble (Fig. 10) has already been well documented in many confined tube experiments [18, 19]: it is characterized by an abrupt axisymmetric expansion of the vortex at a given streamwise station, which encloses an ovoid region of slow recirculating flow. The downstream wake of the bubble is unsteady and non axisymmetric. The cone state (Fig. 11) takes the form of a conical sheet flowing over an open region of nearly stagnant fluid. Shear instabilities are observed to grow on the sheet which breaks

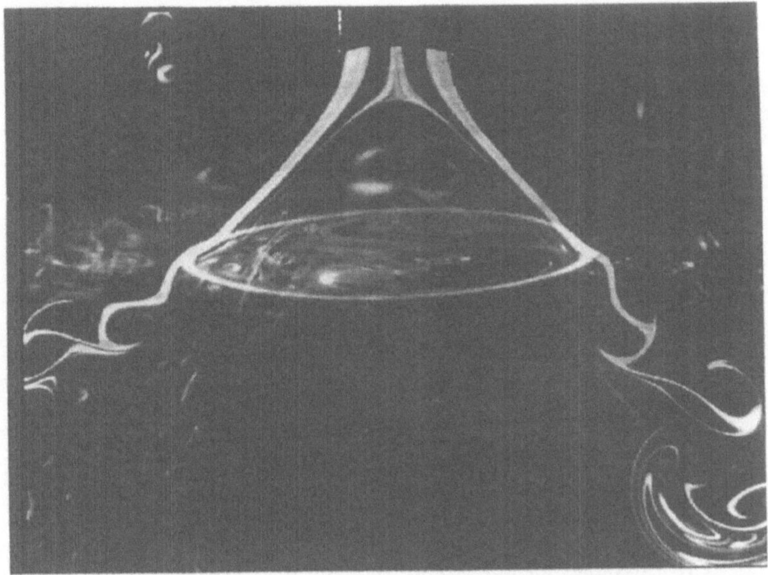

Figure 11. Flow visualisation of cone for $Re = 606$, $S = 1.37$, D_2 [4].

up into vortices further downstream.

At larger Reynolds numbers, two related spiral bubble and spiral cone states have been observed, that are qualitatively similar to their steady relatives, except for a precessing motion of the stagnation point around the axis in the same direction as the upstream vortex flow.

A tentative flow state diagram is given in the Re–S control parameter plane on Figure 12. The symbols Sc_a and Sc_d denote the threshold values for appearance and disappearance of breakdown as the swirl is increased and decreased respectively. As previously stated, breakdown is seen to take place when S exceeds a critical value of the order of 1.3-1.4, whatever the value of the Reynolds number.

4.2. VORTEX BREAKDOWN CRITERION

The critical value $Sc = 1.3$–1.4 for breakdown is seen to be reasonably close to the CI/AI transition value $q = 1.61$ for the Rankine vortex with no external flow ($a = 0$). Alternatively, one may derive a criterion for the existence of a stagnation point in a flow of infinite extent that is based on elementary considerations. Following [4], consider a free vortex with a stagnation point and opening into a cone (Figure 13). According to the Bernoulli equation applied along the vortex axis

$$P_0/\rho + U^2(0, x_0)/2 = P_1/\rho, \qquad (12)$$

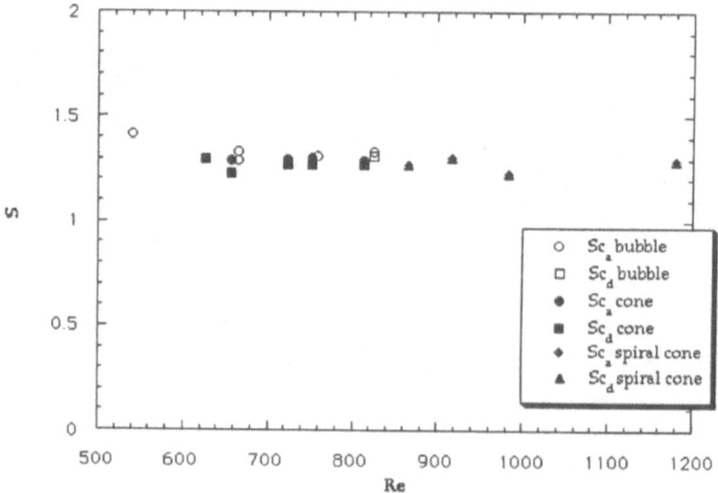

Figure 12. Critical values for appearance Sc_a and disappearance Sc_d of breakdown in (Re, S) parameter space for the D_1 nozzle [4].

where P_0 is the pressure on the vortex axis well upstream of the stagnation point at station x_0, ρ is the density, $U(0, x_0)$ is the upstream axial velocity on the vortex axis at x_0 and P_1 the pressure at the stagnation point. Far upstream of the stagnation point, the radial pressure gradient is balanced by the centrifugal force, so that

$$P_0 - P_\infty = -\int_0^\infty \rho \frac{W(r, x_0)^2}{r} dr \,, \tag{13}$$

where P_∞ is the ambient pressure at infinity in the plane $x = x_0$.

If there exists a path (line with arrows on Fig. 13) around the cone which links a point at rest at infinity to the stagnation point without encountering vorticity, then $P_1 = P_\infty$. By eliminating the pressure difference $P_0 - P_\infty$ from (12) and (13), one then arrives at the simple relation

$$\frac{\displaystyle\int_0^\infty \frac{W^2(r, x_0)}{r} dr}{U^2(0, x_0)} = \frac{1}{2} \,. \tag{14}$$

In the particular case of the Rankine vortex with no axial external flow ($a = 0$) the dimensional azimuthal velocity given in (7) reads:

$$W(r, x_0) = \Omega_c r, \quad r < R; \qquad W(r, x_0) = \frac{\Omega_c R^2}{r}, \quad r > R. \tag{15}$$

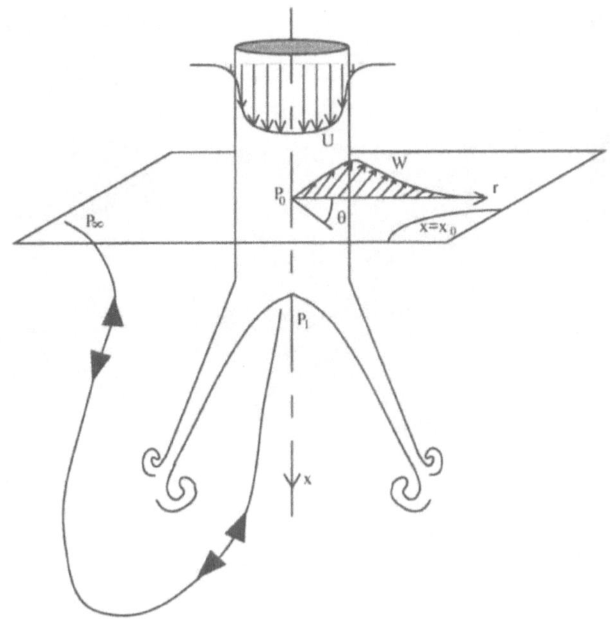

Figure 13. Schematic configuration of cone vortex breakdown.

Upon substituting (15) into (14), one readily obtains the simple breakdown criterion:

$$\frac{W_{\max}(x_0)}{U(0,x_0)} = \frac{1}{\sqrt{2}}, \tag{16}$$

where $W_{\max}(x_0) = \Omega_c R$ is the maximum azimuthal velocity. Bearing in mind that $W(R/2,x_0) \approx W_{\max}(x_0)$, the predicted critical value of the swirl, as defined in (11), giving rise to vortex breakdown is therefore:

$$S_c \equiv \frac{2W_(R/2,x_0)}{U(0,x_0)} = \sqrt{2} = 1.41\,. \tag{17}$$

This prediction is remarkably close to the experimental value $S_c = 1.3 - 1.4$, and significantly closer than the CI/AI transition value $q = 1.61$ obtained for the Rankine vortex with no axial external flow [1]

[1]As discussed in [4], different operational definitions of the swirl parameter S may be used which lead to more or less satisfactory agreement between experiments and theory.

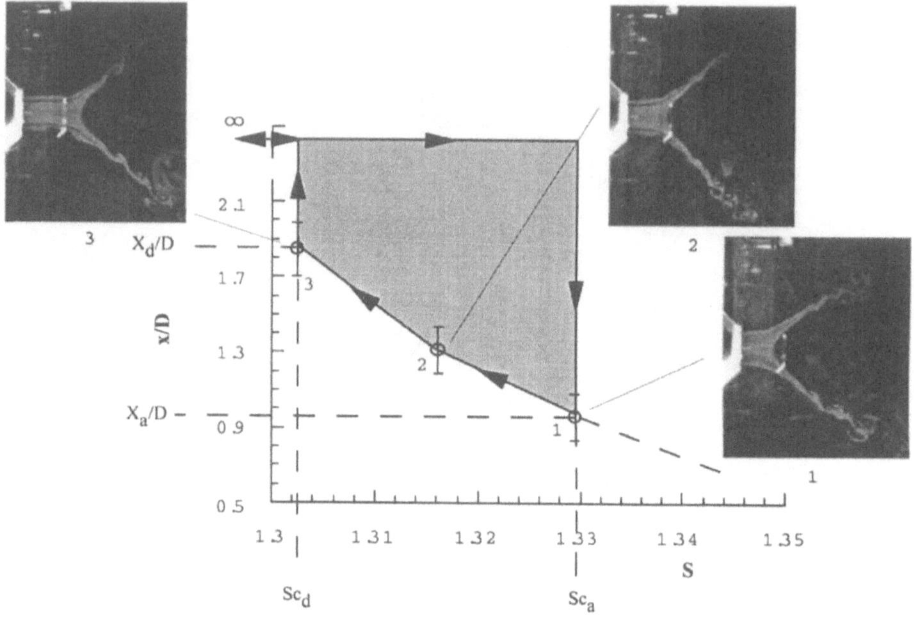

Figure 14. Hysteresis loop of cone in $(S, x/D)$ plane. x/D denotes scaled axial coordinate of stagnation point; $Re = 752$, D_1 [4].

4.3. HYSTERESIS PHENOMENA AND SENSIVITY TO BUOYANCY EFFECTS.

Hysteretic behaviour at the threshold of breakdown has been established by following the changes in the stagnation point location x/D as the swirl parameter S is increased and then decreased. An example involving the cone is displayed on Figure 14: when S is increased, breakdown arises at $S = Sc_a$, $X = X_a$. As S is increased beyond Sc_a, x/D decreases, i.e. the stagnation point moves upstream. Conversely, when S is decreased, the stagnation point moves downstream and still persists as S crosses Sc_a. Breakdown only disappears when S decreases down to $Sc_d < Sc_a$, i.e. when the stagnation point moves beyond X_d. Thus the no-breakdown and cone states coexist within the range $Sc_d < S < Sc_a$. A similar coexistence occurs for the bubble type breakdown.

Numerous experimental observations further support the idea that small temperature inhomogeneities, as small as 0.1° C, have a determining influence on the kind of breakdown that is observed. Such tiny temperature differences induce buoyancy forces which are large enough to select a cone or bubble state at the same

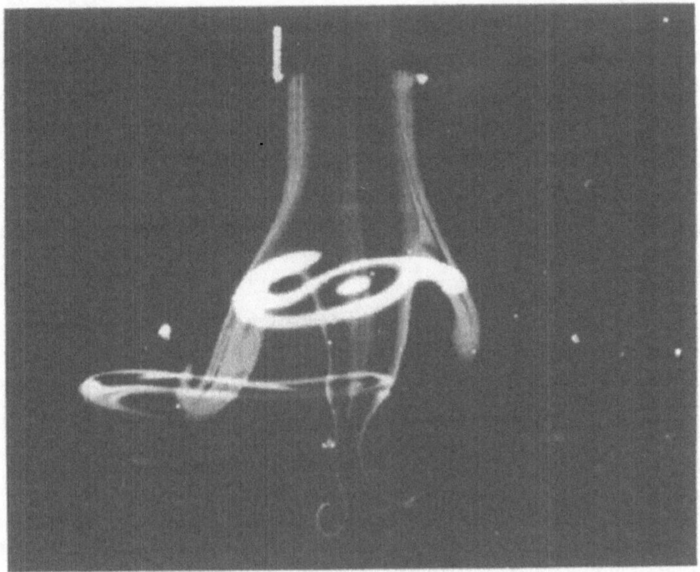

Figure 15. Flow visualisation of helix configuration for $Re = 606$, $S = 1.41$, D_2. Two laser sheets are simultaneously produced in a meridional plane and in a slanted plane from the horizontal [4].

Reynolds number and swirl parameter values. In other words, a reliable state diagram cannot be generated and the example shown on Fig. 12 is purely illustrative of the results that are obtained for a particular case. Bubble and cone are in fact observed at the same nominal flow conditions by applying suitable disturbances or modifying the temperature settings.

4.4. PRECURSOR HELIX STATE

As the rotation rate increases from zero towards Sc_a, helical perturbations distort the swirling jet into a steady helix configuration, as shown on Figure 15. This helical shape, which is dominated by the $m = +2$ mode is present in all experimental runs below onset of breakdown. This feature is in qualitative agreement with the theoretical CI/AI transitional mode encountered in the Rankine vortex model with zero axial external flow ($a = 0$) as q increases below 1.61. More detailed calculations on measured velocity profiles need to be undertaken in order to confirm this relationship.

5. Concluding remarks

The theoretical investigations presented in sections 2-3 have demonstrated that sufficiently high swirl promotes the onset of absolute instability in jet and wake flows without requiring the presence of counterflow. The swirling jet experiments have established that vortex breakdown takes place in the form of a bubble or cone above a critical swirl parameter value that is in good agreement with a simple criterion based on the existence of a stagnation point within the flow. More detailed analysis of the instability properties of *measured* velocity profiles need to be undertaken in order to demonstrate the relationship, if any, between CI/AI transition and vortex breakdown onset.

This research has been supported by the French Ministry of Defense under DRET Grant #92-098.

References

1. Delbende, I., Chomaz J.-M. & Huerre P. Absolute/convective instabilities in the Batchelor vortex: a numerical study of the linear impulse response. *J. Fluid Mech.* (in press).
2. Olendraru, C., Sellier, A., Rossi, M. & Huerre, P. Absolute/convective instability of the Batchelor vortex. *C. R. Acad. Sci. Paris* **323** (1996), IIb, 153–159.
3. Loiseleux, T., Chomaz J.-M. & Huerre P. The effect of swirl on jets and wakes: linear instability of the Rankine vortex with axial flow. *Phys. Fluids* (in press).
4. Billant, P., Chomaz J.-M. & Huerre P. Experimental study of vortex breakdown in swirling jets. Submitted to *J. Fluid Mech.*
5. Huerre, P. & Monkewitz, P. A. Local and global instabilities in spatially developing flows. *Annu. Rev. Fluid Mech.* **22** (1990), 473–537.
6. Benjamin, T. B. Theory of the vortex breakdown phenomenon. *J. Fluid Mech.* **14** (1962), 529–551.
7. Hall, M.G. Vortex breakdown. *Annu. Rev. Fluid Mech.* **4** (1972), 195–217.
8. Leibovich, S. The structure of vortex breakdown. *Annu. Rev. Fluid Mech.* **10** (1978), 221–246.
9. Escudier, M. P. Vortex breakdown: observations and explanations. *Prog. Aerospace Sci.* **25** (1988), 189-229.
10. Althaus, W., Brücker, C. & Weimer, M. Breakdown of slender vortices. In: Green S. (ed) *Fluid Vortices*, Kluwer Academic Pub., Norwell, MA, 373–426, 1995.
11. Batchelor, G. K. Axial flow in trailing line vortices. *J. Fluid Mech.* **20** (1964), 645–658.
12. Howard, L. N. & Gupta, A. S. On the hydrodynamic and hydromagnetic stability of swirling flows. *J. Fluid Mech.* **14** (1962), 463–476.
13. Briggs, R.J. Electron-Stream Interaction with Plasmas. MIT Press, Cambridge, Mass, 1964.
14. Bers, A. Space-time evolution of plasma instabilities -absolute and convective. In: Rosenbluth M.N. & Sagdeev R.Z. (eds) *Handbook of Plasma Physics*, North-Holland, Amsterdam, 451–517, 1983.
15. Krishnamoorthy V. Vortex breakdown and measurements of pressure fluctuations over slender wings. Ph.D thesis, Southampton University, 1966.
16. Batchelor G. K. & Gill A. E. Analysis of the stability of axisymmetric jets. *J. Fluid Mech.* **14**, (1962) 529–551.
17. Spall, R.E., Gatsky, T.B. & Grosch C.E. A criterion for vortex breakdown. *Phys. Fluids.* **30** (1987), 3434–3440.
18. Sarpkaya, T. On stationary and travelling vortex breakdowns. *J. Fluid Mech.* **45** (1971), 545–559.
19. Faler, J.H. & Leibovich, S. Disrupted states of vortex flow and vortex breakdown. *Phys. Fluids* **20** (1977), 1385–1400.

TURBULENT VORTEX BREAKDOWN: EXPERIMENTS IN TUBES AT HIGH REYNOLDS NUMBERS

T. SARPKAYA and F. NOVAK

Naval Postgraduate School
Monterey, CA 93943 U. S. A.

1. Introduction

Trailing vortices, swirling flows in pipes, vortical flows above sweptback wings at large angles-of-attack, flows in closed containers with a rotating lid, and columnar vortices in atmosphere may experience breakdown: *The transformation of a slender vortex into three-dimensional forms.* Where, how, and under what circumstances does this transformation occur in *viscous* vortical flows constitute the essence of the breakdown problem. Neither a stagnation point, nor a region of reversed flow, nor the bridging of laminar–turbulent states is necessary.

The difficulties experienced in describing the nature, identifying the occurrence, and predicting the characteristics of *laminar* vortex breakdowns in tubes and over delta wings have been reviewed by Althaus, Brücker, and Weimer (1995) and will not be repeated here. It suffices to note that after forty years of observations, measurements, and numerical experiments, the phenomenon remains largely in the qualitative, descriptive realm of knowledge. There are neither exact solutions nor universally accepted theoretical models which capture the essential physics, weave 'understanding' into large amounts of numerical, experimental, and observational records, and offer methods of prediction. Theories based on the inviscid-flow assumption have become a rival faith to physical and numerical experiments.

Vortex breakdown in most technological applications (e.g., combustion, aerodynamics) occurs in turbulent swirling flows at high $\gamma = \Gamma/\nu$ and $Re^* = 2U_oR_c/\nu$ (where Γ is the circulation, U_o is a characteristic velocity, R_c is the core radius, and ν is the kinematic viscosity), and the applicability of the information deduced from laminar flow studies to the prediction of breakdowns in turbulent flows is highly questionable. Even though the literature is abound with numerical and experimental studies of swirling turbulent flows (see, e.g., Rhode, Lilley, and McLaughlin, 1982), most works are dedicated either to the development of simple criteria for the identification or position of the vortex breakdown over delta wings or to the improvement of the efficiency of combustion. None of these studies attempted to delineate the topology of the vortex

E. Krause and K. Gersten (eds.), IUTAM Symposium on Dynamics of Slender Vortices, 287-296.

breakdown, an ingredient necessary to the enhancement or to the alleviation of the consequences of vortex breakdown.

Sarpkaya (1995a, 1995b) was the first to show that vortex breakdown in non-cavitating swirling flows in tubes at high Reynolds numbers (Re* up to about 13,000) is significantly different from the other well-known types (basically, double helix, bubble and spiral). The present experiments have shown that at sufficiently high values of γ and Re*, the core initiates a kink, followed by a single spiral, at a point dictated by the prevailing flow and boundary conditions. The nascent spiral adjacent to the kink bursts into turbulence immediately after its inception, while rotating rapidly (at about 500 rev/s), expending gradually, and merging rapidly in a very short distance (within about 10 to 20 mm) with the turbulent flow field downstream. At higher values of γ and Re*, the core first bifurcates into two cores. The two spirals wind in the same sense but in the opposite direction to their surroundings. The remainder of the breakdown (beyond about the first 10 to 20 mm) transforms into a nominally axisymmetric cone of swirling turbulent flow, as seen in photographs taken with exposure times as small as 6 nano-seconds. At still higher values of γ and Re* (about 13,000), one observes only an intense twisting of the core and/or its bifurcations and a highly turbulent conical flow field. The 'bubble' and the 'core-recovery' regions, seen in the usual path of structural transformations at low Reynolds numbers (Sarpkaya, 1971), are completely *bypassed* at sufficiently high values of γ and Re*.

Délery and Molton (1994) examined the topology of the flow resulting from the breakdown over a delta wing at a chord-based Reynolds number of Re = 1.46×10^6 and concluded, on the basis of their velocity measurements, that there is "a reversed flow region, of relatively small extent, which rapidly contracts and disappears, the streamwise mean velocity component becoming everywhere positive again at a short distance from the breakdown point." These experiments were confined to relatively lower γ and Re* values and the turbulence characteristics of the flow field were not investigated.

The numerical studies by Krause (1990); Spall and Gatski (1991); Breuer and Hänel (1993), among others, dealt with axisymmetric or three-dimensional *laminar* flows at relatively low Reynolds numbers, partly because the experimental information for comparison and code tuning was available only at comparable Reynolds numbers and partly because there has not yet been a turbulence model capable of dealing with non-isotropic turbulence in swirling flows, strongly influenced by streamline curvature and centrifugal acceleration. The available modeling approaches (DNS, LES, and Reynolds- or Favre-averaged Navier-Stokes equations: RANS or FANS) have their limitations and it is rather unlikely that standard turbulence models will help to predict the existence of large scale structures in turbulent breakdowns observed in technologically important applications. Spall and Gatski (1995) used the three-dimensional, unsteady, incompressible RANS equations in conjunction with two different turbulence models to simulate a 3-D turbulent breakdown at a Reynolds number of 10^4 (based on vortex core radius) in an unconfined longitudinal vortex, unimpeded by competing and complicating influences such as combustion (i.e., Hogg and Leschziner, 1989). However, the calculated flow field was not in conformity with the observations of Sarpkaya (1995b).

It is evident from the foregoing that one needs archival-quality physical and numerical data and physics-based phenomenological models to sort out the dominant factors for the 'reason d'etre' of the type of vortex breakdown observed at high Reynolds and circulation numbers of technological importance. This process is somewhat complicated by the well-known fact that both the real and simulated breakdowns are highly sensitive to all boundary conditions (not just the upstream conditions). It is, therefore, of paramount importance that the position of the breakdown be rendered as insensitive as possible to boundary conditions through the use of judiciously selected tube geometries. This, in turn, requires a clear demonstration of the fact that the results are not specific to a single geometry and the means of fixing the location of the breakdown did not alter the physics of the phenomenon and the characteristics of the flow field.

2. Experimental Apparatus and Procedure

The original experimental equipment consisted of a Plexiglas water tank, 32 adjustable swirl vanes, a slightly diverging pipe, a constant head reservoir, a rotameter, and the necessary piping system. Its characteristics have been described in some detail by Sarpkaya (1971, 1995a). Here only the modifications made to it for its adaptation to the present investigation are described briefly. The small constant head tank was replaced by a 5 Hp pump which circulated the water between the apparatus and a 15 m^3 reservoir. The tube and vane assembly was placed in a 165 cm long, 66 cm diameter, circular, stainless-steel chamber and the single flow meter was replaced by three larger meters of total capacity of about 15 liter/s. There was no cavitation anywhere in the test tube at any Reynolds-number reported herein.

As previously noted by Sarpkaya (1971), and by others over the ensuing years, breakdowns do not normally remain stationary regardless of the quality of the control of the flow and angular momentum. They dart back and forth, with amplitudes dependent on the magnitudes of the primary controlling parameters (γ and Re*), as well as on those beyond the capacity of the experimenter to control. As the core radius decreases with increasing γ and Re*, the total pressure near the axis decreases to very low values and, if allowed, even give rise to cavitation, i.e., the position of the breakdown becomes sensitized to small pressure changes in the surrounding stream. This sensitivity is aggravated, particularly in high speed swirling flows, by the fact that a small pressure perturbation produces a much steeper rise along a streamline near the axis (Hall, 1961). Unfortunately, it is not possible to accurately prescribe the characteristics of these low-frequency axial oscillations. Their statistical description does not allow one to introduce into the numerical model the same perturbations that occur during the physical experiments. Furthermore, accurate velocity and turbulence measurements are not possible everywhere along the breakdown due to random unsteadiness superimposed on the entire flow field. The alternative is to render the position of the breakdown as insensitive as possible to boundary conditions through the use of judiciously selected tube geometries.

In the present study, experiments were first conducted with the original diverging tube described in Sarpkaya (1971) and the velocity and turbulence measurements were confined to the regions sufficiently upstream and downstream of the darting 'stagnation" point. Subsequently, a converging-diverging inset was introduced into the existing pipe. The inset started at $x = 100$ mm downstream from the start of the divergence of the original tube, converged at an angle of about 4 degrees over an axial distance of about one diameter (D) and then diverged over a distance of about 1.5D, blending smoothly into the existing pipe. The narrowest section of the inset was carefully rounded. This inset rendered the breakdown position practically insensitive (within ± 1 mm) to boundary conditions and to unknown or unknowable disturbances.

The mean velocities and turbulence intensities were measured in forward-scattering mode with a three-component Laser Doppler Anemometer (Dantec) and a 10 W Coherent laser. Bragg-cells were used to shift the frequency of one beam of each pair by 40 MHz in order to provide directional sensitivity and to prevent fringe bias. The crossing of the six beams was achieved using a 25 μm pin hole and the built-in beam steerers. The probe volume (approximately 50 μm in diameter and about 150 mm in length) was positioned at the required location by use of a remotely driven x-y-z traversing unit with a resolution of 0.02 mm. Very small quantities of light scattering particles (5 μm round Polyamid particles of specific gravity of 1.01) were introduced into the large reservoir. No data were collected during the start-up period (at least one hour) of the operation of the system during which the count rate of bursts in coincidence mode reached at least 1000, with a validation rate of 95 to 99%. At least 5,000 samples for the calculation of statistical moments and 10^6 samples for the spectra were collected at each measurement location. The velocity-bias-correction has been properly incorporated into the analysis.

The flow visualization was made through the use of TIO_2 particles and fluorescent dyes with high Schmidt numbers. A thin monochromatic laser light sheet (about 250 μm thick and 250 mm wide, with a wave length of 532 nm) was provided by a Nd-YAG laser (part of a DPIV system) with a pulse duration of 6 ns. The images were recorded in digital form through the use of either the Redlake Imaging Motion Scope or the Kodak Ektapro high speed video systems. From time to time, the said video systems were used in combination with a suitable back lighting at shutter speeds of 1/10,000 s and 1/20,000 s to check the effects of the exposure time. Neither system, at their shortest exposure times, provided images as clear as those obtained through the use of the laser light sheet provided by the Nd–YAG laser, with a 6 ns exposure time.

3. Results and Discussion

We will present first the visual observations and photographic recordings of the breakdown and then the representative velocity, turbulence, and spectral data for γ = 36,000 and Re* = 6,600 where the characteristic velocity U_o is chosen as the mass averaged axial velocity at x = 0 (the start of the divergence of the original tube). This Reynolds number corresponds, in the present study, to a tube-diameter-based Reynolds number of Re_{td} = 115,000. The entire data, including those at higher Reynolds numbers (Re* up to 13,000, Re_{td} = 250,000), will be available for CFD simulations.

In describing his previous observations, Sarpkaya (1995b) noted that when the Reynolds number was set at Re_{td} = 100,000 the entire breakdown resembled a cone, as in Fig. 1, growing almost linearly with distance downstream from a virtual origin. This and other photographs were made from a regular video recording, using a videographic copier. For Re_{td}= 100,000, they showed a small, bubble-like structure in the immediate vicinity of the stagnation point. The downstream end of the 'bubble' joined the rest of the wake in a very short distance. This and other photographs naturally gave rise to a number of important questions. Could the conical shape originate from a rapid precessing of a spiral-like breakdown? Could a smaller exposure time in flow visualizations clarify the internal structure of the flow? How small need the exposure time be? These have underlined the necessity for video recordings at exposure times small enough to capture the temporal evolution of the internal structures and for detailed velocity and turbulence measurements using DPIV and a 3-D LDV system.

Figures 2a and 2b show two sample photographs of the breakdown at γ = 36,000 and Re* = 6,600 (Re_{td} = 115,000), taken under conditions similar to those of Fig. 1.

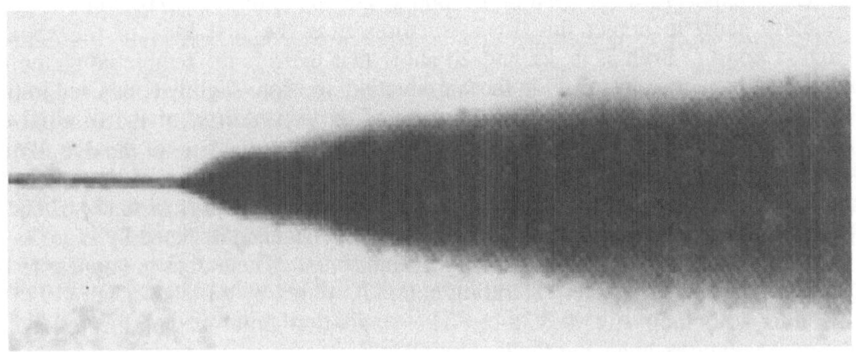

Figure 1. Vortex breakdown in a tube for Re_{td} = 100,00 (exposure time = 1/30 s).

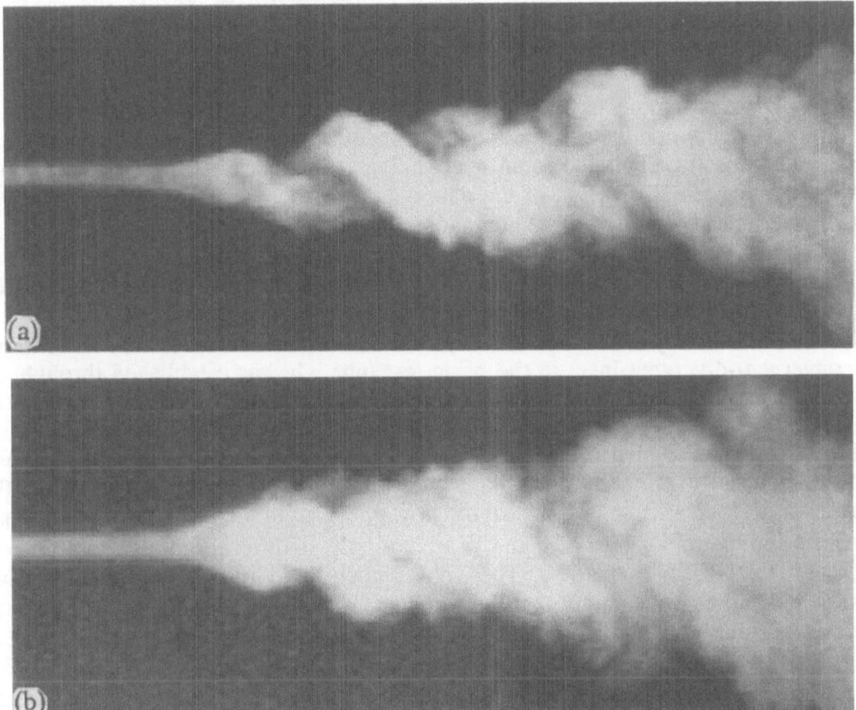

Figure 2. Vortex breakdown in a tube for γ = 36,000, Re* = 6,600, and Re_{td} = 115,00 (exposure = 6 ns). The spiraling about each other of the two branches of the original core is clearly visible in both the top and bottom pictures.

The only major difference is in their exposure times: 1/30 s for Fig. 1 and 6 ns for Figs. 2a-b. Other exposure times such as 1/2,000 s, 1/5,000 s, or even 1/10,000 s were not sufficiently short to depict the flow structures near the first kink or the apparent stagnation point. Both Figs. 2a and 2b show that there is no 'bubble' after the first kink. As we have foreshadowed in the Introduction, the core slightly bends and initiates either a single or a *double spiral*, as a result of the *bifurcation* of the original core (reminiscent of the double helix), as seen in Figs. 2a-b (the diameter of the dye filament upstream of the breakdown is 0.6 mm). Almost immediately, the spirals burst into fine scale turbulence and begin to merge with the rest of the turbulent flow field. There are, to be sure, numerous small variations to the basic mechanism depicted by Figs. 2a and 2b. At times, a half spiral is followed by a major burst. There does not appear to be a stagnation point on the nascent spiraling core. As noted by Althaus et all. (1995), in connection with their Fig. 9.3.3a-b, "The stagnation point is not located on the centerline as often assumed, but rotates around it."

At higher Reynolds numbers, the mechanism of the structural change becomes even more interesting and the breakdown becomes more like an axisymmetric cone. Figures 3a-c show three examples of highly turbulent conical structures for $\gamma = 70{,}000$, $Re^* = 13{,}000$, and $Re_{td} = 250{,}000$. The rate of revolution at the inception of the cone is about 1,500 rev/s and decreases with distance further downstream. The 'bubble' and the 'core-recovery regions', seen in the usual path of structural transformations at low Reynolds numbers (Sarpkaya, 1971), are completely bypassed at sufficiently high values of γ and Re^*.

The LDV measurements were made at numerous sections (5 mm apart) along the entire tube to determine the distributions of velocities, rms values of turbulence, shear stresses, and spectra. This led to an overwhelming number of data points and plots. Here only a very small fraction of the data obtained for $\gamma = 36{,}000$ and $Re^* = 6{,}600$ could be presented because of space limitations.

Figures 4a-b show the normalized u and w components of velocities and their rms values at $x/D = 1.8$ and $x/D = 4.0$. These two sections are quite representative partly because they bracket the axial station at which the breakdown occurred ($X_{bp} = x/D = 2.8$) and partly because they enable one to compare the measurements made with and without the convergent-divergent inset in the basic test tube. It was established through the comparison of the data at various stations such as these that the inset is indeed an advantageous device. It stops the darting of the breakdown point and, in doing so, enables one to make measurements at sections very close to the 'stagnation' point without materially affecting the characteristics of the flow anywhere along the breakdown. It is clear from Fig. 4a that the flow upstream of the breakdown point (X_{bp}) is jet like and the rms values of u and w are not only significant but also significantly different in the vicinity of the core. These observations hold true for all other turbulence stresses and show convincingly that the turbulence for the circumstances encountered herein is indeed highly non-isotropic. Figure 4b shows equally interesting results. The flow becomes wake like, not only at $x/D = 4$, but also at all other sections with $x/D > X_{bp}$. The second observation to be made is that the rms values of u and w are practically indistinguishable. This is almost true for other turbulent stresses as well and tends to show that as the angular velocity of the wake decreases, turbulence becomes increasingly more isotropic. An additional important point to be made in connection with Fig. 4b is that $x/D = 4$ is also the section at which the axial temporal mean velocity becomes zero. There is indeed a small backflow region inside the rapidly spiraling breakdown cone between $x/D = 3$ and 4. It has a lateral extent of about 2 mm. At larger Reynolds

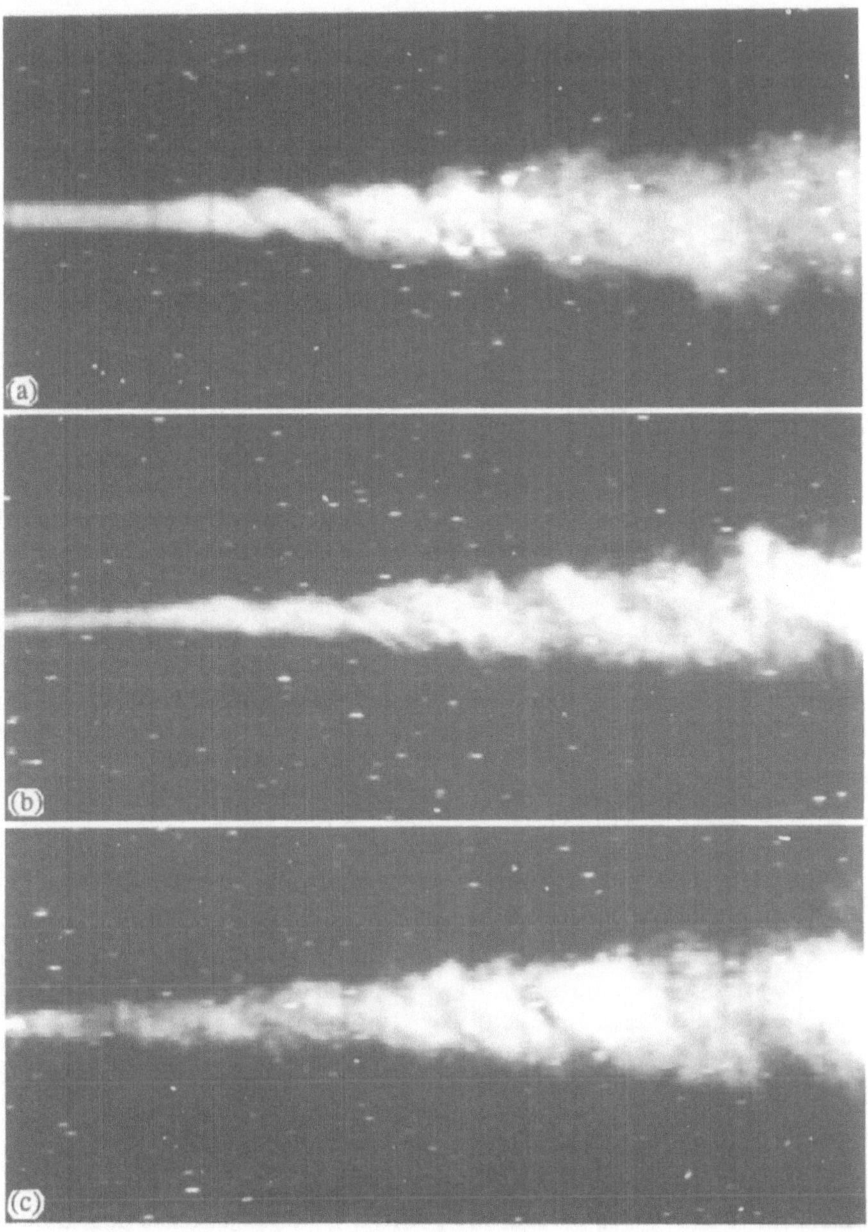

Figure 3. Vortex breakdown in a tube for Υ = 70,000, Re* = 13,000, and Re_{td} = 250,00 (exposure = 6 ns). Particles are TIO_2. The conical shape and the highly turbulent nature of the transformation are seen in all three pictures.

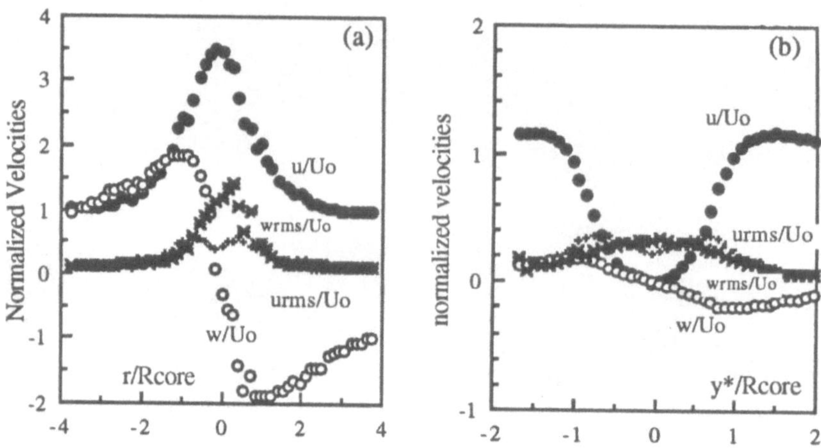

Figure 4. The normalized u and w components of velocities and their rms values at: (a) x/D = 1.8, and (b) x/D = 4.0.

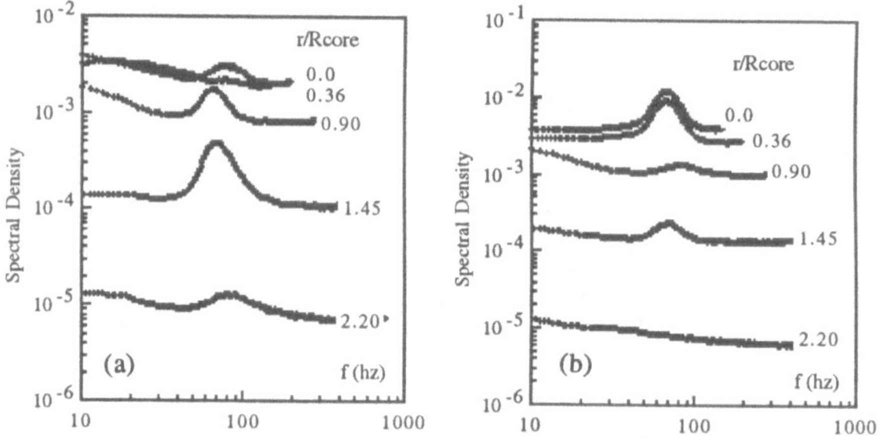

Figures 5. The power spectra of (a) axial and (b) tangential velocity fluctuations at x/D = 3.15, for various values of r/R_c,

numbers, this backflow region becomes practically indistinguishable due to intense turbulent mixing.

The spectra of the velocity components were measured at all sections, upstream and downstream of X_{bp}. The results have shown that at all sections upstream of the breakdown point there is no single characteristic frequency at which the spectrum exhibits a peak. However, all sections downstream of X_{bp} show interesting spectral characteristics. We have chosen here two sections for discussion: x/D = 3.15 and x/D = 4. Figures 5a and 5b show the u and w spectra at x/D = 3.15, for various values of r/R_c,

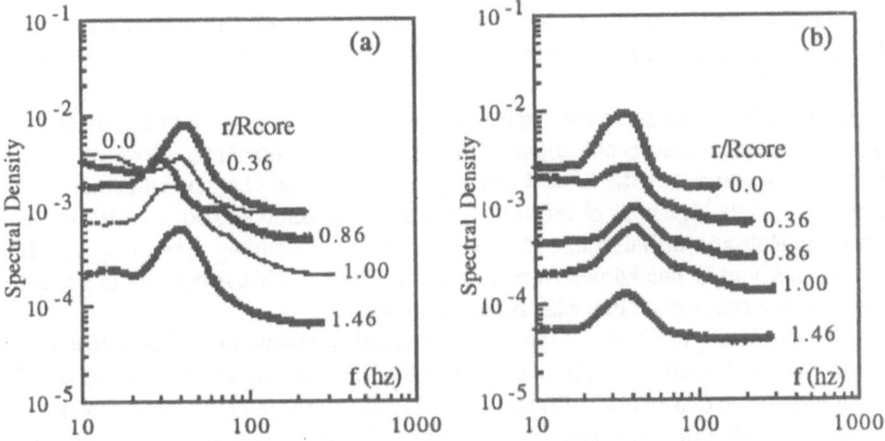

Figures 6. The power spectra of (a) axial and (b) tangential velocity fluctuations at x/D = 4, for various values of r/R_C,

and Figs. 6a and 6b show the corresponding spectra at x/D = 4. The fundamental observations which may be made on the basis of this and other data not shown here are that the higher the frequency at which the spectrum exhibits a peak, the closer is the section to the breakdown point, the w-spectrum is relatively more energetic than the u-spectrum, and, most significantly, the peaks occur at essentially the same frequency regardless of the radial extent of the point at which the spectrum is measured. This becomes increasingly so as the axial position of the section increases. In other words, the turbulence becomes increasingly more isotropic as one moves further downstream. The same point was made earlier in connection with the discussion of the rms values of u and w. It is, therefore, no surprise that the u- and w-spectra in Figs. 6a and 6b exhibit nearly identical energy levels. Observations using very small exposure times as well as measurements have shown that the turbulent wake contains relatively weak coherent structures. In other words, the wake beyond a very short distance from the breakdown point does not show any footprints of a spiraling breakdown. The entire wake becomes a swirling jet and exhibits in many ways the characteristics of a swirling jet. The peak frequencies observed at all x/D values were nearly identical to $w_{max}/(2\pi r)$, indicating that the most energetic eddies were at the mean interface of the swirling cone with its surroundings. In fact, a close examination of the turbulence data has shown that the said interface, the position of the maximum azimuthal velocity, and the maximum turbulent stresses occur almost simultaneously.

4. Conclusions

This paper described experiments on various types of vortex breakdown in *non-cavitating* swirling flows in a slightly diverging cylindrical tube for values of the governing parameters (γ and Re*) up to 70,000, and 13,000 , respectively.

The results refute the conjectures that the circumstances of breakdown are insensitive to the Reynolds number and the local turbulence properties. These two factors have a strong influence on the evolution of the flow. Of all the known forms, the spiral emerges as the most fundamental breakdown form. All other forms may be regarded as transient states affected by various types of instabilities. The nearly axisymmetric form has served to excite imagination, to test some numerical schemes, and to produce numerous 'explanations' of the breakdown phenomenon. However, at very high values of γ and Re* the breakdown acquires forms and characteristics never seen before: Extremely high rates of revolution, onset of core-bifurcation or core-trifurcation, intense nonisotropic turbulence, and a conical shape, resembling a swirling jet. It is clear that the vortex breakdown over a delta wing is not what it appears to be in the model tests but rather more like what is described herein.

If there is any hope of making realistic numerical simulations of turbulent vortex breakdowns, all boundary conditions, in particular the velocity and turbulence profiles and the turbulence dissipation upstream of the breakdown, need to be known with great precision. In this effort one cannot emphasize strongly enough the need for a robust turbulence model which can deal with nonisotropic turbulence in swirling flows subjected to streamline curvature and strong radial pressure gradients.

5. Acknowledgments

This research is being supported by the National Science Foundation under Grant No. CTS-9612528. This support is gratefully acknowledged.

6. References

Althaus, W., Brücker, C.H., and Weimer, M. (1995) Breakdown of slender vortices, in S. Green (ed.), *Fluid Vortices*, Kluwer Academic Publishers, Dortrecht, Chap. 9.

Breuer, M. and Hänel, D. (1993) A dual time-stepping method for 3-D viscous incompressible vortex flows, *Comput. Fluids*, **22**, 467-484.

Délery, J. and Molton, P. (1994) Topology of the Flow Resulting from Vortex Breakdown over a Delta Wing at Subsonic Speed, *Acta Mechanica [Suppl]*, **4**, 297-304.

Hall, M.G. (1961) A theory for the core of a leading-edge vortex, *J. Fluid Mech.* **11**, **209-228**.

Hogg, S. and Leschziner, M.A. (1989) Computation of highly swirling confined flow with a reynolds stress turbulence model, *AIAA J.* **27**, 57-63.

Krause, E. (1990) The solution to the problem of vortex breakdown, *Lecture Notes in Physics*, **371**, 35-50.

Rhode, D.L., Lilley, D.G., and McLaughlin, D.K. (1982) On the Prediction of Swirling Flowfields Found in Axisymmetric Combustor Geometries, *J. Fluids Eng.* **104**, 378-384.

Sarpkaya, T. (1971) On stationary and traveling vortex breakdowns,*J. Fluid Mech.* **45**, 545-568.

Sarpkaya, T. (1995a) Vortex breakdown and turbulence. *AIAA Paper* 95-0433.

Sarpkaya, T. (1995b) Turbulent vortex breakdown, *Phys. Fluids* **7**, 2301-2303.

Spall, R.E and Gatski, T.B. (1991) Computational study of the topology of vortex breakdown, *Proc. R. Soc. Lond. A*, **435**, 321-337.

Spall, R.E. and Gatski, T.B. (1995) Numerical Calculations of Three-Dimensional Turbulent Vortex Breakdown, *Int. J. Numer. Methods Fluids* **20**, 307-318.

VORTEX BREAKDOWN AS A CATASTROPHE

Vladimir SHTERN and Fazle HUSSAIN
Department of Mechanical Engineering
University of Houston, Houston TX 77204-4792, USA

Abstract

By studying swirling *viscous* jets, we develop a new explanation of vortex breakdown, show how solution non-uniqueness appears through cusp and fold catastrophes as the Reynolds number *Re* increases, and obtain analytical solutions for $Re \to \infty$. Although inviscid theories also involve fold catastrophe, they have a strong limitation: the dependence of the head H and circulation Γ_d on stream function Ψ is undetermined inside a recirculatory zone. Analytical continuation and stagnation zone models used to resolve this indeterminacy appear inadequate for swirling jets: $H(\Psi)$ and $\Gamma_d(\Psi)$, which we obtain here by the inviscid limit, differ from those based on the inviscid theories. We therefore suggest a new model consistent with the limiting transition. Also, we analyze turbulent vortex breakdown with a conical wake which has been recently observed at large *Re*.

1. Introduction

1.1. *Hysteretic vortex breakdown.* Swirling flows in nature and technology have striking features that need thorough analysis, particularly for control. One problematic feature is vortex breakdown (VB) which can lead to abrupt transitions between different flow states, occurring at the same values of control parameters. Figure 1 shows the VB visualization above a delta wing [1]. Vortex sheets separating from the leading edges roll up above the

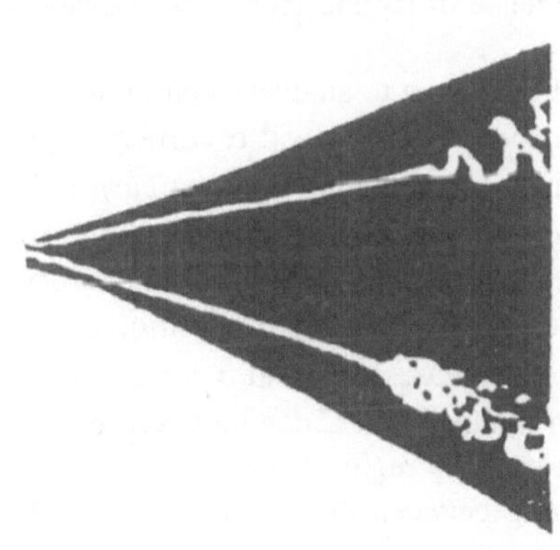

Figure 1. Spiral and bubble-like vortex breakdown (VB) over a delta wing [1].

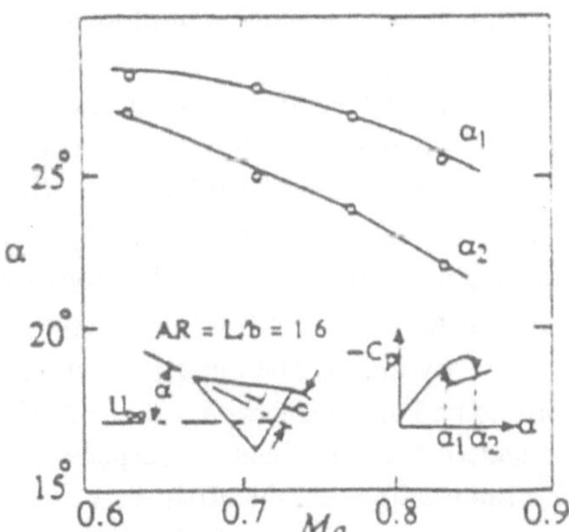

Figure2. Hysteretic transitions in VB as the angle of attack a varies [4].

E. Krause and K. Gersten (eds.), IUTAM Symposium on Dynamics of Slender Vortices, 297-306.

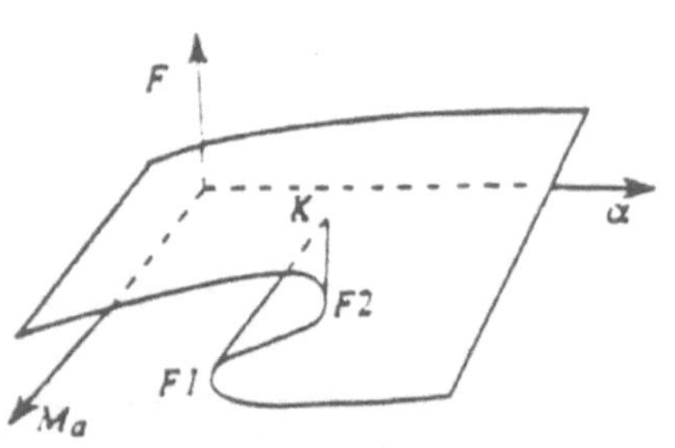

Figure 3. Explanation of the hysteresis in Fig. 2 in terms of cusp K and fold (F1, F2) catastrophes.

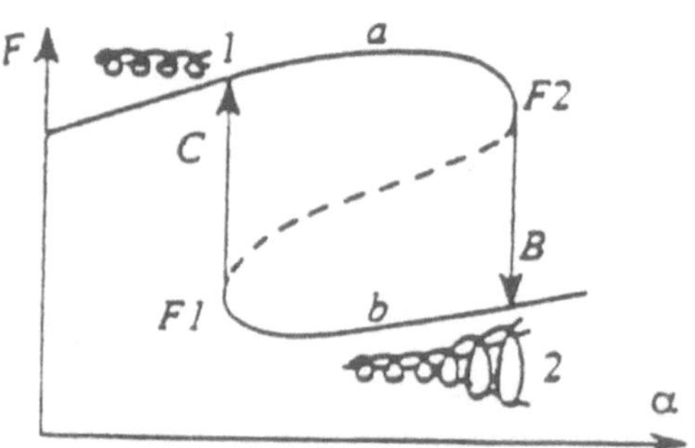

Figure 4. Hysteretic vortex breakdown B and consolidation C as transition between consolidated 1 and open 2 vortex flows.

wing and form longitudinal vortices. The cores of these vortices abruptly expand downstream, forming a 'bubble' or spiral structure as shown in Fig. 1; such an abrupt change is VB [2]. Here we argue that VB is a catastrophe based on experimental evidence of sudden shifts in VB location [3]. Figure 2 presents the experimental data [4] on hysteretic transitions above a delta wing. The insets show the flow schematic and the dependence of pressure coefficient C_p on the angle of attack α at a fixed Mach number *Ma*. The value of C_p jumps at $\alpha = \alpha_2$ as α increases and at $\alpha = \alpha_1$ as α decreases. There are two stable regimes with different C_p (*i.e.* bi-stability) in the range $\alpha_1 < \alpha < \alpha_2$. The jumps correspond to abrupt shifts in the VB location — downstream at $\alpha = \alpha_1$ and upstream at $\alpha = \alpha_2$. The main plot shows the dependence of α_1 and α_2 on *Ma* (however, compressibility is not crucial for VB).

The catastrophe theory is a proper mathematical technique to analyze such jumps [5]. Figure 3 schematically shows the dependence of lift force *F* on *Ma* and α corresponding to Fig. 2. Surface $F(\alpha, Ma)$ has *folds F1* and *F2* which meet and terminate at *cusp* point *K*. Figure 4 shows a section Ma = const of $F(\alpha, Ma)$, i.e. the schematic of the *hysteresis* loop in Fig. 2. Branches *a* and *b* correspond to stable solutions while the dashed curve shows unstable solutions. As one moves along *a* as α increases, a jump transition occurs as shown by line *B*. The opposite jump *C* occurs as α decreases. If *a* and *b* correspond to consolidated and open vortex filaments (Fig. 4), then jumps *B* and *C* can be interpreted as VB and vortex consolidation, respectively. In the bi-stability region (*i.e.* between *F1* and *F2*), finite-amplitude disturbances can cause switching between the stable regimes. The distance between the dashed curve and *a* or *b* at a fixed α is the threshold. If the disturbance amplitude is less than the threshold, then the flow returns to the initial state; otherwise the flow transits to the other stable regime.

Such bi-stability occurs not only in the leading-edge vortices but also in many other practical flows, *e.g.* in tornadoes [6-8] and vortex devices [9,10]. The view that VB is a jump transition is disputable — there is no consensus on the VB mechanism [2,11]. To argue the interpretation of VB as catastrophe, we compare this view with competing approaches.

1.2. *Review of vortex breakdown theories.* The descriptive definition, based on experiments and supported by some theorists [2,12], claims that VB is an **abrupt change** in the core of a slender vortex (see Fig. 1; similar observations have been made in tubes [13-15]). In mathematical terms, the slenderness condition ($a\partial/\partial s$ << 1, where a is the core radius and s is the longitudinal coordinate) is abruptly violated. In slender filaments, vortex dynamics is well predicted by 'parabolized' equations, *e.g.* boundary layer or quasi-cylindrical [16] approximations of the Navier-Stokes equations (NSE). However, elliptic equations, *e.g.* the full NSE, must be applied for the VB region.

VB is often associated with ***flow separation***: a change in flow field topology via the appearance of a *recirculatory zone* (RZ, e.g. the 'bubble' in Fig. 1). Separation in VB is *internal* (i.e. occurs away from walls), but it has common features with *near-wall* separation. In both cases, the streamwise component of the pressure gradient must be positive for separation to occur [16], and the spanwise component of vorticity changes its sign. Some researchers include flow separation in the VB definition [17].

The oldest theory of VB addresses propagation of **inertial waves** in swirling flows. The theory considers a vortex filament as a wave guide and VB as a phenomenon similar to shock waves and hydraulic jumps. A key word is flow *criticality* [18]. If the long-wave speed is larger than the flow velocity, then the flow is sub-critical; otherwise it is supercritical. Both flow types can exist at the same control parameters as *conjugate states* [19]. The theory claims that transitions from super- to sub-critical states can occur by jumps, and that VB is such a jump. The recent version [20,21] of the theory also claims a possibility of jump transitions between different supercritical flow states. So far the wave theory deals with an inviscid fluid.

There were attempts to explain VB in terms of **stability** [22,23]. Symmetry breaking, such as transformation of an axisymmetric vortex into a helical one, relates to instability. The helical instability of swirling jets was studied in Refs. [24-26]. A limitation of these studies is the quasi-parallel approximation, while nonparallel effects are significant in practical swirling flows. The instability does not necessarily lead to VB. It is shown for a flow in a sealed cylinder that VB can occur without any instability and that swirling flow instability can occur without VB [27].

The **fold catastrophe** view of VB was first suggested by Trigub [28] and then developed in Refs. [2,29-31] for inviscid swirling flows. There are also numerical simulations [32-33] of a viscous swirling flow in a diverging pipe showing hysteresis.

1.3. *Discussion of the approaches.* The *descriptive* definition of VB is limited as there is no quantitative criterion for the 'abrupt change'. Catastrophe theory provides an exact definition of such an abrupt change. Although flow *separation* and VB often occur together, these events are not identical. First, internal separation can occur in swirl-free vortical flows as well. Second, separation can occur in swirling flows without abrupt changes in the vortex core. A well-known example is a flow in a sealed cylindrical container driven by a rotating bottom disk where RZ develops starting with zero size as *Re* increases [34]. We show below that flow separation occurs even in a creeping flow without an abrupt change. We view flow separation as VB when separation occurs with an abrupt increase in the core size.

Until now, the *wave theory* has been developed for inviscid fluid and quasi-parallel flows only, while there is a conical wake downstream of VB in most practical applications (Fig.1). The main drawback of the wave theory is that it cannot explain the jump vortex consolidation (C in Fig. 4). The *stability* approach is potentially very powerful. It covers the wave theory, because the inertial waves can be interpreted as propagating neutral disturbances. The critical state supporting an infinitesimal *standing wave* is a marginally stable flow. However, to describe hysteretic VB, one needs a theory for strongly *nonlinear* instability of nonparallel swirling flows. Such a theory needs to be developed because most prior works address linear stability using the quasi-parallel approximation only. The *inviscid theories* treating VB as *fold catastrophe* have a strong limitation: functions $H(\Psi)$ and $\Gamma_d(\Psi)$ are undetermined inside RZ. To overcome this indeterminacy, researchers apply some conjectures to $H(\Psi)$ and $\Gamma_d(\Psi)$ (e.g. the analytical continuation or stagnation zone approaches). Our results for the limiting relations $H(\Psi)$ and $\Gamma_d(\Psi)$ as $Re \to \infty$ differ from those based on these conjectures. It is necessary to study the appearance of hysteresis through a *cusp catastrophe* as Re increases.

Numerical methods, though a powerful tool for the VB study, have their own limitations. There still seem to be no k-ε or LES models which satisfactorily simulate turbulent swirling flows with strong anisotropy and RZ. Direct numerical simulations (DNS) of NSE is effective when VB occurs for not too large Reynolds numbers [32-33]. Other limitations are: (i) the DNS results are very sensitive to boundary conditions and even to grid mesh [27]; (ii) DNS, being expensive, are not appropriate to cover a wide region of control parameters, and one can miss important effects occurring at uncovered parameter values; (iii) asymptotic features or conjectures on flow behavior at infinity are needed to decrease the computation region for open flows and (iv) DNS cannot show the asymptotic relations as $Re \to \infty$. Therefore, numerical methods are the most effective when they are accompanied with analytical or semi-analytical studies. Here, we undertake such a study addressing conically similar viscous jets.

2. Swirling jet in an infinite fluid

2.1. *Problem formulation.* Consider steady flows of a viscous incompressible fluid admitting the representation:

$$v_r = -\nu\psi'(x)/r; \quad v_\theta = -\nu\psi(x)/(r\sin\theta), \quad v_\phi = \nu\Gamma(x)/(r\sin\theta),$$
$$p = p_\infty + \rho\nu^2 q(x)/r^2, \quad \Psi = \nu r\psi(x), \quad x = \cos\theta. \tag{1}$$

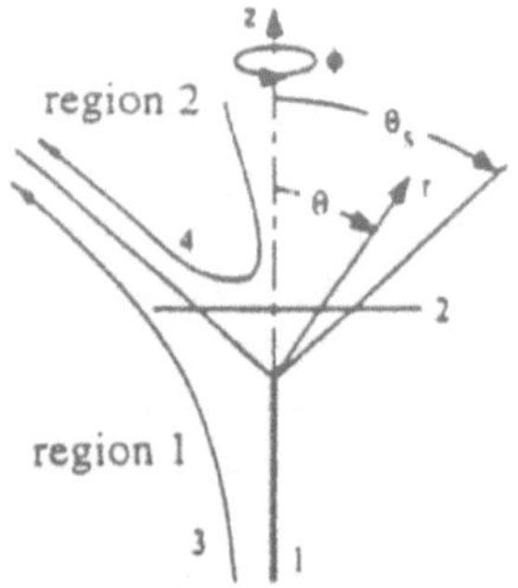

Figure 5. Schematic of the problem.

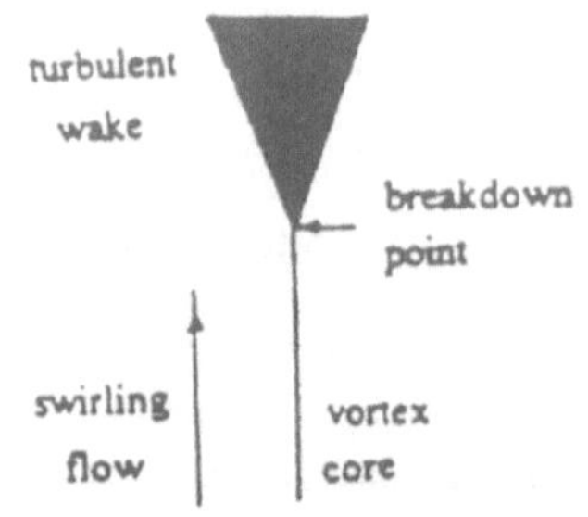

Figure 6. Conical turbulent VB [15].

Here (r, θ, ϕ) are the spherical coordinates, r is the distance from the origin, θ and ϕ are polar and azimuthal angles (Fig. 5); v_r, v_θ, v_ϕ and Ψ are the velocity components and the Stokes stream function, ν is the kinematic viscosity, ρ is the density, and the prime denotes differentiation with respect to x. Substitution of (1) reduces NSE to a system of ordinary differential equations (ODE) for the dimensionless functions ψ, Γ and q. For detailed analyses of this ODE, see Refs. [35,36]; here we discuss only new results and interpretation.

Consider a swirling jet in an infinite fluid as a model of a vortex filament with VB. Fig. 5 shows the problem schematic. A consolidated part of the filament (e.g. in Fig. 1) is idealized as half-line vortex 1 characterized by swirl Reynolds number Re_s, which is the dimensionless circulation ($= \Gamma(-1)$), and $J_0 = J/(2\pi\rho\nu^2)$, where J is the flow force acting on the normal plane, $z = \text{const} > 0$ (line 2 in Fig. 5). The downstream part of the vortex can be either consolidated (without RZ) or 'destroyed' (with RZ). Fig. 5 shows the destroyed vortex, and curves 3 and 4 represent typical streamlines of the meridional motion in the two flow cells separated by conical surface $\theta = \theta_s$. This problem models vortices above delta wings, tornadoes [36] and conical VB as shown in Fig. 6, which is the schematic of experimental observations [15].

2.2. *Analytical solution for small Re_s and J_0.* In the limiting case where $Re_s \to 0$ and $J_0 \to 0$, there is the analytical solution,

$$\Gamma = Re_s(1-x)/2,$$
$$\psi = \tfrac{1}{4}Re_s^2\{(1+x)\ln[2/(1+x)] - (1-x^2)/2\} + Re_a(1-x^2)/2, \qquad (2)$$

where Re_a is the dimensionless velocity on the axis, $x = 1$; Re_a serves here as a free parameter replacing J_0. For $Re_a \geq 0$, $\psi \geq 0$ in $-1 < x \leq 1$ and, therefore the flow is ascending and consolidated. For $Re_a < 0$, the flow is descending near the axis, $x = 1$. Therefore, flow separation from the axis occurs as Re_a passes zero, and RZ develops for $Re_a < 0$. At $Re_a = 0$, ψ is proportional to Re_s^2, and the corresponding value of $J_0 = J_{0s}$ is also proportional to Re_s^2. It is convenient for the presentation of our results to use the Long parameter, $M = 2\pi J_0 Re_s^{-2}$. The M value corresponding to separatiom, $M = M_s$, has the nonzero limit $M_{s0} = 0.482$ as $Re_s \to 0$. The flow is one-cell and ascending for $M > M_{s0}$, but two-cell (e.g. as shown in Fig. 5) for $M < M_{s0}$.

The analytical solution (2) demonstrates the important feature: flow separation (i.e. the appearance of RZ) for infinitesimal Re_s and J_0. We do not consider separation in such an extremely slow flow as vortex breakdown since nothing 'abrupt' occurs in the physical or parameter space. We view vortex breakdown as a result of a catastrophe (fold bifurcation) and of hysteretic transitions, which occur only for rather large Re_s and J_0.

2.3. *Numerical results.* For moderate Re_s and J_0, the problem has been solved numerically. Figure 7(a) shows the results on the parameter plane (M, Re_s). Curve S separates regions of consolidated and two-cell flows (the insets show the meridional motion). As $Re_s \to 0$, S approaches the above asymptote, $M = 0.482$. As $Re_s \to \infty$, S (branch Sl) approaches the asymptote, $M = M_{sL} = 3.79$ [37]. The solution becomes non-unique for $Re_s > Re_K$ corresponding to cusp point K ($M_K = 3.00$, $Re_K = 11.5$). There are three solutions in the region between folds Fl and $F2$ (Fig. 8) and the solution is unique

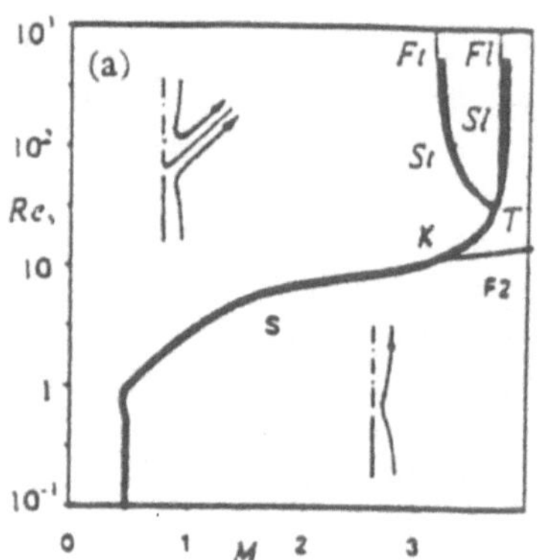

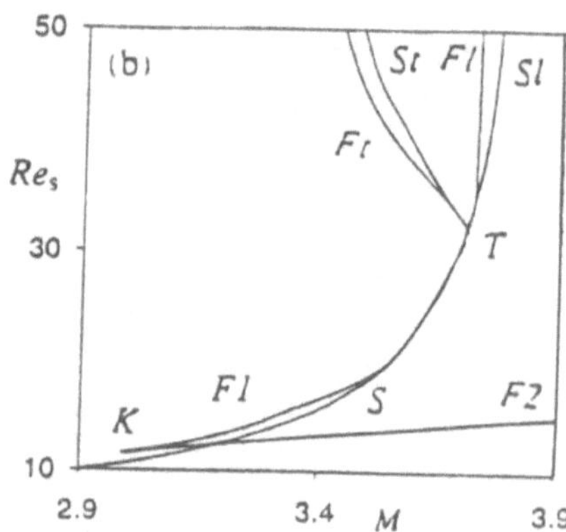

Figure 7. (a) Map of flow patterns and folds. (b) Enlarged vicinity of points K and T.

outside this region. To reveal that curves S and Fl do not merge, figure 7(b) shows the enlarged vicinity of K: Fl and $F2$ meet and terminate at cusp K. Figure 8 illustrates how folds Fl and $F2$ appear as Re_s increases: curves 1-3 show v_m ($= zv_{\phi max}/\nu$, $v_{\phi max}$ is the maximal swirl velocity on a plane z = const) versus J_0 at $Re_s = 11$ (below K in Fig. 7), 11.5 (at the K level) and 12 (above K), respectively. We do not consider flow separation at S for $Re_s < Re_K$ as VB. Jump transitions, corresponding to VB and vortex consolidation, appear together with Fl and $F2$ (B and C in Fig. 8).

2.4. *Analytical solution for large Re_s.* Consider a two-cell flow to compare $H(\Psi)$ and $\Gamma_d(\Psi)$ inside and outside RZ. As $Re_s \to \infty$, the flow (outside a boundary layer near the separating surface, $x = x_s = \cos\theta_s$) is governed by the analytical inviscid solutions [36]:

$$\Gamma \equiv Re_s \text{ in } -1 \le x < x_s \text{ (region 1)}, \quad \Gamma \equiv 0 \text{ in } x_s < x \le 1 \text{ (region 2)},$$
$$\psi_1 = \tfrac{1}{2}Re_s\{(1+x)[3x_s+1-(3+x_s)x]/(1+x_s]\}^{1/2}, \quad \psi_2 = -\tfrac{1}{2}Re_s(1-x)[(1+x_s)/(1-x_s)]^{1/2},$$
$$q_1 = -\tfrac{1}{4}Re_s^2[3x_s+1-(1-x_s)x]/[(1+x_s)(1-x^2)], \quad q_2 = -\tfrac{1}{4}Re_s^2(1+x_s)/[(1+x)(1-x_s)], \quad (3)$$

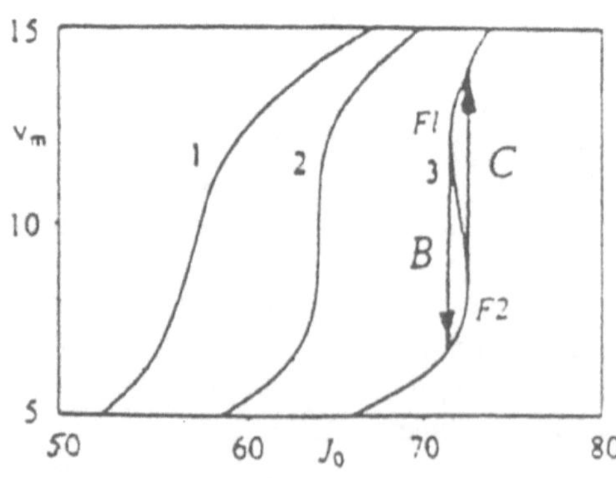

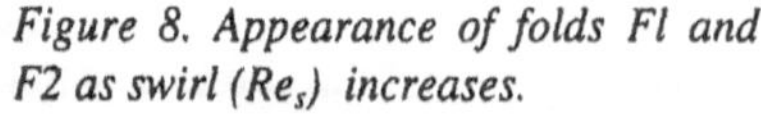

Figure 8. Appearance of folds Fl and F2 as swirl (Re_s) increases.

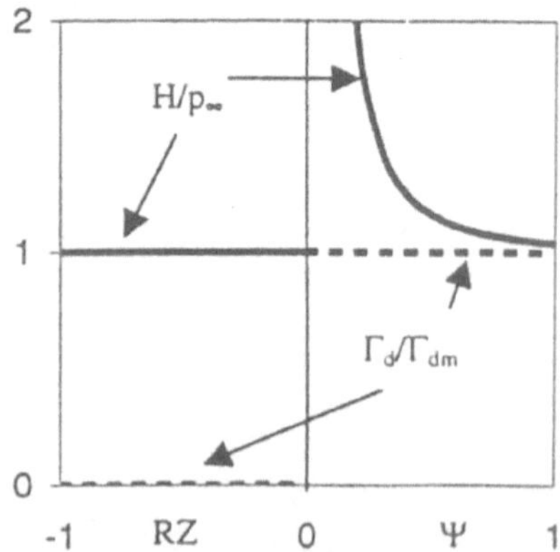

Figure 9. Head H and circulation Γ_d versus stream function Ψ as $Re_s \to \infty$.

where subscripts 1 and 2 indicates the solutions for the regions 1 and 2 (Fig. 5).

Circulation Γ_d (= $\nu\Gamma$) is zero in RZ (region 2, where $\Psi < 0$ and the flow is irrotational) and uniform ($\Gamma_d \equiv \nu Re_s$) outside RZ (region 1, where $\Psi > 0$ and the flow is vortical). The surface $x = x_s$ serves as a sink, and $\Psi = \pm\frac{1}{2}\Gamma_d r \sin\theta_s$ at $\theta = \theta_s \pm 0$. This sink results from entrainment of a strong annular jet which develops near $x = x_s$ as $Re_s \to \infty$. At $x = x_s$, v_r and v_θ change their signs while v_ϕ, being nonzero outside RZ, drops to zero within RZ. In region 1, only the azimuthal vorticity is nonzero: $\omega_\phi = 4\Gamma_d^3\Psi^{-2}(1-x_s^2)^{-1/2}$.

The head, $H = p+\rho v^2/2$, is uniform and equals p_∞ in region 2, while $H = p_\infty + \frac{1}{2}\rho\Gamma_d^4\Psi^{-2}$ in region 1. Both functions $H(\Psi)$ and $\Gamma_d(\Psi)$ have jumps at the RZ boundary in the inviscid limit (Fig. 9). Swirl is absent inside RZ in agreement with the stagnation-zone model. However, the meridional motion *does occur* in RZ and is of the magnitude as that of the meridional motion in region 1; this fact contradicts the stagnation-zone model. Since (3) is an exact solution (validated in [36]), we conclude that neither the analytical continuation nor the stagnation zone model is adequate for this particular flow. Since this example may be exceptional, consider another problem — describing vortex-wall interaction — where boundary conditions significantly differ from those for solution (3).

3. Vortex-wall interaction.

3.1. *General features.* The problem on interaction of the half-line vortex with a rigid wall, also belonging to the conically similar class (1), has striking features and important practical applications [38-40, 35]. Figure 10 shows the problem schematic. A half-line vortex (the thin cylinder in Fig. 10) of circulation $\Gamma_d = \nu Re_s$ locates on the positive z-axis (where also force $F_z = 4\pi A\rho v^2/r$ acts [40]) and induces a swirling flow above a no-slip wall at $z = 0$. The helical curve shows a typical streamline for the case $F_z \geq 0$. Depending on the dimensionless parameters Re_s and A, the flow is ascending, two-cellular and descending. Figure 11 is a map of the flow patterns (the insets show the meridional

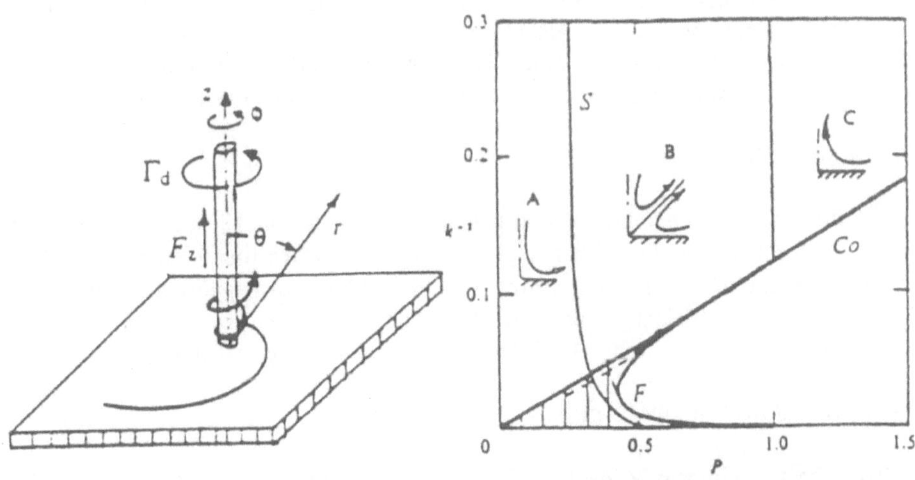

Figure 10. Schematic of the problem. *Figure 11. Map of flow patterns.*

motion) on the parameter plane (k, P), $k = Re_s/2$ and $P = 1+4ARe_s^{-2}$ [40]. Curve F (which terminates at $P = 1$) corresponds to the fold catastrophe and line Co corresponds to a 'collapse' (v_z becomes infinite on the axis and the similarity solution ceases to exist) [35]. There are three solutions in the dashed region, a single solution above Co, and no solution to the right of F. As P increases, the flow pattern changes at separation curve S from that shown by inset A to that shown by inset B; i.e. RZ appears near the wall. Consider the two-cell flow to find the limiting relations $H(\Psi)$ and $\Gamma_d(\Psi)$ as $Re_s \to \infty$.

3.2. *Analytical solution in the inviscid limit.* As $Re_s \to \infty$ while the separation angle θ_s is fixed, $P = (1-x_s)/(1+x_s)$ and the limiting inviscid solutions are:

$$\Gamma \equiv 0 \text{ for } 0 \le x < x_s \text{ (region 1)}, \ \Gamma \equiv Re_s \text{ for } x_s < x \le 1 \text{ (region 2)},$$
$$\psi_1 = Re_s P^{1/2} x, \quad \psi_2 = -Re_s\{x_s(1-x)[(2-x_s)x - x_s]/(1-x_s^2)\}^{1/2},$$
$$q_1 = -\tfrac{1}{2}Re_s^2 P/(1-x^2), \quad q_2 = -\tfrac{1}{2}Re_s^2[1-x_s^2+2x_s(1-x)]/[(1-x^2)(1-x_s^2)] \qquad (4)$$

Again, Γ is a step function, but Γ is nonzero near the axis in (4) in contrast to (3); this difference is due to the fact that the axis is a source of swirl here. However, as in (3), Γ is zero in RZ (which is region 1 now). The head is uniform and equals p_∞ in RZ where the flow is irrotational, and $H = p_\infty + \tfrac{1}{2}\rho\Gamma_d^4\Psi^{-2}x_s^2(1+x_s)^{-2}$ in region 2 where $\omega_\phi = -\Gamma_d^3\Psi^{-2} P^{1/2}x_s(1+x_s)^{-1}$. Thus, the prior inviscid models are invalid for (4) as well.

Common features of flows (3) and (4) are: (a) swirl stagnates in RZ while the meridional motion does not and (b) p, v_r^2, and v_θ^2 are continuous across the RZ boundary, so that the jump in H occurs only due to the jump in v_ϕ^2. An inviscid model consistent with the limiting transition must involve the conditions: (a) swirl is absent in RZ and (b) H inside RZ is smaller by $\tfrac{1}{2}\rho v_\phi^2$ than H outside RZ.

No inviscid model, however, is valid for flows with wide regions of strong turbulence. Such regions are observed in swirling jets and flows with RZ. Turbulent mixing in these regions can be modeled by an eddy viscosity which increases with Re_s, making the inviscid approach invalid. To study how turbulence influences VB, we consider below a simple model where the eddy viscosity is a step function of θ.

4. Turbulent VB

4.1. *Turbulence modeling.* Let us return to the jet studied in §2 and introduce eddy viscosity ν_t generalizing the Schlichting model for a swirl-free round jet, $\nu_t = bJ^{1/2}$ [41], and the Squire model for a swirling vortex, $\nu_t = a|\Gamma_d|$ [42,43], by supposing that $\alpha = \nu_t/\nu = [b^2J_0 + a^2Re_s^2]^{1/2}$. Coefficients a and b are chosen to be consistent with those values used by Schlichting [41] and Albring [44] for the particular cases, $Re_s = 0$ and $J_0 = 0$:

$$\alpha = [J_0/614 + (0.004Re_s)^2]^{1/2}. \qquad (5)$$

The effective viscosity ν_t involves also ν and, therefore, α must be larger than 1. We use (5) if it yields $\alpha > 1$ or else we take $\alpha = 1$. In the case $\alpha = 1$ (which occurs for small J_0 and Re_s), the flow is laminar in the entire region. In the case $\alpha > 1$, we suppose that ν_t is a step function of θ [45], i.e. that the flow is laminar in $-1 \leq x < x_t$ and turbulent in $x_t < x \leq 1$. To define the boundary, $x = x_t$, between the turbulent and laminar regions, we impose the condition, $v_r(x_t) = 0.2v_{rm}$ [46], where v_{rm} is the maximal radial velocity.

4.2. *Numerical results.* Along separation curve S in Fig. 7, the flow is laminar (i.e. (5) yields $\alpha < 1$) for small Re_s up to the point T ($Re_s = 32$) where S splits into two branches. Branch Sl shows the continuation of the laminar ($\alpha = 1$) solution while branch St corresponds to turbulent flow separation. The fold curve Fl also splits unto Fl and Ft. As $Re_s \rightarrow \infty$, the asymptotic values of M_s are $M_{st} = 3.23$ and $M_{sl} = 3.79$ for St and Sl, respectively. The fact that $M_{st} < M_{sl}$ implies that turbulence suppresses flow separation. For example, for a fixed value of the flow force, flow separation occurs for a larger circulation in the turbulent flow than that in the laminar flow. As another example, let M be fixed in the range $M_{st} < M < M_{sl}$ and let Re_s increase (along a vertical line in Fig. 7). For small Re_s, the flow is laminar and consolidated. Flow separation occurs at the point where curve S is crossed. For $Re_s > 32$, the flow becomes turbulent. Crossing curve St indicates the disappearance of RZ. As Re_s further grows at the fixed M, the flow remains consolidated, having no reversed motion near the axis. This result agrees with Sarpkaya's observation [15]: the separation 'bubble' disappears as Re increases. Turbulent mixing tends to make a velocity field uniform, thus preventing the reversed flow, i.e. the 'bubble' appearance. As far as we know, no earlier theoretical study has explained this striking experimental fact.

5. Conclusion.

Our results for swirling viscous jets support the view that vortex breakdown is a fold catastrophe and reveal serious drawbacks of prior inviscid theories. Using an eddy viscosity model, we show that turbulence opposes flow separation leading to the existence of a conical turbulent wake downstream vortex breakdown without a reverse flow inside the wake.

This research has been supported by the NSF Grant CTS – 9622302.

6. References

1. Lambourne, N.C. & Bryer, D.W.: The Bursting of Leading Edge Vortices — Some Observations and Discussion of the Phenomenon, *Aeronautical Research Council Report and Memoranda* 3282, 1962.
2. Saffman, P.G.: *Vortex Dynamics*, Cambridge University Press, 1992.
3. Lowson, M.V.: Some experiments with vortex breakdown, *J. of Roy. Aeronautical Soc.* **68** (1964), 343.
4. Muylaert, I.M.: Effect of compressibility on vortex bursting on slender delta wings, *VKI Project Report* 1980-21.
5. Arnol'd, V. I.: *Catastrophe Theory*, Springer Verlag, 1984.
6. Burggraf, O.R. & Foster, M.R.: Continuation or breakdown of tornado-like vortices, *J. Fluid Mech.* **80** (1977), 685-704.
7. Lugt, H.J.: Vortex Breakdown in Atmospheric Columnar Vortices, *Bul. Amer. Meteorol. Soc.* **70** (1989), 1526-1537.
8. Shtern, V. & Hussain, F.: Hysteresis in a swirling jet as a model tornado, *Phys. Fluids* A **5** (1993), 2183-2195.
9. Spotar', S.Yu. & Terekhov, V.I.: Two spontaneously alternating regimes of vortex flow above a plane, *J. Appl. Mech. Techn. Phys.* (1987) No 2, 68-70.
10. Goldshtik, M. A.: Viscous flow paradoxes, *Ann. Rev. Fluid Mech.*, **22** (1990) 441-472.

11. Atthaus., W., Brücker, Ch. & Weimer, M. Breakdown of slender vortices, In: Green S.I. (ed.) *Fluid vortices*, 373-426, Kluwer, 1995
12. Berger, S. A.: Ellipticity in the vortex breakdown problem. In: *Instability, Transition, and Turbulence*, 96-106, ICASE/NASA LaRC Series, Springer Verlag, 1992.
13. Sarpkaya, T.: Vortex Breakdown in Swirling Conical Flows. *AIAA Journal*, **9** (1971), 1792-1799.
14. Faler, J. H. & Leibovich, S.: Disrupted States of Vortex flow and Vortex Breakdown. *Phys. Fluids* **20** (1977), 1385-1400.
15. Sarpkaya, T.: Turbulent vortex breakdown. *Phys. Fluids* **7** (1995), 2301-2303.
16. Hall, M.: Vortex Breakdown. *Ann.. Rev. Fluid Mech.* **4** (1972), 195-218.
17. Leibovich, S.: The structure of vortex breakdown, *Ann. Rev. Fluid Mech.*, **10** (1978), 221-246.
18. Squire, H.B.: Rotating fluids, *Surveys in Mechanics*, Cambridge University Press, 139-161, 1956.
19. Benjamin, T.B.: Theory of vortex breakdown *J. Fluid Mech.* **14** (1962), 593-629.
20. Escudier, M. E. & Keller, J. J.: Vortex breakdown: a two-stage transition. AGARD CP No. 342 *Aerodynamics of vortical type flows*, paper 25, 1983.
21. Keller, J.J., Egli, W. & Exley, J.: Force- and loss-free transitions between flow states, *ZAMP* **36** (1985), 854-889.
22. Ludwieg, H.: Zur Erklarung der Instabilitat der uber angestellten Deltaflugeln auftretenden freien Wirbelkerne, *Zeitschrift für Flugwissenschaften* **10** (1962), 242-249.
23. Leibovich, S.: Vortex Stability and Breakdown: Survey and Extension, *AIAA Journal* **22** (1983), 1192-1206.
24. Foster, M. R.: Nonaxisymmetric instability in slowly swirling jet flows. *Phys. Fluids A* **5** (1993), 3122-3135.
25. Panda J. & McLaughlin, D. K.: Experiments on the instabilities of a swirling jet. *Phys. Fluids* **6** (1994), 263-276.
26. Khorami, M. R. & Triveli, P.: The viscous stability analysis of Long's vortex. *Phys. Fluids* **6** (1994), 2623-2630.
27. Gelfgat, A.Yu., Bar-Yoseph, P.Z. & Solan, A.: Stability of confined swirling flow with and without vortex breakdown. *J. Fluid Mech.* **311** (1996), 1-36.
28. Trigub, V.N.: The problem of breakdown of a vortex line, *PMM USSR* **49** (1985), 166-171.
29. Buntine, J.D. & Saffman, P.G.: Inviscid swirling flows and vortex breakdown, *Proc. R. Soc. Lond. A* **448** (1995), 1-15.
30. Goldshtik, M. & Hussain, F.: The nature of vortex breakdown, *Phys. Fluids* **9** (1997), 263-265.
31. Wang, S. & Rusak, Z.: The dynamics of a swirling flow in a pipe and transition to axisymmetric vortex breakdown, *J. Fluid Mech.* (1997), **340**, 177-223.
32. Beran, P. S. & Culick, F. E. C.: The role of non-uniqueness in the development of vortex breakdown in tubes, *J. Fluid Mech.* **242** (1992), 491-527.
33. Lopez, J. M. On the bifurcation structure of axisymmetric vortex breakdown in a constricted pipe, *Phys. Fluids* **6** (1994), 3683 - 3693.
34. Escudier, M. P. Observations of the flow produced in a cylindrical container by a rotating endwall. *Experiments in Fluids* **2** (1984), 189-196.
35. Goldshtik, M.A. & Shtern, V.N.: Collapse in conical swirling flows, *J. Fluid Mech.* **218** (1990), 483-508.
36. Shtern, V. & Hussain F.: Hysteresis in swirling jets, *J. Fluid Mech.* **309** (1996), 1-44.
37. Long, R.R.: A vortex in an infinite fluid, *J. Fluid Mech.* **11** (1961), 611-623.
38. Taylor, G.I.: The boundary layer in the converging nozzle of a swirl atomizer, *Quart. J. Mech. Appl. Math.*, **3** (1950), 129-139.
39. Goldshtik, M. A.: A paradoxical solution of the Navier-Stokes equations, *J. Appl. Math. Mech.* **24** (1960), 913-929.
40. Serrin, J.: The swirling vortex. *Phil. Trans. R. Soc. London Ser. A* **271** (1972), 325-360.
41. Schlichting, H.: *Boundary-Layer Theory*, McGrow-Hill N.Y., 1979.
42. Squire, H. B.: The growth of vortex in turbulent flow", *Aero. Quart.* **16** (1965), 302-306.
43. Govindaraju, S. P. & Saffman, P. G.: Flow in a turbulent trailing vortex, *Phys. Fluids* **14** (1971), 2074-2080.
44. Albring, W.: *Elementarvorrgänge fluider Wirbelbewegungen*, Akademie-Verlag, Berlin, 1981.
45. Goldshtik, M.A. & Shtern, V.N.: Conical flows of fluid with variable viscosity, *Proc. Roy. Soc. London*, **419 A** (1988), 91-106.
46. Shtern, V., Herrada, M. & Hussain, F.: A model of turbulent vortex breakdown, *AIAA pap. 97-1842.*

TURBULENT VORTEX BREAKDOWN: A NUMERICAL STUDY

ROBERT E. SPALL
Department of Mechanical and Aerospace Engineering
Utah State University
Logan, UT 84322-4130

THOMAS B. GATSKI
Aerodynamic and Acoustic Methods Branch
NASA Langley Research Center
Hampton, VA 23681-2199

Abstract
Three-dimensional and axisymmetric solutions to the Reynolds-averaged Navier-Stokes equations have been obtained for turbulent vortex breakdown within a slightly diverging tube. Inlet boundary conditions were derived from experimental data for the mean flow and turbulence kinetic energy provided by Sarpkaya (Private Communications). The performance of both two-equation and full differential Reynolds stress models is evaluated. The experimentally determined location of breakdown was well predicted by the differential Reynolds stress model. However, the two-equation models failed to predict the occurrence of breakdown. Failure of the two-equation models is attributed to their inability to accurately account for Reynolds stress anisotropies. Further comparisons between the experimental data of Sarpkaya and the Reynolds stress model results are presented and discussed.

1. Introduction

The majority of numerical work concerning vortex breakdown may be divided into two categories: 1) those for both confined and unconfined flows containing inflow/outflow boundaries, and 2) those that consider flows within cylindrical containers driven by rotating endwalls. In terms of category (1), which represent breakdown flows most similar to those found in practical applications, the majority of the existing numerical work investigates low Reynolds number, laminar, axisymmetric (c.f., Grabowski and Berger (1976)) or three-dimensional flows (c.f., Spall and Gatski (1991), Breuer and Hanel (1993)). An interesting feature of the axisymmetric cases are the existence of nonunique solutions over specific ranges of the flow parameters (c.f., Beran and Culik 1992). However, these nonunique bubble-type solutions have not been observed in three-dimensions; in fact the

E. Krause and K. Gersten (eds.), IUTAM Symposium on Dynamics of Slender Vortices, 307-319.

existence of a long-term stable bubble-type breakdown structure has yet to be demonstrated for fully three-dimensional simulations (c.f., Spall 1996, Tromp and Beran 1997), other than perhaps at very low Reynolds numbers, or within closed cylindrical containers with rotating endwalls.

One of the primary contributions of the numerical works has been to provide insight into the internal structure of vortex breakdown. However, in most technological applications, the vortex breakdown arises within a turbulent swirling flow, and the applicability of laminar results to turbulent breakdown is questionable. In fact, Sarpkaya (1995a, 1995b) presented experimental results for vortex breakdown in non-cavitating, high Reynolds number (up to 225,000) swirling flows and considered the resulting breakdown fundamentally distinct from the various forms of laminar breakdown. These high Reynolds number breakdowns are characterized by the lack of a distinct bubble and appear to be the most robust of all the breakdown forms.

The only existing numerical work appearing in the literature aimed at studying the internal structure of turbulent vortex breakdown (outside of the combustion community, where the breakdown is referred to as a central toroidal recirculation zone, and where geometries typically include such complicating factors as dilution jets and rapid expansions, c.f. Hogg and Leschziner (1989)) are those of Bilanin et al. (1977) and Spall and Gatski (1995). The work of Bilanin et al., in which the Reynolds-Averaged Navier-Stokes (RANS) equations were solved, followed much along the lines of the steady, laminar axisymmetric calculations of Grabowski and Berger (1976). Spall and Gatski also solved the RANS equations (employing the algebraic stress model of Gatski and Speziale (1993)), presenting results for the unsteady, 3-D turbulent breakdown of an unconfined longitudinal vortex. Their results showed some qualitative agreement with Sarpkaya's experimental results (i.e., relative steadiness of the mean flow, robustness, and a lack of asymmetries) but in the absence of common lateral and inflow boundary conditions, a closer comparison between the results was deemed unwarranted.

Several distinct approaches to modeling these high Reynolds number swirling flows exist, ranging from solutions to the Reynolds-averaged Navier-Stokes equations, to large eddy simulations (LES), to direct numerical simulations (DNS). However, to date, DNS and LES simulations utilizing spectral schemes have been limited primarily to geometrically simple configurations, and for the case of DNS, the restrictions include relatively low Reynolds numbers (costs scale as the Reynolds number cubed). Higher-order finite-difference techniques enjoy more flexibility in terms of geometries and boundary conditions, but with the lack of spectral accuracy, their built-in low-pass filter may tend to confuse the issue of resolved scales versus subgrid-scale motions in LES calculations. One must conclude that in the foreseeable future, it is unlikely that LES or DNS approaches will be available as computational tools to be utilized in the investigation and solution of turbulent flow problems in most *technologically important* applications of swirling flows such as the Ranque-Hilsch tube, wake-vortex alleviation, submarine non-acoustic stealth, and flame stabilization.

Hence, we are motivated in the present study to employ the RANS equations to study numerically vortex breakdown in confined high Reynolds number turbulent swirling flows. The geometry is modeled after the axisymmetric diverging tube test section employed in the experimental work of Sarpkaya (1995a, 1995b). Several features contribute to the complex nature of this flow, including the existence of a mild adverse pressure gradient (diverging tube), strong streamline curvature, and an internal separation point (at breakdown). This geometry is somewhat simplified from that of, for instance, most combustors; however it represents an excellent test case to ascertain the capabilities of the RANS equations and associated closure models for strongly swirling flows. The long-term goal of this work then, is to develop RANS closure models that are suitable for engineering analysis of strongly swirling flows.

2. Numerical Method and Turbulence Models

The pressure-based finite-volume code Fluent (Fluent, Inc., Lebanon, NH) has been utilized to solve both axisymmetric and three-dimensional formulations of the Reynolds-averaged Navier-Stokes equations. Cylindrical-polar velocity components were employed to reduce numerical diffusion. Interpolation to cell faces was performed using a QUICK (Leonard 1979) differencing scheme. Pressure-velocity coupling was based upon the SIMPLEC procedure (c.f., Patankar 1980).

A differential Reynolds stress model (DRSM), and both standard and Renormalization Group (RNG) based $K-\varepsilon$ models have been employed for turbulence closure. It is well known that in regions of high strain rate the standard $K-\varepsilon$ model produces excessive levels of turbulence kinetic energy, leading to high values of the isotropic turbulent viscosity. As a result, the model may overpredict the radial diffusion of momentum in strongly swirling flows, leading to an overly rapid decay of maximum swirl velocities. In some cases the RNG based models have been shown to produce results superior to the standard $K-\varepsilon$ model for flows with high streamline curvature and strain rate (c.f., Yakhot et al. 1992). The RNG model is similar in form to the standard $K-\varepsilon$ model except for the addition of a rapid strain term $(\bar{R})$ in the dissipation equation which is written in cartesian tensor form as:

$$\frac{D\varepsilon}{Dt} = \frac{\partial}{\partial x_j}\left(\frac{\nu_t}{\sigma_\varepsilon}\frac{\partial\varepsilon}{\partial x_j}\right) + 2\nu_t C_{\varepsilon 1}\frac{\varepsilon}{K}S_{ij}^2 - C_{\varepsilon 2}\frac{\varepsilon^2}{K} - \bar{R} \tag{1}$$

where,

$$\bar{R} = \frac{C_\mu \eta^3 (1-\eta/\eta_0)}{1+\beta\eta^3}\frac{\varepsilon^2}{K} \tag{2}$$

In addition, $\nu_t = C_\mu K^2/\varepsilon$, $\eta = SK/\varepsilon$, and $S = (2S_{ij}S_{ij})^{1/2}$ (where S_{ij} is the mean strain rate). Note that in regions of large strain the sign of $\bar{R}$ changes, with the effect of decreasing ν_t. This decrease in ν_t may then limit the radial diffusion of momentum and perhaps compensate for that shortcoming of the standard $K-\varepsilon$ model (for which $\bar{R} \equiv 0$). An additional feature of the RNG model is that no empirical constants appear in the equations. Theoretical analysis yields $C_\mu = 0.084$, $C_{\varepsilon_1} = 1.42$, $C_{\varepsilon 2} = 1.68$, $\sigma_\varepsilon = 0.72$, $\beta = 0.012$ and $\eta_0 = 4.38$ (c.f. Yakhot et al. 1992) which may be compared with the values employed for the standard $K - \varepsilon$ formulation of $C_\mu = 0.09$, $C_{\varepsilon_1} = 1.44$, $C_{\varepsilon 2} = 1.92$, and $\sigma_\varepsilon = 1.3$. The values of the coefficients are quite similar, with the exception of $C_{\varepsilon 2}$, which is considerably decreased in the RNG model.

We note that in the past, other modifications to the standard $K - \varepsilon$ model have been applied to the dissipation rate equation in an effort to improve predictions for swirling flows. These are primarily (ad-hoc) Richardson number modifications to the sink or source terms, and have met with only limited success (c.f., Srinivasan and Mongia 1980).

The advantage of Reynolds stress models is that they inherently account for the effects of streamline curvature, body forces and rotation. In the present work, the DRSM closure assumptions are based upon the work of Launder et al. (1975) and Gibson and Launder (1978). In particular, the pressure-strain term, Φ_{ij}, consists of the linear return-to-isotropy model for the "turbulence" portion as,

$$\phi_{ij1} = -C_1 \frac{\varepsilon}{K}\left(\overline{u_i' u_k'} - \frac{2}{3}\delta_{ij} K\right) \tag{3a}$$

and the isotropization of production model for the mean-strain part,

$$\phi_{ij2} = -C_2\left(P_{ij} - \frac{1}{3}\delta_{ij} P_{kk}\right), \tag{3b}$$

(where $\Phi_{ij} = \phi_{ij1} + \phi_{ij2}$, and P_{ij} is the production term). The range of values for C_1 and C_2 employed in the past are usually defined by the relation $(1 - C_2)/C_1 = 0.23$, with the most commonly assigned values (and those employed in this work) given as 1.8 and 0.60, respectively. The work of Gibson and Younis (1986) in which $C_1 = 3.0$ and $C_2 = 0.3$ (which reduces the importance of the mean strain component) is an exception. These values were found to improve the predictive capability for mildly swirling jets. Calculations employing the Gibson-Younis values are also presented in the Results section. In the present work, the wall reflection terms responsible for redistribution of the normal stresses near the wall were not included, however this should not be problematic since we are interested in the vortex core region where contributions from these terms would be negligible. Finally, the dissipation term was modeled by an isotropic dissipation rate while diffusion was modeled by a gradient approximation.

For each of the models, standard equilibrium wall functions were used to implement the duct wall boundary conditions. This eliminated the need for an overly fine grid near the wall, which again, is far removed from the primary area of interest—the vortex core region.

3. Geometry and Boundary Conditions

The course of the numerical study has been guided by experimental data provided to these investigators by Professor Sarpkaya (Private Communication). In particular, the geometry represents a model of the experimental diverging tube apparatus of Sarpkaya (1995a, 1995b). The diverging portion of the computational domain extends from $x = 0$ to $x = 7.25r_0$, where r_0 is the tube radius at the inlet boundary. A section of constant radius, $r = 1.154r_0$, then extends to $x = 41.6r_0$ where zero gradient outflow boundary conditions were imposed.

Experimental data, in terms of the mean flow and turbulence kinetic energy profiles, were utilized to derive the inlet boundary conditions. The inlet distributions for the mean velocities, which represent least-squares fits to the experimental data, are shown in Fig. (1a) (where U and W represent axial and azimuthal velocities, respectively). In these figures and those to follow, velocities have been scaled with respect to the mean axial velocity at inlet, and lengths have been scaled with respect to the inlet tube radius (r_0).

Specification of the inlet turbulence quantities, and in particular the dissipation rate, is problematic. The only experimental data available to the authors at this time is the radial distribution of the turbulence kinetic energy. Hence, in the case of the DRSM calculations, the stress distribution was assumed isotropic, with the normal stresses taken as $2k/3$. One may approximate ε via an eddy-viscosity hypothesis when strain rates and Reynolds stresses are known. However, in the present case the stresses are unavailable, and in addition, the eddy viscosity is likely to be highly anisotropic. Hence, the alternative employed was to specify the dissipation rate at the inlet through a relation of the form:

$$\varepsilon = C\frac{K^{3/2}}{l} \tag{4}$$

where the proportionality factor needs to be specified. If one employs a ratio $C/l = 26$, then the dissipation rate determined by Eq. (4) appears to approximate reasonably well the dissipation rate profiles determined through analysis of preliminary experimental data (Sarpkaya, Private Communication). If the length scale l is taken as the vortex core radius, then this factor is roughly an order of magnitude above what one might employ if standard boundary layer assumptions were employed to determine C as $C_\mu^{3/4}$. To assess the effect of the inlet dissipation rate on the solutions, a limited number of calculation are also presented for which the inlet dissipation was computed by decreasing C/l an order of magnitude to 2.6. It should also be noted that varying ε at the inlet boundary also has the effect of altering the effective viscosity through the relationship $\nu_t = (C_\mu K^2)/\varepsilon$. The inlet distributions employed for the turbulence kinetic energy and dissipation rate are shown in Fig. (1b).

4. Results

The axisymmetric calculations, which are discussed first, were performed on a 298 (axial) x 140 (radial) grid with significant stretching toward the duct centerline, duct outer wall, and (axially) near the breakdown region. At the inlet, approximately 25 grid points were contained within the vortex core radius. A grid resolution study, performed by halving the number of grid points in each direction, did not significantly alter the results. All calculations were performed at a Reynolds number, based upon mean axial velocity and tube diameter at the inlet, of 130,000.

Shown in Fig. (2) is the axial velocity distribution along the vortex centerline ($r = 0$) for each of the models, and for the two inlet distributions of ε discussed above. Available experimental data points outside the breakdown region are also shown. Although experimental data points are lacking from the figure, a small region reversed flow (the breakdown region) occurs between $2.5 < x < 4.0$ (Sarpkaya, Private Communication). The only model to reproduce this reversed flow was the DRSM (for which the higher inlet dissipation rate was specified). The DRSM solution resulting from the lower value of inlet dissipation did produce qualitatively similar results, however the recirculation zone was not predicted. One interesting observation is that the DRSM solution employing the higher dissipation rate showed a pronounced break in the slope of the profile between the inlet boundary and the first minimum in the axial velocity, whereas the other calculation did not. The observation of a break in the slope is consistent with the experimental data.

The effects of inlet dissipation on the two-equation models is much more pronounced. With the lower dissipation level specified at inlet, each model produced a result that is far from the breakdown state. However, with the increased dissipation level, the RNG model produced a minimum axial velocity of approximately 0.09, which may be considered near the breakdown state. The trend of the $K - \varepsilon$ model followed that of the RNG, however the minimum value along the vortex centerline was limited to approximately 0.3. Beyond $x \approx 8$ the predictions of each of the two-equation models, regardless of the specification of the inlet dissipation rate, are quite nearly identical. A similar observation may be made for the two DRSM calculations.

In terms centerline axial velocities then, none of the models was able to predict accurately the behavior of the mean flow in the region of the breakdown. The DRSM was however, able to do a reasonable job of predicting the axial velocity both upstream and downstream of the breakdown location, and did reveal a region of reversed flow. We further explore the regions upstream and downstream of breakdown in the series of figures that follow. In particular, mean axial and azimuthal velocities computed using the higher level of dissipation rate at the inlet boundary are examined at one station upstream of breakdown and two stations downstream of breakdown. In addition, DRSM calculations with the Gibson-Younis values of C_1 and C_2 are also included. For the remaining figures, experimental data points represent the average of values taken at $\pm r$, for a given axial location. The averaging was performed due to small asymmetries in the experimental profiles.

Shown in Figs. (3a,b) are velocity profiles at $x = 0.833$ (upstream of the breakdown location). As revealed in Fig. (3a), both the RNG and $K - \varepsilon$ models have considerably over predicted the radial diffusion of azimuthal momentum. Consequently, the accompanying increase in pressure along the vortex centerline leads to an overly rapid decay of the mean axial velocity profiles, as shown in Fig. (3b). The DRSM (with coefficients $C_1 = 1.8$, and $C_2 = 0.6$, which hereafter will be referred to as DRSM1) has done a good job of preserving the azimuthal velocity profile, and accordingly, the mean axial velocity profile. The results of the DRSM calculation with $C_1 = 3.0$, and $C_2 = 0.3$ (hereafter referred to as DRSM2) indicate that the rate of diffusion of azimuthal momentum has been over-predicted, leading to premature decay of the axial velocity along the centerline. Although not shown, we note that the DRSM1 results in which the lower level of dissipation rate was specified at the inlet did not match the experimental data as well, over predicting the rate of diffusion of azimuthal momentum.

Mean velocity profiles at $x = 5.0$, slightly downstream of the breakdown location, are shown in Figs (4a,b). At this location, each of the models has over-predicted the radial diffusion of azimuthal momentum. In the case of the two-equation models, the radius of the viscous core is over twice that revealed by the experimental data. For the DRSM1 model, the predicted radius is approximately 60% greater than the experimental data, whereas the predicted maximum azimuthal velocity is approximately 25% below the data. The DRSM2 model predicts a rotation rate within the vortex core nearly identical to that of the DRSM1 model, however the peak azimuthal velocity is further under-predicted. Outside the viscous core, the each of the models does a good job of matching the experimental data. In terms of the mean axial velocities, each of the two-equation models have over-predicted its rate of recovery within the core, while the DRSM models have under-predicted this quantity.

The final station for which experimental data was available is given at $x = 8.3$ (see Figs. (5a,b)). Consistent with results for the upstream stations, the two-equation models have considerably over-predicted the vortex core radius, and have also over-predicted the rate of recovery of the axial velocity profile. The DRSM results are much better, particularly with regards to the axial velocity profiles. In terms of the azimuthal velocity, each of the DRSM models gives qualitatively similar results, under-predicting the maximum azimuthal velocity by approximately 25%.

We further delineate the differences between the DRSM1 and two-equation models by comparing contours of mean axial velocities, as shown in Figs. (6a-c). The structure of the resulting flowfields as predicted by the two-equation models are quite similar. The primary difference occurs in the region bounded by $1.5 \leq x \leq 2.5$, where the RNG model predicts a lower mean axial velocity along the vortex centerline. However, neither model predicts a region of recirculating flow. Away from this region, the flowfields predicted by the two models are nearly identical. The flowfield predicted by the DRSM1 model is considerably different. We observe a distinct region of strong axial deceleration just upstream of the breakdown point, which, consistent with experimental results (Sarpkaya, Private Communication), is followed by a short region of reversed flow. Although detailed experimental data is not available, results describing the bounding breakdown "envelope" are also reasonably consistent with the flow visualization studies of Sarpkaya (1995a, 1995b).

As discussed in the previous section, one explanation for the failure of the $K-\varepsilon$ model may be due to its highly dissipative nature in regions of high strain rate. That is, the resultant high levels of apparent viscosity cause the swirl velocities decay so rapidly that, despite the adverse pressure gradient induced by the tube divergence, the flow remains supercritical (i.e., small disturbances are swept downstream) and no breakdown occurs. We investigate this hypothesis by examining contours of constant apparent viscosity, $C_{\mu}K^2/\varepsilon$, shown for each of the models in Figs. (7a-c). Recall that the RNG model includes a rapid strain term in the dissipation rate equation, and employs different values for the coefficients (specifically, $C_{\varepsilon 2}$) such that it tends to predict lower levels of turbulence kinetic energy than does the standard model. Consequently, the decreased levels of turbulence kinetic energy predicted by the RNG model result in notably decreased levels of apparent viscosity. For instance, over the region in which we expect the breakdown to form $(2 \leq x \leq 3)$ the RNG model predicts levels of ν_t approximately 2.5 times below that of both the standard $K-\varepsilon$ and DRSM1 models. In fact, the distribution of ν_t for the $K-\varepsilon$ and DRSM1 models is quite similar. Yet, this difference in the apparent viscosity between the two-equations models is not manifested as a notable difference between the respective mean velocity profiles, as revealed in earlier figures. Hence, it does not seem that the failure of the two-equation models should be attributed solely to excessive levels of apparent viscosity. Rather, it is more likely that the failure of these models should be attributed to the Boussinesq assumption and a resulting inability of the models to properly account for Reynolds stress anisotropies.

In addition to the steady, axisymmetric calculations described above, one set of unsteady, three-dimensional calculations employing the DRSM1 model was performed. The primary purpose of these calculations was to assess the suitability of the dual assumptions of steadiness (of the mean flow) and axisymmetry—recall that for calculations of laminar breakdown, at other than very low Reynolds numbers, axisymmetric bubbles invariably become unstable and transition to spiral-type breakdown. A 168 x 62 x 62 body fitted grid was utilized for these calculations, which were carried out over sufficient time to ensure that no unsteadiness of the mean flow would develop. Representative results, in terms of contours of mean axial velocity in the x-y and x-z planes, are shown in Fig. (8). No indication of transition to spiral breakdown was observed. One explanation for this is that solutions to the RANS equations (at least in 2D) are equivalent to laminar solutions for a specific prescription of a variable viscosity. At these low apparent Reynolds numbers disturbances acting on the mean flow are likely damped. These results are consistent with the experimental results of Sarpkaya (1995a, 1995b) in which the mean flow appeared quite robust, axisymmetric and steady.

5. Summary

A differential Reynolds stress model was found to predict quite well the location of the onset of turbulent vortex breakdown in a slightly diverging pipe. This result was encouraging in that only complete experimental profiles for the mean flow and turbulence kinetic energy were available as inlet boundary conditions. In addition, the existence of a small recirculation region agrees qualitatively with experimental data. Calculations performed by employing Reynolds stress model constants $C'_1 = 3.0$ and $C'_2 = 0.3$ as recommended in Gibson and Younis (1986) for mildly swirling flows, did not improve the predictive capabilities of the model. Results for a fully time-dependent, three-dimensional Reynolds stress model calculation were consistent with the axisymmetric calculations in that the flow evolved to a steady state and transition to spiral breakdown did not occur.

As expected, the performance of the 2-equation $K-\varepsilon$ models was inadequate, and their use for vortex breakdown-like flows must be discouraged. However, rather than attributing the failure of the 2-equation models on excessive diffusivity, it is concluded that the real problem likely lies in the failure of the 2-equation models to accurately predict the Reynolds stress anisotropies. Future work will include an implementation of the complete experimentally determined Reynolds stress distribution at the inlet boundary, with the analysis concentrating on more quantitative aspects of the DRSM predictions, such as a comparison between experimental and numerical results for the Reynolds stresses upstream and downstream of breakdown.

Acknowledgments

Support from NASA Langley Research Center and the National Science Foundation is gratefully acknowledged. Professor Turgut Sarpkaya is thanked for sharing his experimental data with the authors.

References

Beran, P.S. and Culik, F.E.C., (1992) "The Role of Non-Uniqueness in the Development of Vortex Breakdown in Tubes," J. Fluid Mech., Vol. 242, pp. 491-527.

Bilanin, A.J., Teske, M.E. & Hirsh, J.E. (1977). Deintensification as a consequence of vortex breakdown. In *Proceedings of the Aircraft Wake Vortices Conference*, Report No. FAA-RD-77-68.

Breuer, M. and Hanel, D. (1993) "A dual time-stepping method for 3-D viscous incompressible vortex flows," Comput. Fluids, Vol. 22, pp. 467-484.

Gatski, T.B. and Speziale, C.G. (1993) "On Explicit Algebraic Reynolds Stress Models for Complex Turbulent Flows," J. Fluid Mech., Vol. 254, pp. 59-78.

Gibson, M.M. and Launder, B.E. (1978) "Ground Effects on Pressure Fluctuations in the Atmospheric Boundary Layer," J. Fluid Mech., Vol. 86, pp. 491-511.

Gibson, M.M. and Younis, B.A. (1986) "Calculation of Swirling Jets Using A Reynolds Stress Closure," Phys. Fluids, Vol. 29, pp. 38-48.

Grabowski, W.J. and Berger, S.A. (1976) "Solutions of the Navier-Stokes Equations for Vortex Breakdown," J. Fluid Mech., Vol. 75, pp. 525-544.

Hogg, S. & Leschziner, M.A. (1989). Computation of highly swirling confined flow with a Reynolds stress turbulence model. AIAA J. 27, 57-63.

Launder, B.E., Reece, G.J., and Rodi, W. (1975) "Progress in the Development of a Reynolds-Stress Turbulence Closure," J. Fluid Mech., Vol. 68, pp. 537-566.

Leonard, B.P. (1979) "A Stable and Accurate Convective Modeling Procedure Based on Quadratic Upstream Interpolation," Comput. Methods Appl. Mech. Eng., pp. 59-98.

Patankar, S.V. (1980) Numerical Heat Transfer and Fluid Flow, Washington, D.C., Hemisphere Publishing Corp.

Sarpkaya, T. (1995a) "Vortex Breakdown and Turbulence," AIAA 95-0433.

Sarpkaya, T. (1995b) "Turbulent Vortex Breakdown," Phys. Fluids, Vol. 7, pp. 2301-2303.

Spall, R.E. and Gatski, T.B. (1991) "Computational Study of the Topology of Vortex Breakdown," Proc. R. Soc. Lond. A Vol. 435, pp. 321-337.

Spall, R.E. and Gatski, T.B. (1995) "Numerical Calculations of Three-Dimensional Turbulent Vortex Breakdown," Int. J. Numer. Methods Fluids, Vol. 20, pp. 307-318.

Spall, R.E. (1996) "Transition From Spiral- to Bubble-Type Vortex Breakdown," Phys. Fluids, Vol. 8, pp. 1330-1332.

Srinivasan, R. & Mongia, H.C. (1980). Numerical computation of swirling recirculating flows. NASA CR-165196.

Tromp, J.C. and Beran, P.S. (1997) "The Role of Nonunique Axisymmetric Solutions in 3-D Vortex Breakdown," Phys. Fluids, Vol. 9, pp. 992-1002.

Yakhot, V., Orzag, S.A., Thangam, S., Gatski, T.B. and Speziale, T.B. (1992) "Development of Turbulence Models for Shear Flows by a Double Expansion Technique," Phys. Fluids A, Vol. 4, pp. 1510-1520.

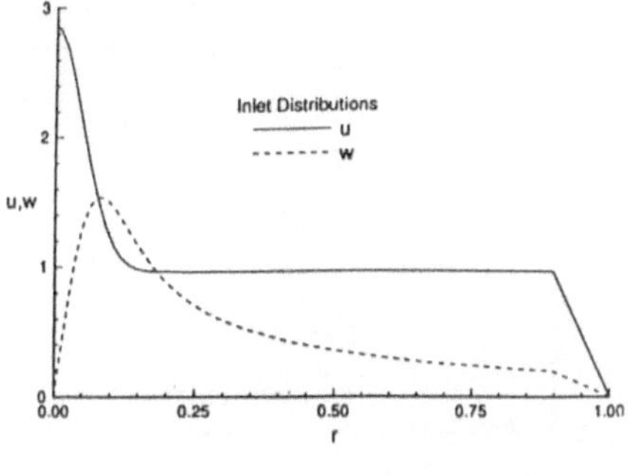

a) Mean velocity profiles.

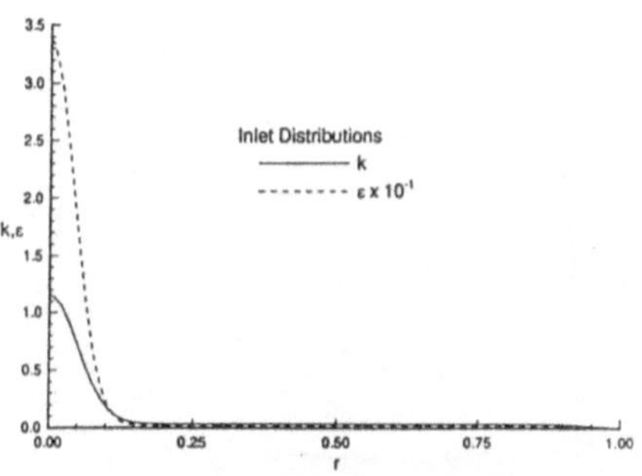

b) Turbulence quantities.

Figure 1. Experimental profiles utilized as inlet boundary conditions.

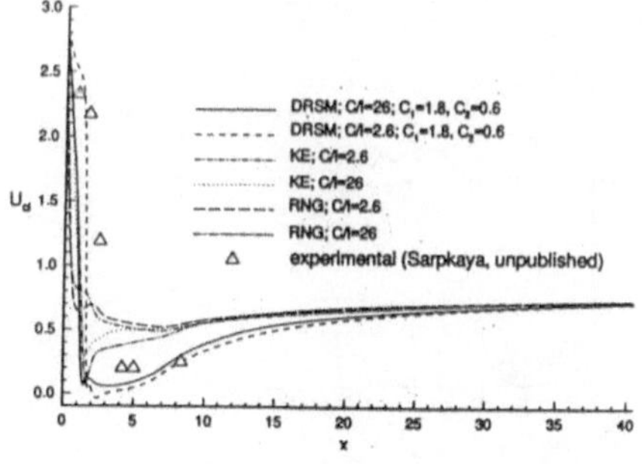

Figure 2. Model predictions of centerline axial velocity.

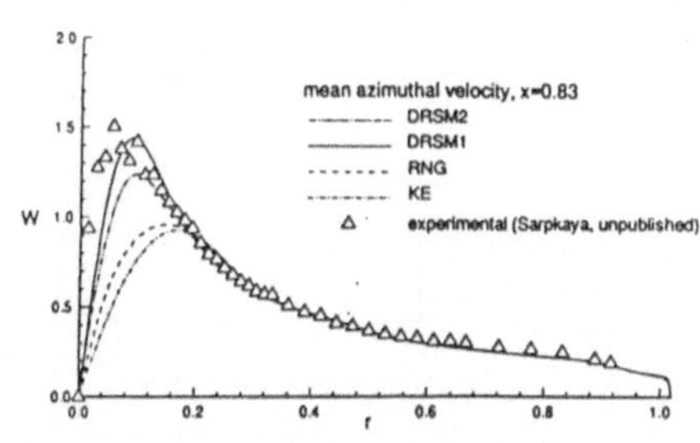

a) Mean azimuthal velocity.

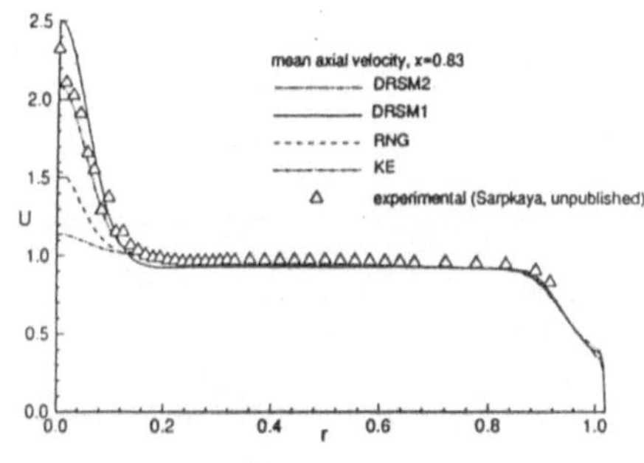

b) Mean axial velocity.

Figure 3. Comparison between model predictions and data at $x = 0.83$

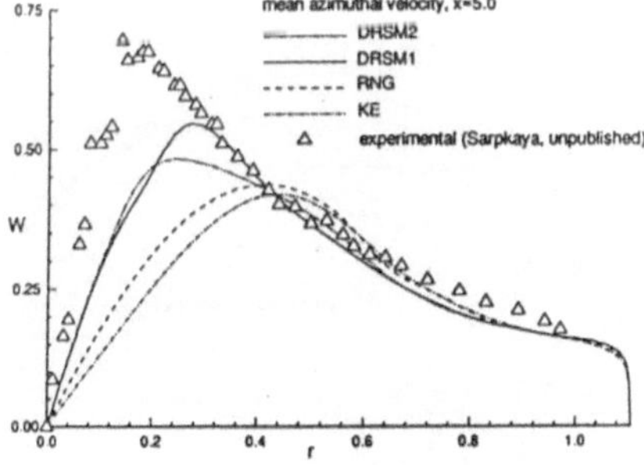

a) Mean azimuthal velocity.

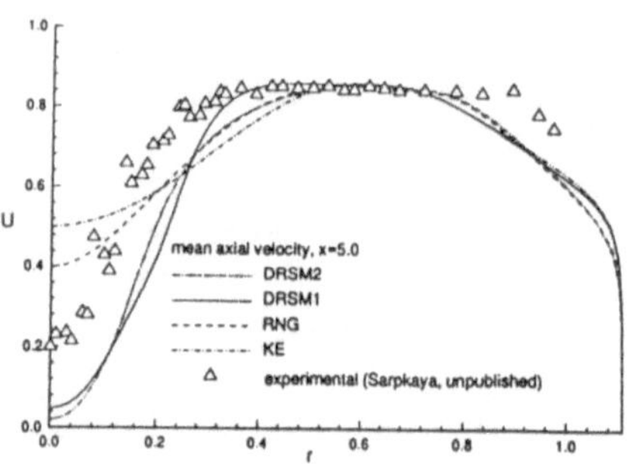

b) Mean axial velocity.

Figure 4. Comparison between model predictions and experimental data at $x = 5.0$.

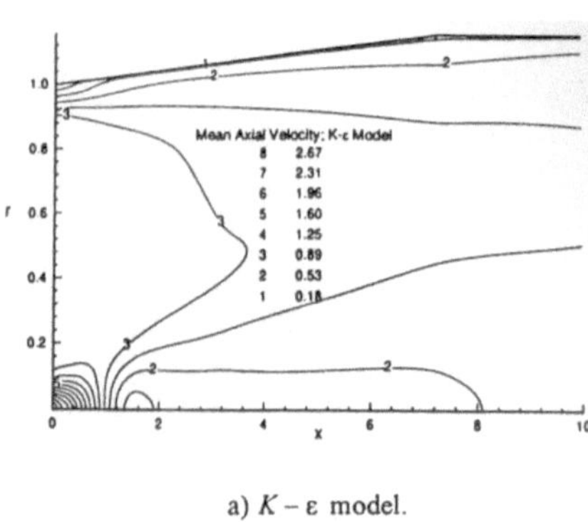

a) $K - \varepsilon$ model.

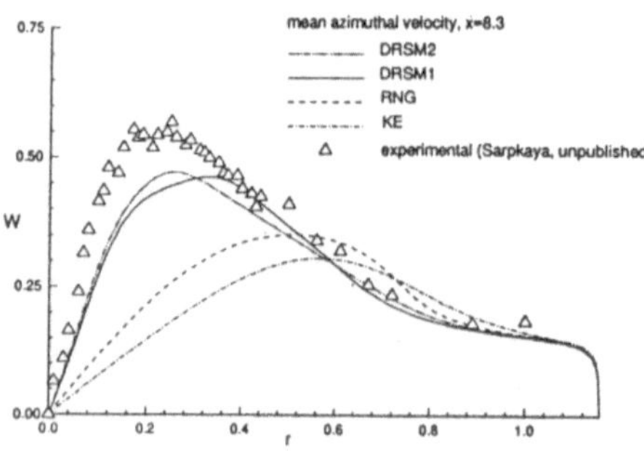

a) Mean azimuthal velocity.

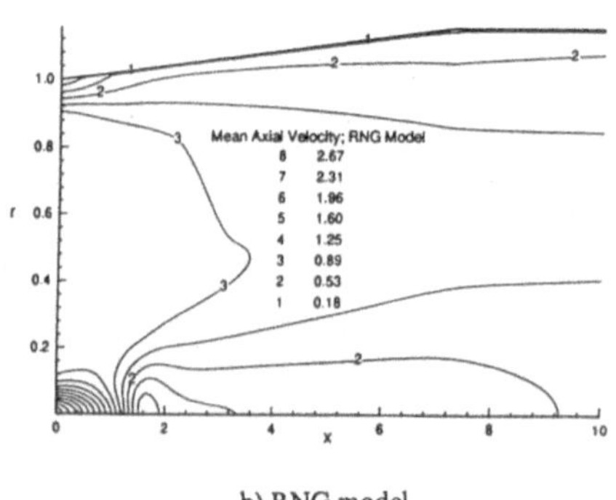

b) RNG model.

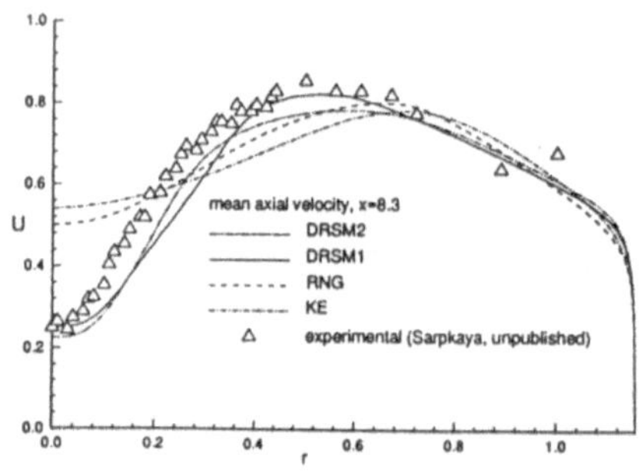

b) Mean axial velocity.

Figure 5. Comparison between model predictions and experimental data at $x = 8.3$.

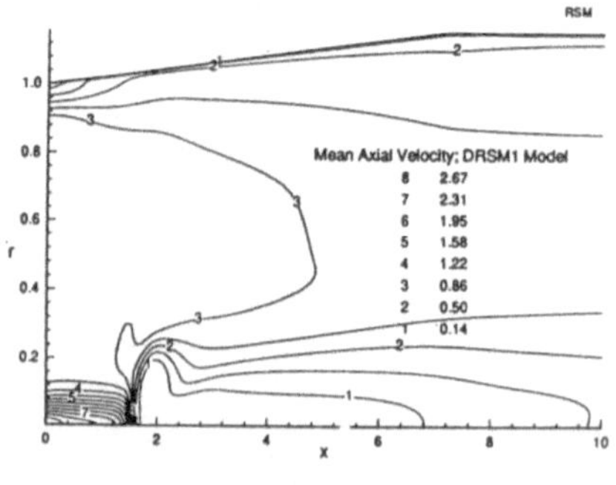

c) DRSM1 model.

Figure 6. Contours of constant mean axial velocity.

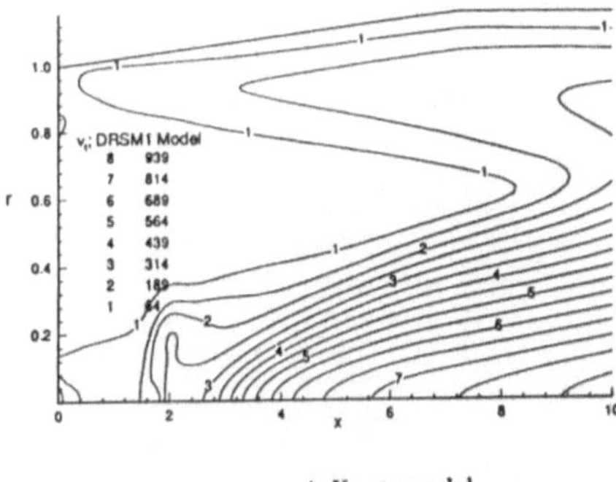

a) $K-\varepsilon$ model.

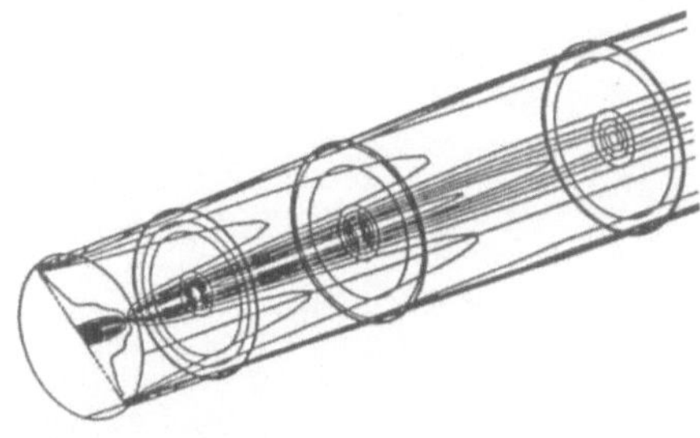

Figure 8. Contours of mean axial velocity for fully three-dimensional, unsteady calculations employing DRSM1.

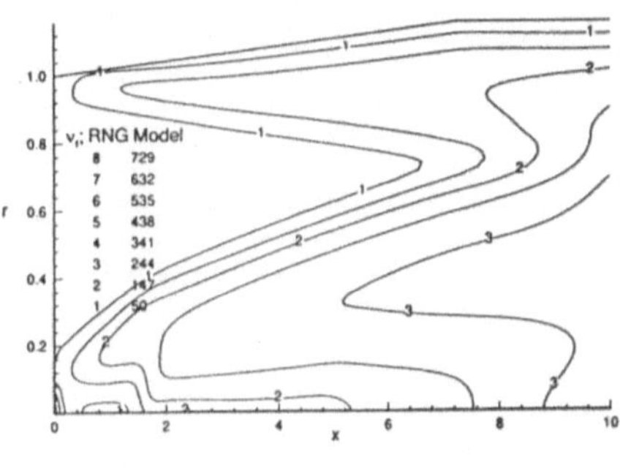

b) RNG model.

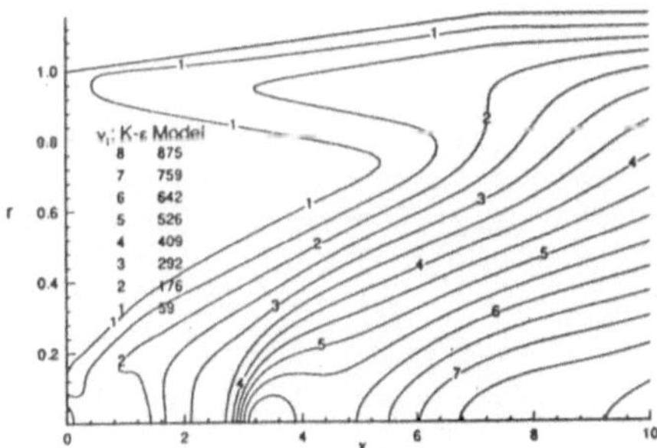

c) DRSM1 model.

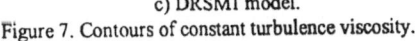

Figure 7. Contours of constant turbulence viscosity.

BREAKDOWN OF SPINNING TUBE FLOWS

CH. BRÜCKER
Aerodynamisches Institut der RWTH Aachen
Wüllnerstr. zw. 5 u. 7
D-52062 Aachen, Germany

Abstract. Breakdown of the turbulent flow in a spinning tube with a diffuser at the end was studied by flow visualization and Digital Particle-Image-Velocimetry (DPIV). The swirl of the flow was introduced at the entrance of the rotating tube with a honeycomb package. The results show, that for the range of parameters investigated, breakdown is initiated at critical Rossby-numbers of Ro≈0.6 similar to those observed in other experiments.

The bursted part of the vortex is of slender conical shape containing near stagnant flow. Its downstream end is characterized by a jump-like contraction near the entrance to the diffuser. The flow then evolves into a jet, enhancing the swirl near the axis. In this region asymmetric instabilities and wavy flow patterns could be observed. Perturbations caused by them travel upstream but do not change the near-axisymmetric shape of the bursted part of the vortex.

1. Introduction

The evolution of vortex breakdown in turbulent flows is still not completely understood [9]. Due to the sensitivity of swirling flows against boundary conditions [7] and external disturbances, experiments are difficult to perform. Most of the experimental studies of vortex breakdown used flow visualization in tubes with guide vanes. In these experiments the circumferential velocity component is always coupled with the axial component. Quantitative results can only be obtained if the velocity profiles are measured. This renders a comparison of results obtained in different set-ups difficult and makes a classification of axisymmetric and non-axisymmetric breakdown almost impossible. With the much improved PIV-technique in the recent years, first instantaneous velocity measurements of the entire flow field in the bursted part of the vortex were obtained, see the review in [1]. The results clearly demonstrated the differences between bubble and spiral-type breakdown. In subsequent time-resolving measurements reported in [3] the transition to spiral-type breakdown was investigated.

In the present article results of turbulent breakdown of spinning tube flows are reported. Previous experiments in a similar set-up reported in [11] were limited to low

E. Krause and K. Gersten (eds.), IUTAM Symposium on Dynamics of Slender Vortices, 321-330.

Reynolds-number flows. The question to be investigated in the present analysis was how turbulent flow in a spinning tube would respond to adverse pressure gradients, generated with a rotating diffuser, and if instabilities similar to those causing the transition from bubble- to spiral-type breakdown affect the spinning flow.

2. Experimental set-up

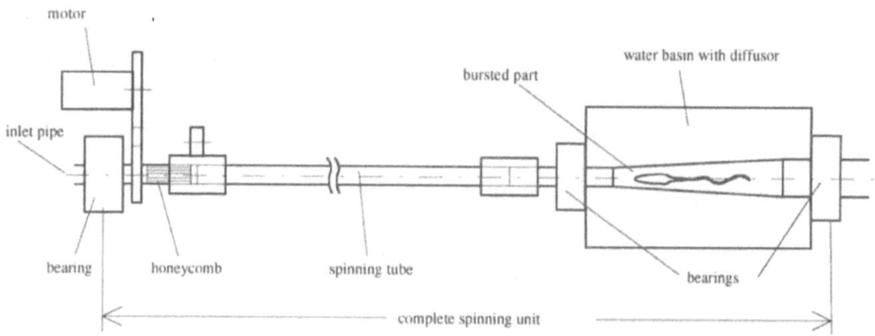

Figure 1. Schematic of the flow channel for studying vortex breakdown in a spinning tube flow

A flow channel with a spinning tube was designed to generate a swirling flow with solid-body rotation (Fig. 1). Water was used as working fluid. Typically it requires 300 tube diameters and more [8] to establish a fully developed swirl. Such a long tube however is difficult to realize with a typical tube radius suitable for flow observation. A honeycomb package was therefore placed at the entrance of the tube of 8 cm in diameter to generate the swirl. A subsequent spinning tube with a length of 20 diameters was chosen in order to damp out any flow disturbances. A diffuser with a length of $l = 3.75$ tube diameters and an opening angle of $\alpha = 1.9°$ was mounted at the end of the tube causing an adverse pressure gradient to promote breakdown. The complete spinning unit was connected upstream to a large water tank with a capacity of 50 m^3. The flow rate into the channel could be controlled with a valve at the channel exit. For DPIV measurements, small tracer particles with a mean diameter of 30 µm (Vestosint, Hüls GmbH, Germany) and a specific gravity of 1.01 g/cm^3 were added to the water upstream of the honeycomb. The flow was visualized with dye entering through a small injection needle placed at the axis of the tube one tube diameter upstream of the diffuser. The maximum flow rate and tube rotation used in the experiments were 120 l/min and 1000 rpm, respectively. The Reynolds-number and the Rossby-number, the two governing similarity parameters, could be varied since the experimental set-up offered the possibility to change the flow rate and the spinning velocity independently.

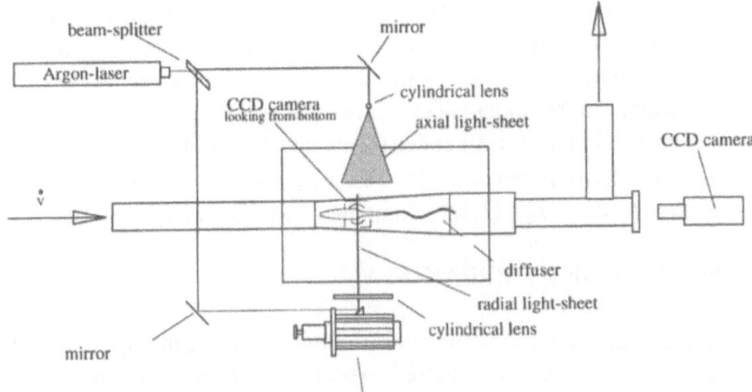

Figure 2. Arrangement of the optical components for flow visualization and DPIV measurements

For flow field measurements, the axial plane and cross-sections in the diffuser were illuminated with thin light-sheets generated with a 2 Watt Ar+ laser beam as depicted in Fig. 2. While the axial light-sheet coincided with the axial plane of the tube in all experiments, the radial cross-sections were rapidly scanned in the axial direction with a 3-D beam scanning device, yielding pictures of the radial flow in 15 planes over the entire breakdown region. The video recordings of the axial and radial cross-sections were processed with the method of Digital Particle-Image-Velocimetry (DPIV). In this manner the time-dependent evolution of the velocity field in the axial plane and the radial cross-sections could be obtained.

3. Results

3.1 INFLOW CONDITIONS

Profiles of the axial and tangential velocity components in the entrance to the diffuser

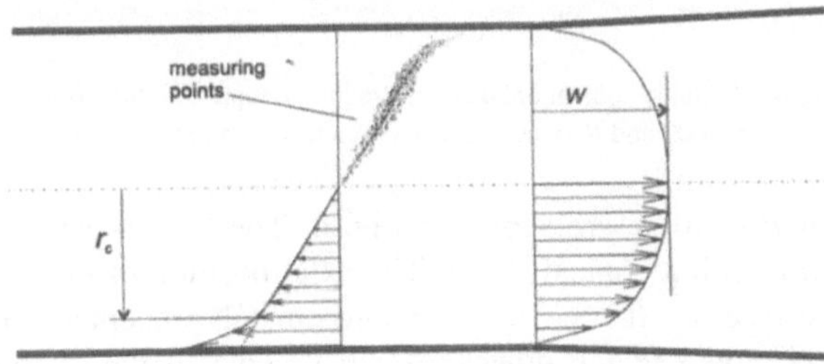

Figure 3. Sketch of the representative velocity profiles of the swirling flow at the entrance to the diffuser after spin-up (from DPIV measurements)

were measured with the method of DPIV. The tube was abruptly started from rest to the final spinning velocity in order to avoid falsification due to breakdown. The results shown in Fig. 3 demonstrate, that solid body rotation is achieved over 80% of the radius. This result was found to be independent from the spinning velocity and flow rate. Therefore the reference length for all experiments was always taken to be $r_C = 0.8\ R_{tube}$. The characteristic similarity parameters, the core Reynolds-number $Re_C = W \cdot 2r_C / \upsilon$ and the Rossby-number $Ro = W / r_C \cdot \Omega$ were computed with the maximum axial velocity W.

3.2 STATIONARY CONICAL BREAKDOWN

Flow visualization studies for Reynolds- and Rossby-numbers ranging up to $Re_C = 8.000$ and $Ro = 0.7$, respectively, showed for all experiments with steady inflow conditions that breakdown occurs at Rossby numbers lower than $Ro \approx 0.6$. Some typical flow visualization pictures of the flow at a core Reynolds-number of $Re_C \approx 8.000$ and a

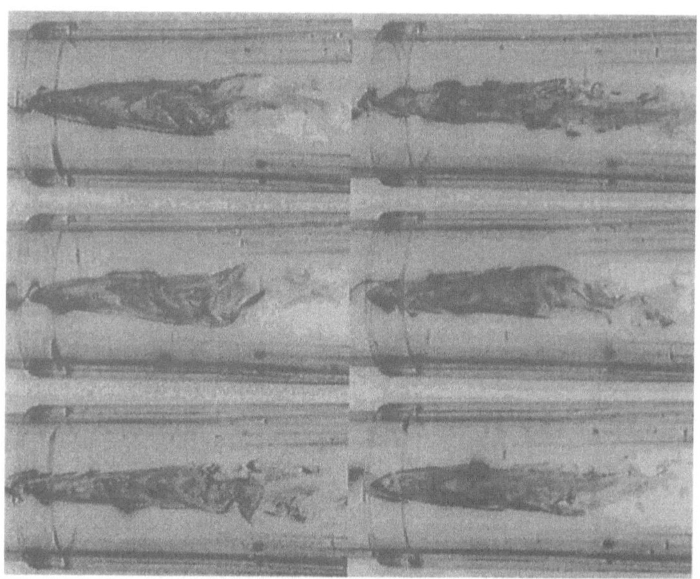

Figure 4. Stable conical breakdown structure of spinning tube flows at $Re_C \approx 8000$ and $Ro \approx 0.35$. shown for a time-span of 15 seconds

Rossby-number of $Ro \approx 0{,}35$ are shown in Fig. 4. The breakdown structure is almost nearly axisymmetric with a conical shape. The axial position of the forward stagnation point and the surface of the cone are approximately independent of time while fluctuations can be noted in the downstream part. Due to the disruption of the dye filaments further downstream it is difficult to recognize the flow structure there.

For this breakdown mode, DPIV measurements were carried out in the breakdown region and further downstream. The relative error of the velocity measurements amounts to ± 0.1 cm/s which is also the limit of the lowest velocities that can be detected. The upper limit is about 20 cm/s for the conventional video-technique used. Fig. 5 depicts the radial profiles of the axial and circumferential velocity components in the breakdown region obtained from the measurements in addition to a typical flow visualization picture in the bursted part of the vortex. The measurements were averaged symmetric to the axis.

The length-to-diameter ratio of the bursted part of the vortex is approximately 4.7. It exceeds the typical value of 1.4-1.7 found in tube flows with guide vanes at lower Reynolds-numbers [1] by a factor of 3. Its conical shape in the front resembles those breakdown modes found by Sarpkaya [9] in turbulent pipe flows for Re > 100.000, based on the tube diameter [9]. If it is assumed that in those experiments the core radius is of the order of $r_C \approx 0.1\ R_{tube}$, a core Reynolds-number of $Re_C \approx 10.000$ is obtained which compares to the experiments presented here.

The mean velocity profiles show that the cone consists of a recirculation region with low axial velocities near the axis which are one order of magnitude lower than the typical axial velocity of 10 cm/s upstream of the bursted part of the vortex. From all measurements the fluctuation velocities could be determined. In the cone they are of order of the measurement error which is ± 0.1 cm/s. Further downstream, the rms fluctuation values were less than 2% of the average axial and tangential velocity.

At the downstream end of the cone, the outer flow surrounding the cone converges towards the axis. The contraction occurs rather suddenly at the entrance to the diffuser, see Fig. 5. It causes a jet-like flow with high axial velocity and swirl of which peak values on the axis are increased by a factor of 2 compared to the corresponding data of the vortex upstream of the bursted part. The distribution of the circumferential vorticity obtained from the mean flow field (not shown here) indicates that negative circumferential vorticity was generated in the bursted part of the vortex while downstream of the contraction, positive circumferential vorticity evolves. It corresponds to the region of accelerated flow near the axis leading to the jet. More detailed DPIV measurements with a close-up view of the downstream end of the cone are planned to analyze the mass exchange at the cone base.

3.3 BREAKDOWN AFTER SPIN-UP

For certain inflow conditions the fluctuations downstream of the cone increase in strength and trigger also the cone base to oscillate around the axis while the front of the cone remains stable (Fig. 4). The fluctuations originate in the region where the jet is formed. In the experiments with steady inflow conditions, visualization in this region was not possible due to the disruption of the dye filaments. In spin-up experiments, these fluctuations could be visualized by injection of the dye shortly before the swirling flow entered the diffuser. The colored fluid was then trapped into the bursted part of the

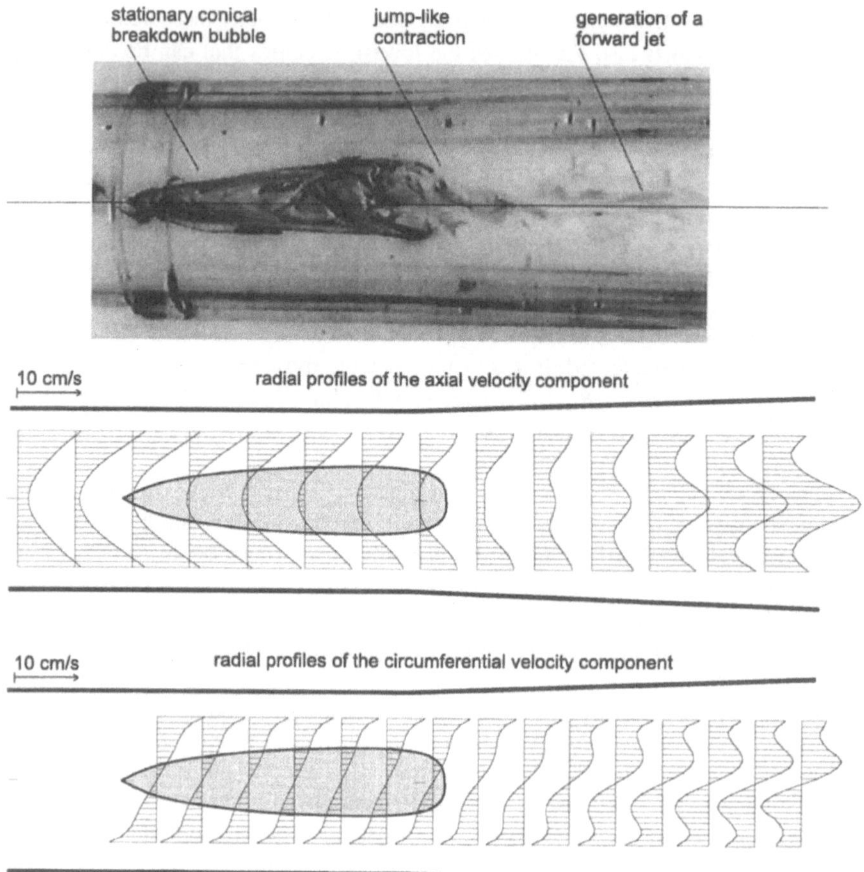

Figure 5. Flow picture of the stable conical breakdown compared to the radial profiles of the axial and circumferential velocity component in the breakdown region ($Re_C \approx 8000$, $Ro \approx 0{,}35$)

vortex and downstream without being disrupted.

3.3.1 *Temporary breakdown*

In case of the spin-up experiments breakdown could be observed also for Rossby-numbers exceeding Ro=0.6 in comparison to experiments where breakdown was initiated at constant flow rates by incremental increase of the spinning velocity. With increasing final spinning velocity in the spin-up experiments, temporary breakdown first occured at Rossby-numbers of $Ro \approx 0.7$. For values lower than $Ro \cong 0.6$ similar to the critical value given in [10] the vortex undergoes permanent breakdown. An example of the temporary breakdown after spin-up of the tube is shown in Fig. 6. First, an axisymmetric recirculation region is generated which travels upstream and grows in size.

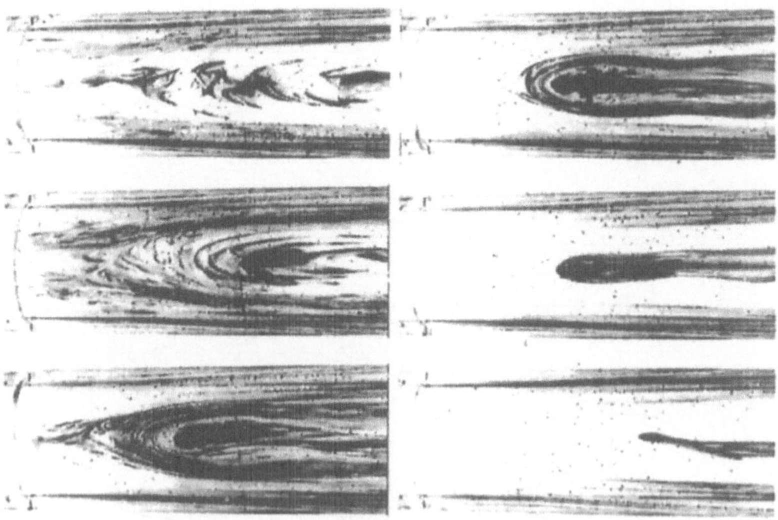

Figure 6. Breakdown development and flow retardation after spin-up of the tube from rest ($Re_C \approx 4480$, the Rossby-number after spin-up is $Ro \approx 0{,}65$)

This process does not continue but stops and is reversed as can be seen in Fig. 6 where the dye surrounding the recirculation region starts to decay and the backflow region vanishes at later times. From the time-resolving recordings it is seen that the fluid trapped inside the recirculation region starts to leave it downstream, caused by the acceleration of the flow at the downstream end of the bubble. With increasing spinning velocity, the upstream travelling recirculation region survives yielding a permanent breakdown state.

3.3.2 *Upstream travelling breakdown*

The instantaneous structure of the upstream travelling breakdown region after spin-up for $Re_C \approx 4480$ and $Ro \approx 0.55$ is shown in Fig. 7 by means of the projection of the instantaneous streamlines in the axial plane and the radial profiles of the axial and circumferential velocity component. Because of the almost axisymmetric nature of the flow travelling upstream, the measurements could be averaged symmetric to the axis to represent the axisymmetric main flow. The front of the travelling cone is smoother and the diameter of the cone is larger compared to the steady conical breakdown. At the downstream end of the cone, the flow reverses with a sharp bend, enters the cone base and turns again into the downstream direction near the axis. From time-resolving measurements the upstream propagation speed of forward stagnation point was determined to 1 cm/s, which is reduced by a factor of 5 compared to the characteristic axial velocity of the flow. In order to obtain the flow pattern as seen from an observer in the bursted part of the vortex, the upstream propagation speed of the stagnation point was subtracted from the instantaneous axial velocity component in the entire flow field (not shown here). The relative flow motion so obtained resembles the structure depicted in Fig. 5 for the steady breakdown characterized by a single recirculation cell and the jet

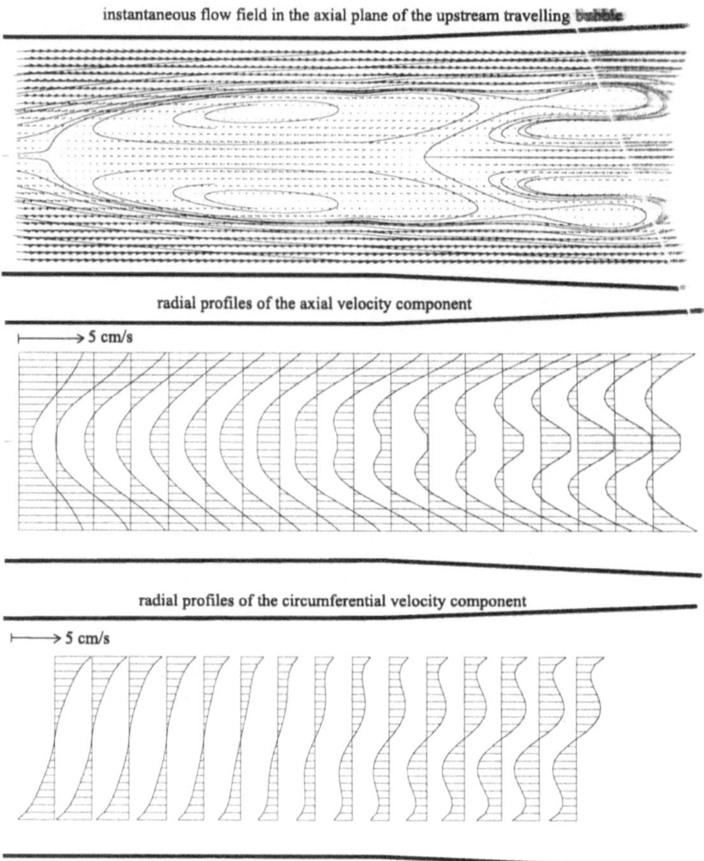

Figure 7. Instantaneous flow structure in the axial plane and profiles of the axial and circumferential velocity component for the upstream travelling breakdown after spin-up ($Re_C \approx 4480$, $Ro \approx 0.55$)

downstream. This demonstrates that the structure of the breakdown region after initiation remains unchanged until it reaches its final axial position at later times.

3.3.3 *Onset of instability after spin-up*

Further downstream of the travelling cone, asymmetric instabilities evolve during the travelling period which were studied by flow visualization with dye, see Fig. 8. The flow pictures for $Re_C \approx 4480$ and $Ro \approx 0.48$ clearly demonstrate wavy patterns downstream from the bursted part of the vortex with an amplitude increasing in time. They travel upstream with the bursted part of the vortex without changing the shape of the cone. Fig. 9 shows the evolution of one section of the wave pattern in the axial plane of the diffuser for laminar and turbulent and turbulent flow conditions at $Re_C \approx 800$ and $Re_C \approx 4480$, respectively. For turbulent flow, the dye filaments disintegrated in this region rapidly, due to the increased mixing in this region. Flow measurements with the method of DPIV

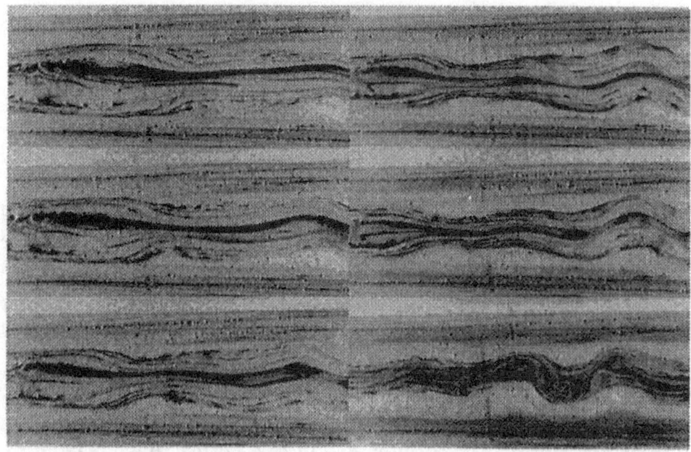

Figure 8. Wave evolution in the downstream region of the cone in a time-span of 15s after breakdown generation in a spin-up process ($Re_C \approx 4480$, $Ro \approx 0.48$)

yielded the instantaneous flow patterns, reconstructed from the experimental data. They are shown on the right hand side of Fig. 9. In the cross-sections, the radial profiles could

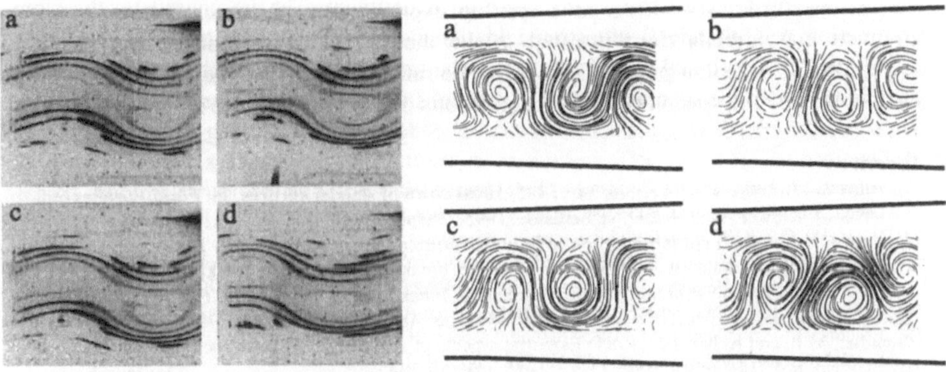

Figure 9: Temporal evolution of a section of the wave pattern over 18 s displayed to 4 moments.
Left: flow visualization at $Re_C \approx 800$, $Ro \approx 0{,}08$ Right: sectional streamline pattern at $Re_C \approx 4480$, $Ro \approx 0{,}48$

not be measured with the conventional video-technique used due to the high shear and the large velocity gradients present there.

The results obtained in the axial plane clearly show a helical pattern which remains nearly unchanged over time in both experiments. The wavelength of the helix is $l \approx 1.2$ tube diameters. They differ markedly from the long spiral waves observed in rotating Hagen-Poisseuille flows in the transitional regime [6]. In order to clarify as to whether or not this steady behavior is due to a standing wave or a three-dimensional moving structure, more detailed experiments are required including particle tracking analysis. This has not been done yet but is planned for the future.

4. Conclusion

In this article experiments of the breakdown of the turbulent vortex in spinning tube flows are presented up to Reynolds-numbers of $Re_C = 8000$. The characteristic length scale of the vortex, the core radius, is almost equal to 0.8 R_{tube} for the range investigated. Since the flow rate and the spinning velocity can be controlled independently, the characteristic similarity parameters, the Reynolds-number and the Rossby-number, can be varied easily.

The bursted part of the turbulent vortex is a near-axisymmetric conical recirculation region with a ratio of $L/D \approx 4\text{-}5$. For comparable Reynolds-numbers other conical breakdown structures were found also in free vortex flows [4] and in tube flows with a guide-vane type of swirl generator [9]. In the experiments described here, the cone ends with a jump-like contraction (compare also [5]). It is located near the entrance of the diffuser and forces the flow back towards the axis. A jet-like flow with high swirl near the axis evolves further downstream. The contraction intensifies the axial vorticity component and decreases the characteristic vortex size about a factor of 2 compared to the corresponding data upstream of the bursted part of the vortex. Usually asymmetric waves develop there as shown in a spin-up experiment. The jet caused by the flow contraction was already demonstrated in the theoretical considerations by Batchelor (compare Fig. 7.5.1d and page 547 in [2]). The rather sudden contraction and the jet are attributed here to the special boundary conditions of the vortex in the spinning tube.

References

[1] Althaus, W., Brücker, Ch., Weimer, M. (1995) Breakdown of slender vortices. In: *Fluid Vortices* (ed. Green, S.I.), Kluwer Acad. Publ., Netherlands, 1995, Chapt. IX, 373-426

[2] Batchelor, G.K. (1967) An introduction to fluid dynamics. Cambridge University Press

[3] Brücker, Ch. and Althaus, W. (1995) Study of vortex breakdown by particle-tracking-velocimetry (PTV). Part 3: Time-dependent structure and development of breakdown modes. *Exp. in Fluids* **18**, pp. 174-186

[4] Delery, J.M. (1994) Aspects of vortex breakdown. Prog. Aerospace Sci. 30, pp. 1-59

[5] Escudier, M.P. and Keller, J.J. (1985) Essential aspects of vortex breakdown. Proc. Int. Coll. on Vortex Breakdown, RWTH Aachen, Febr. 11-12, 1985, pp. 119-144

[6] Imao, S., Itoh, M., Yamada, Y. and Zhang, Q. (1992) The characteristics of spiral waves in an axially rotating pipe. *Exp. in Fluids* 12, pp. 277-285

[7] Krause, E. (1990) The solution to the problem of Vortex Breakdown. *Lecture Notes in Physics* **371**, 33-50, Springer Verlag

[8] Pedley, T.J. (1969) On the instability of viscous flow in a rapidly rotating pipe. *J. Fluid Mech.* **35**, pp. 97-111

[9] Sarpkaya, T. (1995) Turbulent vortex breakdown. *Phys. Fluids* **7**(10), pp. 2031-2303

[10] Spall, R.E.; Gatski, T.B.; Grosch, C.E. (1987) A criterion for vortex breakdown. *Phys. Fluids* 30, pp. 3434-3440

[11] Suematsu, Y., Ito, T. and Hayase, T. (1986) Vortex breakdown phenomena in a circular pipe. Bulletin of the JSME 29 (258), pp. 4122-4129

REVIEW OF THE AACHEN WORK ON VORTEX BREAKDOWN

W. ALTHAUS AND M. WEIMER
Aerodynamisches Institut der RWTH Aachen
Wüllnerstr. zw. 5 u. 7
D-52062 Aachen, Germany

Abstract

Results of the numerical simulations and of the experimental investigations of vortex breakdown are presented. This paper focuses on the development of physical understanding of the breakdown process. Simulations of axisymmetric flows suggested that the Navier-Stokes equations for time-dependent, three-dimensional flows should be solved to describe the breakdown process. The simulation of a free vortex showed bubble- and spiral-type breakdown and the transition between these states. The comparison with the experimental visualization showed a good agreement, thus the analysis of the physical mechanisms of breakdown were based on the numerical results. The initiation and growth of the bubble are described by a feedback mechanism. The shedding of high axial vorticity structures is explained. The growing asymmetry of the internal bubble structure and the subsequent transition to spiral-type breakdown are described. Simulating vortex breakdown in a slightly diverging tube showed stable bubble-and further downstream spiral-type breakdown. The location and internal structure of the bubble corresponds to experimental results. The results of the application of flow visualizations and Particle-Image-Velocimetry are presented

1. Introduction

The breakdown of leading-edge vortices above a delta wing were first observed by Peckham and Atkinson, [1]. According to Benjamin, [2], and Sarpkaya, [3], vortex breakdown is described as "an abrupt change in the structure of the core of a swirling flow". A more restrictive, and generally accepted, definition by Leibovich, [4], defined vortex breakdown as "a disturbance characterized by the formation of an internal stagnation point on the vortex axis, followed by a reversed flow in a region of limited axial extend".

Several theories were proposed which tried to predict the conditions that lead to vortex breakdown, e. g. the concept of hydrodynamic instability by Ludwieg, [5], the model of the transition between two dynamically conjugate states of flow by Benjamin, [2], or the concept of a two-stage-transition by Escudier and Keller, [6]. All these theories can only give statements about the flow state up- and downstream of the breakdown region, but

E. Krause and K. Gersten (eds.), IUTAM Symposium on Dynamics of Slender Vortices, 331-344.

they cannot provide any information about the breakdown region itself. Additionally, they do not consider the influence of boundary conditions on breakdown. The experimental investigations, e. g., by Faler and Leibovich, [7], showed that the internal structure of the breakdown region is dominated by a double vortex ring structure and asymmetric, time-dependent motion inside. Therefore, Krause, [8], proposed 1979 to solve the Navier-Stokes equations for axisymmetric, time-dependent flows to analyse vortex breakdown. This idea initiated the research at the Aerodynamisches Institut (AIA) and the Institut für Luft- und Raumfahrt, (ILR) at the RWTH Aachen. The research was sponsored by the German Research Association (DFG) in the frame of the collaborative research program "Vortical Flows in Aeronautics" (SFB 25) since 1980. The paper focuses on the development of physical understanding of the breakdown process and is structured more or less chronologically.

2. Results

2.1. FIRST SOLUTIONS OF NAVIER-STOKES EQUATIONS

Hartwich, [9], simulated the flow around a delta wing. His results showed streamwise vortices undergoing breakdown with the breakdown point travelling up- and downstream. The spatial resolution of his computations were, due to the lack of computer capacity, too low to resolve details of the breakdown region. A free vortex was simulated by Shi, [10]. He solved the full Navier-Stokes equations for axisymmetric, time-dependent, incompressible, and laminar flow. The unsteady flow was dominated by an upstream travelling vortex ring. The influence of the inflow conditions on the simulation of a free vortex was investigated by Menne, [11]. He stated that the Navier-Stokes equations for three-dimensional, time-dependent flows have to be solved for computing vortex breakdown. The breakdown of a vortex in a slightly diverging tube was computed by Liu and Menne, [12], using the assumption of nearly axisymmetric flow. Their results show an upstream travelling breakdown region with a secondary vortex similar to the experimental results by Faler and Leibovich, [7]. Small non-axisymmetric disturbances initiated at the inflow section had a distinct influence on the internal structure of the bubble. The numerically visualized flow by particle traces looks very similar to experimentally visualized bubbles, Fig. 1. Particles enter the bubble from behind, and downstream of the bubble the flow exhibits a spiralling motion.

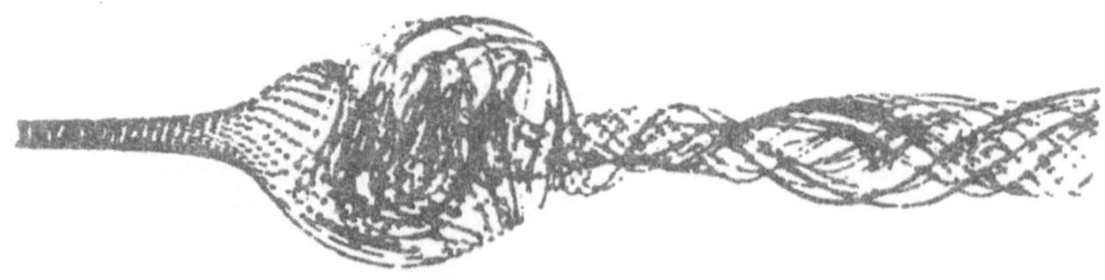

Figure 1. Particle traces for bubble-type breakdown, [12]

2.2. SIMULATION OF A FREE VORTEX

With the growing computer capacity the restriction to axisymmetric or nearly axisymmetric flows, respectively, were dropped. Breuer, [13], simulated a free vortex by solving the Navier-Stokes equations for time-dependent, three-dimensional flow. His results showed the initiation and growth of an upstream travelling bubble which closely resembles the visualized flow structures by Escudier, [14], see Fig. 2. Unfortunately, the upstream migration of the bubble did not stop, as in the experiments, but reaches the inflow section. Thus, the solution did not converge any longer. One can prevent this by adapting the lateral boundary conditions with time. Krause, [16], constructed an iterative procedure for finding appropriate boundary conditions which were applied by Breuer and Hänel, [17]. The basic idea was that the local accelerations of the axial and radial velocity have to vanish asymptotically in a region in the immediate vicinity of the inflow section. With this technique the breakdown region was successfully stabilized inside the domain of integration. Again, for Re=200, based on the vortex core diameter, the initiation and growth of an upstream travelling bubble with one or two vortex rings inside it were simulated, [13]. The upstream migration stopped and, for a certain

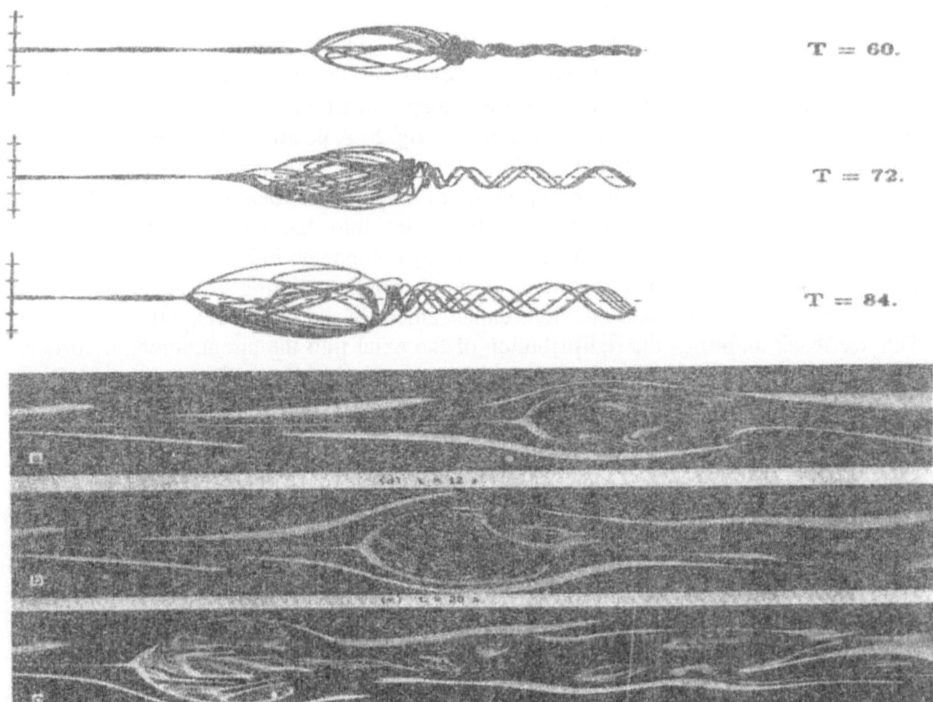

Figures 2. Above: numerical simulation of unstable bubble-type breakdown, [15].
Bottom: Experimental visualization of unstable bubble-type breakdown, [14]

duration, the numerical streaklines showed bubble-type breakdown. However, with increasing time of simulation, the bubble became more and more asymmetric and evolved, finally, to the spiral form of breakdown. Even a periodic transition between both states was observed, similar to experimental observations by Faler and Leibovich, [7], and Althaus and Krause, [18]. Thus, the simulations were repeated for Re=500, as it was in the experiment, [18], and the results compared with the experimental flow visualizations, see [19], [20], and [21].

Fig. 3 shows the initiation and growth phase of bubble-type breakdown. The straight vortex core with a slight swelling enlarges to an umbrella-like structure downstream of the stagnation point. With increasing time, this structure grows further and, finally, evolves into an axisymmetric bubble with an aft vortex ring-like structure. Fig. 4 shows the subsequent transition to spiral-type breakdown. The nearly axisymmetric bubble becomes more and more asymmetric and evolves into the spiral form of breakdown. The comparison of the complete spatial and temporal development, which is documented on video, shows a good agreement.

Thus, vortex breakdown can be reliably simulated and the underlying physical mechanisms were investigated based on the numerical results.

2.3. ANALYSIS OF PHYSICAL MECHANISMS

A feedback mechanism for the initiation of bubble-type breakdown was developed by Althaus et al., [19] and [21], which is based on the redistribution of axial into circumferential vorticity and the corresponding induction of an additional velocity component against the main flow direction, see fig. 5. A positive axial pressure gradient leads to a deceleration of the axial velocity component. Conservation of mass requires radial outflow and from the conservation of angular momentum and the vorticity transport equation, a redistribution of the axial into the circumferential vorticity component follows. This circumferential vorticity component induces an additional axial velocity component against the main flow direction, resulting in an even stronger deceleration of the axial, and therefore enhancement of the radial, velocity components. This feedback amplifies the redistribution of the axial into the circumferential vorticity component which may lead to the formation of a stagnation point. This positive feedback must be balanced by a counteracting mechanism. The radial flow away from the axis leads to an acceleration of the axial velocity component at large radial distances. This acceleration decreases the effect of the positive pressure gradient, thus reducing the radial velocity component. The conservation of angular momentum and the vorticity transport equation require a redistribution of the circumferential into the axial vorticity component, leading to a attenuation of the vortex ring-like structure. This diminishes the induction of the additional axial velocity component, thus increasing the axial velocity. If these positive and negative feedback mechanisms balance each other, a dynamical equilibrium state is attained with a steady position of the bubble. The redistribution of vorticity can be seen very clear in Fig. 6., where the gray-scaled pressure distribution in the midplane of the integration domain with superimposed vortex lines during the growth phase of the bubble is shown. The pressure distribution shows a high degree of axial symmetry. The vortex lines clearly indicate the redistribution from axial into circumferential vorticity as described by the feedback mechanism. In the bubble region

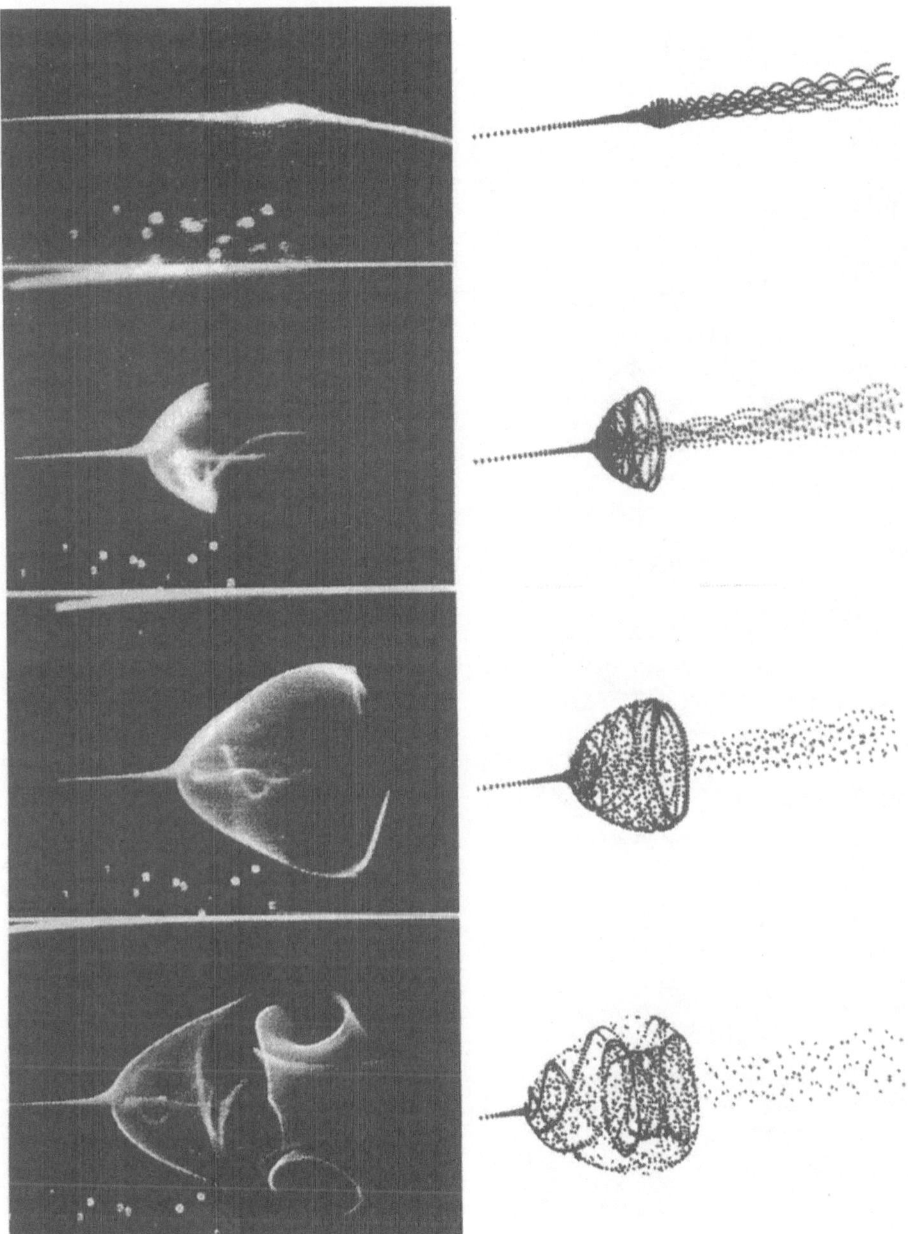

Figures 3. Experimental and numerical flow visualization of the initiation of bubble-type breakdown. Left side: experimental streaklines. Right side: numerical streaklines

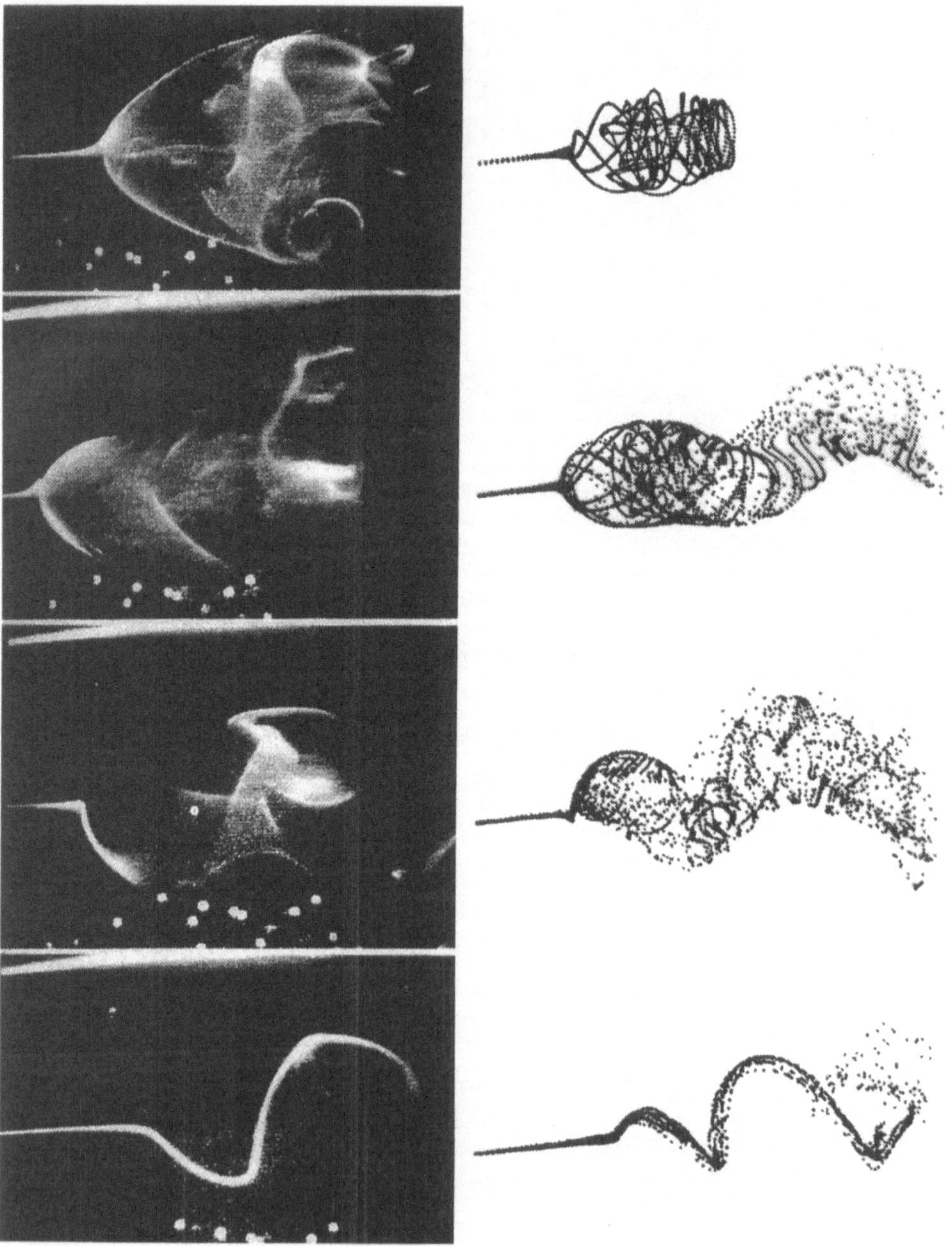

Figures 4. Experimental and numerical flow visualization of transition of bubble-to spiral-type breakdown. Left side: experimental streaklines. Right side: numerical streaklines

the vortex lines become steeper and steeper and accumulate at the rear end of the bubble where they are nearly perpendicular to the main flow direction. They represent the vortex ring-like structure which is due to induction responsible for the generation of the stagnation point and the upstream migration of the bubble.

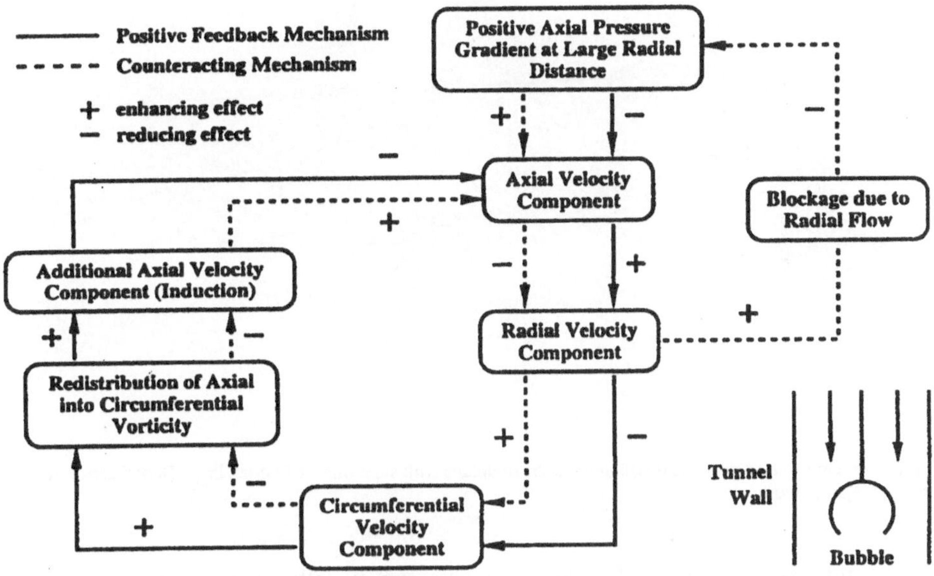

Figure 5. Feedback mechanism for the initiation of bubble-type vortex breakdown.

The vortex ring-like structure feeds the bubble with particles from the outer flow which have a certain circumferential velocity. Due to the conservation of angular momentum these particles enlarge their circumferential velocity and produce immediately downstream of the bubble a region with high axial vorticity as it is clearly indicated by the nearly abrupt change of direction of the vortex lines. The high axial vorticity produces a strong pressure drop in the vicinity of the axis which weakens the induction effect of the vortex ring-like structure. With increasing time this process enhances and the high axial vorticity structure moves downstream and is, finally, shed away. This process is repeated twice but each time it becomes significantly weaker.

During the initiation and growth phase the mass inflow into the bubble is compensated by the growing volume of the upstream migrating bubble. When the bubble has reached its final upstream position, the mass inflow cannot be compensated by the growing bubble volume any longer. But now it is compensated by the shedding of the above described structures with high axial vorticity. This shedding occurs three times, then the compensation of the mass inflow is achieved by a tilting of the vortex ring-like structure as supposed by Sarpkaya, [3], and experimentally verified by Brücker and Althaus, [22].

As a consequence, the stagnation point moves away from the axis and rotates around it. The axial symmetry is lost and with increasing time the pressure distribution becomes

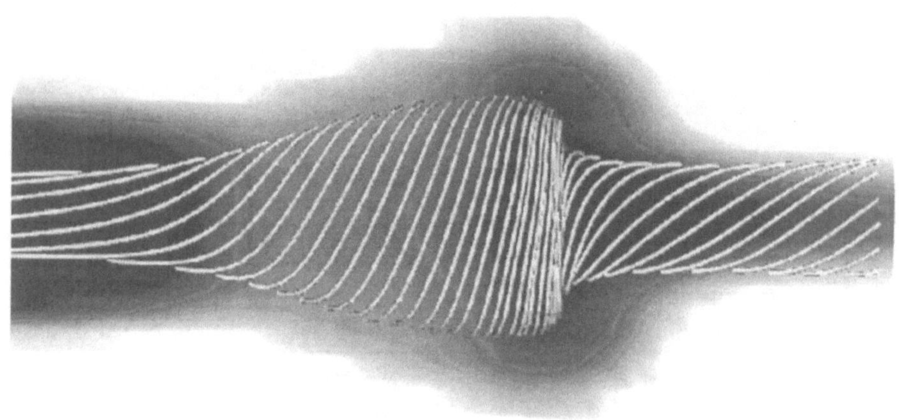

Figure 6. Gray-scaled pressure distribution in the midplane with superimposed vortex lines. Dark means low, bright high pressure.

more and more asymmetric. The stagnation point moves farther away from the axis, too, see Fig. 7. The definition by Leibovich, [4], that vortex breakdown is characterized by a stagnation point on the axis, cannot be maintained in this restrictive manner. The tilting process continues until the stagnation point is located so far away from the axis that the vortex lines clearly show spiral-type breakdown. The inclination of the windings changes with time. Thus, the strength of the inductive effect changes, too, and , dependent on the inclination, there may be backflow between the windings. Correspondingly, for spiral-type breakdown the existence of a stagnation point is not necessary. This observation shows, too, that the definition by Leibovich has to be modified.

2.4. EXPERIMENTAL INVESTIGATION

Helming, [23], generated streamwise vortices by a half model of a wing and investigated breakdown forced by axisymmetric diffusors or nozzles. He visualized the flow with air bubbles or milk and measured velocity profiles with Laser-Doppler-Anemometry. He always found spiral-type breakdown with backflow on the axis and enhanced velocity gradients and turbulence inside the breakdown region. By applying a confusor downstream of the bursted part of the vortex he could regenerate it, see Fig. 8. The velocity profiles showed a strong reduction of the axial and circumferential velocity inside the breakdown region. Vitting, [24] continued the investigations and found always spiral-type breakdown, too. Furthermore, he showed that a regular spiral could burst due to the mutual induction of their windings, Fig. 9. Backstein, [25], induced spiral-type breakdown by weak pressure fields. A pronounced spiral developed only, if the pressure gradient was above a certain value.

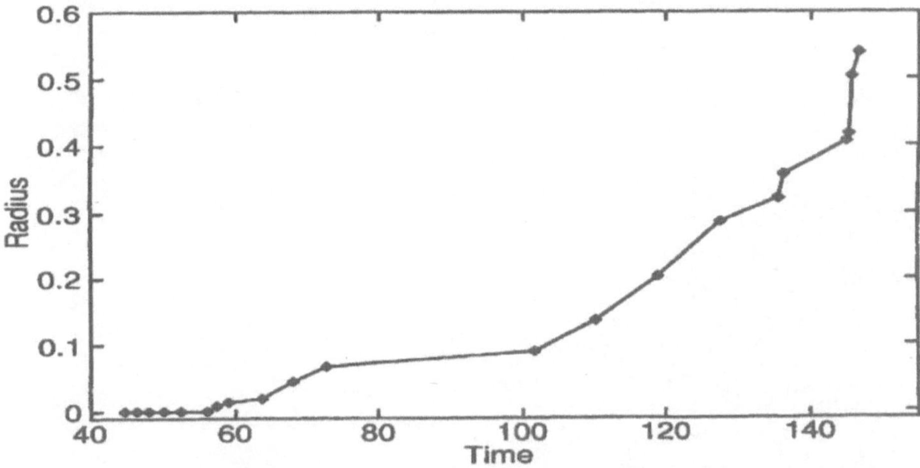

Figure 7. Evolution of the radial distance of stagnation with time for bubble-type breakdown

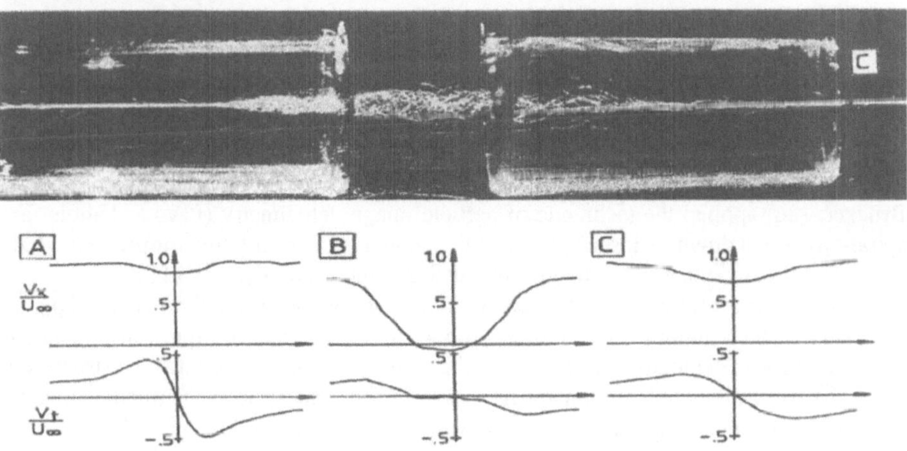

Figure 8. Above: regeneration of the bursted part of the vortex by a confusor. Bottom: axial and circumferential velocity profiles.

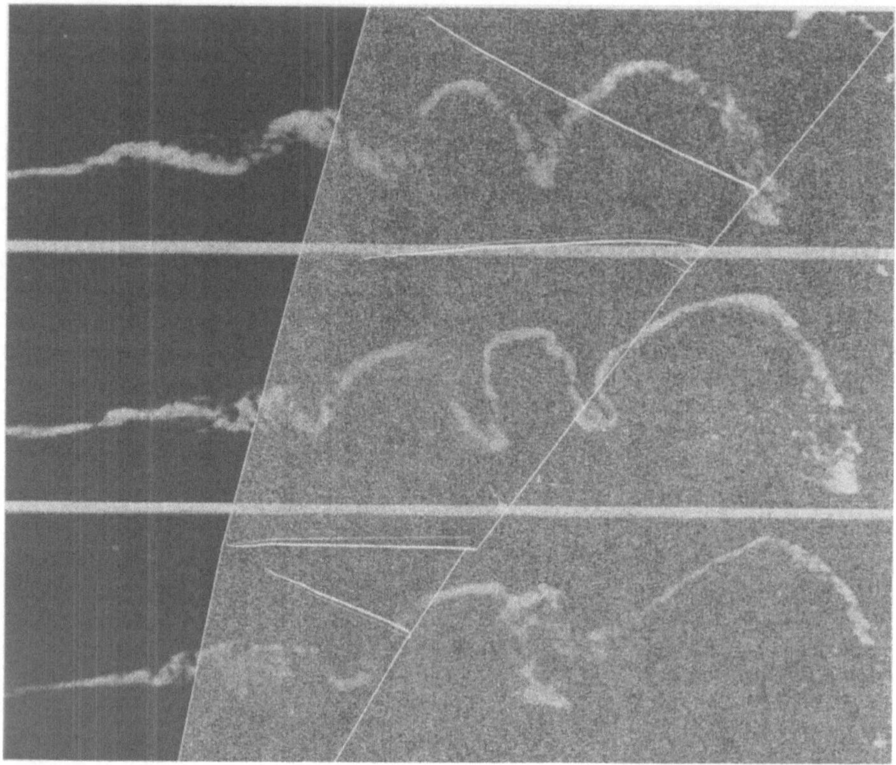

Figure 9. Experimental visualization of the bursting of a regular spiral due to induction

Brücker, [26] applied the technique of particle image velocimetry (PIV) to bubble- and spiral-type breakdown. A laser light sheet is scanned by an oscillating mirror through the flow. Successive records of the illuminated particles are made within known time intervals and from the records the velocity distributions inside the light sheet plane are calculated. The scanning of the sheet allows to reconstruct the quasi-instantaneous, three-dimensional velocity distribution of the flow. A single vortex ring-like structure in the rear part of the bubble dominates its internal flow. This structure is tilted against the centerline and rotates around it. A reconstruction of the three-dimensional shape from the results of the PIV-measurements is shown in Fig. 10, [22]. The investigation of spiral-type breakdown showed backflow between the windings. The stagnation point is not located on the centerline but rotates around it, [27].

2.5. SIMULATION OF VORTEX BREAKDOWN IN A TUBE

Weimer solved the Navier-Stokes equations for time-dependent, three-dimensional, and incompressible flow for Re=3220 in a slightly diverging tube according to the experimental setup by Faler and Leibovich, [7]. He could simulate the initiation and growth of bubble-type breakdown, [28]. The bubble travels upstream during its generation, overshoots and then travels downstream to its equilibrium position. Downstream of the bubble spiral-type breakdown is established. The location and the diameter of the bubble are in good agreement with the results of Faler and Leibovich, [7]. Fig. 11 clearly shows this good agreement between the numerical and experimental

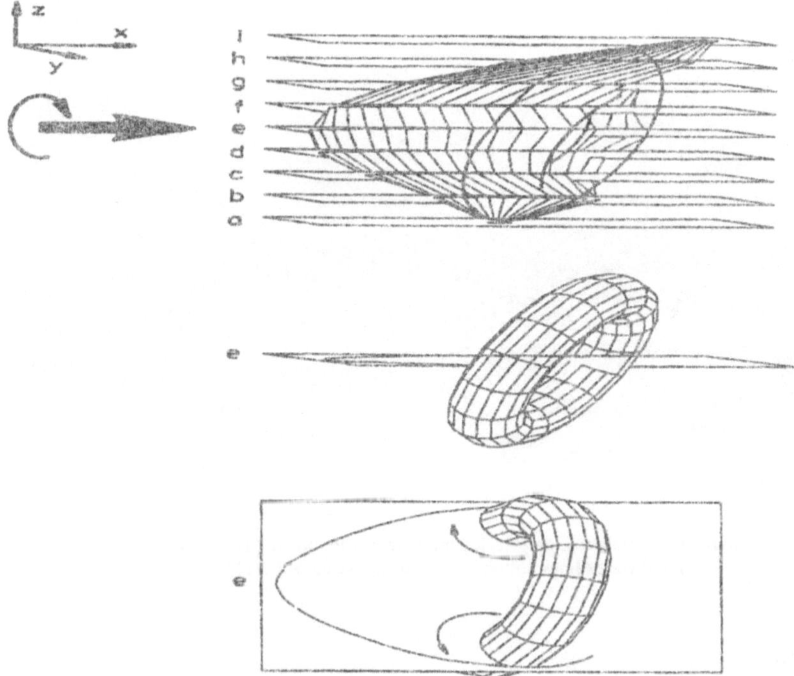

Figures 10. Sketch of the three-dimensional bubble shape including the tilted vortex ring-like structure as reconstructed from measurements

visualization. The picture above shows the gray-scaled pressure distribution in the midplane of the domain of integration with superimposed vortex lines whereas at the bottom the experimental visualization can be seen. The vortex lines as well as the flow visualization indicate first bubble-, and about one bubble diameter downstream, the subsequent spiral-type breakdown.

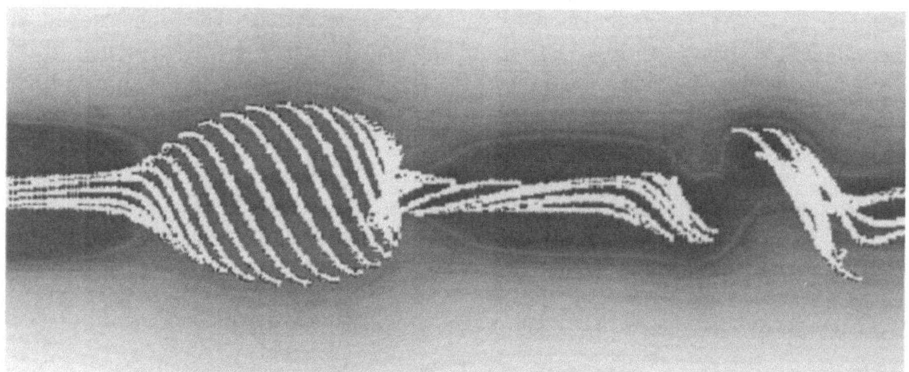

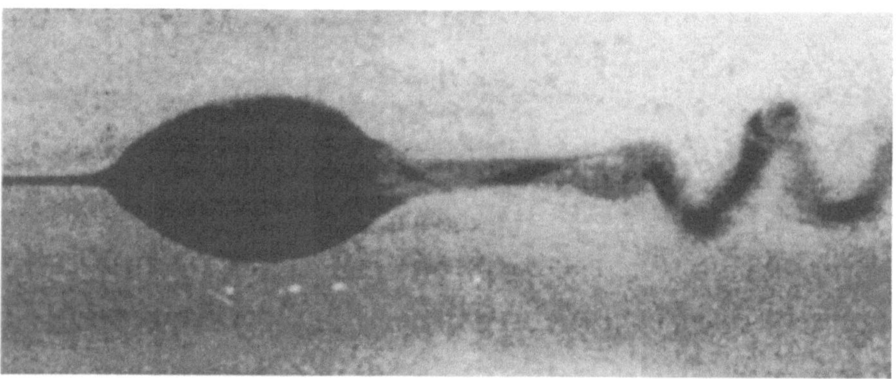

Figures 11. Above: numerical simulation: gray-scaled pressure distribution in the midplane and superimposed vortex lines bottom: experimental visualization, [7].

3. Conclusion

Vortex breakdown is investigated experimentally and numerically since 1980 at the RWTH Aachen. It started with the numerical solution of the Navier-Stokes equations for (nearly) axisymmetric, time-dependent flows. Due to the symmetry condition only bubble-type breakdown was simulated with an upstream travelling bubble. Therefore, a free vortex was simulated by solving the Navier-Stokes equations for three-dimensional, time-dependent flow. The visualization of the numerical results by streaklines showed an upstream travelling bubble which is very similar to experimental visualizations. Contrary to experiments the bubble reaches the inflow section in the simulation and the solution became unstable. A successful stabilization of the breakdown region inside the domain of integration was obtained by applying an iterative procedure to change the lateral boundary conditions with time. With this technique the initiation and growth phase of an upstream travelling bubble, which remains in the domain of integration, was computed. But with increasing time the bubble became more and more asymmetric and, finally, evolves into a spiral. Even a periodic transition, as observed in experiments, could be simulated. Therefore, another simulation and an appropriate experiment were carried out and the spatial and temporal development of bubble- and spiral-type breakdown and the periodic transition between both states were compared. The good agreement showed that vortex breakdown is simulated reliably by solving the Navier-Stokes equations for three-dimensional, time-dependent flow. Based on the numerical results a feedback model was developd which describes the initiation and growth phase of bubble-type breakdown. The shedding away of structures with high axial vorticity could be explained, too. The importance of the balance of the mass flow into and out of the bubble was stressed. Due to the mass balance the vortex ring-like structure is tilted and the stagnation point is located away from the axis. Thus, the definition of vortex breakdown by Leibovich has to be modified. The exchange of mass is according to the model of Sarpkaya, which was verified experimentally, too. The increasing tilting leads to a loss of the axial symmetry and the spiral-type of breakdown evolves.

In the experimental investigations the streamwise vortex was generated either by guidevanes or by a half model of a wing profile. In the former case bubble- and spiral-type breakdown and their periodic transition were visualized in water flow, whereas in the latter case only spiral-type breakdown was observed. The spiral could be regenerated by applying a confusor downstream of it. Furthermore, it was shown that the bursting of the spiral may be due to the mutual induction of the spiral windings. The PIV-technique with a scanning light sheet allows to reconstruct three-dimensional flow structures as, e. g., the tilted vortex ring-like structure. The rotation of the stagnation around the axis and the mass exchange according to the model of Sarpkaya were verified experimentally.

Recently, vortex breakdown in a slightly diverging tube was simulated by solving the Navier-Stokes equations for three-dimensional, time-dependent flow. The visualization of the numerical results showed the initiation and growth phase of an upstream travelling bubble-type breakdown. Downstream of the bubble spiral-type breakdown was observed, which agrees to experimental flow visualizations.

4. Acknowledgement

The author's and their colleagues' projects referred to herein were supported partly by the Deutsche Forschungsgemeinschaft as subprojects of the Sonderforschungsbereich 25 "Wirbelströmungen in der Flugtechnik". The authors thank Prof. E. Krause, Ph. D., who initiated and guided the research on this problem at the Aerodynamisches Institut of the RWTH Aachen, for many intensive and helpful discussions.

References

[1] Peckham, D. H., Atkinson, S. A.: Preliminary results of low speed wind tunnel tests on a Gothic wing of aspect ratio 1. British Aeronaut. Res. Council, **CP 508**, 1957

[2] Benjamin, T. B.: Theory of vortex breakdown phenomenon. J. Fluid Mech., vol. **14**, pp. 593-629, 1962

[3] Sarpkaya, T.: On stationary and travelling vortex breakdowns. J. Fluid Mech., vol. **45**, part 3, pp. 545-559, 1971

[4] Leibovich, S.: The structure of vortex breakdown. Ann. Rev. of Fluid Mech., vol. **10**, pp. 221-246, 1978

[5] Ludwieg, H.: Zur Erklärung der Instabilität der über angestellten Deltaflügeln auftretenden freien Wirbelkerne. Z. Flugwiss. **10,** 242

[6] Escudier, M. P., Keller, J. J.: Vortex breakdown: a two-stage transition. AGARD CP No. **342** Aerodynamics of vortical type flows in three dimensions, paper 25.

[7] Faler, J. H., Leibovich, S.: An experimental map of the internal structure of a vortex breakdown. J. Fluid Mech. **86,** 313-335

[8] Krause, E.: private communication, 1997

[9] Hartwich, P. M. Berechnung von Vorderkantenwirbeln an Deltaflügeln. Diss. Aerodyn: Inst., RWTH Aachen 1983

[10} Shi, X.-G.: Numerische Simulation des Aufplatzens von Wirbeln. Diss. Aerodyn. Inst., RWTH Aachen, 1983

[11] Menne, S.: Rotationssymmetrische Wirbel in achsparalleler Strömung. Diss. Aerodyn. Inst., RWTH Aachen, 1986

[12]. Liu, C. H., Menne, S.: Numerical Investigation of a Three-dimensional Vortex Breakdown. The Fourth Asian Congress of Fluid Mechanics, Aug. 21 – 25, Hong Kong

[13] Breuer, M.: Numerische Lösung der Navier-Stokes-Gleichungen für dreidimensionale inkompressible instationäre Strömungen zur Simulation des Wirbelaufplatzens. Diss. Aerodyn. Inst., RWTH Aachen, 1991

[14] Escudier, M.: Vortex Breakdown: Observation and Explanations, Progress in Aerospace Sciences, vol. **25**, NO. 2, pp. 189-229, 1988

[15] Breuer, M.; Hänel, D.: Solution of the 3-D, incompressible Navier Stokes equations for the simulation of vortex breakdown. Proceedings of the Eigth Gamm-Conference on Numerical Methods in Fluid Mechanics, vol. **29**, 42-51, 1990

[16] Krause, E.: The solution to the problem of vortex breakdown. Lecture Notes in Physics **371**, 35 – 50, 1990

[17] Breuer, M., Hänel, D.: A dual time-stepping method for 3-D, viscous, incompressible vortex flows. Computers and Fluids **22**, 467-484, 1993

[18] Althaus, W.; Krause, E.: Flow Visualization of Flows with Concentrated Vorticity, Progress Report Dec. 1991, EC Contract SC1 -0212

[19] Althaus, W.; Brücker, Ch.; Weimer, M.: Breakdown of Slender Vortices. In: Fluid Vortices. Ed. S. I. Green, Kluwer Academic Publishing, Chapter IX, pp. 373-426, 1995

[20] Weimer, M.; Hofhaus, J.; Althaus, W.: Simulation of Vortex Breakdown. Z. Flugwiss. Weltraumforsch. **6**, 1995

[21] Althaus, W.; Krause, E.; Hofhaus, J.; Weimer, M.: Bubble- and Spiral-Type Breakdown of Slender Vortices. Experimental Thermal and Fluid Science,111 (6): 276-284, 1995

[22] Brücker, Ch.; Althaus, W.: Study of vortex breakdown by particle tracking velocimetry (PTV), Part 1: Bubble-type vortex breakdown.Experiments in Fluids **13** (1992), pp. 339-349

[23] Helming, Th.: Untersuchungen zum Aufplatzen und zur Instabilität von Randwirbeln. Diss. Institut für Luft- und Raumfahrt, RWTH Aachen, 1988

[24] Vitting, Th.: Struktur von Flügelrandwirbeln und Maßnahmen zur Wirbelabschwächung. Fortschr.-Ber . VDI Reihe 7 Nr. 185. Düsseldorf: VDI Verlag 1991

[25] Backstein, S.: Wirbelaufplatzen in schwachen Druckfeldern. Diss. Institut für Luft- und Raumfahrt, RWTH Aachen, 1997

[26] Brücker, Ch.: Experimentelle Untersuchung des Wirbelaufplatzens mit der Particle-Image-Velocimetry. Diss. Aerodyn. Institut, RWTH Aachen, 1993

[27] Brücker, Ch.: Study of vortex breakdown by particle tracking velocimetry (PTV) Part 2: Spiral-type Vortex Breakdown. Exp. in Fluids **14** (1993), 133-139

[28] Weimer, M.: Thesis in preperation. Aerodynamisches Institut, RWTH Aachen, 1997

Session 6

Vortex Sound

SOUND GENERATION BY INTERACTIONS OF TWO VORTEX RINGS

K. ISHII AND H. MARU
Department of Computational Science and Engineering
Nagoya University, Nagoya 464-01, Japan

AND

S. ADACHI
RIKEN
Wako 351-01, Japan

Abstract. Vortex sound from collision of two vortex rings at 16 different angles between head-on and perpendicular collisions is numerically analyzed using an aerodynamic sound theory for the far-field acoustic pressure which results from localized unsteady vortex motion. The vorticity field is obtained by solving the three-dimensional incompressible vorticity equation with the vorticity-potential method. The strong sound source region is identified for each collision angle. It is found that there are three regions of collision angles and that different characteristic vortex motion mainly causes the vortex sound in each region. The relation of the vortex motion with the vortex sound is discussed in detail.

1. Introduction

Vortex sound which results from unsteady motion of vorticity field has been studied extensively [1, 2, 3]. For the localized vortex motion with low Mach and high Reynolds numbers, far-field acoustic pressure is given in terms of the time-derivatives of the vorticity field [4, 5, 6, 7, 8, 9].

In the present paper we study sound generation from the collision of two vortex rings and investigate the relation between vortex dynamics and far-field vortex sound. In the preceding paper [10] we applied a vortex sound theory formulated in the form of multipole expansions to the oblique collisions at right angles. The coefficients of the multipole modes of

E. Krause and K. Gersten (eds.), IUTAM Symposium on Dynamics of Slender Vortices, 347-360.

the acoustic pressure are estimated from the numerical simulation on the time-dependent localized vorticity field. We found that the computed main-mode amplitudes of the wave pressure are consistent with the experimental ones [9] not only qualitatively but also quantitatively. We also succeeded in identifying the source region for the wave emission, and we found that the strong acoustic source locates in the region where the vorticity field varies very rapidly.

When changing the collision angle from the head-on collision to the oblique collision at right angles, the relative importance of the vortex ring expansion, the vortex core deformation and the vortexline reconnection changes, and we may identify the effect of each vortex motion on the far-field acoustic pressure. We therefore analyze the vortex collision with many different angles between the head-on collision and the perpendicular collision. In the present study we perform numerical simulation of two vortex ring collision at 16 different angles. Encouraged by the success in the previous work [10], we use the same numerical method for the simulation of the vorticity field.

In Sect. 2 the formulae of the multipole expansion of the vortex sound are presented. In Sect. 3 we show the vorticity-potential method for numerical simulation on the vorticity field. Sect. 4 is devoted to presenting the results and discussion. The last section includes the concluding remarks.

2. Vortex Sound Formula

2.1. FAR-FIELD ACOUSTIC PRESSURE

We consider sound generation by a localized vortex motion with a low Mach number at a high Reynolds number. When Mach number M is sufficiently small ($M \ll 1$), the whole space can be divided into two regions. One is the inner flow region, and the other is the outer wave region. The governing equation in the inner region is shown to be the incompressible Navier-Stokes equation, and the wave equation for the pressure p is obtained in the outer region. Matching the outer solution to the inner solution is carried out in an intermediate region [5, 6, 7, 8, 9].

In the case of unbounded space where there is neither a solid body nor an external force, the far-field expression retaining only the terms of $O(r^{-1})$ for sufficiently large $r = |\boldsymbol{x}|$ is

$$\begin{aligned} p^{(f)}(\boldsymbol{x},t) &= -\rho_0 P_0^{(1)}(t_r)\frac{1}{r} - \frac{\rho_0}{c^2} Q_{ij}^{(3)}(t_r)\frac{x_i x_j}{r^3} \\ &\quad + \frac{\rho_0}{c^3} Q_{ijk}^{(4)}(t_r)\frac{x_i x_j x_k}{r^4} + \cdots \end{aligned} \qquad (1)$$

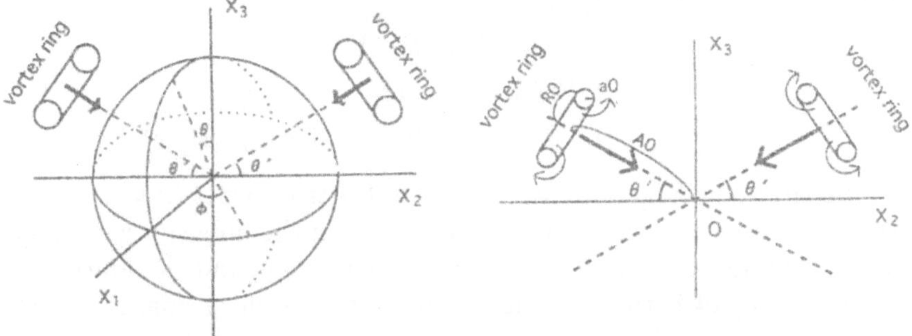

Figure 1. Setup of the collision of two vortex rings and coordinate system.

where superscript (n) denotes the n-th time derivative and $t_r = t - r/c$ the retarded time. The coefficients Q's are

$$Q_{ij}(t) = -\frac{1}{12\pi}\int (y \times \omega)_i y_j d^3y, \tag{2}$$

$$Q_{ijk}(t) = \frac{1}{32\pi}\int (y \times \omega)_i y_j y_k d^3y, \tag{3}$$

where ω is obtained by solving the vorticity equation in the inner region. The terms on the $r.h.s$ of eq. (1) represent monopole, quadrupole and octapole sound respectively. This equation shows that the multipole components of the far-field vortex sound results from time-dependence of the moment of vorticity. Because the first moment of vorticity Q_i is conserved for the vortex motion in free space, the coefficient of the dipole sound becomes zero. The first isotropic term in eq. (1) is related to the rate of energy dissipation as

$$P_0(t) = -\frac{5-3\gamma}{12\pi}\frac{1}{c^2}K^{(1)}(t), \quad K(t) = \frac{1}{2}\int v^2(y,t)d^3y, \tag{4}$$

where K is the total kinetic energy and γ is the ratio of specific heats.

2.2. MULTIPOLE COMPONENTS FOR TWO VORTEX RING COLLISION

In the present study, numerical simulation was performed for the collision of two vortex rings which collides each other with 16 angles between a head-on collision and a perpendicular collision $(0° \le \theta' \le 45°, \Delta\theta' = 3°)$.

The initial geometry is presented in Fig. 1. The centers of vortex rings are located on the (x_2, x_3) plane. Both two vortex rings move in the direction towards the origin. With this geometry the pressure distribution has a symmetry with respect to θ and ϕ. The expression of the pressure observed at a far point (r_{obs}, θ, ϕ), satisfying the symmetry, is expressed as

$$\begin{aligned} p(\theta,\phi,t) &= A_0(t) + A_1(t)P_2^0(cos\theta) \\ &+ A_2(t)P_2^2(cos\theta)cos(2\phi) \\ &+ B_1(t)P_3^0(cos\theta) + B_2(t)P_3^2(cos\theta)cos(2\phi), \end{aligned} \tag{5}$$

where $P_n^k(cos\theta)$ is the Legendre polynomial of the n-th order $(k = 1,\ldots,n)$. The first term represents the acoustic pressure with the directivity of monopole, the second and the third with that of quadrupole, and the fourth and the fifth with that of octapole. The amplitudes of each normal mode A_n, B_n in eq. (5) are related to the time-dependence of the vorticity moments and given by

$$A_0(t) = -\frac{\rho_0}{r_{obs}} P_0^{(1)}(t), \tag{6}$$

$$A_1(t) = -\frac{\rho_0}{c^2 r_{obs}} Q_{33}^{(3)}(t), \tag{7}$$

$$A_2(t) = -\frac{\rho_0}{6c^2 r_{obs}} [Q_{11}^{(3)}(t) - Q_{22}^{(3)}(t)], \tag{8}$$

$$B_1(t) = \frac{\rho_0}{c^3 r_{obs}} [Q_{333}^{(4)}(t) - (1/5)Q_{3kk}^{(4)}(t)], \tag{9}$$

$$B_2(t) = \frac{\rho_0}{c^3 r_{obs}} \frac{1}{30} [\tilde{Q}_{113}^{(4)}(t) - \tilde{Q}_{223}^{(4)}(t)], \tag{10}$$

where the tilde symbol denotes $\tilde{Q}_{113} = Q_{113} + Q_{131} + Q_{311}$ and r_{obs} is a distance between the origin and observation point.

3. Numerical Simulation for Vorticity Field

3.1. VORTICITY-POTENTIAL METHOD

In order to obtain the mode amplitudes A_n, B_n in eq. (5), it is necessary to evaluate the vorticity $\boldsymbol{\omega}$ in the inner region. The governing equations for the vorticity $\boldsymbol{\omega}$ and the induced velocity $\boldsymbol{v}$ are given by

$$\partial_t \boldsymbol{\omega} - \nabla \times (\boldsymbol{v} \times \boldsymbol{\omega}) = \nu \nabla^2 \boldsymbol{\omega}, \tag{11}$$

$$\nabla \cdot \boldsymbol{v} = 0 \tag{12}$$

where ν is a kinetic viscosity. Introducing a vector potential $\boldsymbol{A}$,

$$\boldsymbol{v} = \nabla \times \boldsymbol{A} \tag{13}$$

together with the solenoidal condition $\nabla \cdot \boldsymbol{A} = 0$, the vector potential $\boldsymbol{A}$ satisfies the Poisson equation

$$\nabla^2 \boldsymbol{A} = -\boldsymbol{\omega}. \tag{14}$$

The solution of (14) is given exactly by the Poisson integral

$$A(x,t) = \frac{1}{4\pi} \int \frac{\omega(y,t)}{| x - y |} d^3 y. \tag{15}$$

The three dimensional incompressible vorticity equation is solved with using the vorticity-potential method [11, 12, 13, 14]. After calculating the vector potential (15) on the boundary of the computational region, we solve the Poisson equation (14) on all grid points of the inside of the computational region. We then obtain the vorticity at the next time step following (11) with (13) using the fourth order Runge-Kutta-Gill method. The central finite difference with second order accuracy is used in the numerical scheme.

3.2. PARAMETERS FOR NUMERICAL SIMULATION

Parameters for calculation of vorticity field are presented in TABLE. 1. The three-dimensional problem in the last subsection is solved on a cubic grid of N^3 points. Since the vorticity is localized and the computational region is large enough, the number of operation for evaluating the vector potential can be reduced in the following way. We consider the cases where the magnitude of the vorticity can be assumed to be zero on the boundary. The whole integration domain of the Poisson integral (15) is divided into N_{SD}^3 subdomains. The factor $1/|x - y|$ is replaced with $1/r_i$ where $r_i = |x - y^i|$ is a distance between the boundary point x and the center of the i-th subdomain y^i. The vector potential is evaluated on coarse $\tilde{N}^2$ points on the computational boundary. Then they are interpolated with a surface cubic spline method to give the values at fine N^2 points. With these boundary values the vector potential is obtained by solving the Poisson equation (14) on N^3 points. Comparing the method in which the vector potential is evaluated on N^3 points with eq. (15), the number of numerical simulation is much reduced in the present method [14, 10].

TABLE 1. Parameters for Calculation

Calculation region(in x,y)	-17.5 to 17.5
Calculation region(in z)	-14.0 to 21.0
Number of grid points	$N^3 = 101 \times 101 \times 101$
Space interval	$\Delta x = 0.35$
Time interval	$\Delta t = 0.01$
Number of points of the course grid on the boundary	$\tilde{N}^2 = 13 \times 13$
Number of subcells	$N_{SD}^3 = 33 \times 33 \times 33$

TABLE 2. Parameters for Initial Condition

Radius	$R_0 = 5.0$
Distance from the center of the vortex ring to the origin	$r_0 = 10.0$
Collision angle	$0° \leq \theta' \leq 45°, \Delta\theta' = 3°$
Position of ring A	$(X, Y, Z) = (0.0, r_0 \cos\theta', r_0 \sin\theta')$
Position of ring B	$(X, Y, Z) = (0.0, -r_0 \cos\theta', r_0 \sin\theta')$
Radius of the core	$a_0 = 1.0$
Vortex ring core	Gaussian
Circulation	$\Gamma = 20\pi$
Kinetic viscosity	$\nu = 0.1$
Reynolds number	$\Gamma/\nu = 628$
Velocity of the vortex ring	$U_0 = 3.13$

Parameters for initial condition are presented in TABLE. 2. When there is only one vortex ring and it is very thin ($R_0/a_0 \to \infty$), a vortex ring whose core has Gaussian vorticity distribution is thought to be quasistable. The velocity of the ring with a Gaussian vorticity distribution was given by Saffman [15]:

$$U_0 = \frac{\Gamma}{4\pi R_0}\{\log\frac{8R_0}{a_0} - 0.558\}. \tag{16}$$

The circulation Γ is $\pi\omega_0 a_0^2$. Initial values except a collision angle θ' are common for all simulations.

4. Results and Discussion

4.1. FAR-FIELD VORTEX SOUND FROM VORTEX RING COLLISIONS

In the following figures the time and the mode amplitudes of acoustic pressure are normalized as follows:

$$\tilde{t} = t/(R_0/U_0), \tag{17}$$

$$\tilde{A}_n = A_n/[\rho_0 R_0 U_0^4/(c^2 r_{obs})], \tag{18}$$

$$\tilde{B}_n = B_n/[\rho_0 R_0 U_0^5/(c^3 r_{obs})], \tag{19}$$

where U_0 is an initial speed of a vortex ring given by (16).

The quadrupole mode amplitudes $\tilde{A}_1$ and $\tilde{A}_2$ are plotted in Fig. 2 for six collision angles $\theta' = 0°$, $6°$, $12°$, $21°$, $36°$ and $45°$. The time $\tilde{t} = 0$ corresponds to the time when the center of the vortex ring core would cross the symmetry $x_2 = 0$-plane if there were only one vortex ring. It means that the time $\tilde{t}$ is shifted from the time t by $(r_0 - R_0 \tan\theta')/U_0$, where r_0

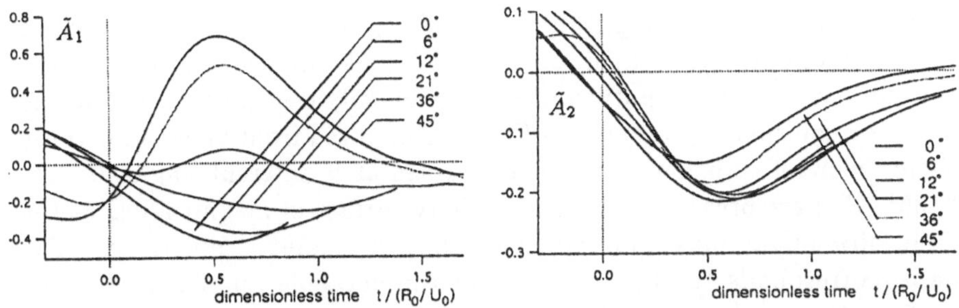

Figure 2. Quadruploe mode amplitude $\tilde{A}_1$ and $\tilde{A}_2$ for six collision angles plotted against the dimensionless scale time.

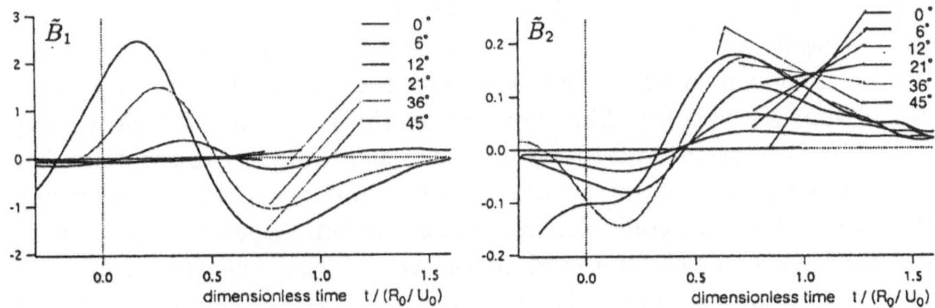

Figure 3. Octaploe mode amplitude $\tilde{B}_1$ and $\tilde{B}_2$ for six collision angles plotted against the dimensionless scale time.

is a distance between the center of the vortex ring and the origin in the initial setup.

For all these angles the amplitude has a maximum around $\tilde{t} = 0.6$. This shows that local vortex motion near the touching region of vortex rings are strongly related to the emission of vortex sound. As the angles θ' becomes larger, the peak value of $\tilde{A}_1$ changes from negative to positive, and it becomes almost zero at $\theta' = 21°$. On the other hand $\tilde{A}_2$ shows similar time dependence for all collision angles.

The octapole mode amplitudes $\tilde{B}_1$ and $\tilde{B}_2$ are plotted in Fig. 3. The amplitude B_1 has two extremes for all collision angles. The height of two peaks becomes rapidly smaller with θ', and $\tilde{B}_1$ is almost zero for $\theta' \leq 21°$. When $\tilde{B}_1$ becomes zero, $\tilde{A}_1$ has a maximum. The magnitude of $\tilde{B}_2$ is smaller than that of $\tilde{B}_1$ by ten times.

4.2. MODE AMPLITUDES AND VORTEX MOTION

The peaks of the mode amplitudes $\tilde{A}_n$ and $\tilde{B}_n$ are thought to be related with dynamics of the vortex ring collision. We look into the relation of these amplitudes and vortex ring dynamics. The vorticity distribution during the vortex ring collision is presented in Fig. 4. Computed front views, top views and side view of collision of two vortex rings at 5 different angles(0°, 6°, 21°, 36°, 45°) are presented. The isovorticity surface of magnitude ω at the level of 40% of the maximum vorticity at each instant is illustrated. The time $\tilde{t}_a$ corresponds to the maximum of the octapole amplitude $\tilde{B}_1$, the time $\tilde{t}_b$ to the maximum of the quadrupole amplitude $\tilde{A}_1$ and the time $\tilde{t}_c$ to the minimum of $\tilde{B}_1$.

The figure of the head-on collision($\theta' = 0°$) shows that whole vortex rings are expanding radially but the increase of the vortex core radius is small. In the oblique collision at right angles($\theta' = 45°$), there are some changes in addition to the radial expansion of the whole vortex ring. In the front view(a) the vortex rings are bending and we can distinguish deformed part from undeformed part. The deformation of the vortex ring is clearly shown in the side view(c). The shoulders of the vortex ring are stretching in x_1- and x_3-direction, and the vortex core is spreading. We can therefore think that the expansion of the whole vortex ring causes the vortex sound in the case of head-on collision. In addition to this expansion, the vortex core deformation at the shoulders of the vortex ring and the bending of the vortex ring are thought to be the cause of the vortex sound emission in oblique collision at right angles.

In the case of $\theta' = 36°$ the deformation of the vortex ring is similar to that in $\theta' = 45°$, but the rate of local deformation is smaller. This difference is thought to bring the difference in the magnitude of the mode amplitudes $\tilde{A}_n$ and $\tilde{B}_n$ for both angles. In the case of $\theta' = 6°$, the deformation in the colliding part of the vortex ring is small and almost same as in $\theta' = 0°$. Comparing the dynamics of the vortex ring collision about the peak of the mode amplitudes for different collision angles, the deformation of the vortex ring which relates to the sound emission is similar to the perpendicular collision ($\theta' = 45°$) when θ' is larger than 21°, while it is similar to the head-on collision for θ' smaller than 21°.

In the case of oblique collisions($\theta' > 0°$), there occurs a phenomenon of vortexline reconnection. In order to show how fast the reconnection proceeds for different collision angles, we present in Fig. 5 the time-dependence of the circulation Γ_1 for the region $S_1(x_1 = 0, x_2 > 0)$ and Γ_2 for $S_2(x_2 = 0, x_1 > 0)$ for five collision angles. In the beginning Γ_2 is zero because vortex rings are far from the (x_1, x_3) plane. Around $\tilde{t} = 0.0$ two vortex rings collide each other on the (x_1, x_3) plane. Some vortexlines which cross S_1

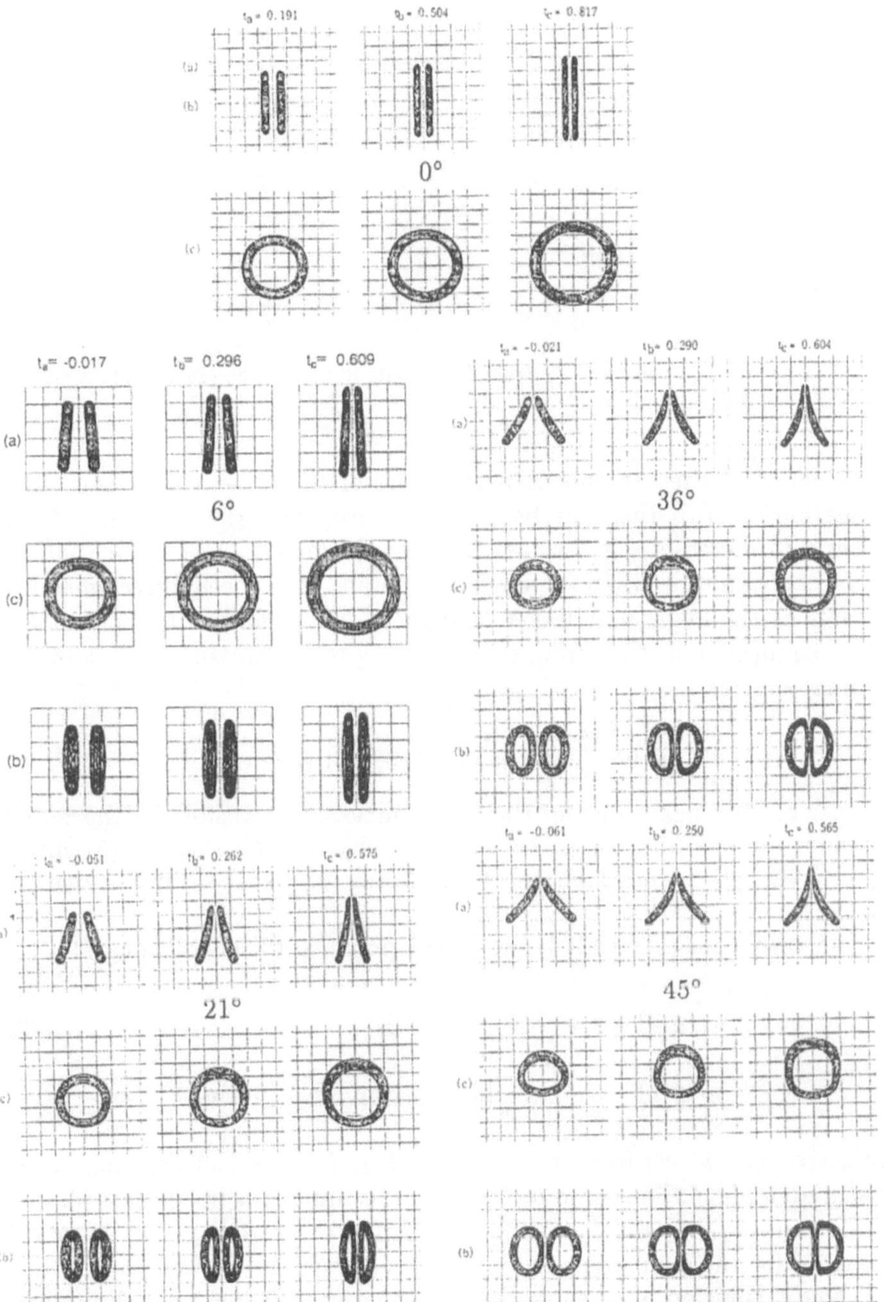

Figure 4. Computed front views(a), top views(b) and side view(c) of collision of two vortex rings at 0°, 6°, 21°, 36° and 45°. The isovorticity surface of magnitude ω with 40% of maximum at each instant is illustrated.

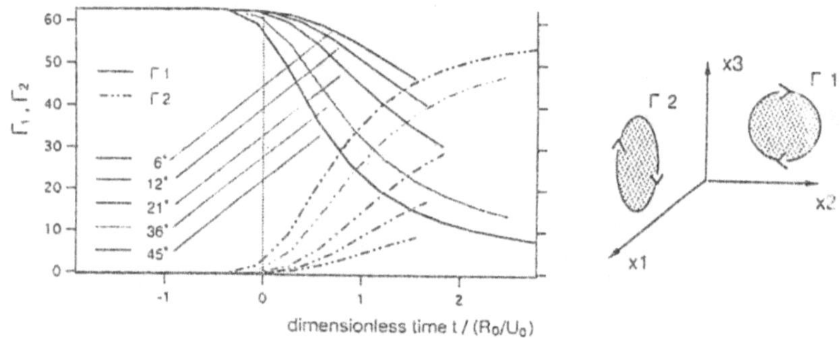

Figure 5. Time-dependence of Γ_1 which is the circulation in the (x_2, x_3) plane and Γ_2 which is the circulation in the (x_1, x_3) plane

disappear, and they reconnect with a part of the other vortex ring. The circulation Γ_1 of S_1 therefore begin to decrease, while the circulation Γ_2 of S_2 begin to increase. Fig. 5 shows that at any tine $\bar{t} > 0$ the circulation Γ_2 has lager magnitude when the angle θ' is larger. The speed of reconnection is faster as the angle θ' is larger, and the time in which a half of vortex-lines reconnects (the time from the beginning to the crossing of Γ_1 and Γ_2) becomes shorter.

4.3. SPATIAL DISTRIBUTION OF VORTEX SOUND SOURCE

In order to investigate further the vortex motion responsible for the sound emission, we look into the sound source of A_n and B_n. Eqs. (2),(3) shows that A_n and B_n depend on the time derivatives of the impulse density $p = \frac{1}{2} y \times \omega$.

The amplitudes A_1 and A_2 are proportional to the space integral of

$$a_1(y,t) = \frac{2}{12\pi} p_3^{(3)}(y,t) y_3, \tag{20}$$

$$a_2(y,t) = \frac{1}{6}\frac{2}{12\pi}[p_1^{(3)}(y,t)y_1 - p_2^{(3)}(y,t)y_2. \tag{21}$$

Projective side views from x_2-direction of a_1 for five collision angles are presented in Fig. 6. The contours for a_n with common magnitude are plotted at three times $\bar{t}_a$, $\bar{t}_b$ and $\bar{t}_c$. In the same figure the vorticity distribution is presented for the comparison with the sound source.

The distribution of a_1 shows that in the case of $\theta' = 6°$, the upper and lower parts of the vortex ring are strong sound source at $\bar{t}_a$, but the sound source spreads over the whole ring at $\bar{t}_b$ and $\bar{t}_c$. In the case of $\theta' = 21°$, $|a|$ is very small at $\bar{t}_a$. But a strong source region begins to appear in the upper

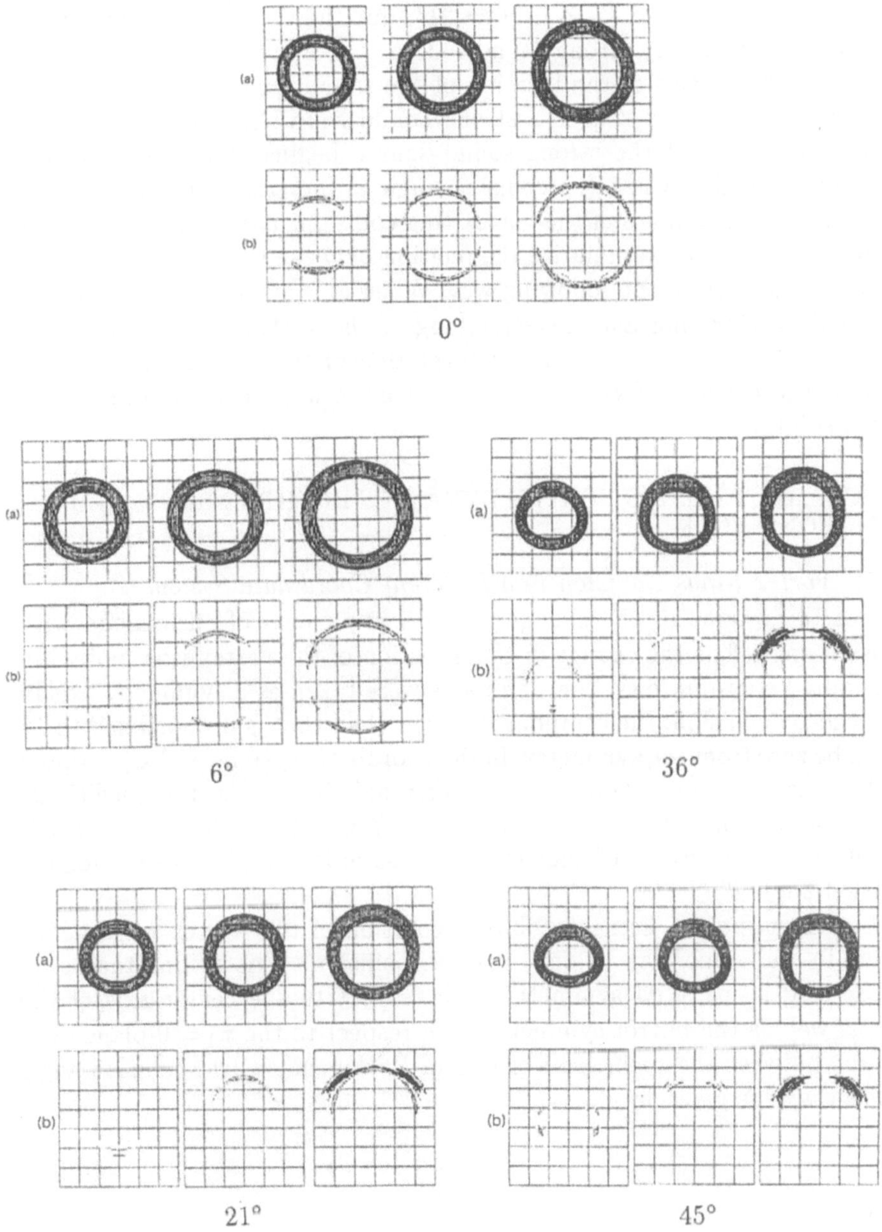

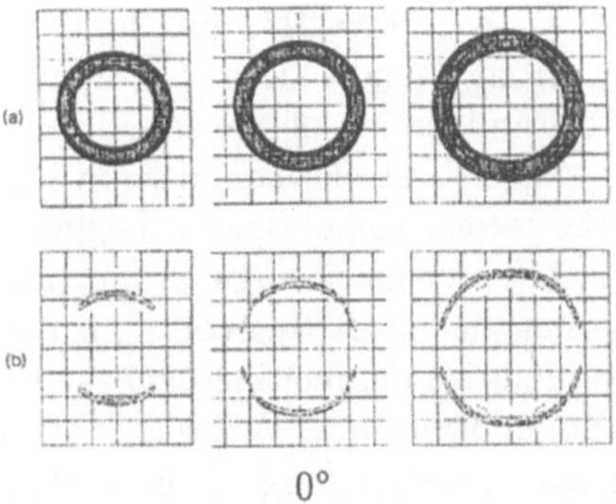

0°

6°

36°

21°

45°

Figure 6. Projective side views of (a) the isovorticity surface and the distribution of (b) a1 for collision angles 0°, 6°, 21°, 36° and 45°. The isovorticity surface of magnitude ω with 40% of maximum at each instant is illustrated.

part of the vortex ring at time $\tilde{t}_b$, and it expands more widely at $\tilde{t}_c$. The shoulder part of the vortex ring has large values. In the case of $\theta' = 36°$ and $45°$, $|a|$ is very small at $\tilde{t}_a$ as in $\theta' = 21°$, but the shoulders of the vortex ring become extremely strong sound source at $\tilde{t}_b$ and $\tilde{t}_c$.

From the analysis of the location of the sound generation, it is shown that the whole vortex ring is a strong sound source in the case of small collision angles, and the strong sound source begins to concentrate into the collision region when the collision angle is increasing. Combining the analysis of the dynamics of the vortex ring collisions and that of the vortex sound source, we can summarize the relation between the vortex sound and the vortex motion in the following way. When θ' is small, the expansion of the whole vortex ring and the stretching of the vortex tube bring vortex sound. As θ' becomes larger, the deformation of the vortex core becomes large in the collision part and its motion gives large contribution to the vortex sound.

4.4. CHARACTERISTIC VORTEX DYNAMICS IN THREE REGIONS OF COLLISION ANGLE

4.4.1. *Vortex Rings Collision in a Different Coordinate System*

One of the origins of the change in the components of each mode with collision angle θ' is the change in the direction of the vortex ring motion in an adopted coordinate system. If x_3-axis were taken as a symmetry axis in the head-on collision, the components other than that is proportional to P_2 would be zero from the symmetry. In the coordinate system (r, θ, ϕ) adopted in the present study, where the symmetry axis is x_2-axis, the coefficients of P_2 and P_2^2 should satisfy the relation $A_1^0 = 2A_2^0$ in the case of head-on collision. The suffix "0" means that these mode amplitudes are for the collision angle $\theta' = 0°$.

Let us suppose a fictitious oblique collision of two vortex rings which evolve in the same way as this head-on collision. This means that each vortex ring keeps axial symmetry with respect to the center axis of the ring as well as the mirror symmetry with respect to the $x_2 = 0$-plane. The relation between the coefficients of P_2 and P_2^2 depends on the collision angle θ' and they are given by

$$A_1' = A_1^0(1 - 3\sin^2\theta'), \quad (22)$$

$$A_2' = A_1^0\frac{\cos^2\theta'}{2}. \quad (23)$$

The prime of A' means that these coefficients are from the fictitious oblique collisions. The amplitude A_1' changes sign at 35.26°, and A_2' approaches to zero as θ' becomes larger. The above relation clearly shows that even if

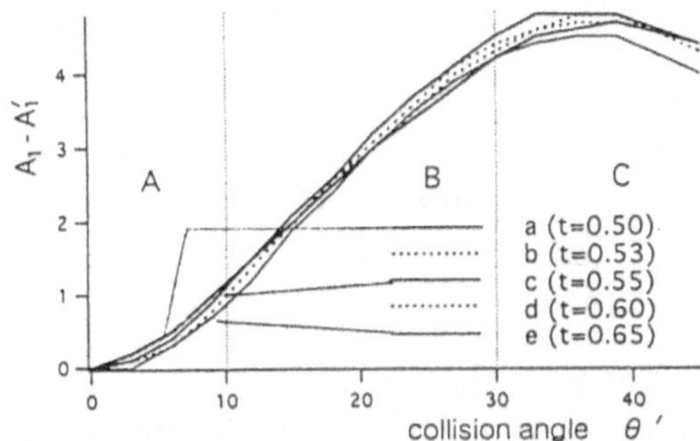

Figure 7. Angle-dependence of difference between A_1 and A_1' at each instant.

the motion of each vortex ring keeps axial symmetry, the mode amplitudes of the emitted sound change as the direction of the initial motion of the vortex rings changes.

It has been shown that the local deformation of the vortex ring give larger contribution to the sound generation as the angle θ' increases. In order to separate the contribution of the vortex ring expansion and the local deformation, we compare the coefficients of quadrupole sound from the real oblique collision and those from the fictitious oblique collision of two vortex rings. In the fictitious collision each vortex ring has the same temporal evolution as a vortex ring in the head-on collision.

Of course we cannot clearly separate the effect of the vortex ring expansion and the local deformation effect, but a measure of the contribution from the local deformation can be obtained from the comparison between the mode amplitudes A and A'. The difference between A_1 and A_1' as a function of collision angle θ' is plotted in Fig. 7 for five different times $a \sim e$ around the maximum of A_1 ($a : \tilde{t} = 0.50$, $b : \tilde{t} = 0.53$, $c : \tilde{t} = 0.55$, $d : \tilde{t} = 0.60$,$e : \tilde{t} = 0.65$). All five curves grow exponentially from 0° and they increase with a constant rate from around 10°. Then they seem to saturate from around 30°. From the angle-dependence of these curves, we conclude that the collision angle is divided into three regions: A($\theta' = 0° \sim 10°$), B($\theta' = 10° \sim 30°$), C($\theta' = 30° \sim 45°$).

4.4.2. *Three regions of Collision Angle*

In both regions B and C, the deformation of the vortex core in the colliding part of the vortex ring gives large contribution. We cannot distinguish between the region B and C from the deformation of votex rings. The division into B and C can be done with Fig. 5 for the vortexline reconnection. Around the peak of $\tilde{A}_1$ ($\tilde{t} = 0.6$), Γ_2 is small at 21°. But Γ_2 for 36° and 45°

is much larger than that for 21°. And the speed of reconnection is much faster at 36° and 45° than 21°. The difference in the time dependence of the vortexline reconnection brings a distinction between the region B and C. When the amount of the vortexline reconnection exceeds a certain value, the vortex sound generation from the vortex core deformation reaches a plateau as a function of the collision angle θ'.

5. Concluding Remarks

The sound source distribution of the multipole components shows that the whole vortex ring is a strong sound source in the case of the angle θ' is small, while the shoulders of the collision part of the vortex ring are strong sound sources when θ' is large. The collision angles are divided into three regions. In each region there is a characteristic vortex motion and it has a relation with the vortex sound as follows:

A. $\theta' = 0° \sim 10°$: The radial expansion of the vortex ring gives large contribution to the vortex sound.

B. $\theta' = 10° \sim 30°$: The effect of the vortex core deformation is large. Its effect is stronger when θ' becomes larger.

C. $\theta' = 30° \sim 45°$: The deformation of the vortex core and the vortexline reconnection occur simultaneously, and thateffect is almost constant for all collision angles in this region.

In the present study, a vortex ring with a slenderness ratio (a ratio of the vortex core radius to the ring radius) 0.2 is adopted. It is known that the slenderness ratio is an important factor for deciding the vortex dynamics. We therefore think it very meaningful to study vortex rings with a smaller slenderness ratio as next subject.

References

1. M.J. Lighthill: Proc. R. Soc. London, Ser A **211** (1952) 564.
2. A. Powell: J. Acoust. Soc. Am. **35** (1964) 177.
3. M.S. Howe: J. Fluid Mech. **71** (1975) 625.
4. F. Obermeier: Acoustica **18** (1967) 238.
5. S.C. Crow: Stud. Appl. Math. **49** (1970) 21.
6. W. Möhring: J. Fluid Mech. **85** (1978) 685.
7. F. Obermeier: J. Sound Vibration **99** (1985) 111.
8. L. Ting and M.J. Miksis: SIAM J. Appl. Math. **50** (1990) 521.
9. T. Kambe, T. Minota and M. Takaoka: Phys. Rev. E **48** (1993) 1866.
10. S. Adachi, K. Ishii and T. Kambe: ZAMM **77** (1997) 716.
11. L.Ting and R.Klein: *Viscous Vortical Flow*, Lecture Notes in Physics 374, Springer-Verlag (1991) 117–174.
12. J.P. Chamberlain and C.H. Liu: AIAA J.**23** (1985) 868.
13. J.P. Chamberlain and R.P. Weston: AIAA paper**84-1545** (1984) .
14. K. Ishii, K. Kuwahara and C.H. Liu: Computers Fluids **22** (1993) 589.
15. P.G. Saffman: Stud. Appl. Math. **49** (1970) 371.

ACOUSTIC SOUND GENERATED BY COLLISION OF TWO VORTEX RINGS

O. INOUE AND Y. HATTORI
Institute of Fluid Science, Tohoku University
2-1-1 Katahira, Aoba-ku, Sendai 980-77, Japan

Abstract. Acoustic sounds generated by head-on and oblique collisions of two vortex rings with an equal strength are simulated numerically. By a finite difference method, the axisymmetric/three-dimensional, unsteady, compressible Navier-Stokes equations are solved directly not only for a near-field but also for a far-field. The sixth-order accurate compact Padé scheme is used for spatial derivatives, together with the fourth-order accurate Runge-Kutta scheme for time integration. The results show that the quadrupolar component of vortex sound is predominant for the case of head-on collision while for the case of oblique collision the octupolar component is also observed.

1. Introduction

Collision of two vortex rings is one of the most fundamental and the most interesting three-dimensional vortical flows. A lot of studies have been made both experimentally and computationally. Acoustic sounds generated by collisions of two vortex rings have also been studied both experimentally and computationally (Minota & Kambe 1986; Kambe, Minota & Takaoka 1993; Adachi, Ishii & Kambe 1997). In computational analyses of vortex sounds, numerical simulations are often separated into two parts; the aerodynamic part and the acoustic part (Kambe *et al.* 1993). First, in the aerodynamic part, near-field flow structures of vortex interactions are simulated by solving the (incompressible) Navier-Stokes equations. Then, in the acoustic part, the far-field sound pressure is calculated theoretically, using the near-field flow quantities obtained in the aerodynamic part. This method saves computational time as well as memory storage if compared with direct numerical simulations, because the flow in the far-field is assumed to be still or uniform, and thus the Navier-Stokes equations are not solved numerically. At

E. Krause and K. Gersten (eds.), IUTAM Symposium on Dynamics of Slender Vortices, 361-368.

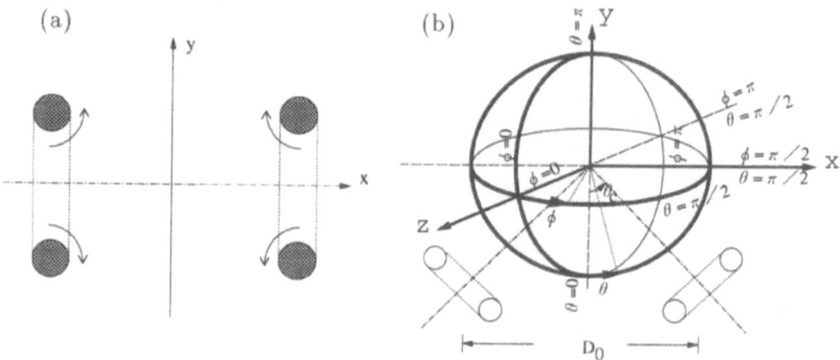

Figure 1. Schematic diagram of flow model. (a) Cylindrical coordinate for head-on collision, (b) Spherical coordinate for oblique collision.

present, however, we can not necessarily say that the applicability of the theory has been well confirmed. Direct numerical simulations of acoustic sounds have become feasible in the recent years (Mitchell, Lele & Moin 1995; Inoue, Hattori & Onuma 1997). In these simulations, the Navier-Stokes equations were solved by using highly accurate schemes both for space and time in order to precisely capture the sound. So far most of the direct simulations of sounds are limited up to two-dimensional flows. The purpose of this paper is to study, using direct Navier-Stokes simulations, the mechanism of a three-dimensional flow field produced by head-on/oblique collision of two vortex rings and to increase our understanding of acoustic sound generated by the collisions.

2. Mathematical Formulation and Numerical Procedure

Schematic diagram of flow model is presented in Figure 1. For head-on collision, the two vortex rings are set initially to move along the x-axis and collide with one another (Figure 1(a)). On the other hand, for oblique collision, the two vortex rings are set initially to move along the paths intersecting at an angle $2\theta_c$ at the origin (Figure 1(b)). The bisecting straight line between the two paths of the vortex centers is taken as the polar axis $\theta = \pi$ (along the y-axis) of the spherical coordinate system. The plane perpendicular to the y-axis is the (x, z) plane on which $\theta = \pi/2$. The plane $\phi = 0$ is taken along the positive z-axis. Initially the vortex centers are included in the (x, y) plane.

For the case of head-on collision, we assume that the flow is axisymmetric with respect to the x-axis. Three different types of vortex ring are considered;

(1) Hill's spherical vortex, $\Gamma = 0.75$,

(2) Gaussian vortex with $r_c = 0.17$, $\Gamma = 0.55$,

(3) Gaussian vortex with $r_c = 0.34$, $\Gamma = 0.69$.

Here the term 'Gaussian vortex' means a vortex ring that has a Gaussian distribution of vorticity, and the symbol r_c denotes the ratio of the vortex core radius to the vortex ring radius. We consider two vortex rings with an equal strength but an opposite sense of rotation to one another. The Mach number M_0 of the vortex rings is defined by $M_0 = U_0/c_0$, and prescribed to be $M_0 = 0.15, 0.3$ and 0.45. Here, U_0 is the initial translational velocity of the vortex rings and c_0 is the sound speed. The circulation, Γ, of the vortex ring is related to U_0 through the following experssions. For Hill's spherical vortex

$$U_0 = \frac{\Gamma}{5R_0}. \tag{1}$$

For Gaussian vortex

$$U_0 = \frac{\Gamma}{4\pi R_0}\left[\ln\left(\frac{8.97}{r_c}\right) - 0.558\right]. \tag{2}$$

The Reynolds number is defined by $Re = U_0 R_0/\nu_0$, and prescribed values of Re are from 250 to 2000. The symbol R_0 denotes the radius of the vortex rings.

The axisymmetric, compressible Navier-Stokes equations are solved by a finite difference method. For spatial derivatives, a sixth-order accurate compact Padé scheme (third-order accurate at the boundaries) proposed by Lele (1992) is adopted. The fourth-order Runge-Kutta scheme is used for time-integration. The non-reflecting boundary conditions (Poinsot and Lele 1992) are used at the boundaries. The computational domain is prescribed to be $[-R_b \leq x \leq R_b,\ \ 0 \leq y \leq R_b]$ with R_b equal to 180. In this paper, lengths are non-dimensionalized by R_0. A non-uniform mesh system is applied in this study. The number of grids is, typically, 2303 (x- direction) $\times$ 862 (y-direction). It has been confirmed that the sound pressure in a far-field is sufficiently larger than numerical errors.

For the case of oblique collision ($2\theta_c = \pi/2, \pi/4$), only Hill's spherical vortex is considered in this study owing to the lack of performance of our supercomputer. The Mach number M_0 is fixed to be 0.15, and the Reynolds number Re is 500.

The three-dimensional, compressible Navier-Stokes equations are solved in the similar way to that for the case of head-on collision. The computational domain is prescribed to be cubic ($|x|, |y|, |z| \leq R_b$), and R_b was prescribed to be 130. The number of grids is 335^3.

3. Results and Discussion

3.1. HEAD-ON COLLISION

A typical example of the head-on collision of two vortex rings is presented in terms of the vorticity in Figure 2, and in terms of the sound pressure in Figure 3, for the case of a Gaussian vortex with $r_c = 0.17, M_0 = 0.15$ and Re=500. The sound pressure $\tilde{p}$ is defined as $\tilde{p} = p - p_\infty$. Hereafter, for simplicity, $\tilde{}$ is omitted. As seen

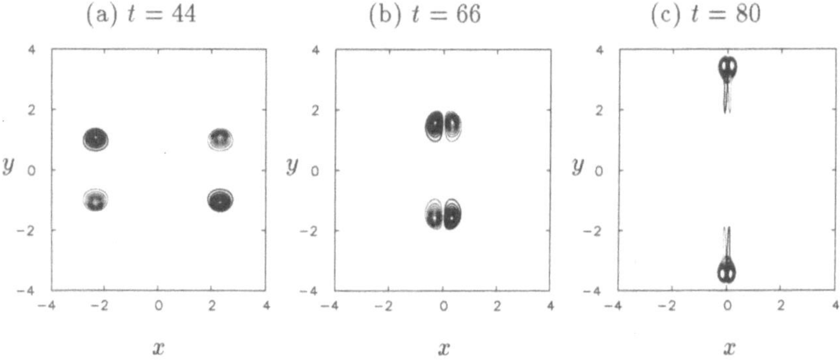

Figure 2. Head-on collision of two vortex rings. Vorticity. Gaussian vortex with $r_c = 0.17$. $M_0 = 0.15$. $Re = 500$.

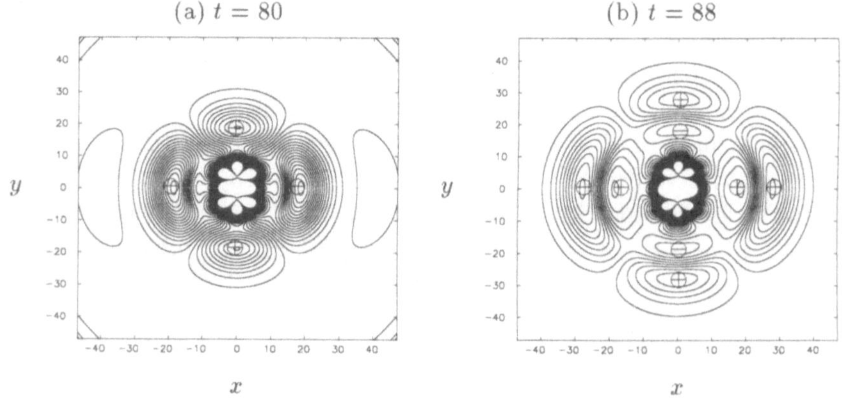

Figure 3. Head-on collision of two vortex rings. Sound pressure. Gaussian vortex with $r_c = 0.17$. $M_0 = 0.15$. $Re = 500$.

from Figure 2, the two vortex rings come closer to one another along the x-axis, collide at $t \simeq 56$, and then stretch along the direction perpendicular to the x-axis. The sound pressure in Figure 3(a) shows that an acoustic wave generated by the collision has a quadrupolar nature; that is, the sound pressure varies circumferentially from positive to negative to positive to negative. Hereafter, this acoustic wave is referred to as the first sound. Then follows the second acoustic wave (the second sound) also having a quadrupolar nature but with opposite alternating signs of circumferential pressure variation to the first sound, as seen from Figure 3(b).

Figure 4 shows the instantaneous directional distributions of the sound pressure

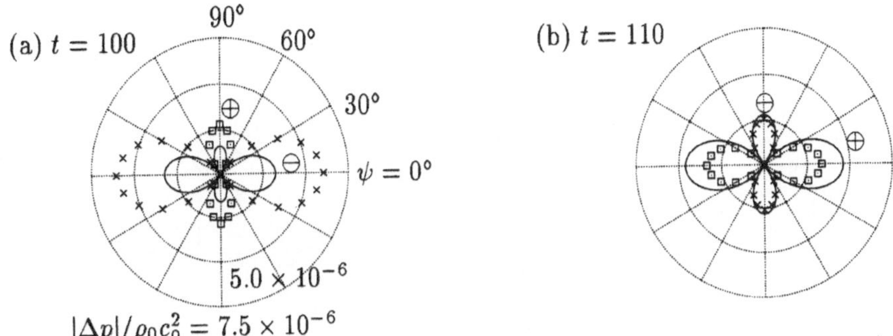

Figure 4. Polar diagram of the sound pressure. Gaussian vortex with $r_c = 0.17$. $M_0 = 0.15$. $Re = 500$.

measured at $r = 40$ where $r^2 = x^2 + y^2$. The radial length from the origin represents the magnitude of the pressure with a linear scale. The symbols □ and × denote positive and negative values of the calculated pressure, respectively. The solid lines are theoretical values of the pressure obtained in the following manner. According to Kambe *et al.* (1993), the sound pressure in the far-field can be expressed, for the case of head-on collision, as follows.

$$p(r, \psi, t) = p_m(r, t) + p_q(r, t)(3\cos^2\psi - 1) \tag{3}$$

Here the angle ψ is defined as $\cos\psi = x/r$. The first term on the right-hand side is usually called as monopole, and the second term as quadrupole. The monopole function $p_m(r, t)$ is obtained after the near-field velocity distributions are known. Similarly the quadrupole function $p_q(r, t)$ is obtained after the near-field vorticity distributions are known. The lines in Figure 4 were obtained from (1), using the values of velocity and vorticity calculated by the present Navier-Stokes simulation. As readily seen from Figure 4, both results show the quadrupolar nature of the sound beautifully. The agreement between the computational and the theoretical results are only qualitatively good.

The calculated peak sound pressures of the first sound measured along the $\psi = 0$ line (x-axis) and also along the $\psi = \pi/2$ line (y-axis) are plotted in Figure 5 for the three types of vortex ring considered in this study. In the figure, the sound pressure along the $\psi = \pi/2$ line is doubled, taking into consideration the ψ-dependence of the quadrupole expressed by Equation (1) and also the computational result that the monopolar component is much smaller than the quadrupolar component. Figure 5 shows that with an increasing distance r the sound pressure falls onto the r^{-1} -line for all the three cases examined. This result indicates that in the far field the sound pressure decays in inverse proportion to the distance r, in agreement with the theoretical prediction (Landau & Lifshitz 1984). We can also see from the figure that the calculated sound pressure at $\psi = 0$ is approximately

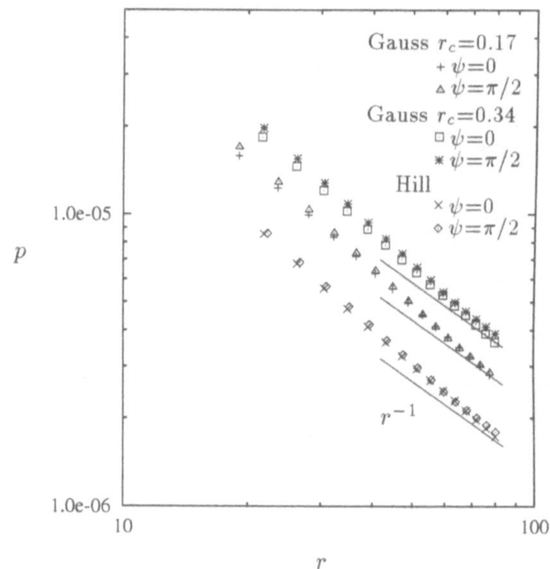

Figure 5. Decay of the sound pressure. The first sound.

twice as large as that at $\psi = \pi/2$. This result is consistent with the theoretical prediction expressed by Equation (1).

3.2. OBLIQUE COLLISION

A typical example of the oblique collision of two vortex rings is presented in Figure 6 for $2\theta_c = \pi/2$. The time-development of the flow field is quite similar to that calculated for incompressible vortex rings (Kida *et al.* 1991; Kambe *et al.* 1993) as well as that observed experimentally for slightly compressible vortex rings with $M_0 = 0.08$ (Kambe *et al.* 1993).

For the case of oblique collision, the sound pressure in the far-field can be expressed as follows (Kambe *et al.* 1993).

$$\begin{aligned} p(r,\theta,\phi,t) &= A0(r,t) \\ &\quad + A1(r,t)P_2^0(\cos\theta) + A2(r,t)P_2^2(\cos\theta)\cos 2\phi \\ &\quad + B1(r,t)P_3^0(\cos\theta) + B2(r,t)P_3^2(\cos\theta)\cos 2\phi \end{aligned} \tag{4}$$

Here, r is the distance from the origin ($r^2 = x^2 + y^2 + z^2$), and P_2^0, P_2^2, P_3^0 and P_3^2 are the Legendre polynomials. The first term on the right hand side shows monopole, the second and third terms show the quadrupole, and the fourth and fifth terms octupole. In Figure 7, the quadrupolar components $A1(t)$ and $A2(t)$

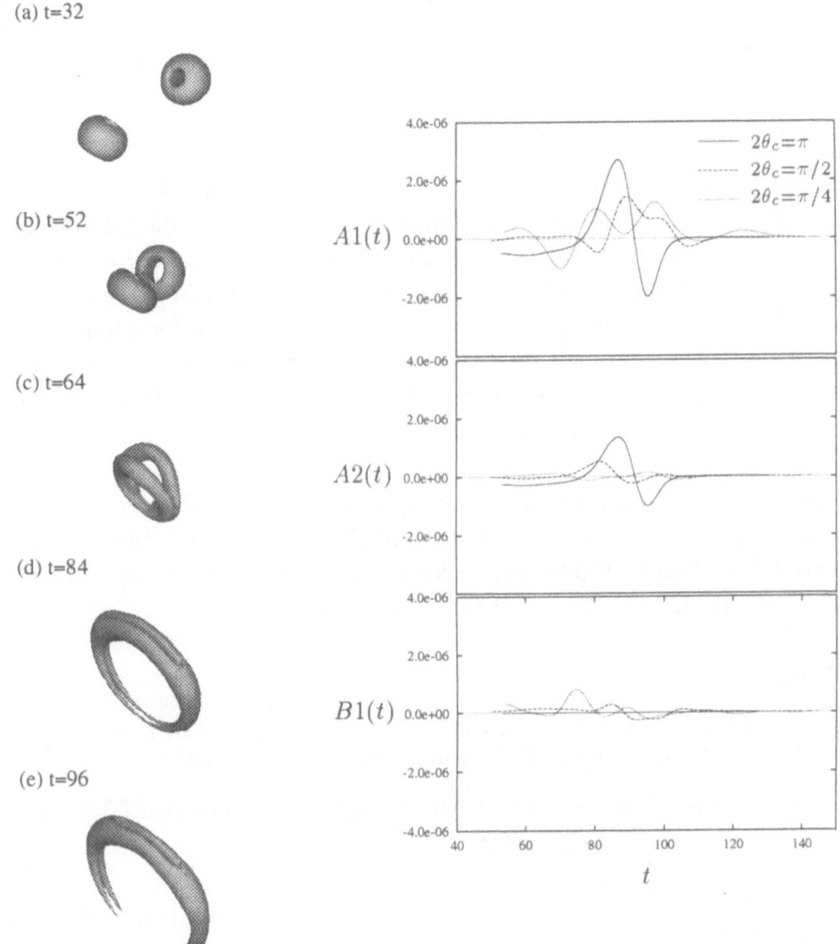

Figure 6. Oblique collision of two vortex rings. Perspective views of the iso-surfaces of the vorticity. Hill's spherical vortex. $2\theta_c = \pi/2$. $M_0 = 0.15$, $Re = 500$.

Figure 7. Calculated mode amplitudes of Equation (2). Measured at $r = 30$. Hill's spherical vortex. $M_0 = 0.15$, $Re = 500$.

and the octupolar component $B1(t)$, all measured at $r = 30$, are plotted for three different initial inclination angles, i.e., $2\theta_c = \pi, \pi/2$ and $\pi/4$. In any of the three cases, the monopolar component $A0(t)$ and the octupolar component $B2(t)$ were small if compared with the other components. We can see from Figure 7 that for the case of head-on collision ($2\theta_c = \pi$) only the quadrupolar components appear and the octupolar components are not observed, but that with a decreasing θ_c the quadrupolar components tend to be suppressed while the octupolar component tends to grow. The quadrupolar components in Figure 7 also suggest that, in contrast to the case of head-on collision where only two sounds are observed, more than two sounds may be generated for the case of oblique collision.

4. Concluding Remarks

Three-dimensional, unsteady, compressible flow fields produced by head-on/oblique collisions of two vortex rings were examined by direct Navier-Stokes simulations, and characteristic features of acoustic sounds generated by the collisions were clarified in some detail. For head-on collision, acoustic sounds having the quadrupolar nature are generated twice. On the other hand, the results for oblique collision suggest that acoustic sounds may be generated more than twice and that the octupolar component is also important. The sound pressure has been confirmed to decay in inverse proportion to the distance in the far-field.

The first author expresses his sincere gratitude to Asako Inoue for her continuous encouragement. Thanks are also given to Mr. Sakari Onuma, Institute of Fluid Science (IFS), Tohoku University, for his technical assistance. Computations were performed with CRAY C916 at IFS, Tohoku University.

References

Adachi, S., Ishii, K. & Kambe, T. (1997) Vortex sound associated with vortexline reconnection in oblique collision of two vortex rings. *ZAMM Z. angew. Math. Mech.*, **5**, 1-5.

Inoue, O., Hattori, Y. & Onuma S. (1997) Acoustic waves generated by shock-vortex interactions. *Proc. 21th Intn'l Symp. Shock Waves*, (in press).

Kambe, T., Minota, T. & Takaoka, M. (1993) Oblique collision of two vortex rings and its acoustic emission. *Phys. Rev. E*, **48**, 1866-1881.

Kida, S., Takaoka, M. & Hussain, F. (1991) Collisions of two vortex rings. *J.Fluid Mech.*, **230**, 583-646.

Landau, L. D. & Lifshitz, E. M. (1984) *Fluid Mechanics*, 2nd ed., Course of Theoretical Physics, Vol.6. Pergamon Press.

Lele, S. K. (1992) Compact finite difference schemes with spectral- like resolution. *J. Comp. Phys.* **103**, 16-42.

Minota, T. & Kambe, T. (1986) Observation of acoustic emission from head on collision of two vortex rings. *J. Sound Vib.*, **111**, 51-59.

Mitchell, B. E., Lele, S. K. & Moin, P. (1995) Direct computation of the sound from a compressible co-rotating vortex pair. *J. Fluid Mech.* **285**, 181-202.

Poinsot, T. & Lele, S. K. (1992) Boundary conditions for direct simulation of compressible viscous flows. *J. Comp. Phys.* **101**, 104-129.

NOISE EMISSION DUE TO SLENDER VORTEX SOLID BODY INTERACTIONS

Ring/Sphere Interaction at High Reynolds Number

O.M. KNIO
Department of Mechanical Engineering, The Johns Hopkins University, 3400 N. Charles St., Baltimore, MD 21218, USA

AND

L. TING
Courant Institute of Mathematical Sciences, New York University, 251 Mercer Street, New York, NY 10012, USA

Abstract. Interactions of a slender vortex ring with and a stationary rigid sphere are analyzed using a 3D, Lagrangian vortex element scheme which discretizes and tracks the filament centerline using smoothed vortex elements. The filament self-induced velocity is obtained from a desingularized Biot-Savart law that reflects the correct asymptotic behavior of the core vorticity distribution. The effect of the sphere is represented in terms of a potential velocity field that is expressed as a line integral along the image of the filament centerline with regular weight functions. It is shown that the acoustic emission due to the interaction between the filament and the sphere essentially consists of dipoles and quadrupoles whose strengths and orientations are determined by the time-evolution of the weighted first and second moments of vorticity, respectively. The scheme is used to analyze the sound generated during the passage of slender vortex rings near a solid sphere.

1. Introduction

Interactions of slender vortices with moving and/or stationary solid bodies play a major role in a large number of applications. Examples include pumps, turbines, propellers, helicopter blades, etc... In most of these examples, detailed computation of the flow and acoustic fields is generally not possible. Thus, it is desirable to consider simplified models which render the analysis tractable, and enable isolation of the essential flow dynamics of their impact on the radiated sound.

Some of the simplest examples of sound generation by slender vortices include the unsteady motion of several vortices in free space and the interaction of a slender

E. Krause and K. Gersten (eds.), IUTAM Symposium on Dynamics of Slender Vortices, 369-378.

vortex with a compact rigid body. In this paper, we consider the simplified example of slender vortex filament interacting with a stationary rigid sphere (Knio et al., 1997). This setting enables us to implement an efficient scheme for simulating the filament motion and for analyzing the associated sound emission.

The motion of the slender vortex in the presence of the sphere is analyzed using a vortex element scheme which we have recently constructed (Klein & Knio, 1995; Klein et al., 1996). The scheme is based on the discretization of the filament centerline into desingularized vortex elements and transport of these elements according to the local centerline velocity. As discussed in section 2, the latter is the sum of the self-induced velocity and the potential velocity due to the presence of the sphere. The self-induced filament velocity is evaluated using a desingularized line Biot-Savart integral which accounts for the evolution of filament core structure (section 3). Meanwhile, the potential velocity due to the presence of the rigid sphere is evaluated using an explicit "image" formula (Knio & Ting, 1997). The far-field acoustic pressure is also analyzed using an explicit formula which shows that sound emission essentially consists of dipoles and quadrupoles whose strengths and orientations are determined by the time-evolution of the weighted first and second moments of vorticity, respectively. In section 4, the numerical model is applied to compute the far-field sound generated by the passage of a slender vortex ring near a stationary rigid sphere. The analysis is restricted to the high-Reynolds-number regime and the 3D computations are used to analyze the effects of filament-sphere separation distance on the far-field sound.

2. Formulation

As illustrated in figure 1, the set-up presently considered consists of a slender vortex ring of radius $\tilde{R}$, circulation $\tilde{\Gamma}$ and initial core size $\tilde{\sigma} \ll \tilde{R}$. The ring is initially located "upstream" of sphere of radius $\tilde{a}$. A Cartesian (x, y, z) coordinate system is used whose origin coincides with the center of the sphere. The axis of the vortex ring initially points along the z-direction and intersects the y-axis at a distance $\tilde{o}$ from the origin.

In the discussion below, we shall adopt the following normalization conventions. The sphere radius $\tilde{a}$ is selected as reference lengthscale, and the circulation of the filament is used to define the following reference velocity $\tilde{U}_{\text{ref}} = \tilde{\Gamma}/\tilde{a}$. With these definitions, the normalized circulation of the filament $\Gamma = 1$, the Reynolds number $\text{Re} = \tilde{\Gamma}/\tilde{\nu}$, and the Mach number $M = \tilde{\Gamma}/\tilde{a}\tilde{c}_0$. Following the same convention, the acoustic pressure is normalized as follows:

$$p = \frac{\tilde{p}}{\tilde{\rho}_0 \tilde{\Gamma}^2 / \tilde{a}^2} \tag{1}$$

Here, $\tilde{\rho}_0$ and $\tilde{c}_0$ are the dimensional density and speed of sound of the undisturbed medium, respectively.

The present study is restricted to interaction events in which the filament remains at all times well separated from the surface of the sphere. We also assume that: (1) the Mach number is so small that the motion surrounding the vortex and the sphere can be treated as essentially incompressible, and (2) the Reynolds number is so large that the thickness of the boundary layer on the surface of the sphere remains much smaller than the sphere radius. Thus, viscous effects near the surface of the sphere are ignored, and the impact of the sphere on the motion of the filament is approximated by the potential velocity field needed to satisfy zero normal velocity at the sphere surface. This leads to a simplified flow model whose construction is summarized in the following section.

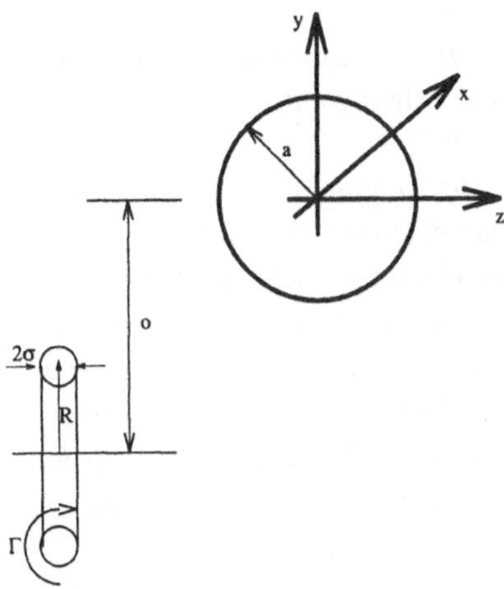

Figure 1. Schematic illustration of the initial flow configuration, showing the locations of the vortex ring and the sphere.

3. Numerical Scheme

The motion of the ring centerline is computed using a vortex element scheme that is consistent with the asymptotic filament evolution equation:

$$\dot{\boldsymbol{X}}(s) = \boldsymbol{v}^{\mathrm{si}}(s) + \boldsymbol{v}^{\mathrm{out}}(s) \tag{2}$$

where $\boldsymbol{v}^{\mathrm{si}}$ denotes the filament self-induced velocity, and $\boldsymbol{v}^{\mathrm{out}}$ is an outer velocity field (Callegari & Ting, 1978). Here, $\boldsymbol{X}$ is the coordinate of the filament centerline and s is the arc length parameter along the centerline. As discussed in the previous section $\boldsymbol{v}^{\mathrm{out}}$ in the present case corresponds to the potential velocity field due to the presence of the sphere.

For a slender filament, the leading-order structure of the core vorticity distribution remains axisymmetric, and the self-induced filament velocity is given by

(Callegari & Ting, 1978):

$$\boldsymbol{v}^{\mathrm{si}}(s) = \frac{\Gamma}{4\pi}\left[\ln\left(\frac{2}{\sigma}\right) + C\right]\kappa(s)\boldsymbol{b}(s) + \boldsymbol{Q}^{f}(\boldsymbol{X}(s)) \tag{3}$$

where $\kappa(s,t)$ and $\boldsymbol{b}(s,t)$ are the curvature and unit binormal at $\boldsymbol{X}$, respectively. The constant $C(t)$ is the time-dependent core structure coefficient, and $\boldsymbol{Q}^{f}(s,t)$ is the so-called finite part of the Biot-Savart integral. In the high-Reynolds-number inviscid limit and in the absence of axial flow, the core structure coefficient evolves according to (Klein & Ting, 1995):

$$C(t) = C(0) + \ln\sqrt{S(t)/S(0)} \tag{4}$$

where $S(t)$ is the total arc length of the filament. When viscous effects are important, the expressions describing the evolution of the core structure coefficient are more involved (Klein et al., 1996).

The numerical scheme used to simulate the above system of governing equations is based on a Lagrangian discretization of the filament geometry into a finite number of regularized vortex elements with spherical overlapping cores. The vortex elements are described in terms of their Lagrangian position vectors, $\chi_i(t)$, $i,\ldots,N$, which are indexed consecutively such that the collection $\{\chi_i(t)\}_{i=1}^{N}$ approximates the filament centerline $\mathcal{L}(t)$ (Knio & Ghoniem, 1990). Based on the Lagrangian variables, a smooth representation of the *regularized* filament vorticity is obtained using the expression (Beale & Majda, 1985):

$$\omega(\boldsymbol{x},t) = \sum_{i=1}^{N}\Gamma\delta\chi_i(t)\, f_\delta(\boldsymbol{x} - \chi_i(t)) \tag{5}$$

where f_δ is a rapidly decaying spherical core function of unit mass, $\delta\chi_i(t)$ is the arc length increment associated with the ith element, and δ is a *numerical* core radius. The above representation yields the following desingularized velocity field:

$$\boldsymbol{v}^{\mathrm{ttm}}(\boldsymbol{x},t) = -\frac{\Gamma}{4\pi}\sum_{i=1}^{N}\frac{(\boldsymbol{x} - \chi_i(t)) \times \delta\chi_i(t)}{|\boldsymbol{x} - \chi_i(t)|^3}\kappa_\delta(\boldsymbol{x} - \chi_i(t)) \tag{6}$$

where $\kappa_\delta(\boldsymbol{x})$ is the velocity smoothing kernel corresponding to f_δ. In the computations, the arc length increments $\delta\chi_i(t)$ are related to the distribution of particle positions using the Lagrangian, spectral collocation procedure described by Klein & Knio (1995). Meanwhile, the numerical core radius δ is related to the physical core structure so that the regularized velocity field $\boldsymbol{v}^{\mathrm{ttm}}$ at the particle positions coincides $\boldsymbol{v}^{\mathrm{si}}$; we set:

$$\delta = \sigma\exp\left(C^{\mathrm{ttm}} - C\right) \tag{7}$$

where C^{ttm} is the numerical core constant which corresponds to the choice of core smoothing function f_δ. In all of the computations presented below, the core function used to regularize the vorticity field is $f(r) = \operatorname{sech}^2(r^3)$, with corresponding velocity kernel $\kappa(r) = \tanh(r^3)$ (Beale & Majda, 1985). For this choice of core smoothing function, the numerical core structure coefficient $C^{\text{ttm}} = -0.4202$ (Klein & Knio, 1995).

The evolution of the filament centerline is determined by tracking the motion of the vortex elements, i.e. by integrating:

$$\frac{\partial \chi_i}{\partial t} = \boldsymbol{v}^{\text{ttm}}(\chi_i(t), t) + \boldsymbol{v}^{\text{out}}(\chi_i(t), t) \tag{8}$$

As mentioned earlier, $\boldsymbol{v}^{\text{out}}$ is the potential flow due to the presence of the sphere. The latter has been analyzed by Knio & Ting (1997) who provide analytical formulas for the "image" potential and the associated velocity field. This analysis, which extends the classical results of Weiss (1944) and Lighthill (1956), shows that the "image" velocity can be expressed as a line integral along the image of the filament centerline with regular weight functions. The result is $\boldsymbol{v}^{\text{out}} = \boldsymbol{v}^{\text{im}}$, with

$$v_j^{\text{im}}(\boldsymbol{x}) = -\frac{\Gamma}{4\pi a} \int_{\mathcal{L}} \left\{ G \frac{\partial \hat{\boldsymbol{r}}}{\partial x_j} + \left[\frac{\partial G}{\partial \lambda} \frac{\partial \lambda}{\partial x_j} + \frac{\partial G}{\partial \mu} \frac{\partial \mu}{\partial x_j} \right] \hat{\boldsymbol{r}} \right\} \cdot \left[\hat{\boldsymbol{\tau}}(s) \times \hat{\boldsymbol{X}}(s) \right] \, ds \tag{9}$$

where $\hat{\boldsymbol{\tau}}$ is the unit tangent vector to $\mathcal{L}$ at $\boldsymbol{X}(s)$, $r \equiv |\boldsymbol{x}|$, $R \equiv |\boldsymbol{X}|$, $\hat{\boldsymbol{r}} \equiv \boldsymbol{x}/r$, $\hat{\boldsymbol{X}} \equiv \boldsymbol{X}/R$, $\mu \equiv \hat{\boldsymbol{r}} \cdot \hat{\boldsymbol{X}}$, $\lambda \equiv a^2/rR$,

$$G(\lambda, \mu) \equiv \frac{\lambda^2}{Z\left[Z + 1 - \lambda\mu\right]}; \; Z(\lambda, \mu) \equiv \sqrt{1 - 2\lambda\mu + \lambda^2} \tag{10}$$

The partial derivatives in the integrand are defined by:

$$\frac{\partial G}{\partial \lambda} = \frac{\lambda}{Z^3(\lambda, \mu)}; \; \frac{\partial G}{\partial \mu} = \frac{\lambda^3 \left[Z^2 + 2Z + 1 - \lambda\mu\right]}{Z\left[Z^2 + Z(1 - \lambda\mu\right]^2} \tag{11}$$

and

$$\frac{\partial \lambda}{\partial x_j} = -\frac{a^2 x_j}{R r^3}; \; \frac{\partial \hat{\boldsymbol{r}}}{\partial x_j} = \frac{\hat{\boldsymbol{j}}}{r} - \frac{x_j \boldsymbol{x}}{r^3}; \; \frac{\partial \mu}{\partial x_j} = \hat{\boldsymbol{X}} \cdot \frac{\partial \hat{\boldsymbol{r}}}{\partial x_j} \tag{12}$$

where $\hat{\boldsymbol{j}}$ is the unit vector along the jth coordinate direction.

The far-field sound emission is estimated using the procedure of Ting & Miksis (1990), which is based on matching the far-field acoustic potential to the potential induced by the slender filament and by its "image". The result is (Knio & Ting, 1997; Knio et al., 1997):

$$p_a^F(\boldsymbol{x}, t) = \frac{M}{4\pi r} \ddot{D}_i(t_r)\hat{x}_i + \frac{M^2}{12\pi r} \dddot{Q}_{il}(t_r)\hat{x}_i\hat{x}_l \tag{13}$$

where $\boldsymbol{x}$ is the observer location, $r = |\boldsymbol{x}|$, $\hat{\boldsymbol{x}} = \boldsymbol{x}/r$, and $t_r = t - Mr$ is the retarded time. The first two terms in (13) represents dipoles with strengths, D_i, while the remaining three represents quadrupoles with strengths, Q_{il}. The dipole strengths D_i are related to the first moments of the filament vorticity and to the weighted first moments of its "image" within the sphere. We identify these two contributions by expressing the dipole strengths as $D_i(t) = d_i(t) - d_i^{\rm im}(t)$ where

$$d_i(t) = \frac{\Gamma}{2} \int_{\mathcal{L}(t)} [X_j \tau_k - X_k \tau_j] \, ds \tag{14}$$

$$d_i^{\rm im}(t) = \frac{\Gamma}{2} \int_{\mathcal{L}(t)} [X_j \tau_k - X_k \tau_j] \left(\frac{a}{|\boldsymbol{X}|} \right)^3 ds \tag{15}$$

for $i = 1, 2, 3$ and i, j, k in cyclic order. Note that the contribution of the filament alone $d_i(t)$ represents the instantaneous impulse associated with the slender vortex filament, and that $d_i^{\rm im}$ is a first moment of vorticity weighted by $(a/|\boldsymbol{X}|)^3$. The quadrupole strengths are expressed as $Q_{il}(t) = q_{il}(t) - q_{il}^{\rm im}(t)$ where

$$q_{il}(t) = \Gamma \int_{\mathcal{L}(t)} [X_j \tau_k - X_k \tau_j] \, X_l \, ds \tag{16}$$

$$q_{il}^{\rm im} = \Gamma \int_{\mathcal{L}(t)} [X_j \tau_k - X_k \tau_j] \, X_l \left(\frac{a}{|\boldsymbol{X}|} \right)^5 ds \tag{17}$$

for $i, l = 1, 2, 3$ and i, j, k in cyclic order. Note that $q_{il}^{\rm im}$ is a second moment of vorticity that is weighted by $(a/|\boldsymbol{X}|)^5$.

In the absence of the sphere the contribution of the image of the filament drops out, and the first and second moments of vorticity reduce to $D_i = d_i$ and $Q_{il} = q_{il}$, respectively, with $\boldsymbol{X}(s,t)$ for the filament alone. In this situation, Moreau's theorem (Moreau, 1948) applies and is used to conclude that the first vorticity moment d_i is time invariant. Consequently, the dipole contribution to the acoustic field vanishes identically, and the acoustic pressure is dominated by the effect of the quadrupoles. Thus, in this case one exactly recovers Möhring's (1978) formula for vortex sound at low Mach number. It also is interesting to note how the present 3D expressions generalize the axisymmetric results of Miyazaki & Kambe (1986), and the predictions of Obermeier (1980).

4. Results and Discussion

To illustrate the implementation of the numerical scheme, we consider a slender vortex ring with $R = 1$, and initial core to radius ratio $\delta/R = 0.04$. The ring is initially "upstream" of the sphere with its centerline belonging to the $z = -4$ plane. We consider several values of the offset; $o = 2.5$, 5.0, 7.5, and 10.

In order to highlight the sound field produced by the interaction of the ring with the sphere, the centerline of the ring is perturbed before the start of the computations. The imposed perturbation has a helical shape with eight waves around the azimuth, and a small amplitude equal to 1% of the undisturbed ring radius. Due to the spinning of the waves around the circumference of the ring, a quadrupolar sound field with the same frequency is generated even in the absence of the sphere (Knio & Juvé, 1996). Its presence enables us to appreciate the dipole-dominated emission which characterizes the ring-sphere interaction.

Figure 2 shows the evolution of the far-field acoustic pressure along the three coordinate axes for an interaction with offset $o = 2.5$ and a Mach number $M = 0.05$. The acoustic emissions along the y and z axes show the occurrence of spikes as the slender ring passes by the sphere. Meanwhile, the emission along the x axis shows the presence of oscillations whose amplitudes rises sharply as the ring approaches the sphere.

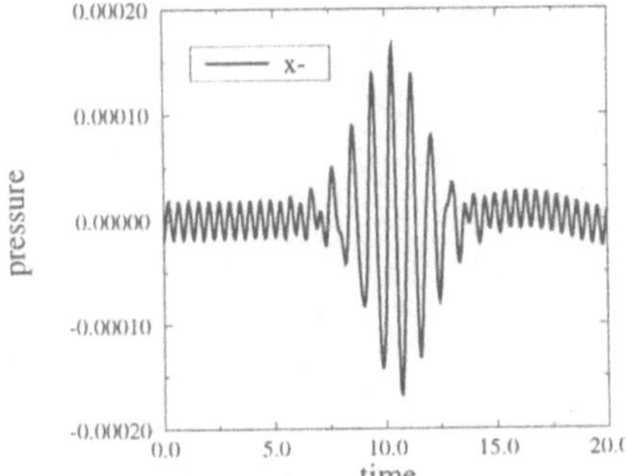

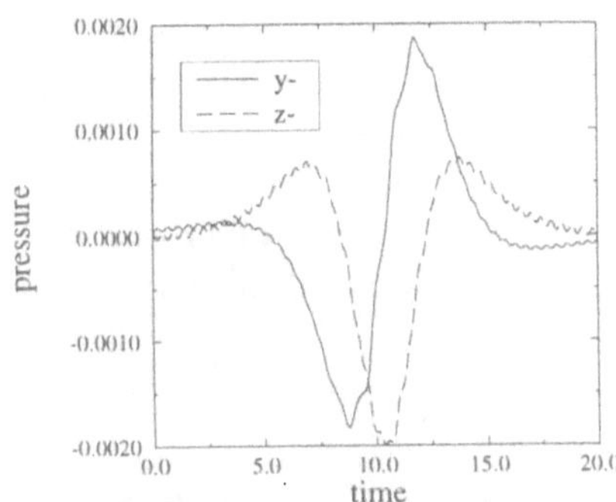

Figure 2. Evolution of the far-field acoustic pressure along the x, y, and z axes for a ring sphere interaction with an offset $o = 2.5$. The plots are generated with $M = 0.05$.

In order to further analyze the character of the noise emission, we generate in figure 3 acoustic pressure signals for a lower Mach number, $M = 0.005$, and examine in figure 4 the evolution of the x and y components of the dipole vector. Comparison of figures 2 and 3 reveals that the amplitude of the acoustic pressure signal along the y and z direction scales linearly with M. Thus, sound radiation along these direction is dominated by the effect of the dipoles (see equation 13). Partial examination of the results in figure 4 and detailed examination of the trajectory of the slender filament (not shown) reveals that the broad features of the evolution of the dipole vector is associated with a tilting of the trajectory of the ring due to the interaction of the sphere. One of the features of the this phenomenon is that there a net change in the impulse associated with the slender filament occurs during the interaction. In particular, figure 4 shows that the y component of the impulse vector does not return to its initial value after the passage is completed.

Meanwhile, the acoustic emission along the x direction exhibits a different character. When the ring is well-separated from the sphere, the pressure signal is dominated by the effect of the quadrupoles. The quadrupole emission, with characteristic $O(M^2)$ far-field pressure dependence (see equation 13), is due in the present to the spinning of the helical wave along the circumference of the ring. As the ring approaches the sphere, the acoustic emission becomes dominated by the effect of the dipoles. As shown in figure 4, The changes of the dipole component along the x-direction has a significantly weaker amplitude than that along the y direction. The evolution of the x component of the dipole vector also exhibits an oscillatory behavior at the same frequency of the helical perturbation. Thus, its origin and behavior are closely coupled to the shape and deformation of the axis of the filament during the interaction. Another distinctive feature of this phenomenon is that the changes in the x-component of the dipole vector decay very rapidly when the distance between the ring and the sphere is large. It is also interesting to simultaneously compare figures 2–4, to observe how the dipoles and quadrupoles affect the far-field emission along the x-direction.

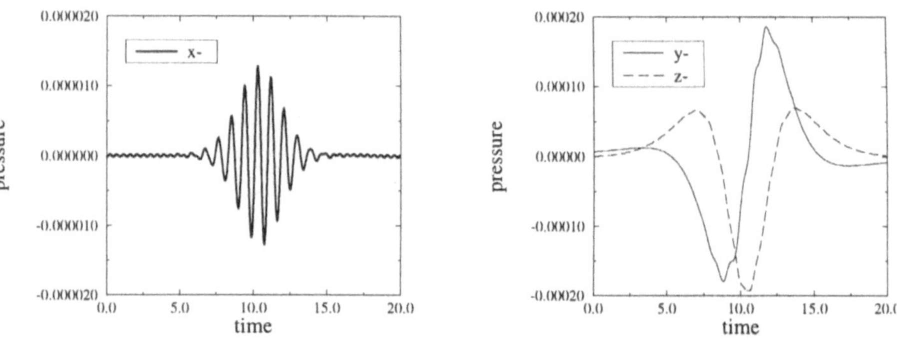

Figure 3. Evolution of the far-field acoustic pressure along the x, y, and z axes for a ring sphere interaction with an offset $o = 2.5$. The plots are generated with $M = 0.005$.

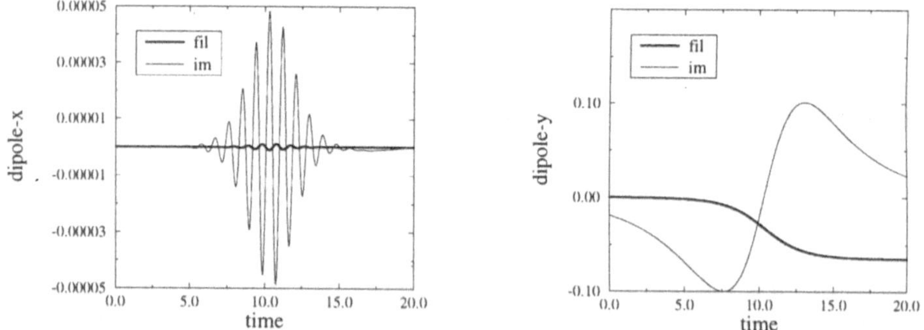

Figure 4. Evolution of the x and y components of the dipole vector for an interaction with $o = 2.5$. The contributions of the filament and the image are identified.

The effect of the initial offset is first examined in figure 5, which depicts far-field pressure signals along the coordinate axes for an interaction with $o = 5$. Figure 5 shows that the emission along the x axis essentially consists of an oscillatory signal with fixed amplitude. Meanwhile, the emission along the y axis appears to be formed by superposing onto a pressure spike a constant-amplitude oscillatory signal. (A similar behavior is observed for acoustic emission along the z direction; the corresponding curve is therefore omitted). Thus, as the offset increases the effect of the dipoles decreases, and the role of the quadrupoles becomes increasingly more important.

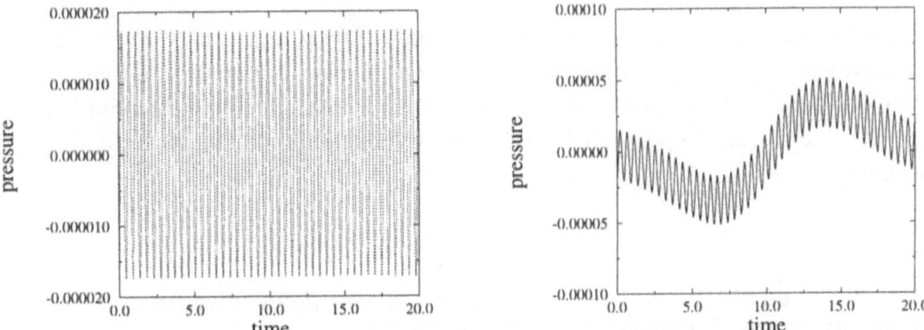

Figure 5. Evolution of the far-field acoustic pressure along the x (left) and y (right) axes for a ring sphere interaction with an offset $o = 5$. The plots are generated with $M = 0.05$.

The dependence of the dipole effect on the filament-sphere separation distance is analyzed by plotting the amplitude of the dipole variation against the offset o in the initial configuration. Results for the x and y components of the dipole vector are shown in figure 6. The evolution of the y component of the dipole vector reveals approximately an o^{-3} relationship, while the evolution of the x component exhibits approximately an o^{-10} dependence.

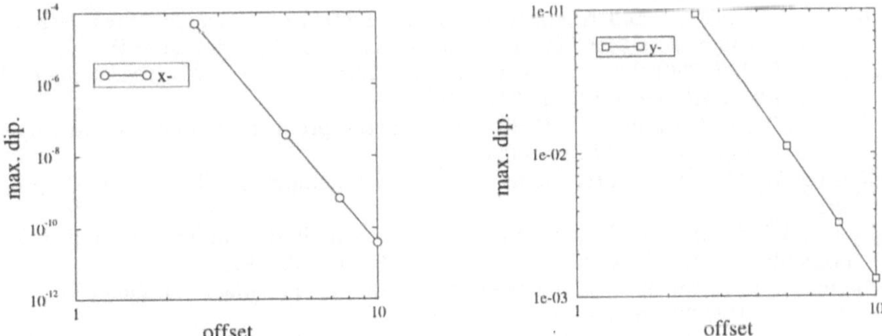

Figure 6. Amplitude of dipole variation versus initial offset.

The results above are consistent with previous observations of the evolution of the dipole vector and the far-field pressure. The behavior of the y component of the dipole vector reflects the analytical expressions of section 3 which lead us to expect that an inverse cube relationship on the separation distance, at least for "non-trivial" "compact" interactions with moderate to large separation distances. The behavior of the x component underscores the potential role of the shape and deformation of the filament centerline in modulating the amplitude and behavior of the dipole emission.

Acknowledgements

Computations were performed at the Pittsburgh Supercomputing Center. Knio's research was supported in part by the National Science Foundation Grant CTS-9706701. Ting's research was partially supported by the Air Force Office of Scientific Research Grant F49620-95-1-0065, and by Alexander von Humboldt-Stiftung, Germany.

References

Beale, J.T., and Majda, A. (1985) High order accurate vortex methods with explicit velocity kernels, *J. Comput. Phys.* **58**, pp. 188–208.

Callegari, A. and Ting, L. (1978) Motion of a Curved Vortex Filament with Decaying Vortical Core and Axial Velocity, *SIAM J. Appl. Math.***35**, pp. 148–175.

Klein, R., Knio, O.M. (1995) Asymptotic Vorticity Structure and Numerical Simulation of Slender Vortex Filaments, *J. Fluid Mech.***284**, pp. 275–293.

Klein, R., Knio, O.M., and Ting, L. (1996) Representation of core dynamics in slender vortex simulations, *Phys. Fluids* **8**, pp. 2415–2425.

Klein, R., and Ting, L. (1995) Theoretical and Experimental Studies of Slender Vortex Filaments, *Appl. Math. Lett.* **8**, pp. 45–50.

Knio, O.M., and Ghoniem, A.F. (1990) Numerical study of a three-dimensional vortex method, *J. Comput. Phys.* **86**, pp. 75–106.

Knio, O.M., and Juvé, D. (1996) On noise emission during coaxial ring collision, *C. R. Acad. Sci. Paris Ser. II* **322**, pp. 591–600.

Knio, O.M., and Ting, L. (1997) Vortical flow outside a sphere and sound generation, *SIAM J. Appl. Math.* **57**, pp. 972–981.

Knio, O.M., Ting, L., and Klein, R. (1997) Interaction of a slender vortex with a rigid sphere: dynamics and far-field sound *J. Acoust. Soc. Am.* (submitted).

Lighthill, M.J. (1956) The image system of a vortex element in a rigid sphere, *Proc. Cambridge Phil. Soc.* **452**, pp. 317–321.

Miyazaki, T., and Kambe, T. (1986) Axisymmetric problem of vortex sound with solid surfaces, *Phys. Fluids* **29**, pp. 4006–4015.

Möhring, W. (1978) On vortex sound at low Mach number, *J. Fluid Mech.* **85**, pp. 685–691.

Moreau, J.J. (1948) Sur deux théorèmes généraux de la dynamique d'un milieu incompressible illimité, *C. R. Acad. Sci. Paris* **229**, pp. 1420–1422.

Obermeier, F. (1980) The influence of solid bodies on low Mach number vortex sound, *J. Sound Vib.* **72**, pp. 39–49.

Ting, L., and Miksis, M.J. (1990) On vortical flow and sound generation, *SIAM J. Appl. Math.* **50**, pp. 521–536.

Weiss, P. (1944) On hydrodynamical images–Arbitrary irrotational flow disturbed by a sphere, *Proc. Cambridge Phil. Soc.* **40**, pp. 259–261.

MULTIPLE SCATTERING OF ACOUSTIC WAVES BY MANY SLENDER VORTICES

MAURIZIO BAFFICO, DENIS BOYER AND FERNANDO LUND
Departamento de Física
Facultad de Ciencias Físicas y Matemáticas
Universidad de Chile
Casilla 487-3, Santiago
CHILE

1. Introduction

Both numerical simulations[1] and laboratory experiments[2] have shown that intense vorticity regions (slender vortices) form in at least some turbulent flows, their precise role being still an open question[3]. One tool that is proving increasingly powerful in the experimental study of vortical structures is acoustic scattering[4]. Most theoretical studies to date have been related to acoustic scattering by one vortical structure. In three dimensions they usually rely on a Born approximation[5], while more accurate analysis is possible in two dimensions with axial symmetry[6]. In this paper, we investigate the scattering of acoustic waves by many vortices both in two and three dimensions. As a first result, we present an application of the study to a two dimensional system of slender vortex dipoles, taken as a simple model of disordered flow.

2. Multiple scattering approach

The main tool we use in our analysis is the multiple scattering approach introduced by Foldy and Lax[7]. In it, the propagation of an acoustic wave through a medium that consists of a large number N of scatterers randomly distributed in the macroscopic volume V, is described as the propagation of an average wave in a homogeneous "effective" medium whose only effect is to introduce a complex index of refraction; its real part gives an effective (phase) speed of sound, and its imaginary part describes the attenuation of the wave due to the energy it loses because of the multiple scattering.

E. Krause and K. Gersten (eds.), IUTAM Symposium on Dynamics of Slender Vortices, 379-387.

In the microscopic scale, the number density $n = N/V$ of scatterers is supposed low, such that they weakly modify the speed of sound of the background medium. Originally these scatterers were thought to be water droplets in air, or air droplets in water; here, we consider non-interacting vortex blobs in a fluid at rest. The following additional assumptions are introduced: the flow generated by vorticity is of low Mach number; the media can be considered as stationary, or frozen, if the typical evolution frequencies associated to the disordered flow, say Ω, are small compared to the sound frequency ν: $\nu \gg \Omega$; moreover, the sound will not disturb the flow field if it is chosen with low amplitude vibrations.

The acoustic wave anywhere away from the scatterers will be a sum of the incident wave that would be present in the absence of scatterers, for example a plane wave of wave vector $\vec{k}$

$$e^{i(\vec{k}\cdot\vec{x}-\nu t)},$$

and the scattered waves from the many vortices. If the scatterers are very many and they do not greatly perturb the plane wave (more on this later), part of the acoustic energy will propagate forward as an attenuated plane wave with a somewhat modified speed of sound:

$$e^{i(\vec{K}\cdot\vec{x}-\nu t)},$$

where $\vec{K} = (K_r + iK_i)\hat{k}$ is a complex wave vector that points in the same direction as the original one, whose real part gives the new speed of sound $c' = \nu/K_r$ and whose imaginary part gives the attenuation length $L = K_i^{-1}$. Note that whatever the scatterers do in time, it is assumed that they do not affect the monochromatic character of the wave.

The way to go about implementing these ideas is the following: A "total" acoustic pressure P is defined as the sum of the plane wave p^0 plus the contributions due to the many scatterings:

$$P = p^0 + \sum_{i=1}^{N} \int G^0 T_i[p_i] \tag{1}$$

where p_i is the acoustic pressure incident on the i-th scatterer, T_i is the so-called T matrix, an operator that completely describes the i-th scatterer in the sense that it says what is the outgoing acoustic pressure for a given incident pressure. It is essential that it be possible to compute this object in order for the method to be practical. G^0 is the free propagator (alias Green function or impulse response), that gives the response of the homogeneous acoustic medium at any point in space and time for a given perturbation that is very localized, both in space and time. It is also assumed that this object is known. The essence of the method lies in the fact that the acoustic presure p_i that is incident on the i-th scatterer is the sum of the plane wave p^0 plus the contributions due to the scatterings from the *other* scatterers:

$$p_i = p^0 + \sum_{j \neq i} \int G^0 T_j[p_j]. \tag{2}$$

In principle, Eqns. (1-2) describe the whole process of wave propagation throug a medium with many scatterers. This, in and of itself, is however not practical when there are very many scatterers. What one looks for, then, is a statement about the average acoustic pressure

$$< P >$$

in which the average is taken over all possible configurations of the scatterers, and hope that it can be written in the form

$$(\nabla^2 + k^2) < P(x) >= - \int \mathcal{M}(x, x') < P(x') > dx' \tag{3}$$

where the kernel $\mathcal{M}(x, x')$ is called the mass operator. Assuming a plane wave form

$$< P(x) >= e^{i\vec{K}\cdot\vec{x}}$$

for the average acoustic wave and replacing this in (3) leads to a dispersion relation:

$$-K^2 + k^2 = -\tilde{\mathcal{M}}(\vec{K}) \tag{4}$$

where $\tilde{\mathcal{M}}$ is the Fourier transform of the mass operator $\mathcal{M}$.

In general, finding the mass operator $\mathcal{M}$ is an impossible task. Progress can be made, however, when the number density, n, of scatterers is low. In this case the mass operator and the T matrix are related through

$$\tilde{\mathcal{M}}(\vec{K}) = n < T_0(\vec{K}, \vec{K}) > +O(n^2) \tag{5}$$

where T_0 is the T matrix for a single scatterer located at the origin and evaluated in the forward scattering direction: both the incoming and the outgoing waves have the same wave vector $\vec{K}$. In the process it has been assumed that scatterers' positions are uncorrelated and uniformly distributed in space, so that the braces in Eqn. (5) now refer to averages only over all possible configurations of a *single* scatterer. A small density means

$$n\sigma_T \ll k$$

where σ_T is the total cross section for scattering by a single scatterer. It turns out that, in Fourier space, the T matrix is proportional to the scattering amplitude F. In three dimensions, for instance:

$$T_0(\vec{k}, \vec{q}) = 4\pi F(\vec{k}, \vec{q})$$

so that, assuming that the new effective wave vector $\vec{K}$ will not differ very much from the original wave vector $\vec{k}$, the following expressions result for the effective speed of sound c' and attenuation length L:

$$\frac{c'}{c_0} = 1 - \frac{2\pi n}{k^2} Re < F(0) >, \tag{6}$$

$$L^{-1} = \frac{1}{2} n < \sigma_T >, \tag{7}$$

where c_0 is the speed of sound in the absence of scatterers, and $F(0)$ is the forward scattering amplitude for a single scatterer. It is related to the total cross section σ_T through the optical theorem

$$\sigma_T = \frac{4\pi}{k} Im F(0) . \tag{8}$$

3. Scattering by a single vortex to order M^2

We now specialize the formalism described in the previous section to the case when the scatterers are vortices. The usual way[5] of dealing with the scattering of an acoustic wave of acoustic pressure p and associated particle velocity $\vec{v}$, by a vortex in a fluid of undisturbed density ρ_0 with vortical velocity $\vec{u}$ ($|\vec{v}| \ll |\vec{u}|$) is to write an inhomogeneous wave equation valid to first order in the Mach number $M = u/c_0$:

$$\nabla^2 p - c_0^2 \partial_{tt} p = -2\rho_0 \nabla_i \nabla_j (u_i v_j) \tag{9}$$

and to solve it by successive approximations in a Born approximation scheme. In this case it is found[5], to leading order, that the scattering amplitude is proportional to the Fourier transform of the component of vorticity that is perpendicular to the scattering plane:

$$F(\theta) \sim \frac{\cos\theta \sin\theta}{1 - \cos\theta} \tilde{\omega}_\perp(\vec{q}) \tag{10}$$

where $\tilde{\omega}_\perp$ is the Fourier transform of the component of the vorticity vector that is perpendicular to the scattering plane, which in turn is defined by the direction of the incident wave vector $\vec{k}$ and the scattering direction $\hat{r}$, the angle between which is the scattering angle θ. The vector $\vec{q}$ is the difference between the scattered and incident wave vectors:

$$\vec{q} \equiv k\hat{r} - \vec{k}$$

and it depends, of course, on the scattering angle θ. In order to find the forward scattering amplitude one has to find the limit of Eqn. (10) when

$\theta \to 0$. The leading term for zero wave vector of a Fourier transform is given by the (space) average of the quantity being transformed. Except in two dimensions and for a vortex with nonvanishing circulation[6], all bounded vorticity distributions have zero average. Taking the next order contribution it is easy to see that the forward scattering amplitude in the first Born approximation is proportional to the dipole moment of the vorticity:

$$F(0) \sim \int \vec{x}\omega_{\perp}(\vec{x})dx. \tag{11}$$

Naturally, it is assumed that the vorticity vanishes sufficiently fast away from the vortex for the scattering problem to make sense; in particular, for integrals such as the last one to converge. The point of this calculation is that the forward scattering amplitude to first order in the Mach number and in a first Born approximation involves a preferred direction. Since it is the average over all possible configurations of the forward scattering amplitude that determines the propagation of the acoustic wave in a medium with many vortices, this average will vanish if all orientations are equally likely, a fact that we will assume to be true since otherwise the flow would have a preferred direction. It can equally be shown that this forward scattering amplitude is still zero if each vortex blob is not at rest but moves at velocity v_c, in a random direction. Consequently, the computation has to be taken to order M^2.

To second order in the Mach number the inhomogeneous wave equation for the acoustic pressure is

$$\nabla^2 p - c_0^2 \partial_{tt} p = s_1 + s_2 + s_3 \tag{12}$$

where

$$\begin{aligned} s_1 &= -2\rho_0 \nabla_i \nabla_j (u_i v_j) \\ s_2 &= -\nabla_i \nabla_j (\rho u_i u_j) \\ s_3 &= -\frac{\delta c^2}{c_0^2} \partial_{tt} p \end{aligned}$$

where ρ denotes the density variations associated with the acoustic pressure p, and δc^2 is the (dimensionless) change in the speed of sound within a given vortex due to non acoustic density changes. This changes are negligible in a liquid such as water but they are non negligible in an ideal gas, in which case they can be calculated to yield

$$\nabla(\delta c^2) = \frac{1-\gamma}{c_0^2}(\vec{u} \cdot \nabla)\vec{u}, \tag{13}$$

where γ is the specific heat ratio ($\gamma \simeq 1$ for water). Actually, there are additional terms of order M^2: they are responsible for the spontaneous generation of sound by an unsteady flow[8]. However, the frequency of this sound is of the order of Ω: it will not be heard by detectors tuned to the much higher frequency ν.

The result for the contribution to order M^2 to the averaged forward scattering amplitude coming from the first Born approximation is

$$\left\langle F^{(1)}(0)\right\rangle = \frac{k^2(\gamma-4)}{4\pi\, c_0^2}\left\langle \int \frac{d^D q}{(2\pi)^D}\left|\hat{k}\cdot\vec{u}(\vec{q})\right|^2\right\rangle \tag{14}$$

with $D = 3$, and the contribution coming from the second Born approximation is

$$\left\langle F^{(2)}(0)\right\rangle = \frac{k^2}{\pi\, c_0^2}\left\langle \int \frac{d^D q}{(2\pi)^D}\frac{(\hat{k}\cdot\vec{q})^2}{q^2-(k+i\eta)^2}\left|\hat{k}\cdot\vec{u}(\vec{k}-\vec{q})\right|^2\right\rangle, \tag{15}$$

where the limit $\eta \to 0$ is understood. Note that $F^{(1)}$ corresponds to a single scattering contribution while $F^{(2)}$ corresponds to a double scattering contribution. Both are of order M^2. The Fourier transform of the velocity, $\vec{u}(\vec{k})$, and of the vorticity, $\vec{\omega}(\vec{k})$, are related through

$$\vec{u}(\vec{k}) = i\frac{\vec{k}\wedge\vec{\omega}(\vec{k})}{|\vec{k}|^2}\ . \tag{16}$$

It should be noted that, strictly speaking, it is not necessary to invoke a multiple scattering approach to obtain the above results, since they involve only the leading order approximation for low vortex number density. It is also possible to take standard, order M, expansions[9], carry them out to order M^2 and replace the velocity corresponding to the desired vorticity distribution. However, we have found the use of multiple scattering ideas helpful in obtaining physical insight into the processes involved.

4. Example: Vortex dipoles in two dimensions

It can be shown that the two dimensional general expressions can be obtained simply by inserting Eqs.(14) and (15) in formulae (6) and (8), and setting $D = 2$. As an example of the type of calculation that becomes possible with the above formalism, we have considered a system of many vortex dipoles in two dimensions: the vorticity is given by

$$\omega = \omega_+ + \omega_- \tag{17}$$

where

$$\omega_\pm(\vec{x}) \equiv \frac{\pm\Gamma}{\pi\epsilon^2}e^{-|\vec{x}\pm\vec{d}|^2/\epsilon^2} \tag{18}$$

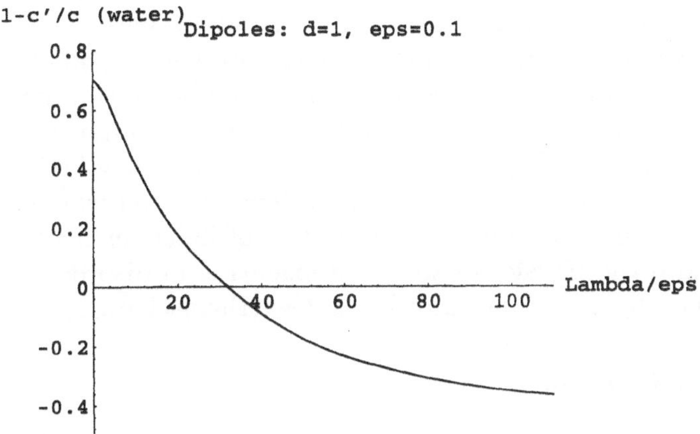

Figure 1. Effective speed of sound for an acoustic wave propagating in a flow consisting of many vortex dipoles in two dimensions, in units of the small quantity $n\Gamma^2/c_0^2$.

This corresponds to two slender vortex disks of opposite circulation $\pm\Gamma$ separated by a distance $2d$ with a Gaussianly distributed vorticity within a core of radius ϵ. This structure is steady if $\epsilon \ll d$. The orientation of the dipole is assumed uniformly distributed over all possibilities. The Fourier transforms that go into (14) and (15) can be computed and then evaluated numerically. Doing this it becomes apparent that the long wavelength behavior of the total scattering cross section goes like λ^{-3}, as it should for Rayleigh scattering by a dipole in two dimensions. The behavior of the effective phase velocity is shown in Figure 1, in units of the small quantity $n\Gamma^2/c_0^2$. It is interesting to note that the phase velocity is smaller than the speed of sound for short wavelengths, and larger than the speed of sound for long wavelengths, with a crossover when the wavelength is at about the size of the dipole.

5. Concluding remarks

We have shown that the properties of an acoustic wave propagating in a disordered flow can be computed in some detail if the flow is modeled as a system of many slender vortices, assuming a weak vortex sound interaction and a low vortex density. The medium has been considered stationary, corresponding to the case where the time scale associated with the acoustic wave is much smaller than the time scales associated with the evolution of the flow. The usual expressions describing wave propagation in a random medium involve averaging over the randomness. that in the present case simply collapses to averaging over the random properties of a single vortex. We have worked out in detail the attenuation length and effective speed of sound for the case of many identical vortex dipoles in two dimensions whose only random variable is the orientation, assumed uniformly distributed. It was found that the effective phase speed of sound may be larger or smaller than the undisturbed speed of sound, depending on wavelength. It is our hope that trying to understand this type of behavior may lead to further insights into the physics of sound propagation in disordered flows, as well as into the physics of the disordered flows themselves.

Acknowledgements

This work was supported by Fondecyt Grants 2950011, 1960892, 3970013, the Andes Foundation, and a Cátedra Presidencial en Ciencias.

References

1. E. Siggia, *J. Fluid Mech.* **107**, 375 (1981); R. M. Kerr, *J. Fluid Mech.* **153**, 31 (1985); Z. She, E. Jackson and S. Orszag, Nature **344**, 226 (1990); A. Vincent and M. Meneguzzi, *J. Fluid Mech.* **225**, 1 (1991); S. Kida and K. Ohkitani, *Phys. Fluids A* **4**, 1018 (1992); M. E. Brachet et. al., *Phys. Fluids A* **4**, 2845 (1992); M. Tanaka and S. Kida, *Phys. Fluids A* **5**, 2079 (1993); J. Jiménez et. al., *J. Fluid Mech.* **255**, 65 (1993); A. Pumir, *Phys. Fluids* **6**, 2071 (1994); T. Passot et. al., *J. Fluid Mech.* **282**, 313 (1995).
2. S. Douady, Y. Couder and M. E. Brachet, *Phys. Rev. Lett.* **67**, 983 (1991); S. Fauve, C. Laroche and B. Castaing, *J. Physique II France* **3**, 271 (1993); P. Abry, S. Fauve, P. Flandrin and C. Laroche, *J. Physique II France* **4**. 725 (1994); O. Cadot, S. Douady and Y. Couder, *Phys. Fluids* **7**, 630 (1995); E. Villermaux, B. Sixou and Y. Gagne, *Phys. Fluids* **7**, 2008 (1995); F. Belin, J. Maurer, P. Tabeling and H. Willaime, *J. Physique II France* **6**, 573 (1996).
3. U. Frisch, *Turbulence*, Cambridge University Press (1995), Ch. 8.
4. P. R. Gromov, A. B. Ezerskii and A. L. Fabrikant, *Sov. Phys. Acoust.* **28**, 452 (1982); C. Baudet, S. Ciliberto and J. F. Pinton, *Phys. Rev. Lett.* **76**, 193 (1991); M. Oljaca, X. Gu, A. Glezer, M. Baffico and F. Lund, *AIAA Paper* 96-0437, (1996); A. Petrossian and J. F. Pinton, *J. Physique II France* **7**. 801 (1997); B. Dernoncourt, J. F. Pinton and S. Fauve, *Physica D*, to appear.
5. A. M. Obukhov, *Dokl. Akad. Nauk. SSSR*, **30**, 616 (1941); M. J. Lighthill, *Proc. Camb. Phil. Soc.* **49**, 521 (1953); R. H. Kraichnan, *J. Acoust. Soc. Am.* **25**, 1096 (1953). T. Kambe and U. Mya-Oo, *J. Phys. Soc. Japan* **50**, 3507 (1981); T. Kambe, *J. Japan Soc. Fluid Mech.* **1**, 149 (1982) (in japanese); M. S. Howe, *J. Sound Vib.* **87**, 567 (1983); A. L. Fabrikant, *Sov. Phys. Acoust.* **29**. 152 (1983); F. Lund and C. Rojas, *Physica D* **37**, 508 (1989); A. T. Skvortsov, *Bull. Russ. Acad. Sciences, Physics Supplement, Physics of Vibrations* **1**, 60 (1993).
6. A. L. Fetter, *Phys. Rev.* **136**, 1488 (1964); P. V. Sakov, *Acoust. Phys.* **39**, 280 (1993); T. Colonius, S. K. Lele and P. Moin, *J. Fluid Mech.* **260**, 271 (1994); R. Bérthet and F. Lund, *Phys. Fluids* **7**, 2522 (1995); J. Reinschke, W. Möhring and F. Obermeier, *J. Fluid Mech.* **333**, 273 (1997).
7. L. L. Foldy, *Phys. Rev.* **67**, 107 (1945). M. Lax, *Rev. Mod. Phys.* **23**, 287 (1951);

Phys. Rev. **85**, 621 (1952). For a recent review, see C. A. Condat and T. R. Kirkpatrick, in *Scattering and Localization of Classical Waves in Random Media*, edited by P. Sheng, World Scientific (1990)).

8. M. J. Lighthill, *Proc. Roy. Soc. A* **211**, 564 (1952); W. Möhring, *J. Fluid Mech.* **85**, 685 (1978); T. Kambe, *J. Fluid Mech.* **173**, 643 (1986).
9. See V. E. Ostashev, *Waves in Random Media* **4**, 403 (1994) and references therein.

VORTEX DYNAMICS IN NONUNIFORM COMPRESSIBLE FLOW

K. EHRENFRIED AND G.E.A. MEIER
DLR-Institut für Strömungsmechanik
Bunsenstr. 10, 37073 Göttingen, Germany

Abstract. The convection of vortices in convergent nozzles and Laval nozzles is investigated experimentally and numerically. All studies are done for the plane case. The experiments are performed in a transonic wind tunnel, where Mach-Zehnder interferometry is used to visualize the density field. For the numerical simulations the Euler equations are solved using a finite-volume method. The experimental and numerical results are compared. The distortion of the vortices and the generation of disturbances is shown with typical examples.

1. Introduction

The convection of vortices and turbulence in flows is often connected with the generation of sound. Well understood and experimentally investigated are the mechanisms of vortex-vortex and vortex-body interactions. In recent investigations it became obvious, that also single vortices in accelerated flows can produce sound of considerable strength.

In [1] and [2] vortex-airfoil interaction was investigated in wind tunnel experiments. The vortices were generated upstream of the test section. Then they passed a convergent section with accelerated flow before they reached the airfoil and the investigated interaction took place. In this case an undisturbed vortex is desirable right before the interaction with the airfoil. But in practice many vortices reach the test section in disturbed form. This can be caused by the generation process of the vortex, when the produced vortex is not exactly symmetric. Also the acceleration of the vortex in the convergent section can cause disturbances, even when the initial vortex was exactly symmetric. When the vortex undergoes an acceleration it is stretched in streamwise direction. The rotation of the vortex prevents

E. Krause and K. Gersten (eds.), IUTAM Symposium on Dynamics of Slender Vortices, 389-398.

a total distortion of its shape and causes the preservation of a more or less circular structure. With respect to further vortex-airfoil interaction experiments it is of practical interest to have more information about a vortex in accelerated flow. Important is to know, how a vortex is distorted and which disturbances are generated in the flow field.

Another question is, how a vortex reacts when it is convected through the throat of a Laval nozzle with a transonic flow regime. At the end of the supersonic region a vortex-shock interaction takes place, in which strong disturbances are generated. But these disturbances cannot propagate in upstream direction. However, vortex-shock interactions were intensively studied elsewhere and are not considered in the present paper. Here the acceleration of the vortex and its passage through the throat is the central point of interest. The vortex reduces the total mass flux during its passage and generates disturbances, which propagate in all directions.

2. Experimental setup

The experiments were performed in a vacuum-driven transonic wind tunnel, which is described in [3]. Atmospheric air is sucked through an inlet into the test section. If necessary, a dehumidifing unit can be installed in front of the inlet to avoid condensation effects. Downstream of the test section the flow is regulated by an adjustable valve before it reaches the vacuum tank.

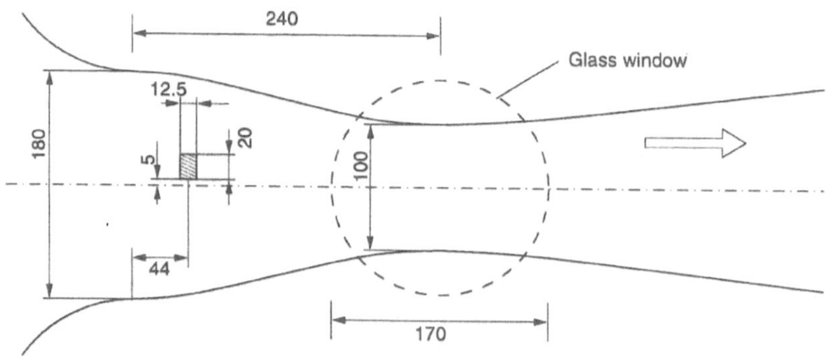

Figure 1. Experimental setup with vortex generator

The experimental setup with a Laval nozzle in the plane test section is shown in figure 1. The throat of the nozzle is located where the flow can be observed through windows in the side walls of the test section. The distance between upper and lower wall in the throat as well as the distance between the side walls is $100mm$. Upstream of the throat a cylinder with rectangular cross-section and sharp edges is fixed between the side walls. This cylinder is used to generate a vortex street, which is then convected through the Laval nozzle. Thus, a series of vortices passes the nozzle, where the dynamics of this process is investigated.

A Mach-Zehnder interferometer is used for the optical visualization of the density field. The pictures are taken by a drum camera with a frame rate of $10kHz$. The flow is investigated under transonic conditions with a supersonic region in the Laval nozzle downstream of the throat. The flow speed is regulated, so that the shock wave, which terminates the supersonic region, remains downstream of the observed region – far outside of the window. Thus, there is no influence from the disturbances further downstream and from the shock wave itself on the flow in the observed region.

3. Experimental results

For the vortex-convection experiments a $20 \times 12.5mm$ cylinder is placed unsymmetrically in the convergent section of the nozzle as indicated in figure 1. The generated vortex street has a period of $\Delta t = 0.8msec$. This means in $0.8msec$ two vortices – a clockwise and a counter-clockwise rotating – are shedded from the cylinder. Due to the acceleration in the nozzle the vortex street is stretched and the distance between consecutive vortices is enlarged. The unsymmetric position of the vortex generator is chosen, so that the upper row of the vortex street is convected approximately along the centerline of the nozzle.

In figure 2 a series of interferograms is depicted. The series shows the passage of a counter-clockwise rotating vortex. The flow direction is from left to right. The time delay between the images is $0.1msec$. At the upper and lower border of the images the walls of the Laval nozzle appear in the interferogram. Thus, the hole throat is visible. The predominantly vertical fringes indicate the density gradient in main flow direction. The superimposed density field of the vortex is visible as closed fringes. The hole density field is disturbed by the vortex street.

Noticeable are disturbances which appear as sharp indentations of the interferometric fringes on the right side above and on the left side below the vortex. These disturbances travel downstream together with the vortex. They are remains of the shear layer which is generated at the vortex generator. This shear layer is partially rolled into the generated vortices, and is convected together with the vortex street. The viscosity causes a rapid damping of the free shear layer and the temperature and entropy increases there. At the same pressure the density is lower in the region with higher entropy. Thus, the former shear layer is visible in the interferograms.

In the following the difference of the density outside of the vortex and

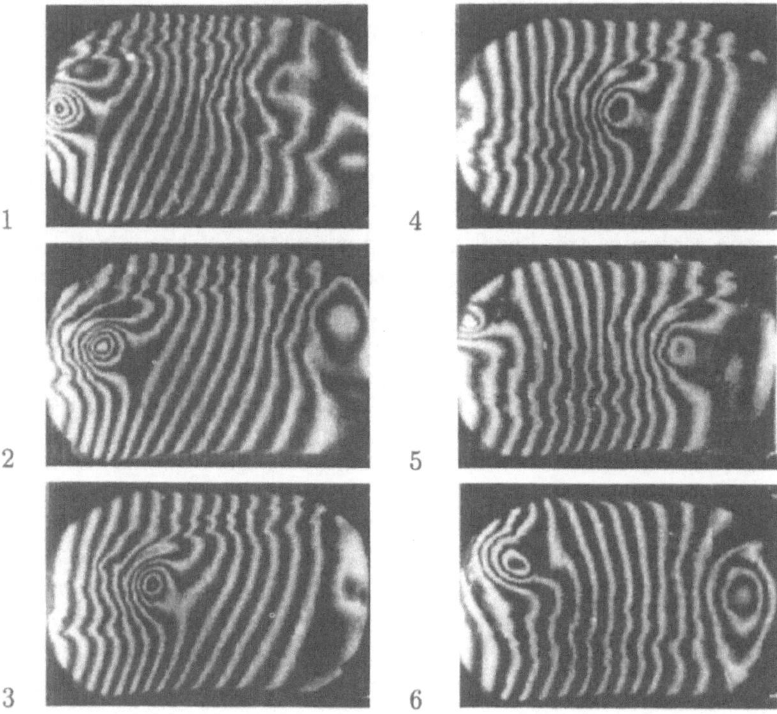

Figure 2. Interferograms of the passage of a counter-clockwise rotating vortex through the throat of the Laval nozzle

the one in the vortex center is denoted as $\Delta\rho_v$. The series of images demonstrate the decrease of the density difference $\Delta\rho_v$ during the passage through the throat. When entering the view-space the vortex appears as compact structure with several concentric rings in the interferogram. The number of rings decreases when the vortex travels downstream. The vortices seem to be weakened during the acceleration into the supersonic region.

4. Numerical method and initial condition

The 2D Euler equations are solved using a finite-volume scheme. To avoid an artificial damping of the vortices, the numerical dissipation has to be reduced. Therefore a second order discretisation together with a relatively fine grid is used. Details of the method are described in [4]. In the following all distances are normalized with the width of the throat, density and pressure are normalized with their stagnation values ρ_s and p_s, and the velocity with $\sqrt{p_s/\rho_s}$. In contrast to the experiment, only a single vortex instead of a vortex street is considered in the numerical investigation. In figure 3 a contour plot of the initial density distribution is shown. The flow direction is from left to right. The computational domain consists of a symmetric Laval nozzle with ducts of constant cross-section attached on both ends. In these regions the normalized cross-section is two. The upstream duct is longer to allow the insertion of a vortex in a region with almost homogeneous flow.

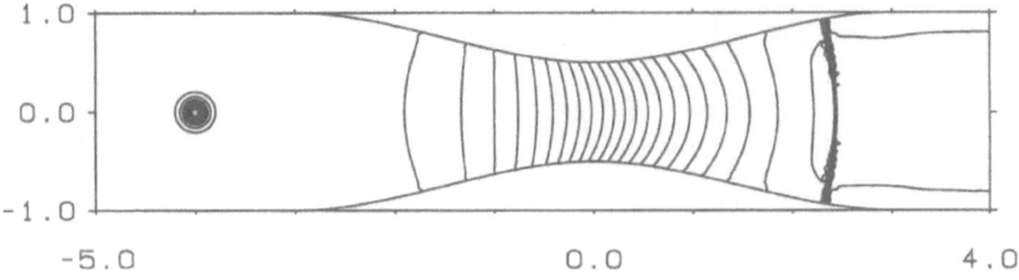

Figure 3. **Contour plot of the initial density field at the beginning of the simulation**

The contour lines show the vortex as well as the smooth density gradient and the shock wave in the Laval nozzle. The shock front appears thinner in the area closer to the axis of the nozzle. In this region the grid is refined to guarantee an accurate computation of the convected vortex. In the given example the core radius of the vortex is 8 times the mesh size.

To generate the initial solution, a steady solution of the transonic nozzle flow is computed. Then a vortex is inserted in the region with almost homogeneous flow. At the beginning the inserted vortex simply is convected downstream without generating any disturbances. In [4] more details about the vortex model and the insertion technique are given. The radial distributions of density, tangential velocity and circulation of the inserted vortex are shown in figure 4. The core radius (defined as the position with maximum tangential velocity) is $R_c = 0.1$. Unlike a Lamb-type vortex our model gives a finite-size vortex. Its velocity field covers only an area up to a limiting radius $R_{end} = 0.5$. This has a numerical advantage with respect to the boundary conditions and the insertion of the vortex in the steady solution. However, also in the experiment the vortices are somehow limited by the wind tunnel walls. But from the experiment only a rough estimation about the density field of the vortices can be obtained. The numerical vortex was chosen to have a density difference $\Delta\rho_v$ in the same range of the experimental vortices. Other quantities are not known at all. Therefore, the numerical vortex is – more or less – an arbitrary model of the vortices in the experiment. Anyway, in the numerical case the nozzle geometry and the setup with a single vortex differs from the experimental situation. Because

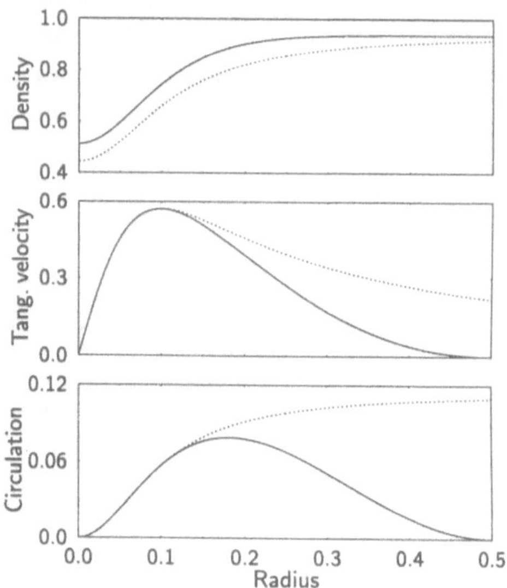

Figure 4. Structure of the used model vortex (solid line) and a Lamb-type vortex (dotted line)

of this the comparison between the two cases can only be of quantitative nature.

5. Numerical results

For comparison with the experimental images artificial interferograms are computed from the numerical solution. Therefor a periodic intensity function was taken, where the density change between one intensity maxima and the next is about 1/20 of the stagnation value. In the left side of figure 5 a series of such images is depicted. Like in the experimental interferograms, the images show the throat region of the nozzle during the passage of a vortex at equidistant time steps. In this case the vortex is rotating clock-wise.

In the right side of the figure the contours of the vorticity $\omega = \operatorname{curl} \vec{u}$ divided by the density ρ is plotted using different grey values. In the 2D case the vorticity-density ratio ω/ρ is constant in a fluid element which is convected in an inviscid and barotropic flow field. This follows from the vortex theorem of Thomson, which says that the circulation along a closed fluid line is preserved in an inviscid and barotropic flow field. Apart from small numerical disturbances the computed flow field is barotropic outside of the shock wave. Thus, the contours of the quantity ω/ρ can be used as an indicator how the vortex is deformed in the flow field.

The numerical result confirms the experimental observation that the

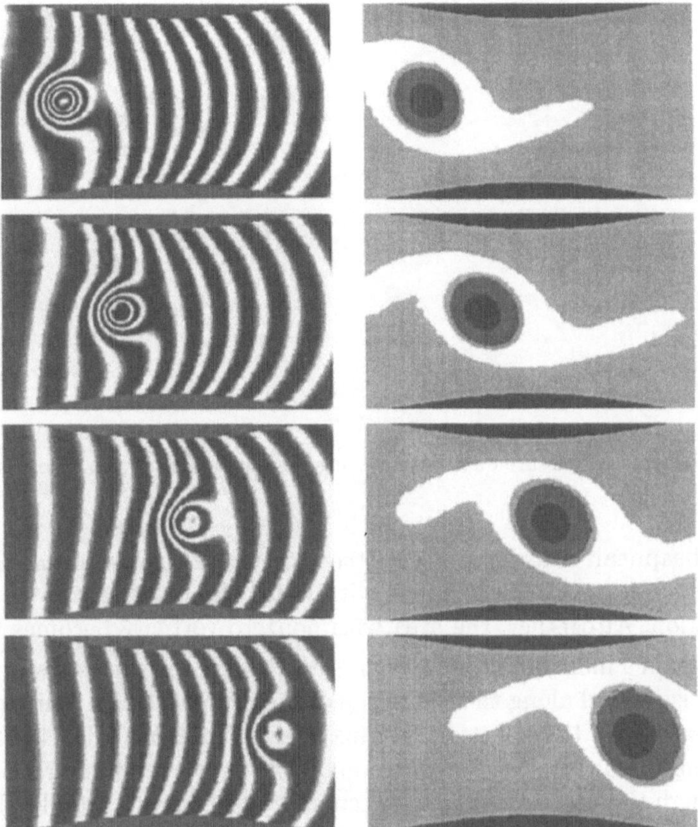

Figure 5. Numerical results: Artificial interferograms of density field (left) and contours of the vorticity-density ratio (right)

vortex density-difference $\Delta\rho_v$ decreases during the acceleration of the vortex. Like in the experiment also the numerical interferograms show the reduction of closed fringes in the vortex. From a density point of view the vortex gets weaker. Additionally the numerical interferograms show distortions of the fringes further away of the vortex center. This indicates disturbances which were generated during the process and are propagating through the flow field. But in the numerical case these effects are much smaller than in the experiment.

The right side of figure 5 shows, how the kinematics of the flow field deformes the vortex. The images indicate a strong distortion of the initially circular ω/ρ field. The outer region has a spiral form and the inner part is elliptical. Of course the involved circulation is preserved and the vortex

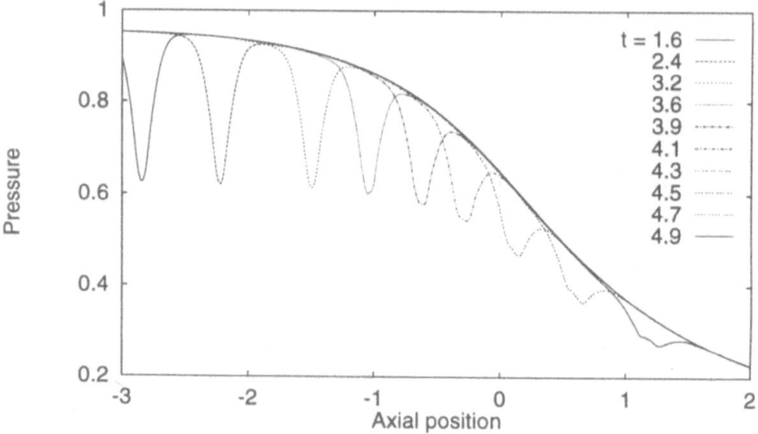

Figure 6. Pressure distribution along the nozzle axis at several times

cannot disappear. Additionally the structure in the ω/ρ contours is growing during the passage. This happens with both, the inner core area and the outer region. Altogether, the area with nonzero vorticity becomes larger.

To obtain a more quantitative representation of the vortex distortion the pressure is plotted along the centerline of the computational domain, which corresponds with the nozzle axis. This is given in figure 6 for several time steps. The throat is located at position 0. The curves represent a symmetric cut through the vortex, because its center travels relative exactly along the nozzle axis. The plot shows the superposition of the pressure field of the vortex with the undisturbed pressure field in the nozzle. Analogous to the density the difference of the pressure outside the vortex and in its center is denoted as Δp_v. The decrease of Δp_v during the passage is clearly visible. This corresponds to the decrease of the density difference $\Delta\rho_v$, which was observed in the experimental and numerical interferograms. In the supersonic region (position > 0) the pressure difference Δp_v is already less than half of its initial value. Further downstream the form of the pressure distribution indicates a strong disturbance of the vortex.

At the beginning of the acceleration the pressure in the center of the vortex remains almost constant, although the vortex has already reached a position in the nozzle with significantly lower undisturbed pressure. Noticeable is that the deviation from the undisturbed curve appears only in a limited area. After the vortex has passed a certain position, the pressure returns there to the initial value. In the given resolution of the plot no propagating pressure waves aside from the vortex are visible.

In the following the generation of pressure disturbances due to the ac-

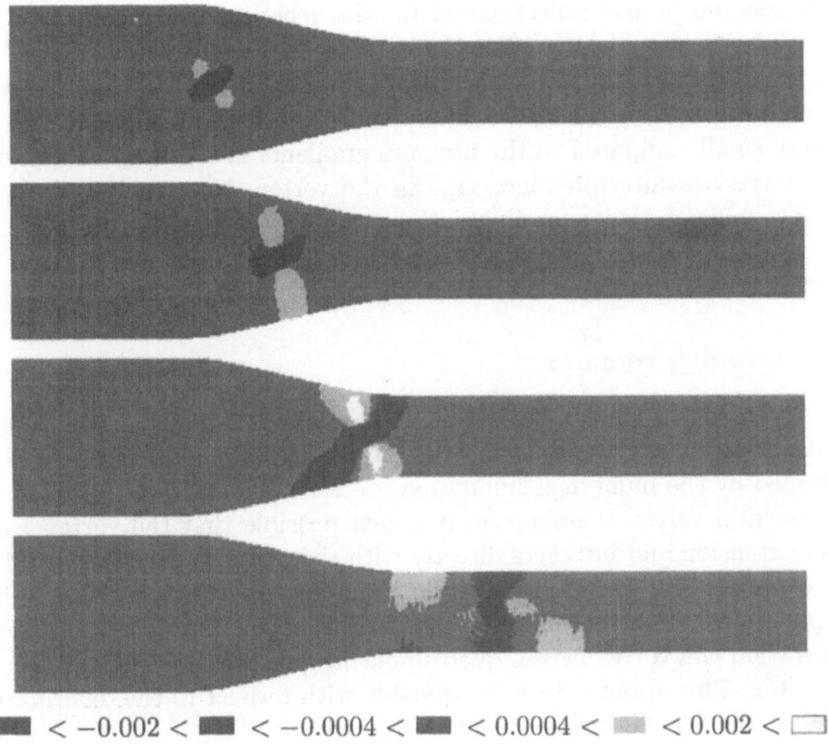

Figure 7. Contours of the pressure difference $p' = p - p_{steady}$

celeration of vortices is demonstrated by a pure subsonic case. The Laval nozzle from figure 3 is modified. The divergent part is replaced by a duct of constant cross-section. Like in the first example, a clock-wise rotating vortex is inserted upstream of the convergent section. The same model is used for the vortex as in figure 4, but the size is reduced. The core radius is only $R_c = 0.05$ and the limiting radius $R_{end} = 0.25$. This small size is taken to avoid a direct interaction of the vortex with the nozzle wall.

In the given example the Mach number in the downstream duct is $Ma = 0.73$. The result of the simulation is presented in figure 7. The images show the solution in the hole computational domain at four equidistant time steps. Plotted are the contours of the pressure difference $p'(x, y)$ between the actual pressure $p(x, y)$ and its steady value $p_{steady}(x, y)$. Dark shades indicate negative p' and bright shades a positive one. Therefore the vortex center appears as a dark spot. During the passage through the convergent nozzle additional pressure fluctuations around the vortex center occur. They show a typical quadrupole pattern. The fluctuations become stronger, and in the last image also reflections of the disturbances from the duct walls are visible. It should be noted, that the plotted contours correspond to relative small values of p' ($\pm 4.0 \times 10^{-4}$ and $\pm 2.0 \times 10^{-3}$ normalized with the stagnation value). Thus, the generated fluctuations outside of the vortex are very small compared to the pressure gradients in the undisturbed flow field or the pressure difference Δp_v in the vortex. Nevertheless, from an acoustic point of view such fluctuations are already interesting. But they are too weak to account for all the disturbances which were observed in the experimental case in figure 2.

6. Concluding remarks

The weakening of the vortex during the passage through the throat of a Laval nozzle, which was first observed in the experimental case, could be confirmed by the numerical simulation. Because the exact structure of the experimental vortex is unknown, it is just possible that the vortex has a larger extension and interacts directly with the nozzle walls. This could be the reason for the stronger disturbances in the experimental density field compared with the numerical solution. As the last example shows, the pure acceleration of a vortex causes quadrupole like pressure fluctuations around the vortex. This appears to be reasonable with respect to the deformation of the vorticity field during the acceleration.

7. Acknowledgments

The described experiment was carried out by A.P. Szumowski. The numerical part of the presented research was supported by the Deutsche Forschungesgemeinschaft (DFG) under contract number ME-344/13.

References

1. Meier, G.E.A., Schievelbusch, U., and Lent, H.-M.: Stoßwellenentstehung bei transsonischer Wirbel-Profil-Wechselwirkung, Z. Flugwiss. Weltraumforsch., Vol. 14 (1990), 327 - 332.
2. Schürmann, O.: Wechselwirkung einer kompressiblen Wirbelstraße mit einem Tragflügelprofil, Mitteilungen aus dem MPI f. Strömungsforschung, Nr. 115 (1994).
3. Szumowski, A.P., and Meier, G.E.A.: Vortex Convection in Nonuniform Compressible Flow, Max-Planck-Institut für Strömungsforschung, Göttingen, Report 4/1988 (1988).
4. Ehrenfried, K., and Meier, G.E.A.: Ein Finite-Volumen-Verfahren zur Berechnung von instationären, transsonischen Strömungen mit Wirbeln, DLR Forschungsbericht 94-33 (1994).

Session 7

Aircraft and Helicopter Vortices

EFFECTS OF COUPLED AND UNCOUPLED BENDING-TORSION MODES ON TWIN-TAIL BUFFET RESPONSE

Osama A. Kandil[1] and Essam F. Sheta[2]
Aerospace Engineering Department
Old Dominion University, Norfolk, VA 23529, USA
and
C. H. Liu[3]
Aerodynamic Methods and Acoustics Branch
NASA Langley Research Center, Hampton, VA 23665, USA

1 Abstract

The effects of coupled and uncoupled bending and torsion modes on flexible twin-tail buffet, due to the unsteady vortex breakdown flow of a 76° sharp-edged delta wing, are considered. This multidisciplinary problem is investigated using three sets of equations on a multi-block grid structure. The first set is the unsteady, compressible, Reynolds-averaged, full Navier-Stokes equations which are used for obtaining the flow-filed vector and the aerodynamic loads on the twin tails. The second set is the aeroelastic equations for coupled and uncoupled bending and torsion modes which are used for obtaining the bending and torsional deflections of the twin tails. The third set is the grid-displacement equations which are used for updating the grid coordinates due to the tail deflections. These sets are sequentially solved accurately in time at each time step. The turbulent model used is the Baldwin-Lomax algebraic model which is modified by Degani and Schiff for cross flow separation. The configuration is pitched at 35° angle of attack and the freestream Mach number and Reynolds number are 0.3 and 1.25 million, respectively. Keeping the twin tails as rigid surfaces, the problem is solved for the initial flow conditions. Next, the problem is solved for the flexible twin tails responses due to the unsteady loads produced by the vortex breakdown flow of the delta-wing leading-edge vortex cores. The configuration is investigated for the effect of coupled and uncoupled bending and torsion modes using two different separation distances of the twin-tail; the inboard and the outboard positions. The computational results are in very good agreement with the experimental data of Washburn, et al.

[1] Professor, Eminent Scholar and Dept. Chair.

[2] Ph.D. Graduate Research Assistant.

[3] Senior Scientist.

E. Krause and K. Gersten (eds.), IUTAM Symposium on Dynamics of Slender Vortices, 401-414.

2 Introduction

In order to maximize the effectiveness of the fighter aircraft that operate well beyond the buffet onset boundary, the design of the new generation of fighter aircraft should account for both high maneuver capabilities and the aeroelastic buffet characteristics at high and wide range of angles of attack. The maneuver capabilities are achieved, for example in the F/A-18 fighter, through the combination of the leading-edge extension (LEX) with a delta wing and the use of vertical tails. The LEX maintains lift at high angles of attack by generating a pair of vortices that trail aft over the top of the aircraft. The vortex entrains air over the vertical tails to maintain stability of the aircraft. At some flight conditions, the vortices emanating from the highly-swept LEX of the delta wing breakdown before reaching the vertical tails which get bathed in a wake of unsteady highly-turbulent, swirling flow. The vortex-breakdown flow produces unsteady, unbalanced loads on the vertical tails and causes a peak in the pressure spectrum that may be tuned to different structural modes depending on the angle of attack and dynamic pressure. This in turn produces severe buffet on the tails and has led to their premature fatigue failure. If the power spectrum of the turbulence is accurately predicted, the intensity of the buffeting motion can be computed and the structural components of the aircraft can be designed accordingly.

Experimental investigation of the vertical tail buffet of the F/A-18 models have been conducted by several investigators such as Sellers, et al. [1], Erickson, et al. [2], Wentz [3], Lee and Brown [4], and Cole, et al. [5]. These experiments showed that the vortex produced by the LEX of the wing breaks down ahead of the vertical tails at angles of attack of 25° and higher producing unsteady loads on the vertical tails, and the buffet response occurs in the first bending mode, increases with increasing dynamic pressure and is larger at $M = 0.3$ than that at higher Mach numbers. Bean and Lee [6] showed that buffeting in the torsional mode occurred at a lower angle of attack and at larger levels compared to the fundamental bending mode. An extensive experimental investigation has been conducted to study vortex-tail interaction on a 76° sharp-edged delta wing with vertical twin-tail configuration by Washburn, Jenkins and Ferman [7]. The vertical tails were placed at nine locations behind the wing. The experimental data showed that the aerodynamic loads are more sensitive to the chordwise tail location than its spanwise location. As the tails were moved laterally toward the vortex core, the buffeting response and excitation were reduced.

Kandil, Kandil and Massey [8] presented the first successful computational simulation of the vertical tail buffet using a delta wing-single flexible vertical tail configuration. The tail was allowed to oscillate in bending modes in laminar flow. Unsteady breakdown of leading-edge vortex cores was captured, and unsteady pressure forces were obtained on the tail. Later on, Kandil, et al. [9-10] allowed the vertical tail to oscillate in both bending and torsional modes. The total deflections and frequencies of deflections and loads of the coupled bending-torsion case were found to be one order of magnitude higher than those of the bending case only. It has been shown that the tail oscillations change the vortex breakdown locations and the unsteady aerodynamic loads on the wing and tail.

The buffet responses of twin-tail model in laminar flow has been studied by Kandil, Sheta and Liu [11]. The twin tails were considered at $\alpha = 30^o$ and for three different spanwise positions of the twin tails. A multi-block grid structure was used to solve the problem. The loads, deflections, frequencies and root bending moments were reduced as the twin tails moved laterally toward the vortex core. The outboard position of the tails produced the least of these responses. In a recent paper by Kandil, Sheta and Massey [12], the buffet response of twin-tail model in turbulent flow was considered at a wide range of angles of attack. The computational results were in good quantative agreement with the experimental data of Washburn, et al [7].

In this paper, we consider the effects of coupled and uncoupled bending and torsion modes on flexible twin-tail buffet response, due to unsteady turbulent vortex breakdown flow of a 76° sharp-edged delta wing, for two different spanwise separation distance of the twin-tail; the inboard position (33% wing span) and the outboard position (78% wing span). The turbulent model used is the Baldwin-Lomax algebraic model which is modified by Degani and Schiff [13] for cross flow separation.

3 Formulation

The formulation consists of three sets of governing equations along with certain initial and boundary conditions. The first set is the unsteady, compressible, Reynolds-averaged full Navier-Stokes equations. The second set consists of the aeroelastic equations for coupled bending and torsional modes. For uncoupled bending-torsion modes, the distance between the elastic axis and inertia axis, x_θ, is set equals to zero in the aeroelastic equations. The third set consists of equations for deforming the grid according to the twin tail deflections. Next, the governing equations of each set along with the initial and boundary conditions are given.

3.1 FLUID-FLOW EQUATIONS

The conservative form of the dimensionless, unsteady, compressible, full Navier-Stokes equations in terms of time-dependent, body-conformed coordinates ξ^1, ξ^2 and ξ^3 is given by

$$\frac{\partial \bar{Q}}{\partial t} + \frac{\partial \bar{E}_m}{\partial \xi^m} - \frac{\partial (\bar{E}_v)_s}{\partial \xi^s} = 0; m = 1-3, s = 1-3 \tag{1}$$

where

$$\xi^m = \xi^m(x_1, x_2, x_3, t) \tag{2}$$

$$\bar{Q} = \frac{1}{J}[\rho, \rho u_1, \rho u_2, \rho u_3, \rho e]^t, \tag{3}$$

$\bar{E}_m$ and $(\bar{E}_v)_s$ are the ξ^m-inviscid flux and ξ^s-viscous and heat conduction flux, respectively. Details of these fluxes are given by Kandil, Kandil and Massey [8].

3.2 AEROELASTIC EQUATIONS

The dimensionless, linearized governing equations for the coupled bending and torsional vibrations of a vertical tail that is treated as a cantilevered beam are considered. The tail bending and torsional deflections occur about an elastic axis that is displaced from the inertial axis. These equations for the bending deflection, w, and the twist angle, θ, are given by

$$\frac{\partial^2}{\partial z^2}\left[EI(z)\frac{\partial^2 w}{\partial z^2}(z,t)\right] + m(z)\frac{\partial^2 w}{\partial t^2}(z,t) + m(z)x_\theta(z)\frac{\partial^2 \theta}{\partial t^2}(z,t) = N(z,t) \tag{4}$$

$$\frac{\partial}{\partial z}\left[GJ(z)\frac{\partial\theta}{\partial z}\right] - m(z)x_\theta\frac{\partial^2 w}{\partial t^2}(z,t) - I_\theta(z)\frac{\partial^2\theta}{\partial t^2}(z,t) = -M_t(z,t) \quad (5)$$

where z is the vertical distance from the fixed support along the tail length, l_t, EI and GJ the bending and torsional stiffness of the tail section, m the mass per unit length, I_θ the mass-moment of inertia per unit length about the elastic axis, N the normal force per unit length, M_t the twisting moment per unit length and x_θ the distance between the elastic axis and inertia axis. When $x_\theta = 0.0$ the bending and torsion modes are dynamically decoupled. The characteristic parameters for the dimensionless equations are c^*, a^*_∞, ρ^*_∞ and c^*/a^*_∞ for the length, speed, density and time; where c^* is the delta wing root-chord length, a^*_∞ the freestream speed of sound and ρ^*_∞ the freestream air density. Details of the solution equations and boundary conditions are given by Kandil, Sheta and Massey [12].

3.3 GRID DISPLACEMENT EQUATIONS

Once w and θ are obtained at the $n+1$ time step, the new grid coordinates are obtained using interpolation equations. In these equations, the twin tail bending displacements, $w^{n+1}_{i,j,k}$, and their displacement through the torsion angle, $\theta^{n+1}_{i,j,k}$ are interpolated through cosine functions. The interpolation equations allow the grid points adjacent to the tail surfaces to move with same deflections as those of the tails and keep the grid points at the computational boundaries fixed.

3.4 BOUNDARY AND INITIAL CONDITIONS

Boundary conditions consist of conditions for the fluid flow and conditions for the aeroelastic bending and torsional deflections of the twin tail. For the fluid flow, the Riemann-invariant boundary conditions are enforced at the inflow and outflow boundaries of the computational domain. At the plane of geometric symmetry, periodic boundary conditions are specified. On the wing surface, the no-slip and no-penetration conditions are enforced and $\frac{\partial p}{\partial n} = 0$. On the tail surface, the no-slip and no-penetration conditions for the relative velocity components are enforced (points on the tail surface are moving). The normal pressure gradient is no longer equal to zero due to the acceleration of the grid points on the tail surface. This equation becomes $\frac{\partial p}{\partial n} = -\rho\bar{a}_t.\hat{n}$, where $\bar{a}_t$ is the acceleration of a point on the tail and $\hat{n}$ is the unit normal.

The initial conditions of the fluid flow correspond to the freestream conditions with no-slip and no-penetration conditions on the wing and tail. For the aeroelastic deflections of the tail, the initial conditions for any point on the tail are that the displacement and velocity are zero, $w(z,0) = 0$, $\frac{\partial w}{\partial t}(z,0) = 0$, $\theta(z,0) = 0$ and $\frac{\partial\theta}{\partial t}(z,0) = 0$.

4 Method Of Solution

The first step is to solve for the fluid flow problem using the vortex-breakdown conditions and keeping the tail as a rigid beam. Navier-Stokes equations are solved using the implicit, flux-difference splitting finite-volume scheme. The grid speed $\frac{\partial\xi^m}{\partial t}$ is set equal to zero in this step. This step provides the flow field solution along with the pressure differences across the tails. The results are used as the initial conditions for the second step, wherein the tails are allowed to deflect. The pressure differences are used to generate the normal force and twisting moment per unit length of each tail. In the second step, the aeroelastic equations are used to obtain the twin tail deflections, $w_{i,j,k}$ and $\theta_{i,j,k}$. The grid displacement equations are then used to compute the new grid coordinates. The metric coefficient of the coordinate Jacobian matrix are updated as well as the grid speed, $\frac{\partial \xi^m}{\partial t}$. The computational cycle consisting of the Reynolds-averaged full Navier-Stokes solver, the aeroelastic equations solver, and the grid displacement solver is repeated every time step.

5 Computational Applications And Discussion

5.1 TWIN TAIL-DELTA WING CONFIGURATION

The twin tail-delta wing configuration consists of a 76°-swept back, sharp-edged delta wing (aspect ratio of one) and dynamically scaled flexible twin tails similar to those used by Washburn, et al. [7]. The vertical tails are oriented normal to the upper surface of the delta wing and have a centerline sweep of 53.5°. A multi-block grid consisting of 4 blocks is used for the solution of the problem. The first block is a O-H grid for the wing and upstream region, with 101X50X54 grid points in the wrap around, normal and axial directions, respectively. The second block is a H-H grid for the inboard region of the twin tails, with 23X50X13 grid points in the wrap around, normal and axial directions, respectively. The third block is a H-H grid for the outboard region of the twin tails, with 79X50X13 grid points in the wrap around, normal and axial directions, respectively. The fourth block is a O-H grid for the downstream region of the twin tails, with 101X50X25 grid points in the wrap around, normal and axial directions, respectively. Figure 1 shows the three dimensional grid topology and a front view blow-up of the twin tail-delta wing configuration.

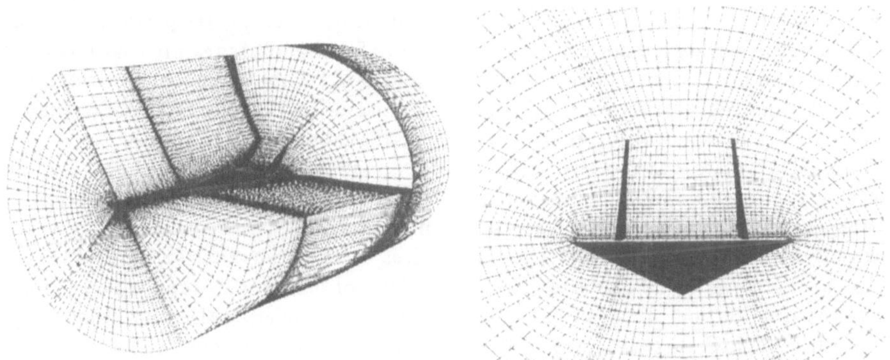

Figure 1: Three-dimensional grid topology and blow-up of the twin tail-delta wing configuration (the tails are in midspan position).

Each tail is made of a single Aluminum spar and Balsa wood covering. The Aluminum spar has a taper ratio of 0.3 and a constant thickness of 0.001736. The chord length at the root is 0.03889 and at the tip is 0.011667, with a span length of 0.2223. The Aluminum spar is constructed from 6061-T6 alloy with density, ρ, modulii of elasticity and rigidity, E and G, of 2693 kg/m^3, $6.896X10^{10}$ N/m^2 and $2.5925X10^{10}$ N/m^2; respectively. The corresponding dimensionless quantities are 2198, 4.595×10^5 and 1.727×10^5; respectively. The Balsa wood covering has a taper ratio of 0.23 and aspect ratio of 1.4. The chord length at the root is 0.2527 and at the tip is 0.058, with a span length of 0.2223. The Balsa thickness decreases gradually from 0.0211 at the tail root to 0.0111 at the tail midspan and then constant thickness of 0.0111 is maintained to the tail tip. The tail cross section is a semi-diamond shape with bevel angle of 20°. The Balsa density, modulii of elasticity and rigidity, E and G, are 179.7 kg/m^3, $6.896X10^8$ N/m^2 and $2.5925X10^8$ N/m^2; respectively. The corresponding dimensionless quantities are 147, 4.595×10^3 and 1.727×10^3; respectively. The tails are assumed to be magnetically suspended and the leading edge of the tail root is positioned at $x/c = 1.0$, measured from the wing apex. The configuration is pitched at 35° angle of attack and the freestream Mach number and Reynolds number are 0.3 and 1.25 x 10^6; respectively.

Keeping the twin tail as rigid surfaces, the unsteady, Reynolds-averaged, full Navier-Stokes equations are integrated time accurately using the implicit, flux-difference splitting scheme of Roe with Reynolds number of 1.25 million and angle of attack of 35°. The initial conditions are obtained after 10,000 time steps with $\Delta t = 0.001$. Next, the results of the coupled and uncoupled bending and torsion modes are presented. For the coupled bending and torsion case, the inertia axis is assumed downstream the elastic axis at $x_\theta = 0.005$. For the uncoupled case, x_θ is set equals to zero.

5.2 UNCOUPLED BENDING-TORSION MODES

Figure 2 shows three-dimensional views of the leading-edge vortex cores particle traces and iso-total pressure surfaces. Figure 3 shows front views for the total pressure contours on the wing surface and in cross flow planes at $x = 1.03$ and $x = 1.22$ of the inboard twin-tail position. The leading-edge vortex cores experience symmetric breakdown upstream of the twin tail. The vortex breakdown location is at about 65% chordstation, and over the wing by 10% chord. The vortices are totally outboard of the twin tail. The cores are moved upward as the flow travels downstream. Smaller size vortex cores appear underneath the primary wing vortex and it becomes larger in size as the flow travels downstream. These are the tail vortices observed by Washburn [7]. The tail vortices exist at the outer surfaces of the tails and they are rotating in the opposite direction to those of the primary wing vortices. Figure 4 shows the distribution of the surface pressure coefficient covering the wing from $x = 0.3$ to $x = 1.0$. Typical turbulent flow distribution are observed, where the largest suction peaks are pronounced at the position of the wing primary vortex cores. Figure 5 shows the distribution of the leading-edge total structural deflection and the root bending moment for the left and right tails for 20 dimensionless time after the initial conditions. The tails deflections are in first, second and third mode shapes. The twin tails are moving opposite to each other in symmetric manner. It is observed that the frequencies of the torsion deflections are almost twice those of the bending deflections while the frequencies of the normal loads are almost the same as those of the twisting moments. The normal forces are out of phase with the bending deflections while the twisting moments are in phase with the torsion deflections.

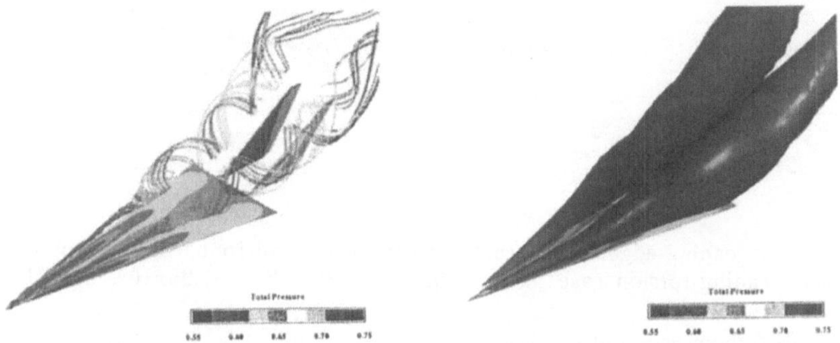

Figure 2: Three-dimensional views showing the total pressure on the surfaces, vortex core particle traces and iso-total pressure surfaces. Uncoupled case after $it = 9,600$, Turbulent flow. Inboard position.

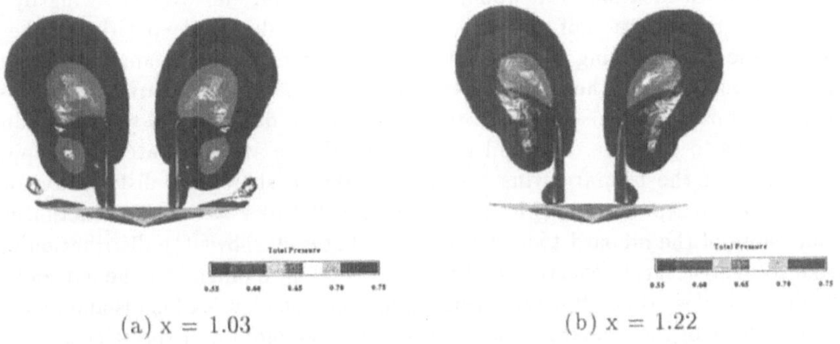

(a) x = 1.03 (b) x = 1.22

Figure 3: Snap shots of total pressure in a cross-flow planes. Uncoupled case after $it = 9,600$, Turbulent flow. Inboard position.

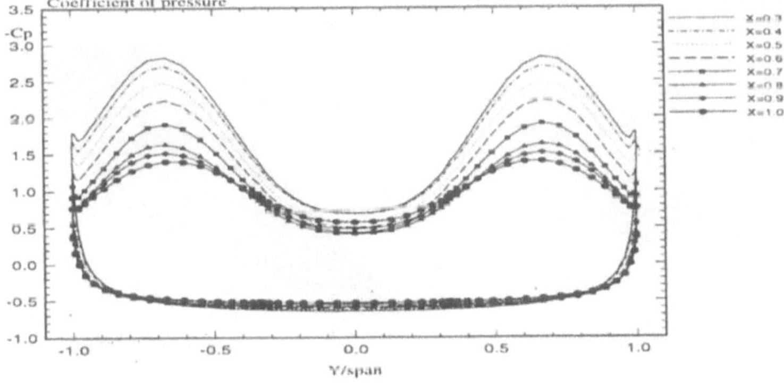

Figure 4: Distribution of Coefficient of pressure. Uncoupled case after $it = 9,600$, Turbulent flow. Inboard position.

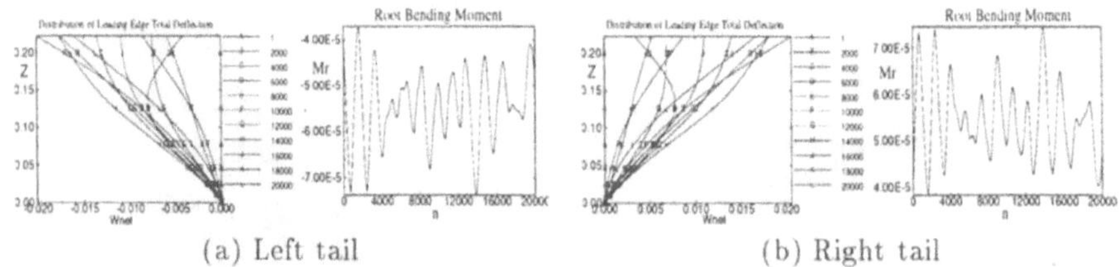

(a) Left tail (b) Right tail

Figure 5: Tail leading-edge total structural deflections and root bending moment for an uncoupled bending-torsion case. $M_\infty = 0.3$, $\alpha = 35^o$, $R_e = 1.25x10^6$, Turbulent flow. Inboard position.

Figure 6 shows three-dimensional views of the leading-edge vortex cores particle traces and iso-total pressure surfaces. Figure 7 shows front views for the total pressure contours on the wing surface and in cross flow planes at $x = 1.03$ and $x = 1.22$ of the outboard twin-tail position. The tails cut through the vortex breakdown flow of the leading-edge vortex cores. The leading-edge vortex cores experience symmetric breakdown upstream of the twin tail. The vortex breakdown location is at about 63% chordstation, and over the wing by 13% chord. The upstream movement of the breakdown location is due to the position of the tails with respect to the primary wing vortex cores which increases the upstream effect. The tail vortices are also outboard of the tails and they are larger in size compared to the case of inboard twin-tail position. This increases the aerodynamic damping on the tails, causing the tails deflections to decrease. The tail vortices are also shown to rotate in the opposite direction to those of the primary wing vortices. Figure 8 shows the distribution of the surface pressure coefficient covering the wing from $x = 0.3$ to $x = 1.0$. The suction peaks are less than those of the inboard twin-tail position. Figure 9 shows the distribution of the leading-edge total structural deflection and the root bending moment for the left and right tails for 20 dimensionless time after the initial conditions. The levels of loads and defections are lower than those of the inboard twin-tail position. The motion of the tails seems more damped than the case of inboard twin-tail. The tails are shown to oscillate in one direction only in first and second mode shapes. It is observed that the bending and torsion deflections are out of phase of the normal force and twisting moment loads, in contrast with the case of inboard twin-tail position.

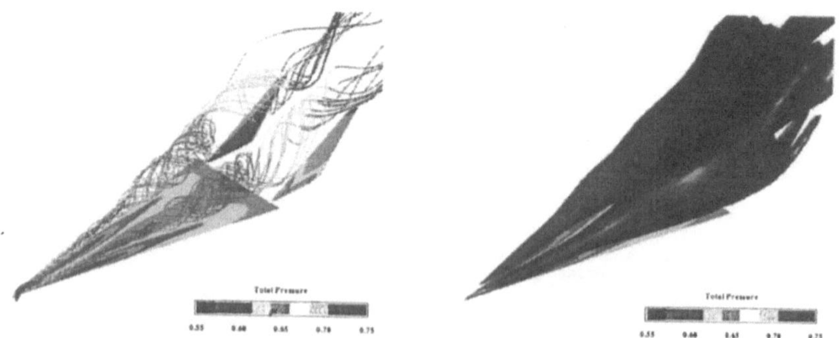

Figure 6: Three-dimensional views showing the total pressure on the surfaces, vortex core particle traces and iso-total pressure surfaces. Uncoupled case after $it = 9,600$, Turbulent flow. Outboard position.

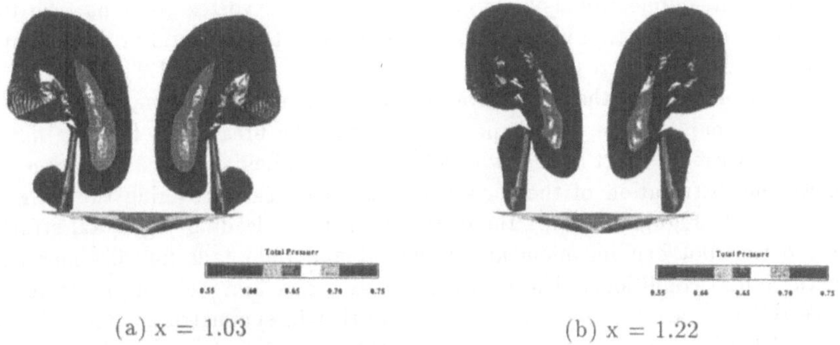

(a) x = 1.03 (b) x = 1.22

Figure 7: Snap shots of total pressure in a cross-flow planes. Uncoupled case after it = 9,600, Turbulent flow. Outboard position.

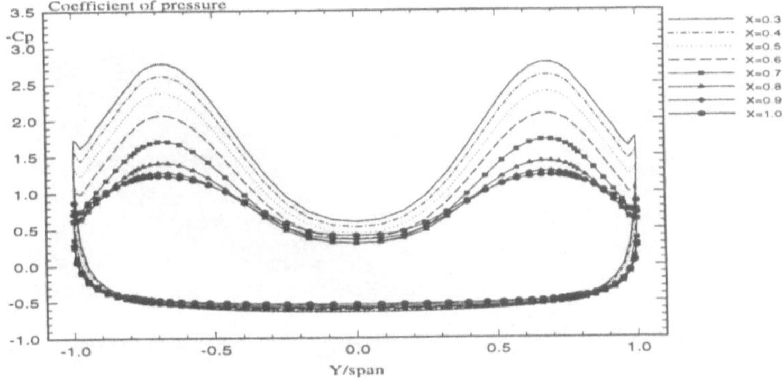

Figure 8: Distribution of Coefficient of pressure. Uncoupled case after it = 9,600, Turbulent flow. Outboard position.

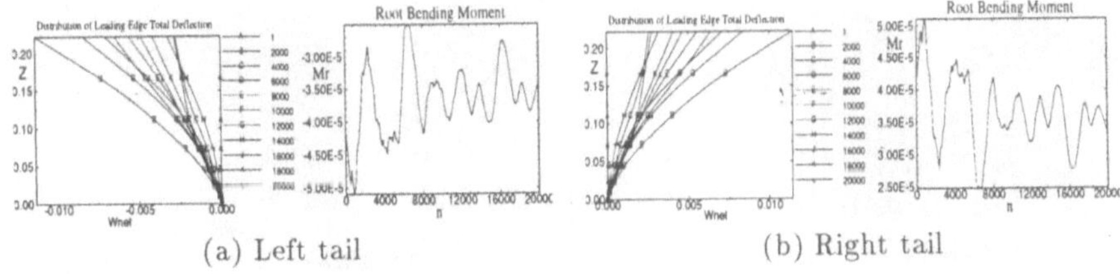

(a) Left tail (b) Right tail

Figure 9: Tail leading-edge total structural deflections and root bending moment for an uncoupled bending-torsion case. M_∞ = 0.3, α = 35°, R_e = 1.25x10^6, Turbulent flow. Outboard position.

5.3 COUPLED BENDING-TORSION MODES

Figure 10 shows three-dimensional views of the leading-edge vortex cores particle traces and iso-total pressure surfaces. Figure 11 shows front views for the total pressure contours on the wing surface and in cross flow planes at x = 1.03 and x = 1.22 of the inboard twin-tail position. Although, the vortex breakdown location is approximately at the same position as the uncoupled case, the shape and traces of the breakdown flow are different which show the upstream effect of the twin-tail coupled bending and torsion motions. Figure 12 shows the distribution of the surface pressure coefficient covering the wing from x = 0.3 to x = 1.0. Figure 13 shows the distribution of the leading-edge total structural deflection and the root bending moment for the left and right tails for 20 dimensionless time after the initial conditions. The tails deflections are in first, second and third mode shapes. The two tails are moving opposite to each other in symmetric manner. The tails deflections and levels of loads are slightly higher than those of the uncoupled case. The frequencies of the bending and torsion deflections, normal forces and twisting moments are slightly different from those of the uncoupled case. The normal forces are out of phase with the bending deflections while the twisting moments are in phase with the torsion deflections.

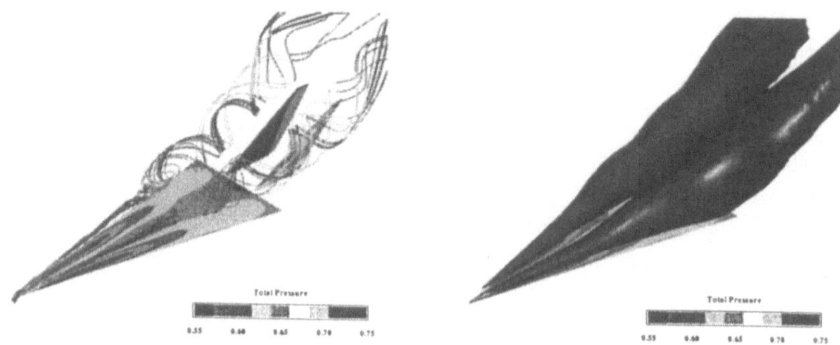

Figure 10: Three-dimensional views showing the total pressure on the surfaces, vortex core particle traces and iso-total pressure surfaces. Coupled case after it = 9,600, x_θ = 0.005, Turbulent flow. Inboard position.

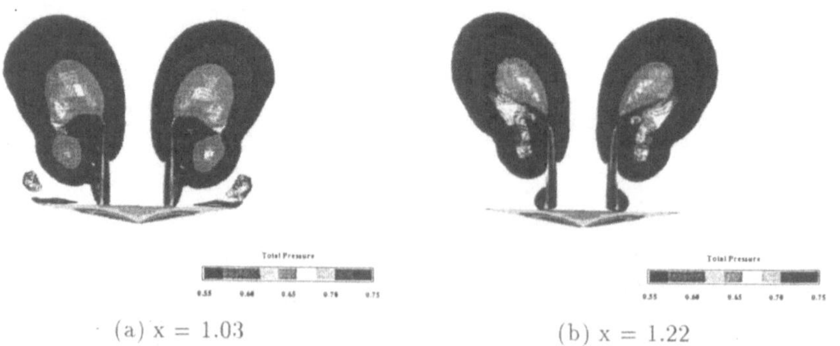

(a) x = 1.03 (b) x = 1.22

Figure 11: Snap shots of total pressure in a cross-flow planes. Coupled case after it = 9,600, x_θ = 0.005, Turbulent flow. Inboard position.

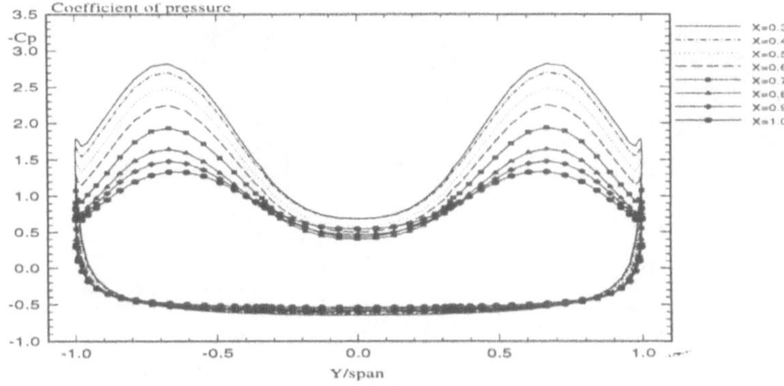

Figure 12: Distribution of Coefficient of pressure. Coupled case after $it = 9,600$, $x_\theta = 0.005$, Turbulent flow. Inboard position.

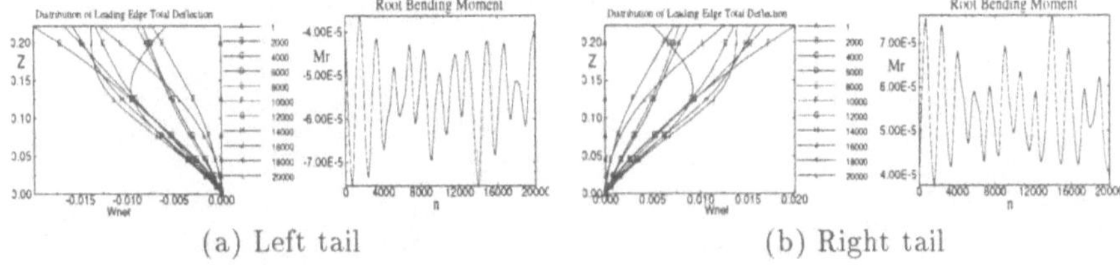

(a) Left tail (b) Right tail

Figure 13: Tail leading-edge total structural deflections and root bending moment for coupled bending-torsion case, $x_\theta = 0.005$. $M_\infty = 0.3$, $\alpha = 35°$, $R_e = 1.25x10^6$, Turbulent flow. Inboard position.

Figure 14 shows front views for the total pressure contours on the wing surface and in cross flow planes at $x = 1.03$ and $x = 1.22$ of the outboard twin-tail position. The shape of the vortex breakdown is more or less similar to that of the uncoupled modes case. This is attributed to the lower levels of tails deflections which minimize the differences between the coupled and uncoupled modes cases. Figure 15 shows the distribution of the surface pressure coefficient covering the wing from $x = 0.3$ to $x = 1.0$. Figure 16 shows the distribution of the leading-edge total structural deflection and the root bending moment for the left and right tails for 20 dimensionless time after the initial conditions. The tails are deflected in one direction only in first and second mode shapes. The tails deflections and levels of loads are slightly higher than those of the uncoupled bending-torsion case of the outboard twin-tail position but still lower than those of the uncoupled and coupled bending-torsion cases of the inboard twin-tail position. The bending and torsion deflections are out of phase of the normal force and twisting moment loads. The frequencies of the bending and torsion deflections are the same as those of the uncoupled bending-torsion case of the outboard twin-tail position.

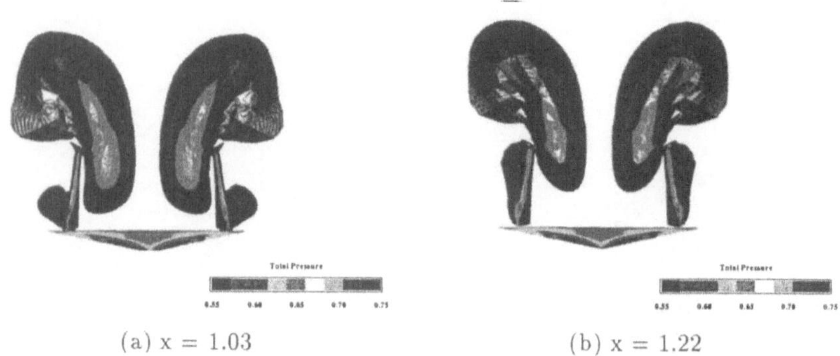

(a) x = 1.03 (b) x = 1.22

Figure 14: Snap shots of total pressure in a cross-flow planes. Coupled case after it = 9,600, $x_\theta = 0.005$, Turbulent flow. Outboard position.

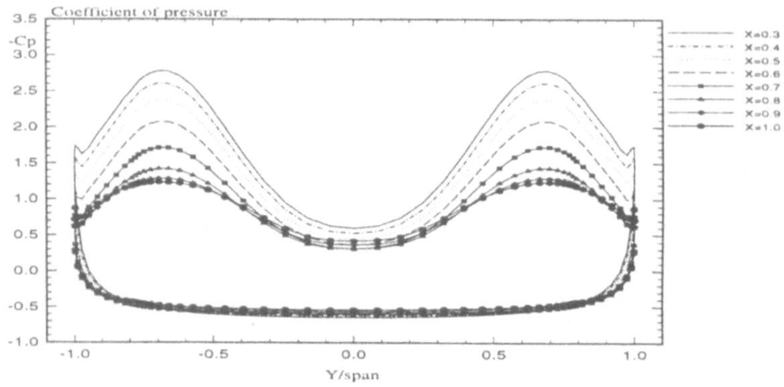

Figure 15: Distribution of Coefficient of pressure. Coupled case after it = 9,600, x_θ = 0.005, Turbulent flow. Outboard position.

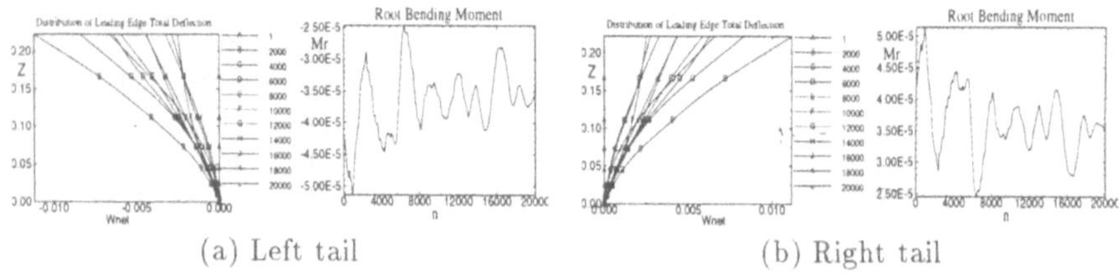

(a) Left tail (b) Right tail

Figure 16: Tail leading-edge total structural deflections and root bending moment for coupled bending-torsion case, $x_\theta = 0.005$. $M_\infty = 0.3$, $\alpha = 35^\circ$, $R_e = 1.25x10^6$, Turbulent flow. Outboard position.

Table 1 shows comparison of the present code (FTNS3D) results with those of Washburn, et al. [7] experimental data, of the mean root bending moment for flexible twin tails, the root mean square root bending moment for flexible twin tails and the lift coefficient with rigid twin tails. The computational results for both coupled and uncoupled bending and torsion modes agree well with Washburn's experimental data. The discrepancies in the results are attributed to the fact that the computational model have some differences than Washburn experimental model. Our wing is a flat plate with zero thickness while Washburn wing is a hummel-type wing (triangular cross section). Although the tails shape is the same, our model assumed magnetically suspended solid material tails while Washburn tails are constructed with spars, an additional Ballast weights and void spaces out in the tails. Our computational model assumed flexible twin-tail, while Washburn experimental model assumed one tail flexible and the other tail rigid. In the experimental work by Washburn [7], the presence of a flexible tail was found to affect the loads and pressures on the other rigid tail.

Parameter	Position	FTNS3D Uncoupled	FTNS3D Coupled	WASHBURN
Mean Root Bending Moment	Inboard	0.157	0.158	0.135
With Flexible Tails	Outboard	0.106	0.1057	0.130
RMS Root Bending Moment	Inboard	0.024	0.0255	0.054
With Flexible Tails	Outboard	0.016	0.015	0.03
Lift Coefficient	Inboard	1.167	1.166	1.16
With Rigid Tails	Outboard	1.1653	1.1653	1.16

Table 1: Validation of FTNS3D Uncoupled and Coupled computational results with Washburn, et al. experimental data [7].

6 Concluding Remarks

The effects of coupled and uncoupled bending and torsion modes on the twin-tail buffet response are investigated for two different spanwise positions of the twin-tail. The inboard position of the twin-tail produces the largest bending and torsion loads and deflections when compared with the results of the outboard position. The coupled bending and torsion modes produce slightly higher deflections and loads than those of the uncoupled modes cases. The frequencies of the torsion deflections are twice those of the bending deflections. It has been shown that the larger the tail deflections are, the higher the upstream effect is on the vortex breakdown flow upstream of the tails. The computational results presented are in good quantative agreement with the experimental data of Washburn, et al. [7].

7 Acknowledgment

For the first two authors, this research work is supported under Grants No. NAG-1-648 and NAG-1-994 by the NASA Langley Research Center. The authors would like to recognize the computational resources provided by the NAS facilities at Ames Research Center and the NASA Langley Research Center.

8 References

1. Sellers, W. L. III, Meyers, J. F. and Hepner, T. E.: LDV Survey Over a Fighter Model at Moderate to High Angle of Attack, SAE Paper No. 88-1448 (1988).
2. Erickson, G. E., Hall, R. M., Banks, D. W., Del Frate, J. H., Shreiner, J. A., Hanley, R. J. and Pulley, C. T.: Experimental Investigation of the F/A-18 Vortex Flows at Subsonic Through Transonic Speeds, AIAA Paper No. 89-2222 (1989).
3. Wentz, W. H.: Vortex-Fin Interaction on a Fighter Aircraft, AIAA Paper No. 87-2474, AIAA Fifth Applied Aerodynamics Conference, Monterey, CA August (1987).
4. Lee, B. and Brown, D.: Wind Tunnel Studies of F/A-18 Tail Buffet, AIAA Paper No. 90-1432 (1990).
5. Cole, S. R., Moss, S. W. and Dogget, R. V., Jr.: Some Buffet Response Characteristics of a Twin-Vertical-Tail Configuration, NASA TM-102749 (1990).
6. Bean, D. E. and Lee, B. H. K.: Correlation of Wind Tunnel and Flight Test Data for F/A-18 Vertical Tail Buffet, AIAA Paper No. 94-1800-CP (1994).
7. Washburn, A. E., Jenkins, L. N. and Ferman, M. A.: Experimental Investigation of Vortex-Fin Interaction, AIAA Paper No. 93-0050, AIAA 31st ASM, Reno, NV, January (1993).
8. Kandil, O. A., Kandil, H. A. and Massey, S. J.: Simulation of Tail Buffet Using Delta Wing-Vertical Tail Configuration, AIAA Paper No. 93-3688-CP, AIAA Atmospheric Flight Mechanics Conference, Monterery, CA August (1993), 566-577.
9. Kandil,O. A., Massey, S. J., and Kandil, H. A.: Computations of Vortex-Breakdown Induced Tail Buffet Undergoing Bending and Torsional Vibrations, AIAA Paper No. 94-1428-CP, AIAA/ASME/ASCE/ASC Structural, Structural Dynamics and Material Conference, SC April (1994), 977-993.
10. Kandil, O. A., Massey, S. J. and Sheta, E. F.: Structural Dynamics/CFD Interaction for Computation of Vertical Tail Buffet, International Forum on Aeroelasticity and Structural Dynamics, Royal Aeronautical Society, Manchester, U.K., June 26-28 (1995), 52.1-52.14. Also published in Royal Aeronautical Journal, August/September (1996), 297-303.
11. Kandil, O. A., Sheta, E. F. and Liu, C. H.: Computation and Validation of Fluid/Structure Twin-Tail Buffet Response, Euromech Colloquium 349, Structure Fluid Interaction in Aeronautics, Institute Fur Aeroelastik, Gottingen, Germany (1996).
12. Kandil, O. A., Sheta, E. F. and Massey, S. J.: Fluid/Structure Twin Tail Buffet Response Over A Wide Range of Angles of Attack, AIAA Paper No. 97-2261-CP, 15th AIAA Applied Aerodynamics Conference, Atlanta, GA, June 23-25 (1997).
13. Degani, D. and Schiff, L. B.: Computation of Turbulent Supersonic Flows Around Pointed Bodies Having Crossflow Separations, J. of Computational Physics 66 (1986), 173-196.

INTERACTION OF WING VORTICES AND PLUMES IN SUPERSONIC FLIGHT

P.M. Sforza
Department of Mechanical, Aerospace, and Manufacturing Engineering
Polytechnic University, Brooklyn, NY 11201

A model of the wake of an aircraft cruising at high speed in the stratosphere is presented. The two major components of the flow field are the trailing vortex wake and the jet exhaust plumes. The prospect of large fleets of supersonic airliners makes it important to accurately predict the dispersion of engine emissions resulting from interaction between the two. Here, synthesis of a number of different jet and vortex studies provides a unified description of aircraft wakes in terms of a length scale bA/C_L, based on the span, aspect ratio, and cruise lift coefficient. The basic premise of the model is that jet plumes, being immersed in the trailing vortex wake downwash, deform into twin vortices typical of jets in a cross-flow. This permits the development of the wake flow field to be assessed with the tools of vortex filament analysis. The wakes of both conventional high subsonic and supersonic aircraft are shown to be accommodated by this approach, as would the wakes of wing-jet combination injectors for scramjet applications. Experimental studies which would aid in the development of more accurate prediction methods are also described.

1. Introduction

1.1 BACKGROUND OF THE VORTEX WAKE PROBLEM

Wing tip vortices became a topic of intense interest and scrutiny in the mid-1960s when they were recognized as a hazard to following aircraft during low-speed terminal operations. Airport safety considerations prompted substantial research efforts led by FAA and NASA through the mid-1970s. During the same period the a race to develop the supersonic transport (SST) was on, and uncertainty about effects of its emissions on the stratosphere catalyzed research on supersonic aircraft vortex wakes. However, because of the rapid demise of the likelihood of a large SST fleet, vortex behavior in the supersonic regime received little sustained attention and is now characterized by a sparse database. Recently renewed interest in supersonic cruise vehicles, both commercial and military, has awakened a concern for fundamental insights regarding

E. Krause and K. Gersten (eds.), IUTAM Symposium on Dynamics of Slender Vortices, 415-424.

their vortex wakes. Several important problems illustrate the need for such information: interaction of tip vortices with shock waves and flow fields of downstream surfaces; entrainment of engine plumes into wing vortices (or fuel jets into scramjet injector vortices); and vortex confinement of entrained material for long times after aircraft passage. In practical applications these can influence stability and control characteristics, alter wake signatures, and increase residence times of possible pollutants, respectively. The object of this paper is to interpret and apply the results of the few past and recent experiments to the second problem, vortex entrainment.

1.2 PAST STUDIES OF ENTRAINMENT BY THE VORTEX WAKE

SST jet plumes stream through a wing wake flow field dominated by trailing vortices. Overcamp and Fay (1973) studied dispersion of SST exhaust products taking into account global characteristics of trailing vortices and the stratospheric atmosphere. They considered the near wake to involve no appreciable interaction and the intermediate wake to be dominated by engulfment of the exhaust plume into the trailing vortex system. Conti, Hoshizaki, Redler, and Cassady (1973), Nielsen, Stahara, and Woolley (1974), and Holdeman (1974) all presented models for predicting behavior of jet exhaust plumes as they interact with the aircraft vortex wake. Farlow, *et al* (1974) presented flight measurements of overall wake dimensions and NO mixing ratios behind a supersonic NASA YF-12. Comparisons to results from the previously mentioned models demonstrated substantial differences between them. Thereafter, for an extended period of almost 20 years, virtually no work along these lines appeared in the literature until Miake-Lye, Martinez-Sanchez, Brown, and Kolb (1991) presented a simplified *ad hoc* model to account for the capture of exhaust effluents by the trailing vortex system. They emphasized differences between supersonic and transonic aircraft wake systems which, they suggested, led to quite different dispersion characteristics. Quackenbush, Teske, and Bilanin (1993) incorporated a Reynolds-averaged turbulent flow model within a large-scale wake code to account for vortex entrainment of SST plumes. Dash and Kenzakowski (1994) presented Navier-Stokes computations for the interaction of a subsonic jet and a low speed tip vortex in response to unpublished experiments carried out at NASA LaRC. Then, in a manner reminiscent of the events of the past, Fahey, *et al* (1995) presented measurements of reactive gases and particles in the wake of a Concorde cruising at supersonic speed in the stratosphere, but in this case no comparisons to existing model predictions were given. Most recently, Gerz and Ehret (1996), Jacquin and Garnier (1996), and Garnier, Jacquin, and Laverdant (1996) presented combined vortex filament and turbulent diffusion models for treating the interaction between the jet plumes and the trailing vortex field. All studies were hampered by lack of a robust experimental database for validating the assumptions utilized, particularly with respect to initializing the plume mixing and reaction problem. As a consequence there is yet no unanimity regarding the mechanisms involved in the entrainment process.

1.3 A MODEL FOR THE PLUME-VORTEX INTERACTION FIELD

The two major components of the flow field are the trailing vortex wake and the jet exhaust plumes, representing the steady state lift and thrust of the aircraft, respectively. For practical lift to drag ratios on the order of 10, the trailing vortex wake represents a significant perturbation to the jet plumes. Recent experimental studies of Mach 2.5 flow past a typical high speed commercial transport (HSCT) planform carried out by Wang, Sforza, and Pascali (1996) show a complex wake system with three different trailing vortices. There is a leading edge vortex from the inboard delta panel of the planform, a junction vortex from the crank in the leading edge, and a true tip vortex from the supersonic leading edge outboard panel. An idealization of the flow field for an aircraft with this planform would involve placing the propulsive jets in the cross-flow of a number of nearby parallel vortices. The basic premise proposed here is that the jet plumes, being immersed in the downwash of the trailing vortex wake, each deform into twin counter-rotating vortices typical of jets in a cross-flow. For inviscid flow, this model of the complete aircraft wake flow field is comprised solely of vortex filaments, thereby providing a simple and consistent means for analyzing the complex arrays of vortices and jet plumes likely to be generated by proposed HSCT designs. Development of a sound basis for such an approach to the interaction between jet plumes and a vortex wake is the subject of the present investigation.

2. The Vortex Wake Flow Field

2.1 NEAR AND FAR FIELD BOUNDARIES

The nature of trailing vortex systems for slender, pointed supersonic planforms was set forth by Spreiter and Sacks (1951) in a comprehensive synthesis of NACAs extensive high speed research. They point out that for low aspect ratio wings small disturbance theory is valid in the vicinity of the wing and the inviscid wake at all Mach numbers and this fact permits them to describe, in some detail, the development of the vortex wake. During this period of high speed research, attention was focused on predicting the downwash field encountered by trailing control surfaces of aircraft or missiles rather than on details of the vortex. On the basis of similarity arguments they show that the downstream distance required for vortex roll-up is given by $x_r = kbA/C_L$ where A is the aspect ratio, b is the span, C_L is the lift coefficient, and k is a constant for any given wing and depends only on the planform shape and the span loading. For elliptical loading k lies between 0.28 for rectangular wings and 0.14 for delta wings. This formulation appears to indicate that the number of span lengths necessary to fully roll up vortices is larger for a conventional jet transport than for an HSCT. However, the parameter $A/C_L = q/(W/b^2)$, where W is aircraft weight and q is the dynamic pressure, tends to be fairly constant for aircraft with similar missions. For large subsonic or supersonic, transport or bomber aircraft in cruise, A/C_L lies between 15 and 20. Thus in most cases one may expect a roll-up distance behind a swept wing to be around 2 or 3 spans for either subsonic or supersonic speeds. Indeed, the experimental measurements of Wang, Sforza, and Pascali (1996) indicate that the vortex wake appears to be fully rolled up within one span of the trailing edge of the tip chord of the Mach 2.5 HSCT planform tested. The results of this section are suggestive in that the use of a modified span $b'=bA/C_L$ as a characteristic length makes the normalized roll-up distance x_r/b' a constant for all geometrically similar wings, a fact exploited in the subsequent analysis.

Well behind a cruising aircraft the flow field is dominated by two parallel trailing vortices. These far field vortices are subject to the sinusoidal instability first described by Crow (1970) and which leads to alteration of the organized pair into a number of elongated rings. The time scale for the onset of this instability corresponds to a downstream distance $x_c = 7.4\ bA/C_L$ or, in terms of the modified span, $x_c/b' = 7.4$. This illustrates that the distance to onset of the Crow instability is the same number of spans for any practical transport aircraft designed for efficient cruise. Note, however, that the elapsed time to the Crow instability will typically be a factor of 10 smaller for the supersonic transport.

2.2 THE INTERMEDIATE FIELD: VORTEX DECAY

In the intermediate region behind a cruising aircraft the two parallel trailing vortices should decay through the effects of viscous dissipation, according to Jacob, Savas, and Liepmann (1997). They report that no accurate method for predicting this decay outside the laboratory exists, although varying degrees of success have been achieved with empirical relations. Iverson (1976) presents a compilation of subsonic laboratory and flight test data for moderate to high aspect ratio wings along with a correlation for the decay of the observed maximum tangential speed. His correlation equation

$$V_m\,[x/U\Gamma]^{0.5} = 5.8 \qquad (1)$$

is applicable provided the vortex Reynolds number, $\Gamma/\nu > 10^6$, where V_m is the maximum tangential speed, Γ is the total circulation of the vortex, and ν is the kinematic viscosity. For elliptically loaded wings $\Gamma = 2C_L Ub/\pi A$ so that

$$V_m/U = 4.63(C_L/A)(x/b')^{-0.5}\,,\ x/b' > 103A^{-2} \qquad (2)$$

with a plateau level of $V_m/U = 0.46C_L$ for lower values of x/b'. This relation is plotted in Figure 1 for practical values of C_L/A along with the location of the roll-up and Crow instability boundaries described previously. This diagram portrays the parameter space for practical subsonic and supersonic transport aircraft designs. Note that onset of the Crow instability appears to occur in the vortex decay region for subsonic transports but well before that for supersonic transports. However, Eq. 2 has not been validated for the supersonic range. Wang and Sforza (1997) note that the experimental database for vortices generated in supersonic flow is sparse, but they also point out that their measurements show the vortex swirl behavior to be much like that found at low speeds. Indeed, their near field results for a rectangular wing with $C_L/A=25$ at M=3.1 show $V_m/U=0.14$, which would tend to fall in line with the upper, subsonic transport, set of plateau curves rather than with the lower set. This suggests that Iverson's relation for the plateau value of maximum tangential speed, or equivalently, for the location of the onset of vortex decay, may be biased toward the high aspect ratio wings which form his entire database. It is clear that more experimental studies of the details of the flow behind supersonic lifting wings is necessary. In the near field, the application of modern diagnostic techniques in wind tunnel simulations can be quite fruitful. For the intermediate field, interference effects make wind tunnel studies impractical. Ballistic range techniques coupled with optical diagnostics would provide much useful data since observations of wake temporal development could be made at a fixed laboratory location. Work of this nature has been carried out on both ballistic and powered projectiles, as described, for example, by Kuo and Fleming (1991).

3. The Jet Plume Flow Field

3.1 THE SUPERSONIC JET IN A SUPERSONIC STREAM

A comprehensive study of the structure of sonic and supersonic jets exhausting into stationary and moving environments was first presented by Love, Grigsby, Lee, and Woodling (1959). One of their motivations was to evaluate the interference flow field caused by propulsive jets. Like the contemporaneous work on inviscid supersonic vortex wakes described previously, concern was focused on near field interference effects on trailing control surfaces caused by the exhaust plumes from wing-mounted jets. The theoretical effort involved method of characteristics solutions for such flows and emphasized the prediction of flow boundaries and shock waves, while the experiments concentrated on schlieren photographs illustrating these same features. Though the turbulent mixing characteristics of low speed jets was then, and is, the subject of multitudinous investigations, the corresponding features of supersonic jets were not studied until the mid-60s when interest in co-flow fuel injectors for scramjets started to grow. Schetz and Swanson (1973) noted that even with this incentive, experiments were still limited to rather low Mach numbers, leading them to carry out studies around M=3.5. Since in scramjet-related experiments the jet Mach number is lower than that of the free stream and the jet is inclined at a substantial angle to the free stream it is difficult to apply any results to the problem at hand. Dash and Kenzakowski (1996) present a modern review of jet flow field modeling which illustrates the dearth of data available in the open literature on supersonic-supersonic jet mixing and a corresponding lack of validated analyses. It is reasonable to assume that experiments, commissioned in support of missile plume technology, provided some data which is presently restricted in circulation. Obviously, there exists a need for fundamental studies of supersonic turbulent mixing and suitable wind tunnel test

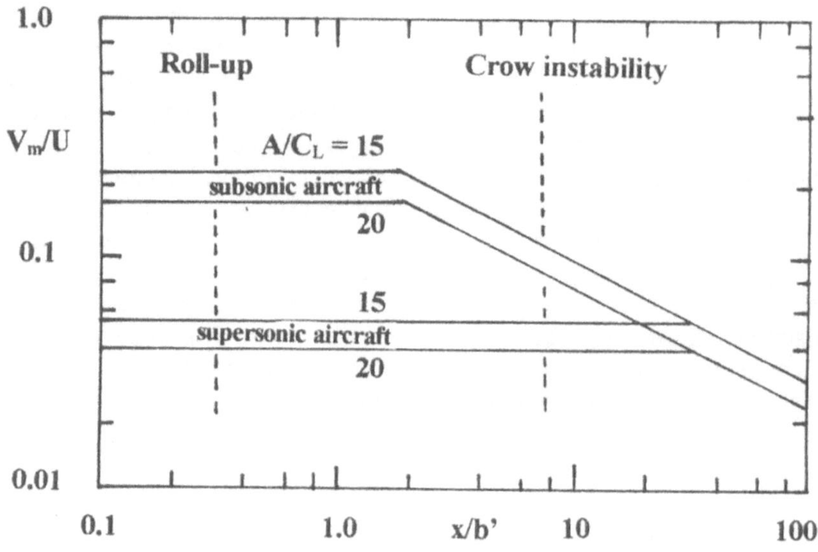

Figure 1. Normalized parameter space for maximum tangential speed of vortex.

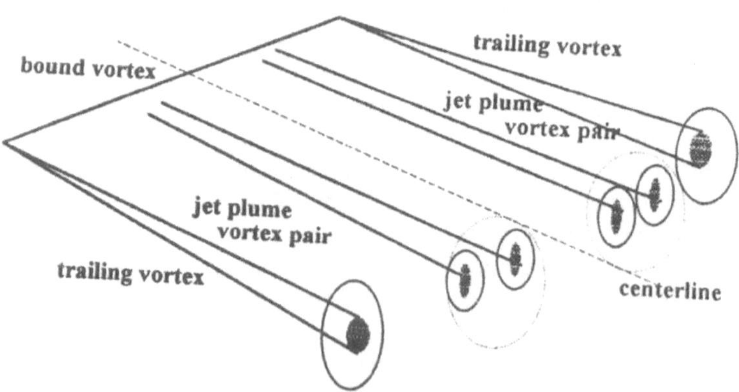

Figure 2. Schematic diagram of the combined vortex flow field model.

rigs are relatively easy and inexpensive to design, install, and operate. The diagnostic techniques to be applied constitute the major challenge for such sorely needed investigations.

3.2 THE SUPERSONIC JET IN A WEAK DOWNWASH FIELD

In the present study the jet plume has velocity U_j oriented parallel to the line of flight, and is exposed to the velocity field of the wing wake. This field is assumed to be composed of three velocity components: the aircraft speed along the line of flight U, the downwash w, and the sidewash v, the last of which is ignored here. Thus, in the slender body inviscid approximation, the jet plumes feel a small cross-flow velocity $w \ll U$ normal to the jet velocity U_j. As a simple analogy, replace the plume by a slender cylinder at angle of attack $\alpha = -\arctan(w/U) \simeq -w/U$. According to Nielsen (1960) such a situation generates a pair of lee-side vortices of circulation $\Gamma \simeq \pi dU\alpha$, or $\Gamma \simeq \pi wd$, where d is the diameter of the cylinder. This corresponds to a pair of vortices below the cylinder rotating as to produce an upwash. A jet in a cross-flow rapidly bends in the downstream direction and simultaneously deforms into a pair of counter-rotating vortices which produce an upwash along the trajectory of the bending jet. This basic flow problem has stirred much controversy regarding the nature of the mechanism responsible for the observed behavior even though *ad hoc* approaches have often provided quite useful correlations, according to Coelho and Hunt (1989). Margason's (1993) review, as well as the other papers collected in an AGARD conference devoted to jets in cross-flow, make no mention of small inclination cases. In addition, the various analytic and semi-empirical approaches mentioned previously do not give consistent results for the circulation to be expected in the vortices. If it is assumed that the jet deforms into a pair of counter-rotating vortices separated by a distance initially equal to the jet diameter d_j and produces an upwash equal to w normal to the line joining the vortex centers, the corresponding jet vortex circulation $\Gamma_j = \pi w d_j$, the same as observed for the slender body at incidence. Since the vortex system has been shown to roll up rapidly, the centerline downwash velocity may be given as $w=(4/\pi)^2(C_L/\pi\Lambda)U$ for a simple wing vortex wake system with only two trailing vortices. This downwash is fairly constant over the spanwise distance $|y|<b/2$, as shown by Spreiter and Sacks (1951). Using this simple approximation for w in the equation for Γ_j leads to a ratio of jet to tip vortex circulation $\Gamma_j/\Gamma=8d_j/\pi b$. The downwash for more complex trailing vortex systems and more accurate values for the jet circulation may readily accommodated in this model. All investigators find the circulation to increase with the momentum flux ratio $\phi^2 = \rho_j u_j^2/\rho w^2$ and Broadwell and Breidenthal (1984) propose a characteristic jet length which, in the present notation, becomes $x_j/b' = 1.72\phi(d_j/b)$. Jacquin and Garnier (1996) present a table of typical parameters for a conventional large jet transport and for the Concorde; using these yields $\Gamma_j/\Gamma = 0.026$ and 0.18, $\phi = 1.25$ and 1.27, and $x_j/b' = 0.0214$ and 0.15, respectively. This suggests that while the circulation ratio developed in the Concorde is much greater than that of the conventional large jet transport, they are both still

small quantities. Furthermore, the momentum function ϕ is the same for both aircraft indicating that the jet vortex circulation in the simple model mainly depends on the jet diameter to span ratio, d_j/b. Finally, the normalized jet characteristic length is on the order of the normalized roll-up distance so that both the wing and jet vortices should be rolled up within the same distance for a given aircraft. The proposed inviscid model for jet plumes immersed in the vortex wake of an aircraft is schematically illustrated in Figure 2. The relatively slow streamwise variation of properties suggests that a turbulent boundary layer analysis could be superimposed on this model to treat the turbulent diffusion process, but that is not addressed here.

4. The Combined Wake Model

The flow field generated by the combined inviscid airplane wake model is illustrated by calculation of a simple generic configuration. Consider an elliptically loaded wing of span b=25.5m with 2 single jet exhausts of diameter d_j=1.75m. The jets are mounted tangent to and under the wing, centered at $|y/b|$= 0.216 and z/b=0.069 (where x points to the nose and z points down). After roll-up the tip vortices are assumed to be located at $|y/b|= b_v/b=(\pi/4)$ and z=0 while the jet plume vortices are assumed to be situated on either side of the horizontal diameter of the jet exit. Normalizing distance with b and circulation with the tip vortex circulation Γ leads to 6 vortex filaments and 2 normalized circulation strengths. Assuming that the initial roll-up location several spans downstream of the wing is sufficient to permit a two-dimensional vortex filament analysis, the development of the flow field under mutual interaction is readily computed. For circulation ratios Γ_j/Γ on the order of 0.1 or less the effect of jet circulation is minimal and the jet vortices move along under the influence of the tip vortices as if passive particles, as shown in Figure 3(a), where the ratio is taken as 0.1and each downstream step is of length $\pi b/4$, the initial vortex span. This behavior occurs because the self-induced field of the jet vortex pairs serves mainly to move them up, against the downwash of the wing. Their effect on other regions of the flow is small because the counter-rotation provides substantial cancellation. The trajectory shown represents a distance of 20 vortex spans, or 15.7 wing spans. Connecting corresponding data points of the two jet vortices provides a measure of the distance between the vortices. Note that as distance downstream increases this separation increases dramatically, suggesting substantial additional transverse strain rate on the jet mixing field, an effect which will certainly affect diffusion. Though the model for the jet circulation ratio suggests a value on the order of 0.1, as used in the calculation, larger values may possibly be encountered in practice. As Γ_j/Γ increases from 0.1, however, the jet vortices begin to influence the motion of the tip vortices, as can be seen in Figure 3(b), which is the same problem carried out for a circulation ratio of 0.5. Here the tip vortex moves outboard and the jet vortices are drastically separated. Such behavior has not been observed yet for aircraft so the model estimate for circulation ratio is probably sound. On the other hand, if trajectories of this sort could be produced they might have application for scramjet injectors. Finally, the number and placement of the jet exhausts will also have an effect on motion in the wake, but the present model is amenable to such parametric studies.

5. Conclusions

An inviscid model for the interaction between the vortex wake of a lifting wing and the jet plumes propelling it has been presented. The background of the problem has been described, relevant past studies reviewed, and experiments needed to fill gaps in relevant areas noted. Introduction of a modified wing span $b'=bA/C_L$ as a characteristic length unified the parameter space for subsonic and supersonic cruise aircraft. The boundaries of vortex roll-up through vortex decay to onset of Crow instability scale with b' and resulting similarities between vortex wakes of subsonic and supersonic cruise transports have been described. Jet plumes, characterized as slightly inclined supersonic jets in a supersonic stream, produce trailing twin counter-rotating vortices. The interaction between the wing vortex wake and the jet plumes was then modeled as one occurring solely between a number of trailing vortex filaments of different strength. Calculations on a simple generic aircraft configuration showed that for the circulation strength expected in aircraft applications, the jet plume vortices merely move with the flow induced by the tip vortices. However, jet plume vortex circulation on the order of that of the tip vortex substantially affects the motion of the tip vortices. The ability to produce such circulation levels in other applications, such as scramjet injectors, may have practical significance.

6. References

Broadwell, J.E. and Breidenthal, R.E. (1984). Structure and mixing of a transverse jet in incompressible flow, *Jl. of Fluid Mechanics*, **148**, pp. 405-412.

Coelho, S.L.V. and Hunt, J.C.R. (1989). The dynamics of the near field of strong jets in crossflows, *Jl. of Fluid Mechanics*, **200**, pp. 95-120.

Crow, S.C. (1970). Stability theory for a pair of trailing vortices, *AIAA Jl.*, **8**, 12, pp. 2172-2179.

Dash, S.M. and Kenzakowski, D.C. (1995). Turbulent aspects of plume aerodynamic interactions, AIAA Paper 95-2373, AIAA/ASME/SAE/ASEE 31st Joint Propulsion Conference and Exhibit.

Dash, S.M. and Kenzakowski, D.C. (1996). Future directions in turbulence modeling for jet flow field simulation, AIAA Paper 96-1775, 2nd AIAA/CEAS Aeroacoustics Conference.

Fahey, D.W. *et al*, (1995). Emission measurements of the Concorde supersonic aircraft in the lower stratosphere, *Science*, **270**, 6 October, pp. 70-73.

Farlow, N.H., Watson, V.R., Lowenstein, M., Chan, K.L., Hoshizaki, H., Conti, R.J., and Meyer, J.W. (1974). Measurements of supersonic jet aircraft wakes in the stratosphere, Proceedings of the 2nd International Conference on the Environmental Impact of Aerospace Operations in the High Atmosphere, American Meteorological Society, Boston, MA, pp.53-58.

Garnier, F., Jacquin, L., and Laverdant, A. (1996). Engine entrainment in the near field of an aircraft, in Proceedings of the International Colloquium CAO on Impact of Aircraft Emissions upon the Atmosphere, Clamart, France.

Gerz, T. and Ehret, T. (1996). Wake dynamics and exhaust distribution behind cruising aircraft, in *The Characterisation and Modification of Wakes from Lifting Vehicles in Fluid*, AGARD CP-584, pp.35-1 to 37-12.

Iverson, J.D. (1976). Correlation of turbulent trailing vortex decay data, *Jl. of Aircraft*, **13**, 5, pp.338-342.

Jacquin, L. and Garnier, F. (1996). On the dynamics of engine jets behind a transport aircraft, in *The Characterisation and Modification of Wakes from Lifting Vehicles in Fluid,* AGARD CP-584, pp. 37-1 to 37-8.

Jacob, J., Savas, O., and Liepmann, D. (1997). Trailing vortex wake growth characteristics of a high aspect ratio rectangular airfoil, *AIAA Jl.*, **35**, 2, pp. 275-280.

Kuo, K.K., and Fleming, J.N. (1991). *Base Bleed - First International Symposium on Special Topics in Chemical Propulsion*, Taylor and Francis, Bristol, PA.

Love, E.S., Grigsby, C.E., Lee, L.P., and Woodling, M.J. (1959). Experimental and theoretical studies of axisymmetric free jets, NASA Technical Report R-6.

Margason, R.J. (1993). Fifty years of jet in cross flow research, in AGARD Conference Proceedings 534.

Miake-Lye, R., Martinez-Sanchez, M., Brown, R., and Kolb, C. (1991). Plume and wake dynamics, mixing and chemistry behind an HSCT aircraft, AIAA paper 91-3158, AIAA Aircraft Design Systems and Operations Meeting.

Nielsen, J.N. (1960). *Missile Aerodynamics*, McGraw-Hill, NY

Nielsen, J.N., Stahara, S.S., and Woolley, J.P. (1974). Ingestion and dispersion of engine exhaust products by trailing vortices for supersonic flight in the stratosphere, AIAA Paper 74-42, AIAA 12th Aerospace Sciences Meeting.

Overcamp, T.J. and Fay, J.A. (1973). Dispersion and subsidence of the exhaust of a supersonic transport in the stratosphere, *Jl. of Aircraft*, **10**, 12, pp. 720-728.

Quackenbush, T.R., Teske, M.E., and Bilanin, A.J. (1993). Computation of wake exhaust mixing downstream of advanced transport aircraft, AIAA Paper 93-2944, AIAA 24th Fluid Dynamics Conference.

Spreiter, J.R. and Sacks, A.H. (1951). The rolling up of the trailing vortex sheet and its effect on the downwash behind wings, *Jl. Aero. Sciences*, **18**, 1, pp.21-32.

Wang, F.Y., Sforza, P.M., and Pascali, R. (1996). Vortex-wake characteristics of a supersonic transport wing planform at Mach 2.5, *AIAA Jl*, **34**, 8, pp.1750-52.

Wang, F.Y. and Sforza, P.M. (1997). Near-field experiments on tip vortices at Mach 3.1, *AIAA Jl*, **35**, 4, pp.750-753.

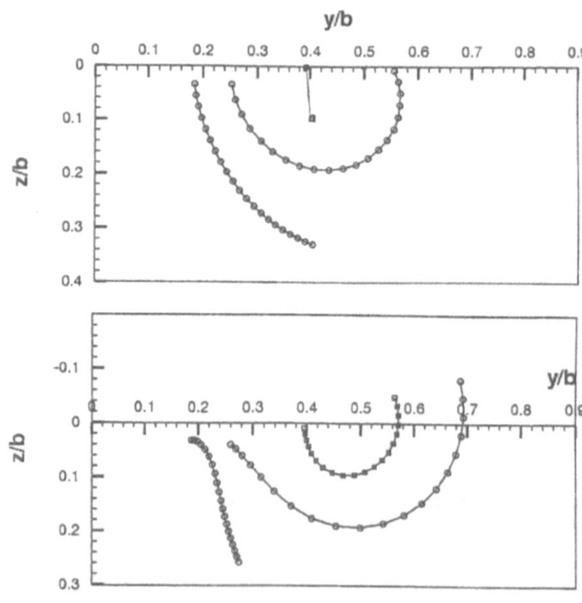

Figure 3. Vortex trajectories for: Γ_j/Γ= (a) 0.1, top; (b) 0.5, bottom.

DYNAMICS OF THE TRAILING VORTICES NEAR THE GROUND

N. KORNEV AND V. TRESHKOV
Marine Technical University St.Petersburg
Lotsmanskaya Str.3, 190008 St.Petersburg, Russia

AND

G. REICHERT
Technical University Braunschweig
Schleinitzstrasse 20, 38023 Braunschweig, Germany

1. Introduction

The vortex behavior near the ground is very important scientific problem for many reasons. At present time, many efforts are being devoted to enable safe improvements in the capacity of the air transportation system by bringing about a reduction in the in-trailspacing of airplanes. The efforts to solve the problem have been directed at investigating of the vortex behavior near the ground and finding ways to reduce the intensity of the vortices. Another important occurrence of vortex-ground interactions is for ekranoplans (wing-in-ground effect crafts). These flying ships move in the operation area of many other marine vehicles and produce the powerful trailing vortices. The study of the vortex behavior near the ground is very important to choose the operating regimes and safe routes of the ekranoplans.

The first part of the paper is devoted to the trailing vortices dynamics in the near vortex wake of a lifting surface near the ground. The problem has been extensively investigated because of its historical importance in the ekranoplan application. A review of some works on the vortex wake of wings near the ground has been written by Hooker [1]. In this paper we describe some specific features of the trailing vortices behavior near the ground.

The second part of the paper focuses on the trailing vortices behavior in the far vortex wake. A summary of the works related to this problem is presented in [2],[3] and [4]. An important aspect of the present paper is the theoretical study of the three-dimensional Crow instability [5] near the

E. Krause and K. Gersten (eds.), IUTAM Symposium on Dynamics of Slender Vortices, 425-434.

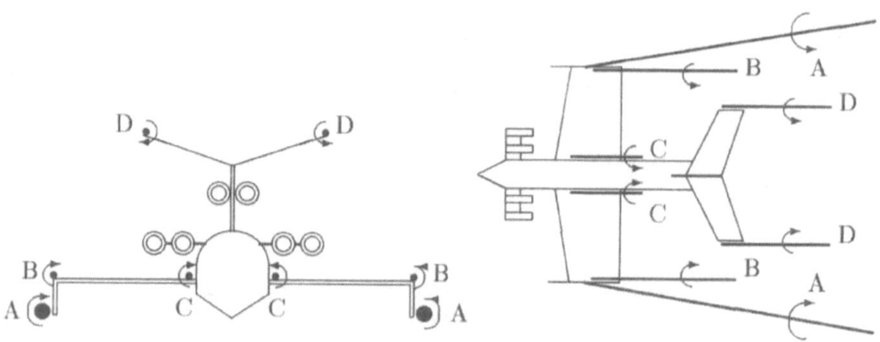

Figure 1. The vortex system of a wingship.

ground. Both the cases of inviscid and viscous flow have been considered. The developed analytical model in inviscid flow predicts three types of the vortex instability near the ground, depending on the dimensionless height of flight. The possible scenario of the vortex trailing instability and decay in ground effect has been proposed as a result of the theoretical analysis.

2. Near Vortex Wake in Ground Effect

Measurements and theoretical investigations shown that the aerodynamic characteristics of a wing and its near vortex wake are not significantly affected by the ground if the height of flight h is larger than the mean aerodynamic chord c. For conventional airplanes the ground has negligible effect on the the near wake, because their smallest height of flight during the take off and landing is greater than the mean chord. Common examples where the influence of the ground on the near vortex wake is important include aerodynamics of the wing-in-ground effects crafts (wingships) and dynamics of ship hull vortices. This section is concerned with the near vortex wake of the wingships.

The vortex wakes of wingships have some specific features which follow from the peculiarities of their wing configuration. As a rule a wingship sheds four counter-rotating vortex pairs shown in Fig.1. Circulations of the vortices D and C are negligibly small compared with circulations of both the upper vortex B and the lower vortex A. At large heights $\overline{h} = h/c > 0.5$ the circulation Γ_B of the vortex B arising on the wing tip is not substantially greater than the circulation Γ_A of the vortex A which is shed from the lower edge of the endplate [6]. As $\overline{h}$ is decreased, the span-load distribution becomes more uniform and Γ_B decreases, whereas the circulation Γ_A is drastically increased. For the case of the small height $\overline{h} < 0.2$ and $A_R \sim 1.0$

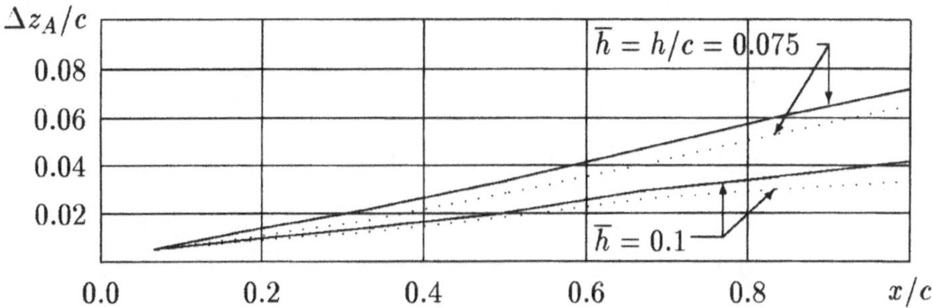

Figure 2. Vortex trajectory data: ...., measurement [7], ——, vortex lattice method [8], h-clearance under trailing edge, $\alpha = 4$ deg.

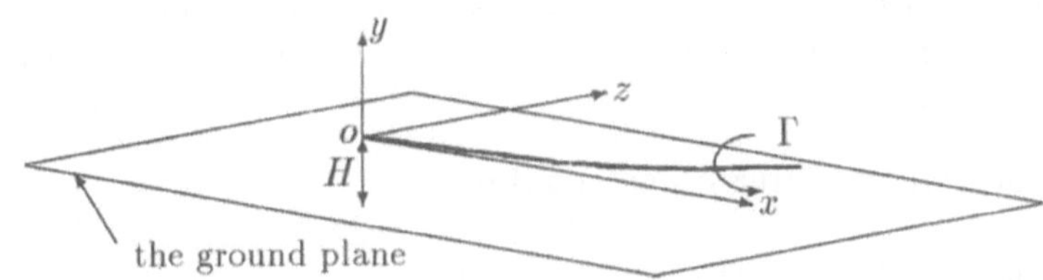

Figure 3. Geometry for investigating the rise of the vortex near the ground.

Γ_A can be four to five times greater than Γ_B [6].

The horizontal motion of a vortex in the outboard direction caused by the interaction with its "mirror" image is the dominant feature near the ground. Whereas the horizontal deflection of the vortex B is negligible at arbitrary h, the vortex A exhibits significant lateral deflection Δz_A. As an example data for the rectangular flat plate airfoil $A_R = 0.5$ equipped with an endplate are shown in Fig.2. The height of the endplate H_{ep} is $0.05c$. As it follows from potential flow theory, in close vicinity to the ground the circulation Γ_A of the vortex A is proportional to $1/h$, and its lateral deflection Δz_A is proportional to $1/h^2$.

In the vertical plane the vortices descend slowly toward the ground due to mutually induced velocities. In close vicinity to the ground the descent is insignificant. Theoretically the trailing vortex may rise up from the ground due to its own self-induced velocity. Let us consider a single vortex beam near the ground (Fig.3). To simplify the explanation, it is convenient to assume that the vortex deformation is small compared to the height H and the vortex core remains small and nearly circular with a radius σ. Using the inverted image method, Biot-Savart's law and the local induction approach

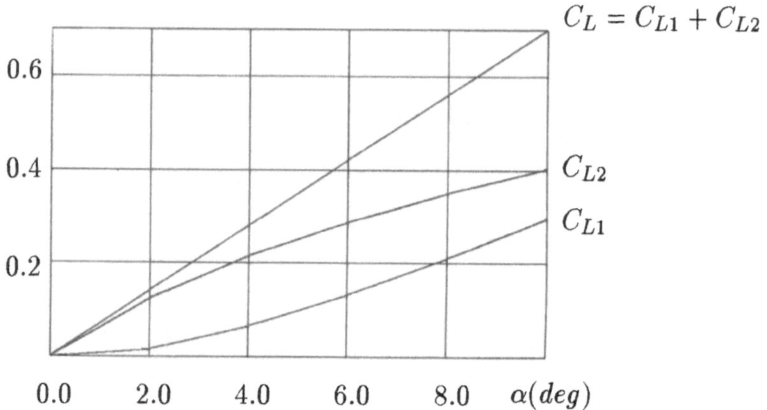

Figure 4. Lift coefficient C_L for the rectangular flat plate airfoil A_R=1: C_{L1}- the part of the lift coefficient caused by the vortex wake deformation, C_{L2}- the part of the lift coefficient obtained with the linear vortex lattice method [8], $\overline{h} = h/c = 0.057$.

we obtain from the stream-line equation

$$\frac{d^2 z}{d^2 x} \sim \frac{\Gamma}{4\pi} \frac{2H}{[x^2 + 4H^2]^{3/2}}, \; z \sim \frac{\Gamma}{8\pi H}(x - 2H + \sqrt{x^2 + 4H^2}),$$

$$y \sim \frac{\Gamma^2}{32\pi^2 H} |\ln \sigma| \frac{x}{\sqrt{x^2 + 4H^2}}$$

The semi-infinite vortex has a positive curvature near the ground and is rising up from the ground. If the vortex radius σ or the height H tend to zero, the rise of the vortex tends to infinity. For the values of H, σ and Γ of practical interest the rise of the vortex due to self-induction is suppressed by the descent due to induction of the opposing vortex.

Visel [7] has shown that the lower trailing vortex of the wingship can experience the breakdown in the close vicinity to the ground. For instance, breakdown was observed throughout the interval of $0 < \overline{h} < 0.075$ for the flat plate airfoil $A_R = 0.72$ equipped with an endplate ($H_{ep} = 0.05c$) at $\alpha = 4$ deg. Since the minimal height of flight of wingships is limited by the seaworthiness requirements in cruise regime ($\overline{h} > 0.1$), in practice the breakdown hasn't been observed yet.

The deformation of the vortices in the near wake is responsible for a number of important effects in the wingship aerodynamics. In the case of a single wing near the ground the vortex deformation is very important for small aspect ratios A_R. The dependence of C_L on α involves competition between two types of nonlinearities. On the one hand, there is the well-known nonlinearity at the large angles of attack $C_{L1} \sim \alpha^{5/3}$ [9]. The coefficient

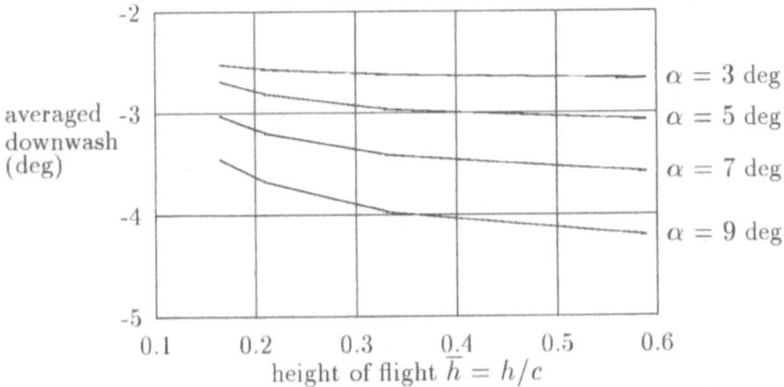

Figure 5. The averaged downwash on the tail unit of the wingship MPE. Vortex lattice method [8].

C_{L1} is obtained via combination of the method of matched asymptotics [9] and the linear vortex lattice method [8]. On the other hand, the ground effect also results in the nonlinearity $C_{L2}(\alpha)$. Combination of these nonlinearities with curvatures of opposite sign can lead to the suppression of the nonlinearity. The dependence C_L on α becomes nearly linear (Fig.4). One of these nonlinearities can be dominant, depending on α, h and the wing configuration.

The vortex wake has a strong quantitative and qualitative effect on the aerodynamics of the wingship with a tail unit. The averaged vortex-induced downwash on the tail unit increases with the decrease of the height of flight (see Fig.5). This result doesn't seem to be trivial. On the one hand, a decrease of h leads to the increase of the averaged downwash due to the increase of the vortex circulation. On the other hand, as expected from the potential flow theory, the downwash should be lowered due to the mutual approach of the vortex wake and its mirror image. The first effect proves to be stronger.

The aerodynamic characteristics and the stability criteria of the wingship with a tail unit must be calculated taking into account vortex deformation in the near wake [8]. The interaction between the trailing vortices and the tail unit of a wingship is stronger than that observed in the case of a conventional airplane. A set of mathematical models of different level of sophistication are used in practice, including semiempirical models and the advanced free-wake lifting-surface method. A summary of these models is presented in [1] and [8].

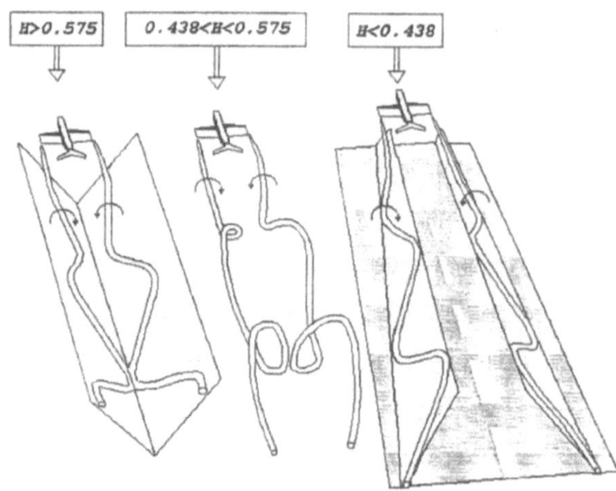

Figure 6. Schematic of the vortex instability near the ground: $1, \overline{H} > 0.575; 2, 0.575 > \overline{H} > 0.438$, and $3, \overline{H} < 0.438$.

3. Dynamics of the Trailing Vortices in the Far Vortex Wake Near the Ground

This section consists of two parts. Basic types of the convective instability of the trailing vortices are predicted in the next subsection using the linear stability theory. The physics of the process in close vicinity to the ground is largely viscous. The nonlinear vortex instability in the viscous flow near the ground is considered in subsec. 3.2.

3.1. INVISCID STABILITY THEORY FOR A PAIR OF TRAILING VORTICES NEAR THE GROUND

In [4] the well-known Crow stability theory [5] has been extended on the ground effect. Consider the basic results for a pair of trailing vortices near the ground. We found three types of vortex instability, depending on the dimensionless height of initial position $\overline{H} = H/b$ of the trailing vortices (see Fig.6).

In the upper region $\overline{H} \geq 0.575$, the instability is similar to Crow instability [5]. In the case $\overline{H} = \infty$, the growing perturbations are planar symmetric waves (see Fig.7). The planes containing any maximally unstable mode are inclined at about $\theta_1 = \theta_2 = 48$ deg. The symmetry of the perturbation waves is violated with a decrease of the height $\overline{H}$. The planes containing the planar sinusoidal waves are fixed at different angles θ_i to the

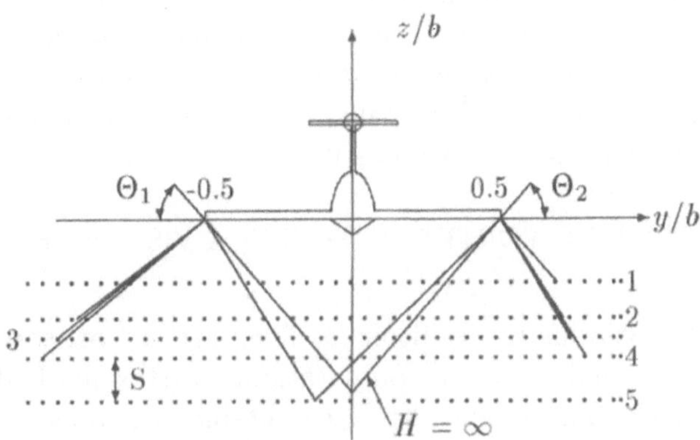

Figure 7. The planes containing maximally unstable modes for different dimensionless height. b is the wing span. The dotted lines show different positions of the ground with respect to the aircraft: 1 $\overline{H}$=0.2, 2 0.313, 3 0.375, 4 0.438, 5 0.575; and S, region of the helical instability.

horizontal. At $\overline{H} = 0.575$, for example, the angle θ_1 is about 57 deg and θ_2 is about 42 deg, as can be seen directly from Fig.7. The sinusoidal instability grows until the cores of the two vortices come into contact. Throughout the interval $0.575 \leq \overline{H} \leq \infty$, the point of contact is located between the vortices below their unperturbed position. At $\overline{H} = 0.575$ the point lies on the ground.

The trailing vortices are bent into helix in the middle region $0.438 \leq \overline{H} \leq 0.575$. The helix amplitude grows along the x-axis, resulting in a helical instability.

The growing perturbations near the ground (lower region, $\overline{H} < 0.438$) are planar waves, which are confined to fixed planes inclined at negative angles to the horizontal. The limiting value of this angle for $\overline{H} \to 0$ is -42 deg. The instability of a trailing vortex in close vicinity to the ground is like the so-called Crow instability [5] for two trailing vortices. In this case, the inverted vortex plays the role of the second trailing vortex. The instability grows until the vortex core comes into contact with the ground. The point of contact is located outside of the wingspan (Fig.7).

With increase of $\overline{H} \to \infty$ the amplification rate α tends to 0.83. Designation α corresponds to that introduced by Crow. In very close proximity to the ground, $\overline{H} \sim 0$, α is proportional to $\overline{H}^{-2}$. The amplification rate is minimal in the region of the helical instability. The dependence of the amplification rate on the vortex core size is rather weak.

As follows from the presented linear analysis in inviscid fluid, the trailing vortices have undergone a growing sinusoidal instability in close vicinity to the ground $\overline{H} < 0.438$ and have come into contact with it. However, in reality the approaching vortices generate the boundary-layer along the ground that separates and causes the primary vortices to follow a complicated trajectory. The next section of the paper is devoted to the nonlinear instability of the trailing vortices in a viscous flow in close vicinity to the ground.

3.2. THE TRAILING VORTEX MOTION IN A VISCOUS FLOW NEAR THE GROUND

To model the viscous mechanisms involved in the vortex-wall interaction, we use a new version of the computational vortex method based on the splitting scheme. This scheme has been obtained rigorously from the initial-boundary value problem for the Navier-Stokes (NS) equation in the context of vortex method [10],[11].

The numerical investigation of the three-dimensional problem were carried out in [12] for sinusoidally deformed vortex at Reynolds number $Re = \Gamma/\nu = 10^7$. The mode of perturbation corresponds to the leading mode predicted by the linear analysis at this height. Assuming that the height of vortex is very small in comparison with the vortex separation, we neglected the influence of the second tip vortex. Figures 12 and 13 [12] show the vortex lines of the flow.

Qualitatively three stages of the perturbation development can be singled out near the ground. The first stage is linear stage when an amplitude of the perturbation grows , however, the perturbed vortex remains sine shaped. The linear analysis presented in the subsec. 3.1 describes this process. At the second stage, the nonlinearity of the vortex behavior becomes stronger, the lowest regions of the vortex approach the ground and are developed stronger than upper ones. The sinusoidal shape of the vortex is violated. The lower regions of the vortex undergo the vortex stretching and the vortex core flattening. The approaching the ground vortex generates the secondary vortex system which causes the primary vortex to follow a complicated trajectory at the third stage of process. Below the lowest regions of the primary vortex the secondary vortex grows more rapidly and strongly. The vortex tongues are formed in the secondary vortex system. At an early stage of the evolution, the new-formed vortex tongue detaches from the ground because of the effect of the primary vortex. The secondary vortex, moving upwards, stops the horizontal motion of the primary vortex.Then the mutual interaction between primary and secondary vortices and the negative curvature-induced velocity on the vortex tongue peak result in moving the vortex tongue again toward the ground and in penetrating of the tongue peak between the upper regions of the primary vortex. As can be seen from Fig.12 and 13 [12], the forming vortex configuration consists of the counter-rotating vortex tubes inclined at non-zero angles. Consequently, there are the conditions for realization of the vortex reconnection. The primary and secondary vortices form the complicated vortex tangle in which the individual vortex tubes are collided, reconnected and annihilated by mutual diffusion of oppositely signed vorticity. The simulation of these processes is possible on the more powerful computer than that used in this work. At the present time, simulations of the 3D vortex motion in a viscous flow near the ground have not been completed yet. Our future works will be devoted to this problem. Interesting results at the small Reynolds numbers are presented in the recent work [13].

4. Conclusions

The complexity of the aerodynamics near the ground originates in many respects from the complicated vortex dynamics in ground effect. The study of the vortex dynamics in the near vortex wake has been extensively performed either experimentally, numerically, or theoretically as a part of the development of wingships. The vortex deformation in the near wake has a strong quantitative and qualitative effect on the aerodynamics, dynamics and stability characteristics of the wingship with a tail unit.

The following possible scenario of the process of the trailing vortex instability in the far vortex wake near the ground, that can be proposed from the results of this investigation, is as follows. The trailing vortex pair produced by a lifting surface undergoes a growing instability caused, for example, by the effects of the lift fluctuations or by flying through atmospheric turbulence. The dominating nature of the instability is inviscid. The corresponding linear stability theory is proposed in this work. In the upper region of the dimensionless height $\overline{H} \geq 0.575$ the growing perturbations are the planar waves, which are confined to fixed planes inclined at different angles to the horizontal. The instability grows until the vortices come into contact and break up into crude rings. At the small heights $\overline{H} \leq 0.438$, the mutual influence of the right and left trailing vortices is weak. The nature of the vortex instability depends mainly on the interaction between the vortex and the ground. The deformed vortex approaches the ground and induces a secondary vortex system. Due to the fact that the location of the perturbed primary vortex is changed with respect to the ground, the generated secondary vortex tubes have a curvature and are advected both by self-induced velocity and the velocity induced by the primary vortex. Due to the self-induction and mutual effects the primary and secondary vortices form the complicated vortex tangle in which the individual vortex tubes are collided, reconnected and annihilated by mutual diffusion of oppositely signed vorticity. As follows from the linear stability analysis, the intermediate zone of $\overline{H}$ ($0.438 \leq \overline{H} \leq 0.575$) is the zone of the helical instability. Most probably that this instability does not exist separately. The lowest regions of the helical vortex will approach the ground and undergo the Crow-like instability in the lower region of $\overline{H} \leq 0.438$.

Acknowledgments

The authors would like to thank Alexander von Humboldt Foundation for support of this work.

References

1. Hooker, S.: A review of current technical knowledge necessary to develop large scale wing-in-surface effect craft, Intersoc. Adv. Mar. Veh. Conf., Arlington, VA,(1989), 367-429.
2. Doligalski,T.L.,Smith,C.R.,and Walker,J.D.A.: Vortex interactions with walls, Annual Review of Fluid Mechanics **26** (1994), 573-616.
3. Zheng,Z.C.,and Ash,R.L.: Study of Aircraft wake Vortex Behavior near the ground, AIAA Journal **34**, No.3 (1996), 580-589.
4. Kornev N.V., and Reichert, G.: Three-Dimensional Instability of a Pair of Trailing Vortices Near the Ground, AIAA Journal **35**, No.9 (1997).
5. Crow,S.C.: Stability Theory for a Pair of Trailing Vortices, AIAA Journal **8**, No.12 (1970), 2172-2179.
6. Treshkov,V.K.: Hydrodynamics of the wing near the ground. Report of the Leningrad Shipbuilding Institute, I-3-A-668,(1971), 1-52 (in Russian).
7. Visel, E.P.: Investigation of the trailing vortices of the wing with a small aspect ratio equipped with an endplate near the ground, Utchenye zapiski TSAGI **2**, No.3 (1971), 12-19 (in Russian).
8. Kornev,N.V.,and Treshkov,V.K.: Numerical Investigation of Nonlinear Unsteady Aerodynamics of the WIG Vehicle, Proceedings of the Intersociety High Performance Marine Vehicle Conference, Arlington, VA, (1992), ws38-ws48.
9. Moltchanov,V.F.: Method of separation of the principal part of nonlinear characteristics of a rectangular wing in ideal fluid flow, Utchenye zapiski TSAGI **11**, No.1 (1980), 12-17 (in Russian).
10. Basin,M., and Kornev,N.V.: Beruecksichtigung der Reibung in der Wirbelmethode, Z. angew. Math. Mech. **77**, No.7 (1997), 1-10.
11. Kornev,N.V., and Basin,M.A.: A Way to split the Navier-Stokes Equations in the Context of Vortex Method, appears in Commun. Numer. Meth. in Eng.
12. Kornev,N.V., and Reichert,G.: Randwirbelzerfall bei Bodeneffektfahrzeugen,Proc. of the German Aerospace Congress **2** (1996), 1043-1052.
13. Chang, T.Y., Hertzberg, J.R., and Kerr, R.M.: Three-dimensional vortex/wall interaction: Entrainment in numerical simulation and experiment, Phys.Fluids **9**, No.1 (1997), 57-65.

MODIFICATIONS OF THE TIP VORTEX STRUCTURE FROM A HOVERING ROTOR USING SPOILERS

JUSTIN W. RUSSELL AND LAKSHMI N. SANKAR
School of Aerospace Engineering
Georgia Institute of Technology, Atlanta, GA 30332-0150

CHEE TUNG
Army Aeroflightdynamics Directorate
NASA Ames Research Center, Moffett Field, CA 94035

Abstract

Numerical studies of the tip vortex structure from a hovering rotor with and without various spoilers are presented. A general multizone unsteady three-dimensional Navier-Stokes solver is developed to determine the flowfield. A scheme that is fifth-order accurate in space and first-order accurate in time is used to improve the capturing of the tip vortex. Velocity data from the core of the vortex is studied at various planes behind the blade trailing edge. Computations of this velocity data for a clean rotor are first compared with experimental results obtained for the same rotor test case. Three different trailing edge spoiler configurations are then investigated to see if the tip vortex structure can be favorably altered.

1. Introduction

With greater emphasis today being placed on noise reduction both for civil and military rotorcraft, studies of the tip vortices produced by rotors become more and more important. This is because a substantial component of the noise generated by a rotor is due to Blade-Vortex Interaction (BVI), which is the effect of the tip vortex created by previously passing blades on a given blade. Preliminary numerical studies [Lee and Smith] have shown that vortices with larger core sizes have a less detrimental effect on the lift of a blade during BVI. Hence, if the tip vortex can be somehow altered so that its core size is substantially increased, then both the BVI noise and the vibratory airloads can be reduced.

Early studies [Tangler] aimed at reducing "blade slap" have shown that passive devices such as a stub/subwing mounted at the rotor tip can be used to improve slap

E. Krause and K. Gersten (eds.), IUTAM Symposium on Dynamics of Slender Vortices, 435-448.

signature with no noticeable degradation in performance. Another rotor has also been tested [Berry and Mineck] with a stub/subwing and a winglet, but a higher torque requirement was found for one test case in hover and therefore these concepts were eliminated from further testing. More recently, devices such as a stub/subwing have been shown to diffuse the tip vortex by as much at 47% with a 9% decrease in drag, and the NASA star tip device has been shown to diffuse the tip vortex by 100% with a 67% increase in drag [Smith and Sigl]. All of these results offer motivation that a passive device, if properly designed, can be used produce significant increases in vortex core size (thus reducing BVI), and in some cases with minimal or no loss in rotor performance.

To properly capture the dynamics of the rotor near-wake, especially those in the vortex core, an accurate solution method is needed. This is particularly true for rotor flows, where the structure of the tip vortex greatly influences the loading on the rotor blade and vice-versa. Higher-order spatial accuracy schemes, such as the one used in the present analysis, have been shown to conserve the vortex structure over large distances [Wake and Choi]. This is necessary if the rotor wake is to properly develop and yield accurate performance predictions. Additionally, this higher-order methodology is needed in order to simulate more complex vortical flows that evolve when passive tip devices are added to simple rotor geometries. It is extremely important to accurately model these vortical flows to be sure that it is the tip device that is altering the vortex core and not the numerics of the scheme.

The present work addresses the aspect of tip vortex modeling (i.e., the details of the velocity field within the core), as well as the study of tip vortex alteration by passive devices. A recent survey [Yu] offers insight into the types of tip alterations that have been tested to reduce BVI noise. However, he notes that only acoustic measurements are taken in these experiments with the tip devices. That is, the physics of the flowfield due to these devices is not well understood. In the case of rotors, the problem is made more complex by the fact that the vortex structure significantly influences the blade loading and vice versa. The present calculations for a clean rotor are compared with experiments [McAlister et al.] to show the effectiveness of the numerical scheme in capturing the dynamics and size of the vortex core. With this established, passive tip devices such as spoilers are then investigated to show their benefits and/or detriments to the rotor and their effects on near-wake characteristics.

1.1. A MATHEMATICAL CRITERION FOR VORTEX STABILITY

Theoretical analyses [Leibovich and Stewartson, Stewartson and Leibovich] have shown that diffusion of the tip vortex core, in general, tends to destabilize the tip vortex. A thorough investigation into these theoretical studies has been done [Russell et al., AHS Forum]. For brevity, only the result is given here. It has been shown that a sufficient condition for vortical flow to be unstable is:

$$|V_Z|\frac{d\Omega}{dr}\left[\frac{d\Omega}{dr}\frac{d\Gamma}{dr}+\left(\frac{dV_x}{dr}\right)^2\right]<0 \tag{1}$$

where $|V_z(r)|$ is the azimuthal (normal) velocity component, $V_x(r)$ is the axial velocity component, r is the radial distance from the vortex axis, $\Omega = |V_z(r)|/r$ is the angular velocity, and $\Gamma = r|V_z(r)|$ is the circulation.

The result of this stability criterion is that if the core of the tip vortex can be diffused (i.e., make $d|V_z(r)|/dr$ small where r is small), this results in an unstable vortex. Hence, the goal behind the use of the trailing-edge spoiler is to decrease the slope of the azimuthal (normal) velocity in the vortex core, especially near the center of the vortex. It will be shown that the higher-order numerical scheme used in the present analysis can capture this variation in slope resulting from the addition of three different trailing-edge spoilers.

2. Mathematical Formulation

The mathematical and numerical formulation behind the present approach has been extensively documented [Hariharan]. For brevity, only the general characteristics of the formulation are described here.

This numerical scheme solves the three dimensional, unsteady, compressible Navier-Stokes equations. The inviscid and viscous flux terms are computed using a cell-vertex finite volume formulation. The inviscid fluxes at the cell faces are computed using Roe's approximate Riemann solver [Roe]. This solver requires flow information on the left and right sides of a cell face for each coordinate direction. In this work, this information is obtained using a fifth-order essentially-non-oscillatory (ENO) scheme [Chakravarthy, Harten and Chakravarthy].

The solution is advanced in time using an implicit three-factor diagonal alternating-direction-implicit (ADI) scheme [Pulliam and Chaussee]. This makes the procedure first order accurate in time. In addition, implicit fourth-order artificial dissipation using a spectral radius scaling factor is used to improve the temporal stability characteristics of the scheme.

3. Numerical Modeling

For all of the simulations described below, the full Navier-Stokes equations are solved in a time-accurate fashion using the algebraic Baldwin-Lomax turbulence model. The complexity of the configuration requires that the flowfield be divided into zones or blocks, as will be discussed later. Data is passed between zonal interfaces to ensure fifth-order spatial accuracy throughout the interior of the computational domain. Note, however, that the scheme drops to first-order accurate near solid surfaces and farfield boundaries. Only one blade of the multi-bladed rotor is solved in hover, with a periodic boundary condition used at the azimuthal boundaries.

3.1. BOUNDARY CONDITIONS

All of the farfield boundaries are approximated by a first-order extrapolation. Two-point averages of the flow properties are used at the zonal interfaces. As previously stated, a simple periodic condition is also used at the azimuthal boundaries simulating the existence of other blades. A no-slip boundary condition is used at all solid surfaces. At these surfaces, density is extrapolated to second-order and pressure to first-order. For the blunt ends of the blade, special zones called "cap" grids are used as shown in Figure 1. These cap grids contain singular faces at the leading edge and trailing edge of their airfoil shape. Here, two-point averages are used just as is done at the leading and trailing edge of the blade surface (due to the H-grid topology).

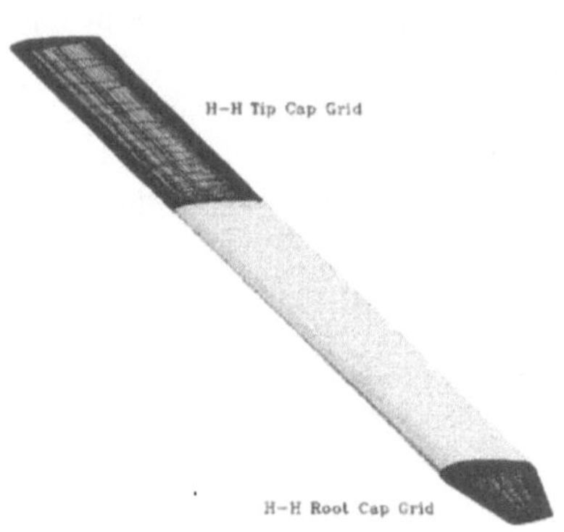

Figure 1: Airfoil-Shaped H-H Cap Grids.

4. Configurations Considered

4.1. CLEAN ROTOR

The reference rotor is a model two-bladed rotor which has been experimentally investigated [McAlister et al.]. The grid used for the model consists of 1,210,330 grid points and has an H-H-O topology. The grid is divided into 6 zones with the first two zones forming the H-H-O topology and encompassing a majority of the computational domain. These two zones allow the root and tip airfoil geometries to extend to the farfield boundaries which minimizes grid "kinks" (a source of numerical error) in the radial direction. In most three dimensional simulations for rotors in both hover and forward flight, the grid in the rotor root and tip regions is simply "pinched off", which physically represents a wedge-shaped end to the rotor root and tip. However, in this simulation we seek to more accurately model the rotor tip geometry and capture the tip vortex. Hence, the rotor root and tip regions are "capped" with two-zone H-H grids that are in the shape of airfoils of the region from which they extend [see Figure 1]. Note that each zonal interface matches grid point for grid point so as to avoid the use of an interpolation scheme.

4.2. ROTOR WITH TRAILING EDGE SPOILER

In order to make a just comparison between the clean rotor results and the results for the rotor with a spoiler, the exact same grid is used in the spoiler simulations. The only difference between the two grids is that the upper and lower H-H-O zones are split vertically at the trailing edge. This yields a total of 8 zones. A schematic diagram of this configuration is shown in Figure 2 below.

To model the three spoilers, a solid surface boundary condition is applied at the trailing-edge zonal interfaces over a range of specified grid points. This yields a spoiler model which is grid-aligned (nearly parallel to the axis of rotation) and of zero thickness. A graphical representation of the trailing-edge spoiler is given in Figure 3 below.

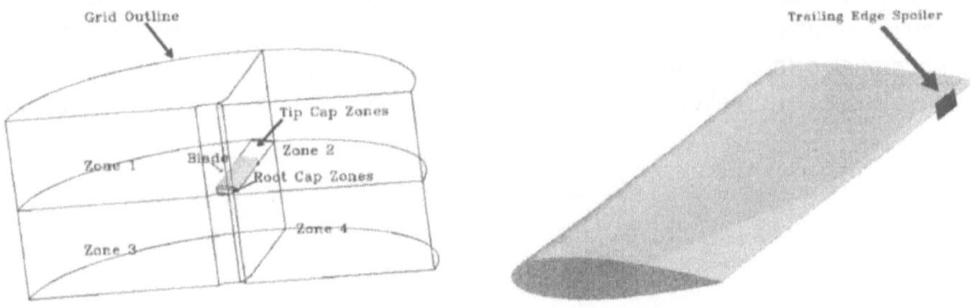

Figure 2: Schematic Diagram of Grid Configuration for Simulations of the Rotor with a Trailing-Edge Spoiler.

Figure 3: Graphical Representation of a Trailing-Edge Spoiler.

Three different spoiler sizes are tested, with the dimensions and locations as shown below in Table 1.

Table 1. Dimensions for the Three Spoilers

Spoiler Number	Height (%chord)	Width (% radius)	Location (% radius)
1	3.9	5.8	87.5 - 93.3
2	5.0	8.8	85.6 - 94.4
3	8.4	13.3	83.5 - 96.8

These dimensions are approximate since the spoiler is actually grid-aligned. The width corresponds to the spanwise dimension and the height is that dimension normal to the flow direction both on the upper and the lower side of the rotor.

5. Results and Discussion

5.1. CLEAN ROTOR

Results for the clean rotor are first presented to validate the numerical procedure for predicting the characteristics of the vortex core. Results are presented at two instances in time, with the solution starting from a zero-flow initial condition. The measurement plane is the same as that in the experiment. The reader should think of the x-component as the chordwise or axial component (positive towards the blade), the y-component as the radial component (positive inboard), and the z-component as the normal component (positive upward).

Figure 4 below compares the three velocity components in the core of the tip vortex with the experimental data. At this point in time, the blade has traveled only one half of a revolution. The usefulness of this figure is to show that the general characteristics of the tip vortex in the x and z directions can be captured with little computational effort. However, the peak of the y-component (not shown) is not captured at all. This makes sense, since the inboard component is greatly dependent on the wake contraction and the effects of the vortex wake from below. In contrast, the x-component is strongly dependent on the wake viscous effects, and the z-component largely dependent on the lift. Notice that the z-component is overpredicted at the peaks. This is due to the fact that the inflow has yet to develop, which leads to higher lift and consequently higher shed circulation. However, also notice that the thickness of the core and the velocity slope are captured well, which are important for determining stability characteristics.

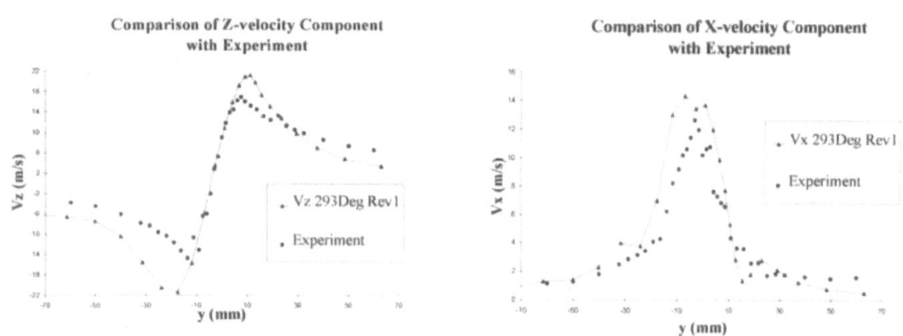

Figure 4: Comparison of Core Velocity Component Magnitudes with Experiment for a Clean Rotor; 3 Chordlengths Downstream of the Blade Trailing Edge after One Half of a Revolution.

For visualization purposes, Figure 5 on the following page gives a particle trace in a blade-fixed frame of reference after one half of a revolution. In addition, Figure 6 (see next page) shows the velocity vectors in the measurement plane. In these figures, the tip vortex has a well-behaved and expected form.

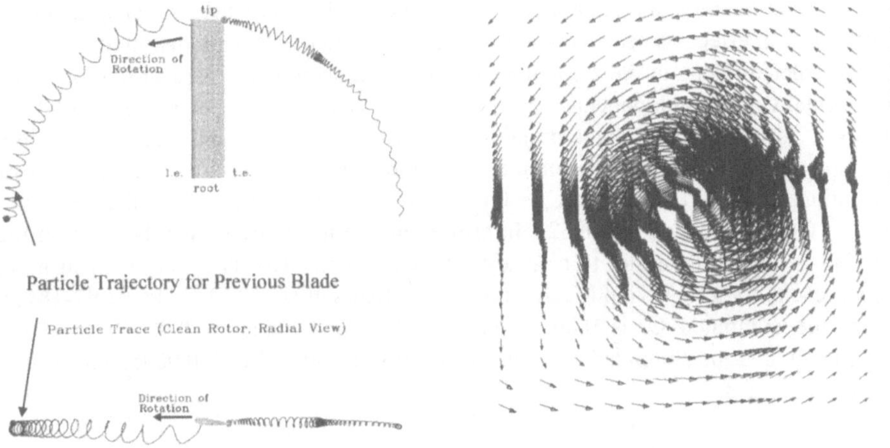

Figure 5: Particle Trace (Evolution of the Tip Vortex) after One Half of a Revolution.

Figure 6: Velocity Vectors Showing the Tip Vortex after One Half of a Revolution; 3 Chordlengths Downstream of the Blade Trailing Edge.

It was expected that simply running the simulation further would improve the results. This was not exactly the case. When beginning from a zero-flow condition, the blade initially sheds a starting tip vortex in the plane of rotation. Hence, once the two-bladed rotor travels one half of a revolution, the second blade hits this starting vortex which causes a large decrease in the lift of the rotor. Consequently, the shed tip vortex becomes weaker, and the predictions worsen. Eventually the inflow reestablishes as the vortices shed by the previous blade are pushed below the blade and the lift recovers. In addition, the wake begins to contract (though rather slowly).

Velocities through the vortex core are presented after one and a half revolutions in Figure 7. The magnitude of the peaks for the x and z-components of velocity are predicted well. The y-component (not shown) is still well underpredicted, however

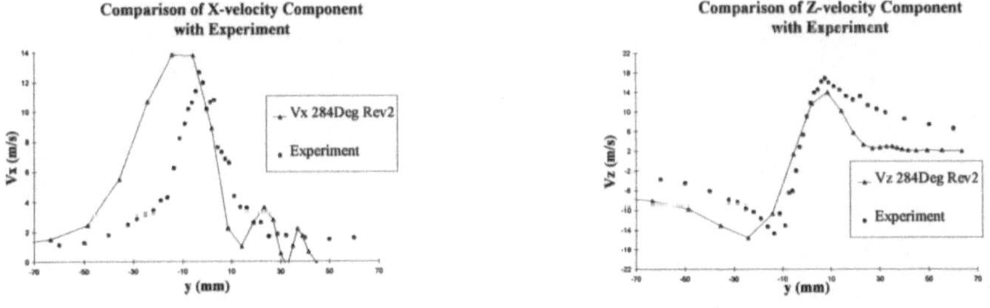

Figure 7: Comparison of Core Velocities 3 Chordlengths Downstream of the Blade Trailing Edge after 1.5 Revolutions.

some inboard velocity is beginning to form, which is encouraging. Also notice that the vortex thickness is overpredicted. This is due to the fact that the vortex has moved inboard (due to the contraction effects of the vortex below) where fewer grid points exist. This, in effect, "smears" the vortex causing a fatter x-velocity distribution and an underpredicted slope in the z-component velocity distribution. The fact that the axial and normal velocity components are captured well within one half of a rotor revolution, however, allows us to study the destabilizing effects on the tip vortex due to a trailing edge spoiler. Finally, the computed thrust and torque coefficients after one and one half revolutions are $C_T = 0.0045$ and $C_Q = 0.00037$ which are lower than McAlister's values of $C_T = 0.0051$ and $C_Q = 0.00052$. Improvements to these results will be made if the solution is advanced further to a steady-state condition. This is clearly seen in more recent results for a higher tip Mach number case [Russell et al., AIAA Paper 97-1845], where the computed C_T is 0.00501 compared with a value of 0.00500 from the experiment, and the computed C_Q value is 0.000531 compared with 0.000500 measured in the experiment.

5.2. SPOILER RESULTS

As previously stated, the most important components of velocity through the core of the vortex (as far as stability of the vortex is concerned) are the axial (x) and normal or azimuthal (z) components. Hence, since the goal is to destabilize and diffuse the tip vortex, we will focus only on these components in the vortex core. Figure 8 below and on the next page shows comparisons of the velocity components for the three spoiler configurations with the clean rotor solutions after one half of a revolution. It is

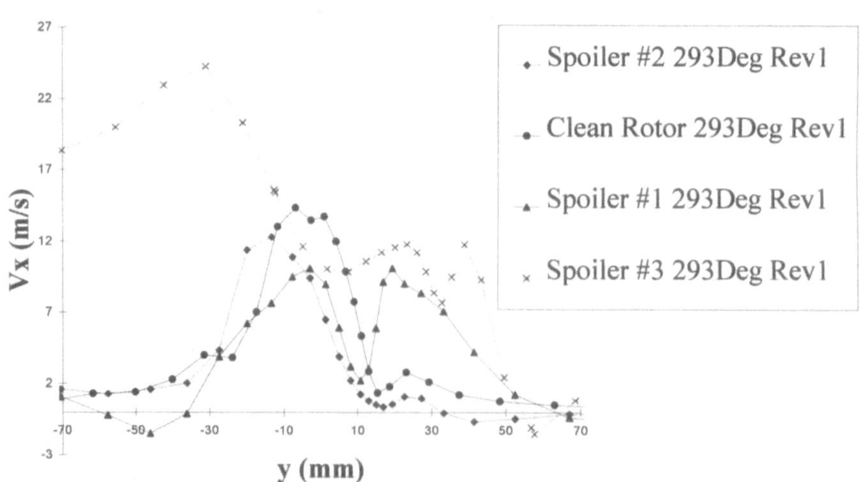

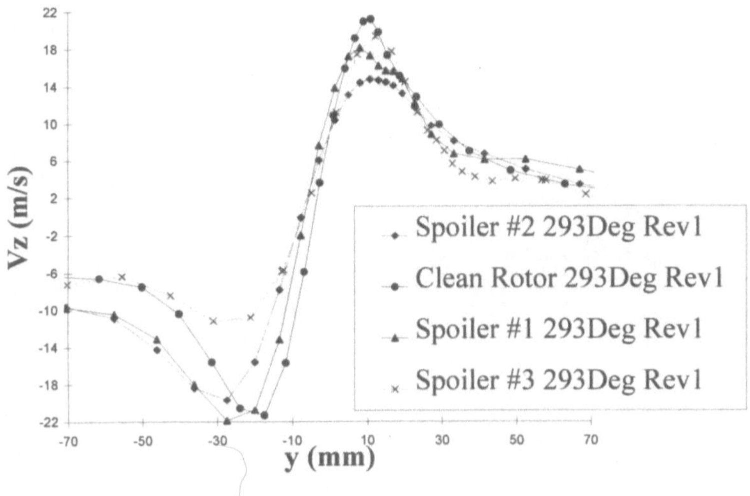

Figure 8: Comparison of the Core Velocity Distributions 3 Chordlengths Downstream of the Blade Trailing Edge.

evident from the normal velocity distribution that the presence of the spoiler does indeed decrease the slope of the normal velocity distribution in the vortex core. Furthermore, this slope decreases in the core region as the spoiler size increases. Hence, with these results, it is seen that a larger spoiler (as much as excess power would allow) is more beneficial in destabilizing the vortex. Evidence of this destabilization is seen in Figure 9, where the velocity vectors show that the overall organization of the tip vortex degrades downstream for the large spoiler. Again referring to the axial velocity distribution in Figure 8, it is evident that dV_x/dr remains relatively large for all spoiler configurations, which is a requirement for vortex destabilization. In addition, the axial

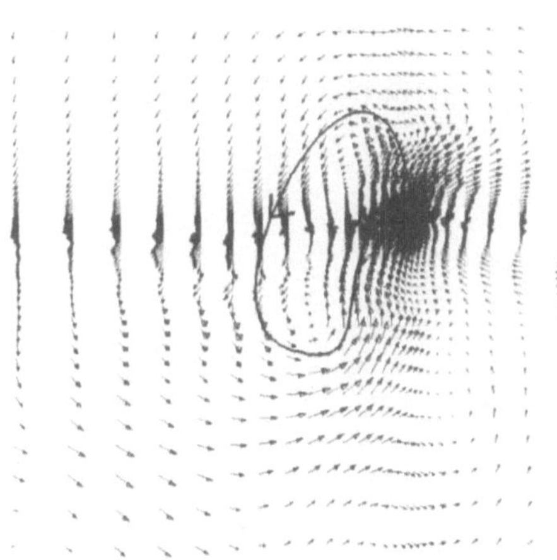

Figure 9: Velocity Vectors in the Measurement Plane for the Large Spoiler (#3).

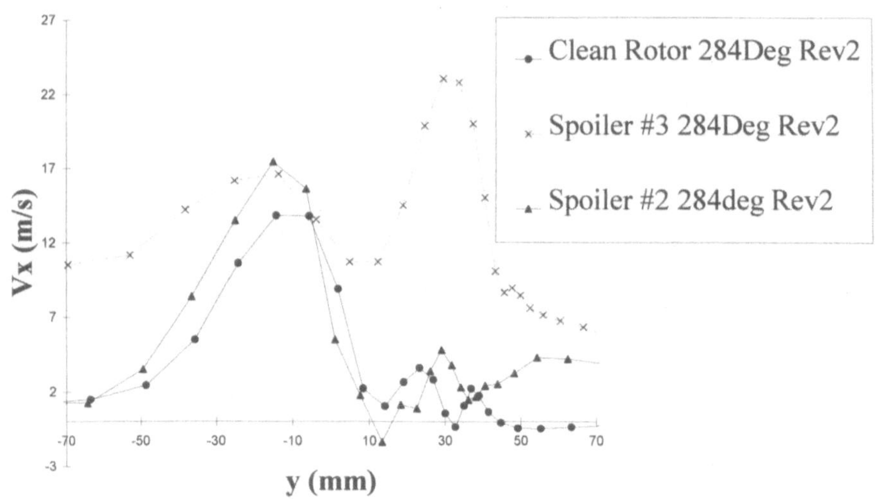

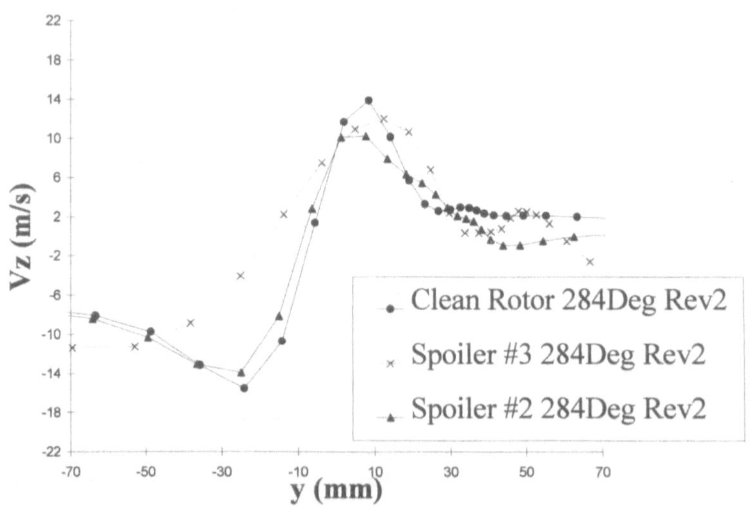

Figure 10: Comparison of Velocity Distributions for the Rotor with and without a Trailing-Edge spoilers 3 Chordlengths Downstream of the Blade Trailing Edge after One and a Half Revolutions.

velocity distribution for the spoiler cases shows that more air is dragged due to the presence of the spoilers, resulting in a significant torque penalty.

The velocity distributions for the rotor/spoiler configurations after one and a half revolutions are shown on the previous page in Figure 10. For comparison purposes, the clean rotor results are also plotted. Notice that the larger the spoiler the less the normal or azimuthal velocity changes in the core, as was seen in the first half revolution. Here, however, the spoiler shows an even more pronounced effect on the slope of the normal velocity through the core. Again, according to equation (1), this has a destabilizing effect on the vortex, which is what we are seeking. Note also that the axial velocity maintains a relatively large slope near the vortex core, with the larger spoiler having a greater "smearing" effect on the distribution.

Overall, the effects of the different spoilers can be seen below in Figure 11 which shows the lift coefficient distribution variations for the rotor with the spoilers. Again, the clean rotor results are plotted for comparison. As expected, the presence of the spoiler causes a decrease in lift in the vicinity of its location. Since lift is proportional to the bound circulation on the rotor blade, conservation of vorticity requires that trailing vorticity must be shed due to the radial change in lift. At the inboard station where the lift first drops due to the presence of the spoiler, it is expected that a significant amount of vorticity will be shed forming a vortex that rotates in the same direction as the tip vortex. Also, outboard of the spoiler where the blade experiences a sharp increase in lift, a counter-rotating vortex is expected to be shed. A schematic diagram of this is shown below in Figure 12.

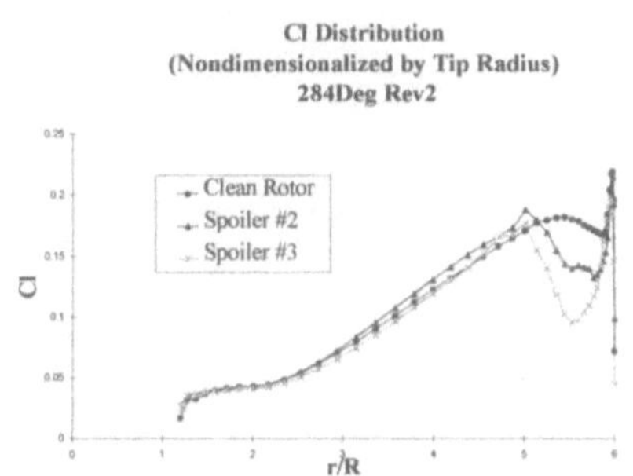

Figure 11: Variation in C_l Distribution for the Rotor Blade with and without Various Spoiler Configurations.

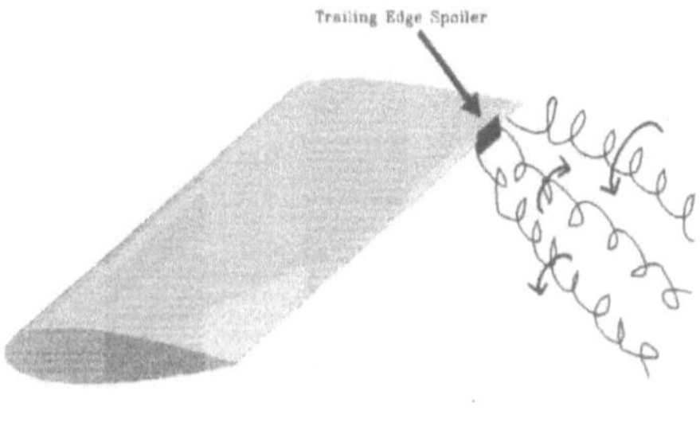

Figure 12: Schematic Diagram of Expected Vortex Shedding Due to the Presence of the spoiler.

These effects do indeed occur and can be seen on the following page in Figure 13, which shows the velocity vectors in a radial plane approximately 0.5 chordlengths behind the blade trailing edge. Further downstream, these vortices tend to interact and diffuse the tip vortex to various degrees. The overall effect of this tip vortex alteration is clearly seen in the particle traces shown in Figure 14 on the next page for both spoiler configurations. Comparing these traces with that shown in Figure 5 for the clean rotor,

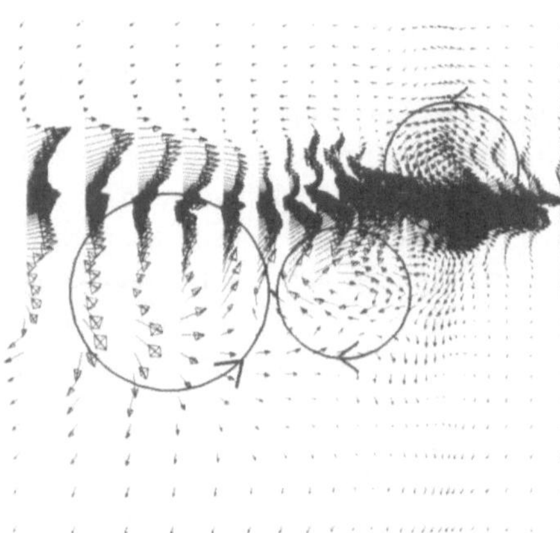

Figure 13: Velocity Vectors in the Near Wake of the Rotor with Spoiler #3 Showing Counter and Co-Rotating Vortex Formations.

it is obvious the destabilizing effect the spoilers have on the tip vortex. From these figures, it is seen that increasing the spoiler size has the effect of "unwinding" the tip vortex. This "unwinding" or diffusion of the tip vortex is what is desired in order to improve the BVI characteristics of the blade.

Finally, the computed thrust and torque coefficients for the two largest spoilers are C_T = 0.0044 and C_Q = 0.00058 (spoiler #2), and C_T = 0.0040 and C_Q = 0.00119 (spoiler #3). As expected, the trends show that the thrust decreases and the torque increases as the spoiler size increases when compared with the values calculated for the clean rotor.

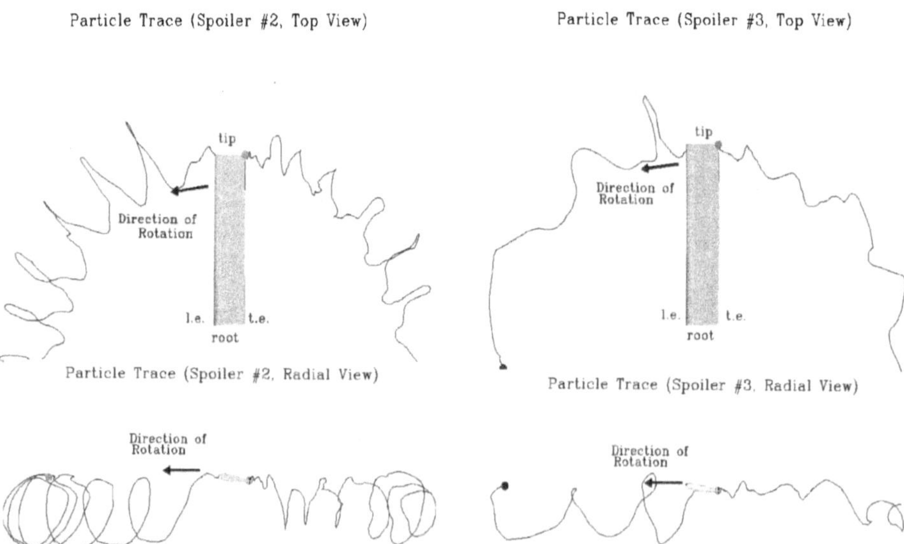

Figure 14: Particle Traces Around the Azimuth for the Rotor with Various Trailing-Edge Spoilers

6. Conclusions

A high-order spatial accuracy scheme is capable of capturing the rotor tip vortex over large distances, provided enough grid points remain distributed through the vortex core. However, many revolutions are potentially needed to allow the rotor wake to develop and to accurately capture the rotor tip vortex with respect to all three coordinate velocity distributions (especially the inboard direction), as well as to yield accurate performance predictions.

It has been shown that reasonable qualitative and quantitative results about the vortex core in the normal and axial directions can be captured within one half of a rotor revolution. This can lead to less computational effort if only general effects on the vortex due to some rotor modification (i.e., implementation of a passive tip device) are desired. With this in mind, a trailing edge spoiler has been shown to yield a beneficial effect with regards to destabilization and/or diffusion of the rotor tip vortex in hover, which leads to better BVI characteristics for the blade. The larger the spoiler, the more the diffusion/destabilization there is to the tip vortex.

Further studies are needed to determine an "ideal" spoiler for a given rotor configuration and flight condition (i.e., the spoiler yielding the most destabilizing effects for the least amount of performance penalty), both in terms of spoiler size and location. Finally, convergence to a steady state is necessary to yield accurate performance predictions for both the clean rotor and the trailing-edge spoiler configurations.

Acknowledgements

The first two authors acknowledge the support of the National Rotorcraft Technology Center (NRTC) for this project, as part of the Georgia Tech Rotorcraft Center of Excellence.

References

Lee, D.J. and Smith, C.A., "Effect of Core Distortion on Blade-Vortex Interaction," *AIAA Journal*, Vol. 29, (9), September 1991.

Tangler, J.L., "Experimental Investigation of the Subwing Tip and Its Vortex Structure," NASA CR 3058, 1978.

Berry, J.D., and Mineck, R.E., "Wind Tunnel Test of An Articulated Helicopter Rotor Model with Several Tip Shapes," NASA TM 80080, December 1980.

Smith, D.E. and Sigl, D., "Helicopter Rotor Tip Shapes for Reduced Blade Vortex Interaction An Experimental Investigation," AIAA Paper 95-0192.

Wake, B.E., and Choi, D., "Investigation of high-order upwinded differencing for vortex convection," *AIAA Journal*, Vol. 34, February 1996.

Yu, Y.H., "Rotor Blade-Vortex Interaction Noise: Generating Mechanisms and its Control Concepts," American Helicopter Northeast Region Aeromechanics Specialist Meeting, Bridgeport, CT, October 1995.

McAlister, K.W., Schuler, C.A., Branum, L., and Wu, J.C., "3-D Wake Measurements Near a Hovering Rotor for Determining Profile and Induced Drag," NASA Technical Paper 3577, August 1995.

Leibovich, S. and Stewartson, K., "A sufficient condition for the stability of columnar vortices," *Journal of Fluid Mechanics*, Vol. 126, July 1982.

Stewartson, K. and Leibovich, S., "On the stability of a columnar vortex to disturbances with large azimuthal wavenumber: the lower neutral points," *Journal of Fluid Mechanics*, Vol. 178, August 1986.

Russell, J.W., Sankar, L.N., Tung, C., and Patterson, M.T., "Alterations of the Tip Vortex Structure from a Hovering Rotor using Passive Tip Devices," American Helicopter Society 53rd Annual Forum, Virginia Beach, Virginia, April 29 - May 1, 1997.

Hariharan, N., "High Order Simulation of Unsteady Compressible Flows Over Interacting Bodies with Overset Grids," Ph.D. Thesis, Georgia Institute of Technology, Atlanta, GA, August 1995.

Roe, P.L., "Approximate Riemann Solvers, Parametric Vectors, and Difference Schemes," *Journal of Computational Physics*, Vol. 39, 1981.

Chakravarthy, C.R., "Some Aspects of Essentially Nonoscillatory (ENO) Formulations for the Euler Equations," NASA CR 4285, May 1990.

Harten, A. and Chakravarthy, C.R., "Multi-Dimensional ENO Schemes for General Geometries," NASA CR 187637, September 1991.

Pulliam, T. H. and Chaussee, D.S., "A Diagonal Form of an Implicit Approximation -Factorization Algorithm," *Journal of Computational Physics*, Vol. 39, 1981.

Russell, J.W., Sankar, L.N., and Tung, C., "High Accuracy Studies of the Tip Vortex Structure from a Hovering Rotor," AIAA Paper 97-1845

MEASUREMENTS OF ROTOR TIP VORTICES USING LASER DOPPLER VELOCIMETRY

J. GORDON LEISHMAN
Department of Aerospace Engineering,
Glenn L. Martin Institute of Technology,
University of Maryland, College Park, Maryland, USA.

1 Introduction

A helicopter rotor wake is dominated by strong vortices that form at each blade tip, e.g. Refs. [1]-[8]. When the helicopter is descending or maneuvering, the blades may closely intersect the tip vortices resulting in a phenomenon known as blade vortex interaction (BVI). BVI produces significant three-dimensional unsteady airloads, which manifest as high rotor vibrations and impulsive noise, e.g. Ref. [6]. The reduction of BVI induced rotor noise and vibrations has become an important goal, and it is currently an area of active research on both theoretical and experimental fronts.

The complete understanding and prediction of the aeroacoustics of BVI stems, in part, from the ability to fully understand the fundamental evolution, convection, diffusion and dissipation mechanisms of the blade tip vortices. Unfortunately, the vortical flows generated by helicopter rotors are very difficult to measure accurately. Some of the key issues limiting such measurements are discussed in Ref. [7]. Yet good quality measurements are essential because all existing computational models of rotor wakes contain empirically measured parameters, e.g. see Ref. [8] and references therein. This may include a description of the velocity field close to the tip vortex core, as well as the viscous diffusion characteristics. Without proper representation of these parameters and how they are affected by rotor operational and geometric factors, the quantitative predictive capabilities of computational models can be severely limited. This is especially important for BVI, which is sensitive to small changes in the tip vortex structure and positions relative to the rotor blades.

Several detailed experiments on tip vortices have been conducted with fixed (non-rotating) wings, e.g. Refs. [9]-[13]. For rotor, the situation is considerably more complicated because of the time-varying (but nominally periodic) flow, the curved nature of the vortices, the close proximity of the tip vortices to each other, and the effects of following blades. The problem is further dependent on geometric parameters such as blade twist and tip shape, as well as operational parameters such as rotor thrust, climb or descent velocity, and advance ratio. Measurements on rotor tip vortices were first performed using hot-wire anemometry, e.g., Refs. [14]-[17]. However, hot-wire probes are limited by their

E. Krause and K. Gersten (eds.), IUTAM Symposium on Dynamics of Slender Vortices, 449-458.

spatial resolution, and proximity concerns make measurements difficult near the rotating blades. Laser Doppler velocimetry (LDV) can alleviate many limitations posed by hot-wire probes, although optical access, seeding issues [7, 18] and the need to ensure the periodicity of the flow [19] are significant operational constraints. Early rotor measurements using one-dimensional LDV systems are reported in Refs. [20]-[23]. Other work with two-component LDV systems is described in Refs. [24]-[28]. More recent rotor studies with three-component LDV include the work of Seelhorst *et al.* [30, 31], McAlister [32, 33] and Leishman *et al.*[34]-[36], which have all begun to detail the nature of the tip vortex flow.

2 Description of the Experiment

Both one and two-bladed teetering-type rotors were used in the present experiments. The rotors were driven by a three-phase constant torque electric motor. The rotor blades had a rectangular planform with a radius of 406 mm and chord of 42.5 mm, with a root cut-out of 20% rotor radius. The airfoil profile was a NACA 2415. The blades were untwisted to keep the tip region well loaded so as to produce a relatively strong tip vortex. The one-bladed rotor was operated at 2,100 rpm, giving a tip speed of 89 ms^{-1}. A collective pitch of 4° was used. The two-bladed rotor was operated at a condition to give the same nominal blade loading and average tip vortex strength as for the one-bladed rotor, this being a tip speed of 85 ms^{-1} with a collective pitch of 5°. For both rotors, the nominal tip Mach number and chord Reynolds number were approximately 0.28 and 2.7×10^5, respectively.

A three-component fiber-optic based LDV system was used. The argon-ion laser beam was projected through a 2 mm aperture, and then to a beam separator containing a dispersion prism. This created a matrix of six beams, with colored pairs of green, blue and violet. One beam of each color was shifted in frequency by using a Bragg cell. The beams were then focused by couplers into single-mode polarization-preserving fiber optic lines and carried to the optics. The transmitting and receiving optics were contained in a pair of fiber optic probes – see Fig. 1. One probe was used for the green and blue beams, and the other for the violet beams. The beams were collimated when they emerged from the fiber optic lines, passed through the transmitting optics and beam expanders, and focused down to a point to form the measurement volume. The beam expanders enhanced the spatial resolution of the measurements and improved the signal-to-noise ratio. Optional lenses of focal length 750 mm or 2100 mm were attached to the beam expanders. Both fiber-optic probes were moved by a three-axis linear traverse.

The LDV measurement volume was defined by three pairs of intersecting ellipsoids (two from the green and blue pairs, and the other from the violet pair) approximately 73 μm in diameter and 0.85 mm long (based on the Gaussian $1/e^2$ points at $\lambda = 488.0\ nm$). Scattered light was collected by the receiving lens and coupled to the receiving fiber. From there, the light signal was carried via fiber-optic cables to a receiver, separated by color, and converted into analog signals by a set of photomultipliers. A three-channel digital processor provided filtering, amplification and digitization. A real-time autocorrelator provided the Doppler frequency, which was converted into a velocity measurement for each channel. The velocity components were then converted into an orthogonal axis system based on the beam crossing angles.

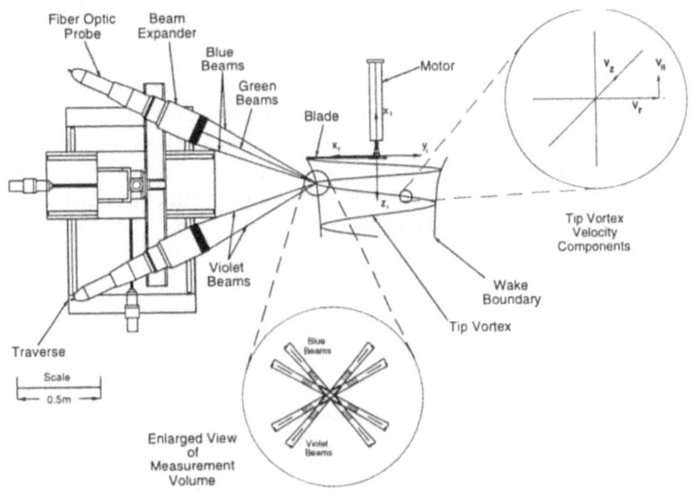

Figure 1. LDV system and rotor, also showing traverse and vortex coordinate systems

The flow field was seeded with atomized olive oil, which provided a calibrated average particle size of 0.6 – 0.8 μm. The seed was directed from the atomizer and introduced into the flow through a series of low velocity nozzles about three rotor radii upstream and to one side of the rotor. Precautions were taken to ensure that the seeder did not inject any artificial velocity or turbulence into the rotor wake.

All measurements were made in coincidence mode, which means that all three LDV processors had to recognize a valid data point within a pre-set coincidence window before accepting the data. This ensured correlated velocity components and that each component of velocity was measured to the same statistical accuracy. For the phase-resolved measurements, an integral shaft encoder provided phase-locked synchronization for the LDV system through a rotating machinery resolver (RMR). The RMR interface allowed the inclusion of phase data synchronous with velocity data from the signal processor. After processing, the data were sorted into 400 discrete azimuth bins (phase resolution of 0.9 degrees). Only data within a short acceptance window were considered valid, so data falling into these bins were heavily biased toward the leading edge of the bin pulse.

Because of the axi-symmetric nature of a rotor wake in hover, tip vortex velocity field data at different wake ages could be measured over one-dimensional radial grids located parallel to the TPP and at different axial planes below the rotor. The tip vortices were convected along the wake boundary and passed through these successive grids. Estimates of the wake boundary from flow visualization were used to determine the initial measurement coordinates, which were centered about the vortex axis at each axial plane. Typically, a traverse grid with a 2 mm spacing was used outside the vortex core region, but this was reduced to a 0.2 mm spacing near and inside the core.

A first set of experiments were run to acquire time-averaged measurements over several complete rotor revolutions at the specified grid points. Then phase-resolved measurements were made, with up to 40,000 coincident data samples per-channel per-grid point being specified. After the vortex age (bin number) was selected when the vortex axis coincided with the measurement grid, a second test was conducted to allow more detailed measurements of the tip vortex structure. In this second test, the measurement time was 'windowed' in an azimuthal sense to ± 25 bins relative to the estimated vortex/grid crossing azimuth. This gave a much larger number of data samples in each bin, significantly improving the statistical accuracy of the measurements.

3 Results

3.1 Tip Vortex Tangential Velocities

The tangential flow field velocities in the tip vortices were obtained by subtracting the time-averaged velocities at the same grid points from the phase-resolved (temporal) measurements [34]-[36]. These residual velocities can be considered the components induced by the tip vortex alone. Fig. 2 shows the residual tangential velocities measured at different wake ages (ζ) for the one-bladed rotor. The one-bladed rotor allowed the tip vortex to be studied for almost one complete rotor revolution prior to the first blade passage and, therefore, without significant mutual interference from other blades and/or tip vortices. This helped provide a baseline for measurements with more blades. In the case of the two-bladed rotor, the tip vortex generated by the previous blade lies in closer proximity to the following blade, somewhat complicating the problem.

The results shown in Fig. 2 are similar to a 'fixed-wing' tip vortex flow, but with some differences. One effect is the velocity peaks on each side of the vortex axis are different; that is, the velocity profile is asymmetric. Furthermore, a non-zero convection velocity is found at the vortex axis (even after the time-average in removed) compared to a corresponding two-dimensional vortex. These effects result from the self-induced velocities produced by the curvature of the tip vortex [34], essentially similar to that obtained with a vortex ring.

It will be seen from Fig. 2 that the largest peak tangential velocities in the tip vortex were obtained at the early wake ages, these being approximately one-third of the rotor tip speed. Results measured at $\zeta = 9.7°$ showed slightly lower peak velocities than at $\zeta = 16.9°$. This is because the measurements are being made closer to the origin of finite length vortex filament, which is still in some state of evolution. At the earliest wake ages the relatively high swirl velocities cause the seed particles to centrifuge out of the core region, making it difficult to obtain a sufficient number of statistically valid samples. In the present work, any data points that held fewer than 25 samples were discarded. As the vortex ages, the seed particles are more easily entrained into the core and correspondingly larger numbers of samples were obtained in the rotational core.

Fig. 2 shows that the tangential velocities quickly diminished in magnitude as the vortex aged. This is a consequence of viscous diffusion. The velocity peaks exist on either side of the viscous core, the radial distance between the peak velocities being defined as the nominal core diameter. Therefore, it can readily be deduced that the viscous core dimensions enlarged with increasing wake age. The exception to this diffusion trend was noted for the first blade passage event, which occured at $\zeta = 360°$ for the one-bladed rotor. Here, it was found that the diffusion trend momentarily reversed, with an increase in peak tangential

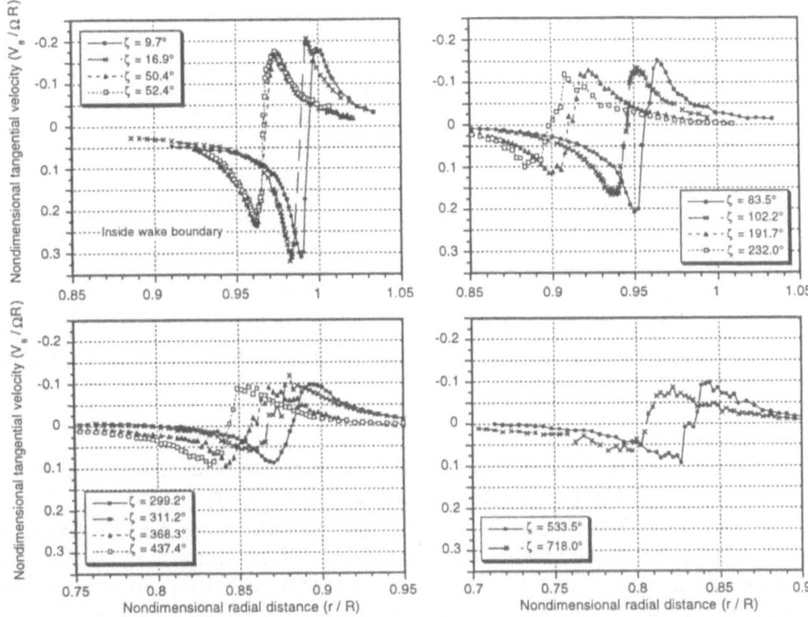

Figure 2. Tangential velocity component in the tip vortex generated by the one-bladed rotor for various wake ages

velocity and a decrease in core size just after the blade passage. This effect, which was first noted in Ref. [34], is most likely a result of vortex stretching resulting from the wake axial (slipstream) velocity gradients produced at the blade passage. Because the preponderance of tip vortex filaments generated by helicopter rotors remain very close to the rotor plane, the consequences of this blade passage effect on the diffusive characteristics of the tip vortex is potentially very important.

3.2 Vortex Core Growth

The vortex core dimensions are, on average, inversely related to the peak swirl velocities in the tip vortex. Therefore, the core size is an indicator as to the intensity of a blade vortex interaction event. The core size also directly affects the intensity of vortex-vortex interactions and the overall rollup of the rotor far wake. The average viscous core radius for tip vortices generated by helicopter rotors has been previously estimated to be of the order of the blade thickness, e.g. [15, 22, 39], which is typically 5-15% of mean blade chord. However, the subsequent growth in the viscous core is relatively undocumented. Even for fixed-wings, the growth trends are hard to measure because wandering effects bias the measurements [37] and can manifest as false diffusion trends. For rotors, aperiodicity effects introduce a similar velocity bias [19], but a further complicating factor is that tip vortices generated by each individual blade strongly

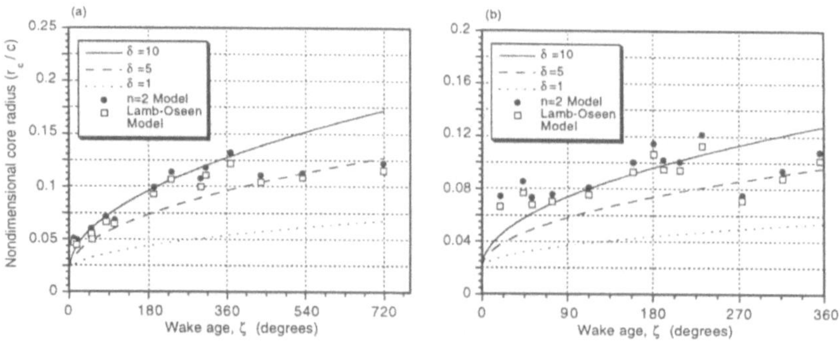

Figure 3. Growth in the viscous core radius as a function of wake age for (a) the one-bladed rotor, and (a) the two-bladed rotor

interact with one another and can exhibit pairing effects. This distorts both the spatial locations of the tip vortices, as well as their velocity profiles.

Figs. 3(a) and 3(b) show the viscous core growth trends for the one and two-bladed rotors, respectively. These estimates of core radius, r_c, with wake age were based on least-squares fits to the tangential velocity profiles using the Vatistas $n = 2$ model [38, 39] and the Lamb-Oseen model. Two data points are, therefore, shown for each wake age. For the one-bladed rotor the viscous core was found to grow rapidly from 4% chord to 12% chord during the initial stages, with the trend reversing at the first blade passage. For the two-bladed rotor, the viscous core was initially much larger but grew less rapidly to about 10% of blade chord before the first blade passage ($\zeta = 180°$). Again, the core growth trend at the blade passage is altered, with a reduction in the core radius close to the value at formation.

Figs. 3(a) and 3(b) also show the growth curves established on the basis of a simple diffusion model. The Lamb-Oseen vortex velocity profile satisfies the laminar Navier-Stokes equations, and the resulting variation of r_c with time can be shown to vary with the square-root of vortex age [40, 41]. To model the actual diffusion rate an average apparent or virtual viscosity coefficient δ can be inferred from the measurements [41]-[43]. This assumption is frequently made in wakes and jets [44]. For rotor applications, the vortex age can be measured in terms of rotor azimuth, ζ, and in air the core growth can be represented by $r_c(\zeta) = r_0 + 0.00855\sqrt{\delta\zeta/\Omega}$, where r_0 is the initial core size, Ω is the angular velocity of the rotor, with ζ being measured in radians. The initial core size was estimated by extrapolating back to zero wake age, this being $r_0/c = 0.025$ for the one-bladed rotor. Results are also shown for different values of δ, where it is apparent that the laminar decay ($\delta = 1$) of the vortex is indeed unrealistically slow. The identical curves are plotted on both Figs. 3(a) and 3(b) to help delineate the differences in the vortex core growth trends between the two rotors. It appears that $\delta = 10$ provides a fairly good approximation to the observed growth trend, at least up to the first blade passage. This is consistent with measurements for fixed-wing vortices, see for example Dosanjh *et al.* [9]. However, it is likely that the value of δ will be considerably larger for full-scale rotors, which will have vortex Reynolds numbers that are at least one order of magnitude greater. These quantitative effects are, however, still unknown.

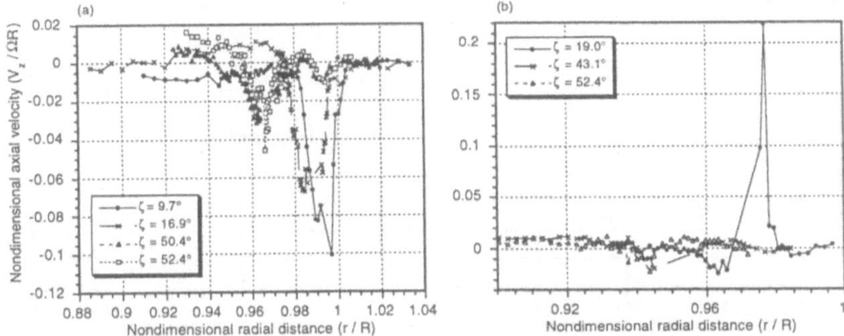

Figure 4. Axial velocity component in the tip vortex at early wake ages for the (a) one-bladed rotor, and (b) the two-bladed rotor

3.3 Axial Velocities

The axial velocities in the vortex core were found to be relatively small in magnitude compared to the swirl velocities. For the one-bladed rotor, the axial velocities were noted to have an almost Gaussian deficit profile, as shown in Fig. 4(a) for three early wake ages. At later ages, the magnitudes were negligible and are not shown. The classical Lamb result predicts that the peak axial velocity should diminish in proportion to $1/\zeta$, which is fairly consistent with observed behavior.

Further examination of the axial velocity profiles in Fig. 4(a) revealed that there were small but significant perturbations just outside the primary vortex core. These were observed at essentially the same radial location relative to the vortex axis at all early wake ages. The reason for this is not completely clear, but it may be because of secondary vortical structures embedded inside the primary vortex. Some evidence of such secondary vortex structures generated by fixed-wings have been reported by Corsiglia *et al.*[10] and McAlister [32, 33]. However, it is likely that there may be several vortical structures that develop from the side-edge of the blade, which then roll into a primary vortex.

Fig. 4(b) shows that for the two-bladed rotor the peak axial velocities were substantially lower, although with one exception. At the earliest wake age of 19° (which is about one chord downstream of the blade trailing-edge), a substantial positive peak was obtained on the outer side of the tip vortex flow. This was not a spurious measurement because it coincided with a large seed particle count. This large axial velocity gradient across the vortex core is likely a result of its rollup process. Unfortunately, more detailed measurements in the side-edge region pose considerable challenges with LDV, including optical access and the necessary entrainment of seed particles.

3.4 Tip Vortex Circulation

The circulation around the tip vortex was calculated by integrating the velocity field around a circular curve that enclosed the entire vortex flow. The asymptotic circulation values were evaluated separately for the outer (outside rotor wake) and inner (inside rotor slipstream) parts of the induced velocity field on a plane perpendicular to the vortex axis, and the average value was then taken.

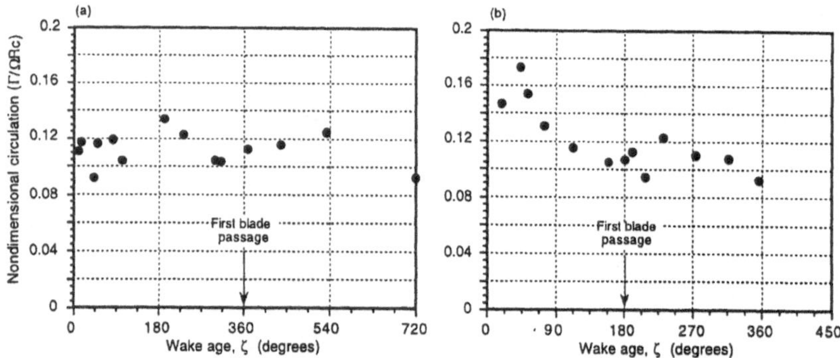

Figure 5. Net tip vortex strength (circulation) as a function of wake age for (a) the one-bladed and, (b) the two-bladed rotor

These data are plotted as a function of wake age in Fig. 5 for the one and two-bladed rotors. For the one-bladed rotor, the net circulation was found to remain nominally constant for all wake ages at a non-dimensional value of $\overline{\Gamma} = (\Gamma/\Omega Rc) = 0.12$. This means that the dissipation of the vortex is negligible even after the blade passage event, which has been shown to cause a significant change in the viscous diffusion trend. For the two-bladed rotor, a somewhat different behavior in the circulation trend was observed. Fig. 5(b) shows that the initial circulation in the tip vortex is higher, reaching a value as high as 0.17 just after formation, but quickly decreasing within a quarter of a rotor revolution to a nominally constant value of 0.12, which is consistent with the one-bladed rotor case. The reason for these differences can be traced to the initial velocity profiles for the one and two-bladed rotors. While the two-bladed rotor showed a slightly lower initial peak-to-peak swirl velocity than for the one-bladed rotor, the core radius was also larger and so gives a higher overall vorticity in the initial stages.

4 Concluding Remarks

An experimental investigation has been made to study blade tip vortex structures in the wake of hovering rotors using three-component laser Doppler velocimetry. Several interesting conclusions have been drawn from the study. First, the swirl or tangential velocity profiles surrounding the tip vortices were found to be asymmetric. This is because of the curved nature of the vortices, which produces a self-induced effect. Second, for the one-bladed rotor the vortex core was found to diffuse approximately with the square-root of wake age up to the first blade passage. For the two-bladed rotor, the initial core size was somewhat larger and the tip vortices did not diffuse as quickly. However, in each case the core sizes after the first blade passage were approximately the same. Third, at the first blade passage event, a change was observed in the tip vortex diffusion because of an increase in tangential velocity and decrease in viscous core size. However, the net circulation about the vortex remained constant throughout. Fourth, the tip vortices from both rotors exhibited deficit type axial velocity profiles in their early stages, although of considerably different magnitudes.

Acknowledgments

This work was partly supported by the U.S. Army Research Office under contract DAAH-04-93-G-0223 and the National Rotorcraft Technology Center under grant NCC 2944. The author would like to acknowledge the contributions of Andrew Baker, Alan Coyne, Mahendra Bhagwat and Ashish Bagai to this work.

References

1. Lehman A. F. (1968) Model Studies of Helicopter Rotor Patterns, *Proceedings of the 24th Annual American Helicopter Society Forum*, Washington DC.
2. Landgrebe, A. J. and Bellinger, E. D. (1971) An Investigation of the Quantitative Applicability of Model Helicopter Rotor Wake Patterns Obtained from a Water Tunnel, UARL K910917-23.
3. Landgrebe, A. J. (1972) The Wake Geometry of a Hovering Rotor and its Influence on Rotor Performance, *Journal of the American Helicopter Society*, **17, 4**, pp. 2-15.
4. Tangler, J. L. (1977) Schlieren and Noise Studies of Rotors in Forward Flight, *Proceedings of the 33rd Annual American Helicopter Society Forum*, Washington DC.
5. Leishman, J. G. and Bagai, A. (1991) Rotor Wake Visualization in Low Speed Forward Flight, AIAA paper 91-3232, *AIAA 9th Applied Aerodynamics Conference*, Baltimore, MD.
6. Schmitz, F. H. (1991) Rotor Noise, in NASA Reference Publication 1258, Chapter 2, *Aeroacoutics of Flight Vehicles: Theory and Practice.*
7. Leishman, J. G. and Bagai, A. (1996) Challenges in Understanding the Vortex Dynamics of Helicopter Rotor Wakes, Paper 96-1957, *27th Fluid Dynamics Conference*, New Orleans, LA.
8. Bagai, A. and Leishman, J. G. (1995) Rotor Free-Wake Modeling Using a Relaxation Technique - Including Comparisons with Experimental Data, *Journal of the American Helicopter Society*, **40, 3**, pp. 29-41.
9. Dosanjh, D. S., Gasparek, E. P. and Eskinazi, S (1962), Decay of a Viscous Trailing Vortex, *The Aeronautical Quarterly*, pp. 167-188.
10. Corsiglia, V. R., Schwind, R. G. and Chigier, N. A. (1973) Rapid Scanning, Three Dimensional Hot Wire Anemometer Surveys of Wing-Tip Vortices, NASA CR-2180.
11. Orloff, K. L. (1974) Trailing Vortex Wind Tunnel Diagnostics with a Laser Velocimeter, *Journal of Aircraft*, **11, 8**, pp. 477-482.
12. Singh, P. I. and Uberoi, M. S. (1976) Experiments on Vortex Stability of Aircraft Wake Turbulence, *The Physics of Fluids*, **19, 12**, pp. 1858-1863.
13. McAllister, K. W, and Takahashi, R. K. (1991) NACA 0015 Wing Pressure and Trailing Vortex Measurements, NASA TP 3151.
14. Simons, I. A., Pacifico, R. E. and Jones, J. P. (1966) The Movement Structure and Breakdown of Trailing Vortices from a Rotor Blade, *Presented at the CAL/USAA Avlabs Symposium*, Buffalo, NY.
15. Cook, C. V. (1972) The Structure of the Rotor Blade Tip Vortex, Paper 3, In: Aerodynamics of Rotary Wings, AGARD CP-111.
16. Caradonna, F. X., and Tung, C. (1981) Experimental and Analytical Studies of a Model Helicopter Rotor, *Vertica*, **5**, pp. 149-161.
17. Tung, C., Pucci, S. L., Caradonna, F. X. and Morse, H. A. (1981) The Structure of Trailing Vortices Generated by Model Helicopter Rotor Blades, NASA TM 81316.
18. Leishman, J. G. (1996) On Seed particle Dynamics in Tip Vortex Flows, *Journal of Aircraft* Vol. 33, No. 4, July/Aug. pp. 823-825.
19. Leishman, J. G. (1997) Measurements of the Aperiodicity in Rotor Wakes, University of Maryland, Dept. of Aerospace Engineering. Aug. 1997.
20. Landgrebe, A. J. and Johnson, B. V. (1974) Measurements of a Model Helicopter Rotor Flow Field with a Laser Doppler Velocimeter, *Journal of the American Helicopter Society*, **19, 3**, pp.39-43.

21. Scully, M. P. and Sullivan, J. P. (1972) Helicopter Rotor Wake Geometry and Airloads and Development of Laser Doppler Velocimeter for Use in Helicopter Rotor Wakes, Technical Report 183, MIT DSR No. 73032, *Massachusetts Institute of Technology Aerophysics Laboratory.*
22. Sullivan, J. P. (1973) An Experimental Investigation of Vortex Rings and Helicopter Rotor Wakes Using a Laser Doppler Velocimeter, Technical Report 183, MIT DSR No. 80038, *Massachusetts Institute of Technology Aerophysics Laboratory.*
23. Thomson, T. L., Komerath, N. M, and Gray, R. B. (1988) Visualization and Measurement of the Tip Vortex Core of a Rotor Blade in Hover, *Journal of Aircraft*, **25, 2**, pp. 1113-1121.
24. Biggers, J. C. and Orloff, K. L. (1974) Laser Velocimeter Measurements of the Helicopter Rotor-Induced Flowfield, *Proceedings of the 30th Annual American Helicopter Society Forum*, Washington DC.
25. Biggers, J. C., Chu, S. and Orloff, K. L. (1975) Laser Velocimeter Measurements of Rotor Blade Loads and Tip Vortex Rollup, *Proceedings of the 31st Annual American Helicopter Society Forum*, Washington DC.
26. Biggers, J. C., Lee, A., and Orloff, K. L. (1977) Measurements of Rotor Helicopter Tip Vortices, *Proceedings of the 33rd Annual American Helicopter Society Forum*, Washington DC.
27. Hoad, D. R. (1983) Preliminary Rotor Wake Measurements with a Laser Velocimeter, NASA TM 83246, March 1983.
28. Elliot, J. W., Althoff, S. L and Sailey, R. H. (1988) Inflow Measurements made with a Laser Velocimeter on a Helicopter Model in Forward Flight, NASA TM-100543.
29. Hoad, D. R. (1991) Helicopter Local Blade Circulation Calculations for a Model Helicopter Rotor in Forward Flight Using Laser Doppler Velocimeter Measurements, *Proceedings of the 47th Annual American Helicopter Society Forum*, Phoenix, AZ.
30. Seelhorst, U., Beesten, B.M. and Butefisch, K. A. (1994) Flowfield Investigation of a Rotating Helicopter Rotor Blade by Three-Component Laser Doppler Velocimetry, *Proceedings of the 75th AGARD Fluid Dynamic Panel Symposium*, Berlin, Germany.
31. Seelhorst, U., Raffel, M,m Willert, H., Vollmers, H., Butefisch, K. A., Kompenhans, J. (1996) Comparison of Vortical Structures of a Helicopter Rotor Model Measured by LDV and PIV, *Proceedings of the 22nd European Rotorcraft Forum*, Brighton, UK.
32. McAllister, K. W., Sculer, C. A., Branum, L. and Wu, J. C. (1995) 3-D Wake Measurements Near a Hovering Rotor for Determining Profile and Induced Drag, NASA TP 3577.
33. McAllister, K. W. (1996) Measurements in the Near Wake of a Hovering Rotor, Paper 96-1958, *27th AIAA Fluid Dynamic Conference*, New Orleans, LA.
34. Leishman, J. G., Baker, A. and Coyne, A. J. (1996) Measurements of Rotor Tip Vortices Using Three-Component Laser Doppler Velocimetry, *Journal of the American Helicopter Society*, **41, 4**, pp. 342-353.
35. Han, Y. O., Leishman J. G. and Coyne, A. J. (1997) Measurements of the Velocity and Turbulence Structure of a Rotor Tip Vortex, *AIAA Journal*, **35, 3**, pp. 477-485.
36. Coyne, A. J., Bhagwat, M. J. and Leishman, J. G. (1997) Investigation into the Rollup and Diffusion of Rotor Tip Vortices using Laser Doppler Velocimetry, *Proceedings of the 53nd Annual American Helicopter Society Forum*, Virginia Beach, VA.
37. Devenport, W. J., Rife, M. C., Liapis, S. I. and Follin, G. J. (1996) The Structure and Development of a Wing-Tip Vortex, *Journal of Fluid Mechanics*, **312**, pp. 67-106.
38. Vatistas, G. H., Kozel, V. and Mih, W. C. (1991) A Simpler Model for Concentrated Vortices, *Experiments in Fluids*, **11**, pp. 73-76.
39. Bagai, A. and Leishman, J. G., 1993) Flow Visualization of Compressible Vortex Structures Using Density Gradient Techniques, *Experiments in Fluids*, **15**, pp. 431-442.
40. Leonard, A. (1980) Vortex Methods for Flow Simulation, *Journal Computational Physics*, **37**, pp. 289-335.
41. Ogawa, A. (1993) *Vortex Flow*, CRC Press Inc.
42. Lamb, H. (1962) *Hydrodynamics*, 6th Edition, Cambridge University Press, pp. 592-593, 668-669.
43. Squire, H. B. (1965) The Growth of a Vortex In Turbulent Flow, *Aeronautical Quarterly*, **16**, pp. 302-306.
44. Schlichting, H. (1979) *Boundary Layer Theory*, 7th Edition, McGraw-Hill.

Mechanics

FLUID MECHANICS AND ITS APPLICATIONS

Series Editor: R. Moreau

21. J.P. Bonnet and M.N. Glauser (eds.): *Eddy Structure Identification in Free Turbulent Shear Flows.* 1993 ISBN 0-7923-2449-8
22. R.S. Srivastava: *Interaction of Shock Waves.* 1994 ISBN 0-7923-2920-1
23. J.R. Blake, J.M. Boulton-Stone and N.H. Thomas (eds.): *Bubble Dynamics and Interface Phenomena.* 1994 ISBN 0-7923-3008-0
24. R. Benzi (ed.): *Advances in Turbulence V.* 1995 ISBN 0-7923-3032-3
25. B.I. Rabinovich, V.G. Lebedev and A.I. Mytarev: *Vortex Processes and Solid Body Dynamics.* The Dynamic Problems of Spacecrafts and Magnetic Levitation Systems. 1994 ISBN 0-7923-3092-7
26. P.R. Voke, L. Kleiser and J.-P. Chollet (eds.): *Direct and Large-Eddy Simulation I.* Selected papers from the First ERCOFTAC Workshop on Direct and Large-Eddy Simulation. 1994 ISBN 0-7923-3106-0
27. J.A. Sparenberg: *Hydrodynamic Propulsion and its Optimization.* Analytic Theory. 1995 ISBN 0-7923-3201-6
28. J.F. Dijksman and G.D.C. Kuiken (eds.): *IUTAM Symposium on Numerical Simulation of Non-Isothermal Flow of Viscoelastic Liquids.* Proceedings of an IUTAM Symposium held in Kerkrade, The Netherlands. 1995 ISBN 0-7923-3262-8
29. B.M. Boubnov and G.S. Golitsyn: *Convection in Rotating Fluids.* 1995 ISBN 0-7923-3371-3
30. S.I. Green (ed.): *Fluid Vortices.* 1995 ISBN 0-7923-3376-4
31. S. Morioka and L. van Wijngaarden (eds.): *IUTAM Symposium on Waves in Liquid/Gas and Liquid/Vapour Two-Phase Systems.* 1995 ISBN 0-7923-3424-8
32. A. Gyr and H.-W. Bewersdorff: *Drag Reduction of Turbulent Flows by Additives.* 1995 ISBN 0-7923-3485-X
33. Y.P. Golovachov: *Numerical Simulation of Viscous Shock Layer Flows.* 1995 ISBN 0-7923-3626-7
34. J. Grue, B. Gjevik and J.E. Weber (eds.): *Waves and Nonlinear Processes in Hydrodynamics.* 1996 ISBN 0-7923-4031-0
35. P.W. Duck and P. Hall (eds.): *IUTAM Symposium on Nonlinear Instability and Transition in Three-Dimensional Boundary Layers.* 1996 ISBN 0-7923-4079-5
36. S. Gavrilakis, L. Machiels and P.A. Monkewitz (eds.): *Advances in Turbulence VI.* Proceedings of the 6th European Turbulence Conference. 1996 ISBN 0-7923-4132-5
37. K. Gersten (ed.): *IUTAM Symposium on Asymptotic Methods for Turbulent Shear Flows at High Reynolds Numbers.* Proceedings of the IUTAM Symposium held in Bochum, Germany. 1996 ISBN 0-7923-4138-4
38. J. Verhás: *Thermodynamics and Rheology.* 1997 ISBN 0-7923-4251-8
39. M. Champion and B. Deshaies (eds.): *IUTAM Symposium on Combustion in Supersonic Flows.* Proceedings of the IUTAM Symposium held in Poitiers, France. 1997 ISBN 0-7923-4313-1
40. M. Lesieur: *Turbulence in Fluids.* Third Revised and Enlarged Edition. 1997 ISBN 0-7923-4415-4; Pb: 0-7923-4416-2

Kluwer Academic Publishers – Dordrecht / Boston / London

Mechanics

FLUID MECHANICS AND ITS APPLICATIONS

Series Editor: R. Moreau

41. L. Fulachier, J.L. Lumley and F. Anselmet (eds.): *IUTAM Symposium on Variable Density Low-Speed Turbulent Flows.* Proceedings of the IUTAM Symposium held in Marseille, France. 1997 ISBN 0-7923-4602-5
42. B.K. Shivamoggi: *Nonlinear Dynamics and Chaotic Phenomena.* An Introduction. 1997 ISBN 0-7923-4772-2
43. H. Ramkissoon, *IUTAM Symposium on Lubricated Transport of Viscous Materials.* Proceedings of the IUTAM Symposium held in Tobago, West Indies. 1998 ISBN 0-7923-4897-4
44. E. Krause and K. Gersten, *IUTAM Symposium on Dynamics of Slender Vortices.* Proceedings of the IUTAM Symposium held in Aachen, Germany. 1998 ISBN 0-7923-5041-3

Kluwer Academic Publishers – Dordrecht / Boston / London